Interdisciplinary Applied Mathematics

Volume 51

More information about this series at http://www.springer.com/series/1390

Merab Svanadze

Potential Method in Mathematical Theories of Multi-Porosity Media

Springer

Merab Svanadze
Ilia State University
Tbilisi, Georgia

ISSN 0939-6047 ISSN 2196-9973 (electronic)
Interdisciplinary Applied Mathematics
ISBN 978-3-030-28024-6 ISBN 978-3-030-28022-2 (eBook)
https://doi.org/10.1007/978-3-030-28022-2

Mathematics Subject Classification: 74F05, 74F10, 74G25, 74G30, 74H45, 74M25, 35E05

This Springer imprint is published by the registered company Springer Nature Switzerland AG.
The registered company address is: Gewerbestrasse 11, 6330 Cham, Switzerland

Dedicated to
Maia, Lika, and Giorgi

"Nature is written in mathematical language"

Galileo Galilei, *The Assayer, 1623*

Preface

The goal of this book is to investigate the three-dimensional problems of the mathematical theories of elasticity and thermoelasticity for multi-porosity materials. The purposes of this work are twofold. The first is to develop the classical *potential method* (boundary integral equation method) in these theories, and the second is to generalize the basic results of the classical mathematical theories of elasticity and thermoelasticity.

The theory of porous media plays a central role in various fields of applied mathematics, continuum mechanics, and engineering and has already been dealt with in numerous books and papers. In the last three decades, there has been interest in investigation of problems of elasticity and thermoelasticity for solids with double and triple porosities. Moreover, in the recent research papers, there are indications that the construction and investigation of the mathematical models of quadruple and quintuple porosity structures of geo- and biomaterials are important and necessary for various fields of engineering and technology.

For clarity, in this book, we investigate the linear mathematical theories of elasticity and thermoelasticity for quadruple porosity materials. On the one hand, each result of this work can be reduced to the corresponding result for the triple (double) porosity materials neglecting fourth level (third and fourth levels) of porosity of materials. On the other hand, in a quite similar manner, the research methods and results of this book can be generalized to quintuple and more stage hierarchical porosity systems. Furthermore, quadruple porosity structures are the ones most likely to represent realistic systems occurring in nature, for example, bone, human brain, oil, gas, and thermal reservoirs.

The quadruple porosity model represents a new possibility for the study of important problems of engineering and mechanics. The intended applications of the mathematical theories of elasticity and thermoelasticity for solids with a quadruple porosity structure are to geological materials such as oil, gas and geothermal reservoirs, rocks and soils, manufactured porous materials such as ceramics and pressed powders, and biomaterials such as bone and human brain.

All the results of this book are unpublished contributions of the author. In this work, the boundary value problems (BVPs) of the linear mathematical theories of elasticity and thermoelasticity for quadruple porosity materials and the BVPs of rigid bodies with quadruple porosity are investigated, and in each theory, the following results are obtained:

- The fundamental and singular solutions of the systems of equations of equilibrium, quasi-static, steady vibrations, and in the Laplace transform space are expressed in explicit form by means of elementary (harmonic, biharmonic, metaharmonic) functions, and their basic properties are established.
- The representations of Galerkin-type solutions of these systems of equations are obtained.
- Green's formulas and integral representations of Somigliana type of regular vector and classical solutions are established.
- A wide class of the internal and external BVPs of equilibrium, quasi-static, steady vibrations, and in the Laplace transform space are formulated.
- The radiation conditions are established and the uniqueness theorems of regular (classical) solutions of the BVPs are proved.
- The surface (single-layer and double-layer) and volume potentials are constructed, and their basic properties are studied.
- On the basis of these potentials, the BVPs are reduced to the equivalent integral equations for which Fredholm's theorems are valid.
- The existence theorems for classical solutions of the above-mentioned BVPs are proved by using the potential method and the theories of singular and Fredholm integral equations.

This book is addressed to a large circle of readers: mathematicians (especially those involved in applied mathematics), physicists (particularly those interested in mechanics and its connections), engineers of various specialities (civil, mechanical, etc., who are scientific researchers), graduate students of these professions, etc. I have assumed that the reader is conversant with the basic elements of the mathematical theory of elasticity, but otherwise, the book is self-contained.

The text is written in the moderately clear form and structured in a stepwise hierarchical fashion from basic to more advanced topics.

This book contains 13 chapters. The text begins with an introductory chapter. Chapters 2–12 can be roughly divided into two parts. The first part (Chaps. 2–8) and the second part (Chaps. 9–12) deal with problems of the linear theories of elasticity and thermoelasticity of quadruple porosity materials, respectively. The objective of the final chapter is to mention several interesting topics in the theory of multi-porosity media for future research.

The author warmly thanks Professor Brian Straughan for his many helpful discussions on the modeling questions of multi-porosity materials and who prompted the author to write this book. I am also grateful to three anonymous referees for useful comments and suggestions.

Finally, I would like to thank Christopher Tominich for his help with editorial matters and the Springer production team for their professional competence and fruitful cooperation in preparing the final version of this book.

Tbilisi, Georgia
September 2019

Merab Svanadze

Contents

Chapter 1
Introduction

This chapter is divided into seven main sections.

In Sect. 1.1, a brief review of the theories of multi-porosity materials is presented.

In Sect. 1.2, the short history of the potential method is introduced.

In Sect. 1.3, the basic notations are given. These notations are used throughout this work.

In Sects. 1.4 and 1.5, the basic equations of thermoelasticity and elasticity of quadruple porosity solids are presented, respectively.

In Sect. 1.6, these equations are rewritten in the matrix form.

Finally, in Sect. 1.7, the stress operators of the considered theories are given.

1.1 Brief Review of the Theories of Multi-Porosity Materials

1.1.1 Single Porosity Materials

Since the publication of fundamental works by Truesdell and Toupin [376] and Truesdell and Noll [375], there has been significant interest in thermodynamic formulations of general theories of porous media.

The prediction of the mechanical properties of porous materials has been one of hot topics of continuum mechanics for more than 100 years. The construction of mathematical models of fluid flow through porous media and the intensive investigation of the problems of porous continua arise by extensive use of porous materials into civil engineering, geotechnical engineering, technology, hydrology, and recent years, medicine and biology (see Bear [36], Coussy [106, 107], Cowin [111], de Boer [123, 125], Ichikawa and Selvadurai [171], Liu [228], Straughan [319, 320], Wang [392], and the references therein).

M. Svanadze, *Potential Method in Mathematical Theories of Multi-Porosity Media*, Interdisciplinary Applied Mathematics 51,
https://doi.org/10.1007/978-3-030-28022-2_1

In addition, most astrophysical, geological, biological bodies, or otherwise are porous materials with micro- and nanostructures (see Cowin [110], Dormieux et al. [133], Straughan [326], and Su et al. [329]). In reality, poroelasticity is a well-developed theory for the interaction of fluid and solid phases of a fluid saturated porous medium. Moreover, poroelasticity is an effective and useful model for deformation-driven bone fluid movement in bone tissue (for details see Cowin [110, 111], de Boer [124], Holzapfel and Ogden [166], and Vafai [383]).

In the investigation of porous materials, mathematical models have always been used. Interest in the models of porous media arose long ago and still exists mainly due to the development of technologically new types of porous materials (see Das et al. [122], Wu [397], and the references therein). There are many different approaches to a theoretical formulation of linear and nonlinear models for materials with single and multiple porosities. Historically, mathematical models of single porosity materials were first models in the theory of porous media (see, e.g., de Boer [123]). A wide information on the construction of mathematical models that describe phenomena of flow and transport in porous media is given in the new books of Bear [36] and Das et al. [122].

There are a number of theories which describe mechanical properties of single porosity materials (see Cheng [78], de Boer [123], and the references therein), and the most well known of them are: (i) Biot's [41] consolidation theory based on Darcy's law (1856) and (ii) Nunziato–Cowin [266] theory based on the volume fraction concept. For details of Darcy's law and the volume fraction concept see Bluhm and de Boer [59] and de Boer [123, 125].

(i) Biot's Consolidation Theory. M.A. Biot in his original paper [41] presented the governing equations of poroelasticity for an isotropic single porosity material based on Darcy's law. In this theory the independent variables are the displacement vector field and the pressure in pore network. On the basis of this theory and the linear model of thermoelasticity [44] he developed the porothermoelasticity theory (the thermal–hydraulic–mechanical theory) in [48]. Nowadays, the problems of Biot's elegant theories of poroelasticity [41] and porothermoelasticity [48] for single porosity materials are intensively investigated (see Chirita [80], Chirita et al. [87], Ciarletta et al. [100], Scalia [299], and the references therein). The basic results and historical information on these theories can be found in the books of Cheng [78], de Boer [125], Ichikawa and Selvadurai [171], Selvadurai and Suvorov [308], Straughan [319], and Verruijt [389]. A comprehensive review of analytical solutions for problems governed by poroelasticity equations is presented by Selvadurai [307]. A review on consolidation theories and their applications in numerous fields are given by Radhika et al. [285].

(ii) Nunziato–Cowin Theory. Nunziato and Cowin [266] introduced a theory for the behavior of single porous deformable materials in which the skeletal or matrix materials are elastic and the interstices are voids (vacuous pores). On the basis of this model the linear theory of elastic materials with single voids is presented by the same authors [112]. Ieşan [174] developed a linear theory of thermoelastic materials with single voids. In the above-mentioned theories for materials with single voids

the independent variables are the displacement vector field, the volume fraction of pores, and the variation of temperature.

Several mathematical models for single porosity materials based on the volume fraction concept are suggested by Aouadi [13], Bowen [60, 61], Chirita et al. [84], Ciarletta and Straughan [94], De Cicco and Diaco [131], Dhaliwal and Wang [132], Ieşan [172, 177], Ieşan and Quintanilla [182], Passarella [272], Ricken and Bluhm [289], and extensively studied by Aouadi [14, 15], Casas and Quintanilla [68], Chirita and Scalia [85], Ciarletta et al. [90, 95–97], Cowin [108], Cowin and Puri [113], D'Apice and Chirita [120], de Boer and Svanadze [129, 357], Ieşan [173], Magaña and Quintanilla [235], Marin [239], Passarella et al. [274], Puri and Cowin [279], Quintanilla [282], Scalia [297], Scarpetta [302], and Tsagareli [377].

The basic results on the Nunziato–Cowin theory may be found in the books of Ciarletta & Ieşan [89], Ieşan [175, 176], Straughan [319], and the references therein.

1.1.2 Double Porosity Materials

A double porosity material is a solid that contains pores on a macro scale referred to as macro (primary) pores and pores on a much smaller scale referred to as micro (secondary) pores. In the past half century, the mathematical models of double porosity media, as originally developed for the mechanics of naturally fractured reservoirs, have found applications in many branches of civil and geotechnical engineering, technology, and, in recent years, biomechanics. Significant progress has been made towards understanding and modeling of flow processes in fractured rock. However, fractured rock may be considered as a multi-porous medium but the most studies have focused naturally fractured reservoirs with double and triple porosities (for details see Aguilera [3], Ichikawa and Selvadurai [171], Nikolaevskij [262], Rezaee [288], Showalter and Walkington [312], and Wu [398]).

In addition, heat transfer in single- and multi-porosity materials has received much attention in science and engineering, for instance, heat extraction in hot dry rock, heat transfer in biological tissue, geothermal and oil–gas reservoirs (see Aifantis [7], Aifantis and Beskos [8], Cushman [114], Nield and Bejan [261], and Straughan [326]).

Moreover, Darcy's classical law is extended for describe fluid flow through undeformable double porosity materials by Barenblatt and Zheltov [24], Barenblatt et al. [25], and Warren and Root [394]. The linear theory for double porosity deformable materials by using Darcy's extended law is developed by Wilson and Aifantis [396]. This theory unifies the earlier proposed models of Barenblatt and coauthors [24, 25] for a rigid body with double porosity and Biot [41] for an elastic solid with single porosity.

In the last three decades more general models (the cross-coupled terms are included in the equations of conservation of mass for the pore and fissure fluid) of the theories of elasticity and thermoelasticity for double porosity materials by using Darcy's extended law are introduced by Bai and Roegiers [20], Berryman and Wang [37], Gelet et al. [149], Khalili et al. [199], Khalili and Selvadurai [200], Masters et al. [243], Pride and Berryman [278], and Zhao and Chen [404]. A general form of the double porosity model of single phase flow in a naturally fractured reservoir is derived from homogenization theory by Arbogast et al. [16]. The theory of viscoelasticity for Kelvin–Voigt materials with double porosity is presented by Svanadze [338]. Double-diffusion models from a highly heterogeneous medium are introduced by Showalter and Visarraga [311].

These mathematical models for double porosity materials are investigated extensively by several authors. In this connection, the linear theory of elastodynamics for anisotropic nonhomogeneous materials with double porosity is developed by Straughan [322]. The properties of the acceleration waves in the nonlinear double porosity elasticity are established by Gentile and Straughan [152]. The fundamental solutions in the theories of elasticity and thermoelasticity for materials with double porosity are constructed by Scarpetta et al. [305], Svanadze [331, 336], and Svanadze and De Cicco [358]. The Coupling effects in double porosity media with deformable matrix are studied by Khalili [198].

The basic properties of plane harmonic waves in the coupled linear theory of elasticity for solid with double porosity are established by Ciarletta et al. [99] and Svanadze [334]. Elastic wave propagation, dispersion, and attenuation properties of Rayleigh and Love waves in a double porosity medium are analyzed by Ba et al. [19], Berryman and Wang [38], Dai and Kuang [116], and Dai et al. [117]. The reflection and transmission of elastic waves from the interface of a fluid-saturated porous solid and a double porosity solid are studied by Dai et al. [118].

The uniqueness theorems for three-dimensional BVPs of the theories of elasticity and thermoelasticity for materials with double porosity are proved by Scarpetta and Svanadze [304] and Svanadze [337]. Uniqueness and a variational principle are established in the theory of consolidation for double porosity materials by Beskos and Aifantis [40]. Using the finite element method a numerical solution of Aifantis' equations of double porosity is obtained by Khaled et al. [197].

The nonclassical BVPs of the linear theories of elastic and thermoelastic double porosity materials are investigated in the series of papers. Namely, the BVPs of quasi-static and steady vibrations and the initial-BVPs are studied by means of the potential method and the theory of singular integral equations by Svanadze [333, 335, 339] and Svanadze and Scalia [359]. The plane BVPs of the elastic equilibrium of porous Cosserat media with double porosity are considered by Janjgava [188] and a concrete BVP for a concentric ring is solved.

Explicit solutions of the BVPs in the theory of consolidation of double porosity materials for the half-plane are constructed by Basheleishvili and Bitsadze [31]. The two-dimensional version of the fully coupled linear equilibrium theory of elasticity for materials with double porosity is considered and the solutions of BVPs for an elastic circle and for a plane with circular hole are constructed in the form of absolutely and uniformly convergent series by Tsagareli and Bitsadze [378]. The explicit solutions of the basic BVPs in the fully coupled linear equilibrium theory of elasticity for the double porosity space with spherical cavity and spherical layer are constructed by Bitsadze and Tsagareli [56, 57]. Explicit solutions in the fully coupled theory of elasticity for a circle with double porosity are presented by Tsagareli and Bitsadze [379] and Tsagareli and Svanadze [381].

Moreover, Svanadze [366–370] constructed the fundamental solutions of the systems of steady vibrations equations in the linear theories of viscoelasticity and thermoviscoelasticity for Kelvin–Voigt materials with double porosity, and proved the existence and uniqueness theorems for classical solutions of the BVPs of steady vibrations.

We may remark that in the above-considered models for materials with double porosity the independent variables are the displacement vector, the pressures in the pores and fissures networks, and the variation of temperature.

Recently, Ieşan and Quintanilla [183] developed the theory of Nunziato and Cowin [112, 266] for thermoelastic deformable materials with double voids structure by using the volume fraction concept. More recently, Ieşan [179] presented a theory of thermoelastic prestressed solids with double voids. A generalized theory of thermoelastic diffusion with double voids is considered by Kansal [192]. In these theories the independent variables are the displacement vector, the volume fractions of pores and fissures, and the variation of temperature.

On the basis of Ieşan and Quintanilla [183] theory the elastodynamic problem of an infinite thermoelastic double voids body with a spherical cavity in the context of Lord–Shulman theory of thermoelasticity with one relaxation time is examined and some numerical results are obtained by Kumar and Vohra [207, 208] and Kumar et al. [210]. The variational principle for Lord–Shulman theory of thermoelastic material with double voids is developed by Kumar et al. [212] and the propagation of plane waves for thermoelastic material with double voids structure with one relaxation time is studied. Exponential decay, existence, and uniqueness of the solutions in the one-dimensional version of thermoelasticity for solids with double voids are established by Bazarra et al. [35].

Plane waves, uniqueness theorems, and existence of eigenfrequencies in the theory of rigid bodies with double voids are investigated by Svanadze [343]. The existence of classical solutions in the external BVPs of steady vibrations of this theory are established by the same author [342]. The basic three-dimensional BVPs of the equilibrium theory of elasticity for materials with a double voids structure are studied by using the potential method and the theory of singular integral equations by Ieşan [178]. The existence and uniqueness of solutions of the BVPs of steady vibrations in the theories of elasticity and thermoelasticity for materials with double voids are studied by Svanadze [347, 349].

Spatial and temporal behavior of solutions of the dynamical problems in the linear theory of thermoelasticity for solids with double voids are established by Arusoaie [17] and Florea [139]. The micropolar model of thermoelasticity for solids with double voids is presented by Marin et al. [242], and the existence, uniqueness, and stability of the weak solutions are proved. The explicit expressions for the fundamental solution of the system of equations in the case of steady vibrations of this theory are presented by Kumar et al. [207]. The existence and stability results for thermoelastic dipolar bodies with double voids are established by Marin and Nicaise [241]. The state space approach to plane deformation in the theory of elasticity for materials with double voids is developed by Kumar and Vohra [206]. The BVPs for isotropic thermoelastic double porous microbeam are considered and numerical results are obtained in [209]. The components of stress and temperature distribution in isotropic, homogeneous, thermoelastic porous medium with double voids structure due to thermomechanical sources are determined by Kumar et al. [211]. A priori estimates for the amplitude of a harmonic vibration in the linear thermoelasticity theory of anisotropic materials with double voids are derived by Florea [140]. The fundamental solution of the system of equations of pseudo-oscillations in the theory of thermoelastic diffusion materials with double voids is constructed by Kansal [193].

The three-temperature model of the linear theory of double porosity thermoelasticity under local thermal nonequilibrium (LTNE) for anisotropic materials is developed by Franchi et al. [142]. In this theory the temperature is allowed to be different in the skeleton, in the fluid in the macro pores, and in the fluid in the micro pores. In addition, in this paper the uniqueness and decay of solutions of the quasi-static problems are established. The linear theory of thermoelasticity for isotropic materials with double porosity under LTNE is considered by Svanadze [355], and the basic nonclassical BVPs are investigated by using the potential method.

1.1.3 Triple Porosity Materials

To the double porosity concept, a number of triple porosity models have been proposed to describe flow through fractured vuggy rocks. Indeed, mathematical formulations of flow through triple porosity media were first introduced three decades ago by Liu [226]. Abdassah and Ershaghi [1] developed a triple porosity model to describe the well test data of fractured reservoirs with a dual matrix structure. Liu et al. [227] proposed a triple porosity model for a reservoir which consists of fractures, rock matrices, and cavities within the rock matrix (as an additional porous portion of the matrix). Wu et al. [398] used the same model to describe flow in a reservoir which consists of matrices, large fractures, and small fractures. A triple porosity reservoir modeled as a parallel resistance network for matrix, fractures, and non-connected vugs by Aguilera and Aguilera [4].

A triple-porosity model is presented by Bai and Roegiers [21] to study the solute transport in heterogeneous porous media. In this paper the transport processes are distinctly different between three levels of porosity (macro-, meso-, and microporosity). In addition, the general governing equations of solute transport for a triple-porosity rigid body are expressed in terms of the solute concentrations in the three levels of porosity and the deformation of a body is neglected.

The several new triple porosity mathematical models for solids with hierarchical macro-, meso-, and microporosity structure are presented by Straughan [324] and Svanadze [344]. The fundamental solutions in the theories of triple porosity elasticity and thermoelasticity are constructed explicitly by Svanadze [345, 354] and their basis properties are established. The same author [350] investigated the

BVPs of steady vibrations in the theory of triple porosity elasticity by means of the potential method and the theory of singular integral equations. The uniqueness and stability of solutions in the theory of triple porosity thermoelasticity are established by Straughan [327]. The BVPs of quasi-static and steady vibrations in the theory of thermoelasticity for solids with triple porosity are studied by Svanadze [346, 348, 351]. Numerical solutions of problems in the triple porosity model for shale gas reservoirs accommodating gas diffusion in kerogen are presented by Sang et al. [294]. On the basis of the triple-porosity/dual-permeability model the numerical results for fluid-flow process are obtained by Wei and Zhang [395] and Zou et al. [405].

Recently, the basic BVPs of equilibrium, quasi-static, and steady vibrations in the linear theories of elasticity and thermoelasticity for materials with triple voids are investigated by Svanadze [352, 353, 356].

1.1.4 Multi-Porosity Materials

There exists considerable investigation involving multi-porosity continua. The theories of multi-porosity materials are an extension of Biot's classical poroelasticity theory (see Biot [41]).

On the basis of Darcy's extended law the mathematical models of elasticity and thermoelasticity for media with multiple porosity are presented by Bai et al. [22] and Moutsopoulos et al. [254]. Indeed, based on the multi-porosity concept (see Aifantis [6]), as an extension to the traditional dual-porosity approach, a multi-porosity/multipermeability model is presented by Bai et al. [22]. Moutsopoulos et al. [254] studied the hydraulic behavior and contaminant transport in multiple porosity media.

Mehrabian and Abousleiman [246] derived the fully coupled N-poroelasticity model and studied the Mandel's problem. The linear theory of multiple porosity and multiple permeability poroelasticity theory is presented for homogenous, isotropic, fluid-saturated materials by same authors [247]. In this paper, numerical results in the time domain are presented for single-porosity, double-porosity, triple-porosity, quadruple-porosity, and quintuple-porosity poroelastic models of organic-rich shale. The analytical solutions for an N-porosity/N-permeability porous medium are derived by Liu and Abousleiman [225] and Mehrabian and Abousleiman [248]. The model proposed by Sheng et al. [310] is more comprehensive, considering not only the reservoir-stimulated characteristics of multi-fractured horizontal wells in shale gas reservoirs but also the mechanisms of transport through media at multiple scales. The poroelastic constants of multiple-porosity solids are established by Mehrabian [245].

A quadruple-porosity model for shale gas reservoirs with multiple migration mechanisms is developed by He et al. [164]. The paper by Patwardhan et al. [275] reviews the methods and experimental studies used to describe the flow mechanisms of gas through quad porosity shale systems, and critically recommend the direction in which this work could be extended. A quad-porosity numerical model based on the production scenario of multi-stage fracturing horizontal well is established by Zhang et al. [403].

Recently, shale-petroleum reservoirs at discovery are characterized by a quad-porosity system by Lopez and Aguilera [233]. In this work, a petrophysical model is built that allows quantification of storage capabilities in shales through determination of adsorbed porosity, organic porosity, inorganic porosity, and fracture porosity. The governing equations that describe the gas mass balance in the quintuple porosity model are presented in detail by Aguilera and Lopez [5, 232].

The general models for anisotropic thermoelastic solids with quadruple and quintuple porosity structures are presented by Straughan [324]. In addition, in this paper the equations of various nonlinear generalizations are introduced and the uniqueness of a solution to the equations for a body with quintuple porosity is established.

In the mathematical models of multi-porosity materials (see Bai et al. [22], Mehrabian and Abousleiman [246, 247], Moutsopoulos et al. [254], and Straughan [324]) the independent variables are the displacement vector, the pressures in the pore networks, and the variation of temperature.

An extensive review of the results on the multi-porosity media may be found in the new book of Straughan [326]. The basic results and historical information on the theory of porous media can be found in the books of de Boer [123, 125], Ichikawa and Selvadurai [171], and Straughan [322, 323].

Nowadays, the investigation of nonclassical mathematical problems of elastic and thermoelastic multi-porosity materials is very important for applied mathematics, continuum mechanics, geosciences, and biomechanics. In this book, a wide class of these problems for quadruple porosity materials is solved by means of the potential method.

1.2 Potential Method: An Overview

The potential method is the most powerful and elegant techniques for solving elliptic BVPs of mathematical physics and mechanics. This method is a theoretical tool for proving the existence of solutions of BVPs and a practical tool for constructing of analytical and numerical solutions.

Classical potential method grew out of mathematical physics, in particular, out of the theories of gravitation and electrostatics. Green [157] used a physical argument to introduce a function, which he called a "potential function", to investigate the problems of electrostatics and magnetics.

Using the fundamental solution of Laplace equation, Green [157] presented the formula for integral representation of a regular function (Green's third identity). This representation formula contains the convolution type three integrals (potentials): two surface (the single-layer and double-layer) potentials and the volume (Newtonian) potential.

As is known from classical potential theory (see, e.g., Günther [161] and Kellogg [195]), if the bounding surface and the boundary conditions satisfy certain smoothness conditions, then the internal and external BVPs of Dirichlet, Neumann, Robin, and mixed type were reduced to the equivalent integral equations by the virtue of the above-mentioned harmonic potentials.

The creation of the theory of integral equations by Fredholm [144] gave a new impetus to the development of potential methods. On the basis of Fredholm's theorems for integral equations and Green's identities the existence and uniqueness of solutions of the BVPs were proved (for internal Neumann problem, the uniqueness is only up to an arbitrarily additive constant). In the diffraction theory, Kupradze [213, 214] established the basic properties of potentials and proved the existence and uniqueness theorems for the BVPs of Helmholtz' equation by means of the potential method.

The potential method plays a pivotal role in the investigation of BVPs of mathematical physics and continuum mechanics. The application of this method to the 3D (or 2D) BVPs of mathematical physics reduces these problems to 2D (or 1D) boundary integral equations of Fredholm's type. Besides, the boundary integrals are very convenient for constructing numerical solutions of BVPs by virtue of the boundary element method. An extensive review of works on this subject can be found in Cheng and Cheng [79].

For historical and bibliographical materials on the potential theory see the classical books of Günther [161], Kellogg [195], and Kupradze [215]. The main results on the application of the potential method in the theory of harmonic functions are given in the book of Hsiao and Wendland [169].

In the classical theories of elasticity and thermoelasticity the formulae of integral representations of a regular vector functions and the surface and volume potentials are constructed by means of the fundamental solutions of these theories (for details see Kupradze et al. [219]). The potentials of the elasticity theory were constructed and applied to BVPs in the works of representatives of the Italian mathematical school E. Betti, T. Boggio, G. Lauricella, R. Marcolongo, F. Tricomi, V. Volterra, and others. The main results in this subject are received by J. Boussinesq, K. Korn, H. Weyl, H. Poincaré, Georgian scientists N. Muskhelishvili, V. Kupradze, T. Gegelia, M. Basheleishvili, T. Burchuladze, and their pupils.

It is well known (see, e.g., Kupradze et al. [219]) that the BVPs of continuum mechanics are reduced to the singular integral equations by using the potential method. In these equations the boundary integrals are strongly singular and need to be defined in terms of Cauchy principal value integrals. The theory of one-dimensional singular integral equations is developed in the monographs of Muskhelishvili [256, 257], and using this theory, the plane BVPs of the elasticity theory are studied. Owing to the works of Kupradze [217], Kupradze et al. [219], and Mikhlin [252], the theory of multidimensional singular integral equations has presently been worked out with sufficient completeness. This theory makes it possible to investigate not only 3D problems of classical elasticity theory, but also problems of the theory of elasticity with conjugated fields.

For an extensive review of the works and basic results on the potential method in the classical theories of elasticity and thermoelasticity, see the books of Burchuladze and Gegelia [64], Kupradze [217], Kupradze et al. [219], and the review paper by Gegelia and Jentsch [147].

In recent years several continuum theories with microstructure have been formulated in which the deformation is described not only by the usual displacement vector field, but by other vector or tensor fields as well. In this connection, the potential method is developed in linear theories of elasticity and thermoelasticity for materials with microstructure, see, e.g., Burchuladze and Svanadze [65], Scalia and Svanadze [299, 300, 360], Scalia et al. [301], and Svanadze [330, 332, 340, 341].

Moreover, employing the potential method and the theory of singular integral equation the BVPs of the linear theories of elasticity and thermoelasticity for double and triple porosity materials are investigated in the series of papers by Ieşan [178], Svanadze [333–335, 339, 342, 346–351], and Svanadze and Scalia [359].

In this book, the classical potential method is developed in the linear mathematical theories of elasticity and thermoelasticity for quadruple porosity materials.

1.3 Notation

Each chapter of this book has its own numeration of formulas. The formula number is denoted by two figures enclosed in brackets; for example, (3.2) means the second formula in the third chapter. Theorems, lemmas, definitions, and remarks are numerated in the same manner but without brackets; for example, Theorem 3.2 means the second theorem in the third chapter.

We denote the vectors (vectors fields), matrices (matrices fields), and points of the Euclidean three-dimensional space $\mathbb{R}^3$ by boldface letter, and scalars (scalar fields) by Italic lightface letters.

Let $\mathbf{x} = (x_1, x_2, x_3)$ be a point of $\mathbb{R}^3$ and let t denote the time variable, $t \geq 0$, $|\mathbf{x}| = \sqrt{\sum_{j=1}^{3} x_j^2}$. The nabla (gradient) and the Laplacian operators will be designated by ∇ and Δ, respectively, i.e.

$$\nabla = \mathbf{D_x} = \left(\frac{\partial}{\partial x_1}, \frac{\partial}{\partial x_2}, \frac{\partial}{\partial x_3}\right), \qquad \Delta = \frac{\partial^2}{\partial x_1^2} + \frac{\partial^2}{\partial x_2^2} + \frac{\partial^2}{\partial x_3^2}.$$

The unit matrix will be always denoted by $\mathbf{I}_m = \left(\delta_{lj}\right)_{m\times m}$ for $m = 2, 3, \cdots$, where δ_{lj} is the Kronecker delta. The Dirac delta function will be denoted by $\delta(\mathbf{x})$.

The inner (scalar) product of two vectors $\mathbf{w} = (w_1, w_2, \cdots, w_m)$ and $\mathbf{v} = (v_1, v_2, \cdots, v_m)$ will be denoted by $\mathbf{w} \cdot \mathbf{v} = \sum_{j=1}^{m} w_j \bar{v}_j$, where $\bar{v}_j$ is the complex conjugate of v_j. If $m = 3$, then the vector product of vectors $\mathbf{w}$ and $\mathbf{v}$ will denote by $[\mathbf{w} \times \mathbf{v}]$. However, any m-dimensional vector $\mathbf{w} = (w_1, w_2, \cdots, w_m)$ will be considered as a single-column matrix, and the product of the matrix of dimension $r \times m$ by the vector $\mathbf{w}$ will be the r-dimensional single-column matrix.

A matrix differential operator is a matrix the elements of which are differential operators. If $\mathbf{A} = (A_{lj})_{m\times r}$ is a matrix differential operator and $\mathbf{B} = (B_{lj})_{r\times s}$ is a matrix function, then $\mathbf{AB}$ will be matrix function $\mathbf{C} = (C_{lj})_{m\times s}$, where the functions C_{lj} are determined by $C_{lj} = \sum_{k=1}^{r} A_{lk}B_{kj}$. If $\mathbf{A} = (A_{lj})_{m\times r}$ is a matrix function and $\mathbf{B} = (B_{lj})_{r\times s}$ is a matrix differential operator, then $\mathbf{AB}$ will be matrix function $\mathbf{C} = (C_{lj})_{m\times s}$, where the functions A_{lk} are the coefficients of the differential operators B_{kj}. The transpose of matrix $\mathbf{A}$ will be denoted by $\mathbf{A}^\top$.

In this book, we always consider that an isotropic and homogeneous elastic solid occupies a region of $\mathbb{R}^3$ and possesses the hierarchical structure with four different porosity (macro-, meso-, micro-, and submicroporosity) scales. We denote the displacement vector by $\hat{\mathbf{u}} = (\hat{u}_1, \hat{u}_2, \hat{u}_3)$, the fluid pressures in the macro, meso, micro, and submicro scales by $\hat{p}_1, \hat{p}_2, \hat{p}_3$, and $\hat{p}_4$, respectively. We shall usually use T_0 to denote the constant absolute temperature of the body in the reference configuration, $T_0 > 0$. The notation $\hat{\theta}(\mathbf{x}, t)$ is employed to denote the temperature measured from T_0.

We assume that the subscripts preceded by a comma denote partial differentiation with respect to the corresponding Cartesian coordinate. Repeated Latin and Greek indices are summed over the ranges (1,2,3) and (1,2,3,4), respectively. A superposed dot denotes differentiation with respect to t, so that, for instance,

$$\hat{\theta}_{,j}(\mathbf{x}, t) = \frac{\partial \hat{\theta}(\mathbf{x}, t)}{\partial x_j}, \qquad \hat{\theta}_{,jj}(\mathbf{x}, t) = \sum_{j=1}^{3} \frac{\partial^2 \hat{\theta}(\mathbf{x}, t)}{\partial x_j^2} = \Delta\hat{\theta}(\mathbf{x}, t),$$

$$\dot{\hat{\theta}}(\mathbf{x}, t) = \frac{\partial \hat{\theta}(\mathbf{x}, t)}{\partial t}, \qquad \ddot{\hat{\theta}}(\mathbf{x}, t) = \frac{\partial^2 \hat{\theta}(\mathbf{x}, t)}{\partial t^2},$$

$$\hat{u}_{j,j}(\mathbf{x}, t) = \sum_{j=1}^{3} \frac{\partial \hat{u}_j(\mathbf{x}, t)}{\partial x_j} = \operatorname{div}\hat{\mathbf{u}}(\mathbf{x}, t),$$

$$a_\alpha \hat{p}_\alpha(\mathbf{x}, t) = \sum_{\alpha=1}^{4} a_\alpha \hat{p}_\alpha(\mathbf{x}, t),$$

where a_1, a_2, a_3, and a_4 are arbitrary numbers.

Now we introduce the well-known classes of functions and surfaces which will be used in what follows (see, e.g., Günther [161], Kupradze et al. [219]).

Let f be the function determined in the domain Ω of $\mathbb{R}^3$. The function f belongs to $C(\Omega)$ ($C^0(\Omega)$), if f is continuous in Ω and we shall write $f \in C(\Omega)$. The function f belongs to $C^k(\Omega)$, where k is a positive integer, if at each point $\mathbf{x}$ of

Ω there exist all partial derivatives of f up to order k and they are continuous in Ω; in this case we shall write $f \in C^k(\Omega)$.

Let $\partial\Omega$ be the boundary of Ω. The function f is continuously extendible at the point $\mathbf{z} \in \partial\Omega$ from Ω, if there exists a limit

$$\lim_{\Omega \ni \mathbf{x} \to \mathbf{z} \in \partial\Omega} f(\mathbf{x}).$$

The function f belongs to $C^k(\bar{\Omega})$ (denoted by the symbol $f \in C^k(\bar{\Omega})$), where $\bar{\Omega} = \Omega \cup \partial\Omega$ and k is an arbitrary nonnegative integer, if $f \in C^k(\Omega)$, f and all its derivatives with respect to the Cartesian coordinates up to order k are continuously extendible at each point of $\partial\Omega$.

The function f belongs to $C^{0,\nu}(\Omega)$ (denoted by the symbol $f \in C^{0,\nu}(\Omega)$), where $0 < \nu \leq 1$ and Ω is the bounded domain in $\mathbb{R}^3$, if the following Hölder's inequality is fulfilled

$$|f(\mathbf{x}) - f(\mathbf{y})| < c|\mathbf{x} - \mathbf{y}|^\nu$$

for any $\mathbf{x}$ and $\mathbf{y}$ in Ω, c is a positive constant chosen for f and not depending on $\mathbf{x}$ and $\mathbf{y}$ in Ω. However, if Ω is the unbounded domain in $\mathbb{R}^3$, then above inequality is fulfilled for $\forall\, \mathbf{x}, \mathbf{y} \in \Omega$ and $|\mathbf{x} - \mathbf{y}| \leq 1$.

The function f belongs to $C^{k,\nu}(\Omega)$ (denoted by the symbol $f \in C^{k,\nu}(\Omega)$), where k is an arbitrary positive integer and $0 < \nu \leq 1$, if f has all derivatives with respect to the Cartesian coordinates up to order k, these derivatives are uniformly continuous in the domain Ω and all k-order derivatives belong to $C^{0,\nu}(\Omega)$. Class $C^{k,\nu}(\bar{\Omega})$ of functions is defined similarly.

If $\mathbf{f}$ is vector or matrix determined in Ω, then $\mathbf{f} \in C^k(\Omega)$ ($\mathbf{f} \in C^{k,\nu}(\Omega), \mathbf{f} \in C^k(\bar{\Omega})$, $\mathbf{f} \in C^{k,\nu}(\bar{\Omega})$) means that every component of it belongs to $C^k(\Omega)$ ($C^{k,\nu}(\Omega)$, $C^k(\bar{\Omega})$, $C^{k,\nu}(\bar{\Omega})$).

Let S be the closed surface surrounding the finite domain Ω^+ in $\mathbb{R}^3$, $\overline{\Omega^+} = \Omega^+ \cup S$, $\Omega^- = \mathbb{R}^3 \backslash \overline{\Omega^+}$ and $\overline{\Omega^-} = \Omega^- \cup S$; $\mathbf{n}(\mathbf{z})$ be the external (with respect to Ω^+) unit normal vector to S at $\mathbf{z}$, $\mathbf{n} = (n_1, n_2, n_3)$, $\frac{\partial}{\partial \mathbf{n}}$ is the derivative along the vector $\mathbf{n}$.

The reader will find the definition of a surface of class C^k or $C^{k,\nu}$, for example, in the books by Kupradze et al. [219] and Günther [161], where k is an arbitrary natural number and $0 < \nu \leq 1$. Indeed, the surfaces of classes C^1, $C^{1,\nu}$, and C^2 are called the smooth surface, the Liapunov surface, and the surface with a continuous curvature, respectively.

1.4 Equations of Thermoelasticity

Within a multi-porosity conceptual framework (see Bai et al. [22], Liu and Abousleiman [225], Mehrabian and Abousleiman [247], Moutsopoulos et al. [254],

Straughan [326]), the governing system of field equations of the linear theory of thermoelasticity for quadruple porosity isotropic materials with macro-, meso-, micro-, and submicropores may be written in the following six sets of equations (see Straughan [324]):

1. *The equations of motion*

$$t_{lj,j} = \rho \left(\ddot{\hat{u}}_l - \hat{F}_l^{(1)} \right), \qquad l = 1, 2, 3, \tag{1.1}$$

where t_{lj} is the component of total stress tensor, ρ is the reference mass density of solid, $\rho > 0$, $\hat{F}_l^{(1)}$ is the component of body force per unit mass.

2. *The equations of fluid mass conservation*

$$v_{j,j}^{(\alpha)} + \dot{\zeta}_\alpha + a_\alpha \dot{e}_{rr} + d_\alpha + \varepsilon_\alpha \dot{\hat{\theta}} = 0, \tag{1.2}$$

where $\mathbf{v}^{(\alpha)} = \left(v_1^{(\alpha)}, v_2^{(\alpha)}, v_3^{(\alpha)} \right)$ $(\alpha = 1, 2, 3, 4)$; $\mathbf{v}^{(1)}$, $\mathbf{v}^{(2)}$, $\mathbf{v}^{(3)}$ and $\mathbf{v}^{(4)}$ are the fluid flux vectors for the macro-, meso-, micro-, and submicropore (first, second, third, and fourth) phases, respectively; e_{lj} is the component of the strain tensor,

$$e_{lj} = \frac{1}{2} \left(\hat{u}_{l,j} + \hat{u}_{j,l} \right), \qquad l, j = 1, 2, 3, \tag{1.3}$$

a_α is the effective stress parameter, ε_α is the constitutive thermal constant, ζ_α is the increments of fluid (volumetric strain) in the α-th pore phase and defined by

$$\zeta_\alpha = b_{\alpha\beta} \hat{p}_\beta, \tag{1.4}$$

$b_{\alpha\beta}$ is the cross-coupling compressibility for fluid flow at the interface between the α-th and β-th pore systems $(\alpha, \beta = 1, 2, 3, 4)$ and measures the compressibilities of the pore systems,

$$d_\alpha = \sum_{\beta=1;\beta\neq\alpha}^{4} \gamma_{\alpha\beta} (\hat{p}_\alpha - \hat{p}_\beta), \qquad \alpha = 1, 2, 3, 4, \tag{1.5}$$

$\gamma_{\alpha\beta}$ is the internal transport coefficient and corresponds to a fluid transfer rate respecting the intensity of flow between the α-th and β-th pore phases, $\gamma_{\alpha\beta} = \gamma_{\beta\alpha} \geq 0$ for $\alpha \neq \beta$. Obviously, (1.5) implies

$$d_\alpha = d_{\alpha\beta} \hat{p}_\beta, \qquad \alpha = 1, 2, 3, 4, \tag{1.6}$$

where

$$\mathbf{d} = \left(d_{\alpha\beta}\right)_{4\times 4}$$

$$= \begin{pmatrix} \gamma_{12}+\gamma_{13}+\gamma_{14} & -\gamma_{12} & -\gamma_{13} & -\gamma_{14} \\ -\gamma_{12} & \gamma_{12}+\gamma_{23}+\gamma_{24} & -\gamma_{23} & -\gamma_{24} \\ -\gamma_{13} & -\gamma_{23} & \gamma_{13}+\gamma_{23}+\gamma_{34} & -\gamma_{34} \\ -\gamma_{14} & -\gamma_{24} & -\gamma_{34} & \gamma_{14}+\gamma_{24}+\gamma_{34} \end{pmatrix}_{4\times 4}. \tag{1.7}$$

Clearly, $\det \mathbf{d} = 0$. Moreover, if we ignore fluid flow between the α-th and β-th pore systems, then $\gamma_{\alpha\beta} = 0$.

3. *The constitutive equations*

$$t_{lj} = 2\mu\, e_{lj} + \lambda\, e_{rr}\delta_{lj} - \left(a_\alpha\, \hat{p}_\alpha + \varepsilon_0 \hat{\theta}\right)\delta_{lj}, \qquad l, j = 1, 2, 3, \tag{1.8}$$

where λ and μ are the Lamé constants, and ε_0 is the thermal expansion coefficient.

4. *Darcy's law for material with quadruple porosity*

$$\mathbf{v}^{(\alpha)} = -\frac{\kappa_{\alpha\beta}}{\tilde{\mu}} \nabla\, \hat{p}_\beta - \mathbf{s}^{(\alpha)}, \qquad \alpha = 1, 2, 3, 4, \tag{1.9}$$

where $\tilde{\mu}$ is the fluid viscosity and $\kappa_{\alpha\beta}$ are the macroscopic intrinsic permeabilities associated with the four pore systems; indeed, $\kappa_{\alpha\beta}(\alpha \neq \beta)$ are the cross-coupling permeabilities for fluid flow at the interface between the four pore systems; $\mathbf{s}^{(\alpha)} = \rho_\alpha \hat{\mathbf{s}}^{(\alpha)}$ (no sum), ρ_α, and $\mathbf{s}^{(\alpha)}$ are the densities of fluid and the external forces (such as gravity) for the α-th pore phase, respectively.

5. *Fourier's law of heat conduction*

$$\hat{\mathbf{q}} = -k \nabla \hat{\theta}, \tag{1.10}$$

where $\hat{\mathbf{q}}$ is the heat flux vector and k is the thermal conductivity of the material.

6. *The heat transfer equation*

$$\operatorname{div}\hat{\mathbf{q}} = -T_0 \dot{\eta} + \rho \hat{s}, \tag{1.11}$$

where η is the entropy per unit mass,

$$\eta = a_0 \hat{\theta} + \varepsilon_0 \operatorname{div}\hat{\mathbf{u}} + \varepsilon_\alpha\, \hat{p}_\alpha, \tag{1.12}$$

a_0 is the heat capacity, $a_0 > 0$ and $\hat{s}$ is the heat source.

Substituting Eqs. (1.3), (1.4), (1.6)–(1.10), and (1.12) into (1.1), (1.2), and (1.11) we obtain the following system of equations of motion in the linear theory of thermoelasticity for materials with quadruple porosity expressed in terms of the displacement vector $\hat{\mathbf{u}}$, the pressures $\hat{p}_\alpha (\alpha = 1, 2, 3, 4)$, and the temperature $\hat{\theta}$:

$$
\begin{aligned}
&\mu \hat{u}_{l,jj} + (\lambda + \mu)\hat{u}_{j,lj} - a_\alpha \hat{p}_{\alpha,l} - \varepsilon_0 \hat{\theta}_{,l} - \rho \ddot{\hat{u}}_l = \hat{F}_l, \\
&k_{\alpha\beta}\, \hat{p}_{\beta,rr} - b_{\alpha\beta}\, \dot{\hat{p}}_\beta - d_{\alpha\beta}\, \hat{p}_\beta - a_\alpha\, \dot{\hat{u}}_{j,j} - \varepsilon_\alpha \dot{\hat{\theta}} = \hat{F}_{\alpha+3}, \qquad (1.13) \\
&k\, \hat{\theta}_{,jj} - T_0 \left(a_0 \dot{\hat{\theta}} + \varepsilon_0\, \dot{\hat{u}}_{j,j} + \varepsilon_\alpha\, \dot{\hat{p}}_\alpha \right) = \hat{F}_8,
\end{aligned}
$$

where

$$
\hat{F}_l = -\rho \hat{F}_l^{(1)}, \qquad \hat{F}_{\alpha+3} = -\operatorname{div} \mathbf{s}^{(\alpha)},
$$

$$
\hat{F}_8 = -\rho\, \hat{s}, \qquad k_{\alpha\beta} = \frac{\kappa_{\alpha\beta}}{\tilde{\mu}},
$$

$$
l = 1, 2, 3, \qquad \alpha, \beta = 1, 2, 3, 4.
$$

Clearly, from (1.13) we get

$$
\begin{aligned}
&\mu \Delta \hat{\mathbf{u}} + (\lambda + \mu)\nabla \operatorname{div} \hat{\mathbf{u}} - \nabla(\mathbf{a}\, \hat{\mathbf{p}}) - \varepsilon_0 \nabla \hat{\theta} - \rho\, \ddot{\hat{\mathbf{u}}} = \hat{\mathbf{F}}^{(1)}, \\
&\mathbf{K}\, \Delta\, \hat{\mathbf{p}} - \mathbf{b}\, \dot{\hat{\mathbf{p}}} - \mathbf{d}\hat{\mathbf{p}} - \mathbf{a}\, \operatorname{div}\dot{\hat{\mathbf{u}}} - \boldsymbol{\varepsilon}\, \dot{\hat{\theta}} = \hat{\mathbf{F}}^{(2)}, \qquad (1.14) \\
&k \Delta\, \hat{\theta} - T_0 \left(a_0 \dot{\hat{\theta}} + \varepsilon_0 \operatorname{div} \dot{\hat{\mathbf{u}}} + \boldsymbol{\varepsilon}\, \dot{\hat{\mathbf{p}}} \right) = \hat{F}_8,
\end{aligned}
$$

where

$$
\hat{\mathbf{F}}^{(1)} = (\hat{F}_1, \hat{F}_2, \hat{F}_3), \qquad \hat{\mathbf{F}}^{(2)} = (\hat{F}_4, \hat{F}_5, \hat{F}_6, \hat{F}_7), \qquad \mathbf{K} = (k_{\alpha\beta})_{4\times4},
$$

$$
\mathbf{a} = (a_1, a_2, a_3, a_4), \qquad \mathbf{b} = (b_{\alpha\beta})_{4\times4}, \qquad \boldsymbol{\varepsilon} = (\varepsilon_1, \varepsilon_2, \varepsilon_3, \varepsilon_4),
$$

$$
\hat{\mathbf{p}} = (\hat{p}_1, \hat{p}_2, \hat{p}_3, \hat{p}_4), \qquad \mathbf{a}\hat{\mathbf{p}} = a_\alpha \hat{p}_\alpha.
$$

In what follows, we consider five special cases of equations of motion (1.14).

- *Equations of steady vibrations.* If $\hat{u}_l$, $\hat{p}_\alpha$, $\hat{\theta}$ and $\hat{F}_j$ are postulated to have a harmonic time variation, that is,

$$\left\{\hat{u}_l,\, \hat{p}_\alpha,\, \hat{\theta},\, \hat{F}_j\right\}(\mathbf{x},t) = \mathrm{Re}\left[\left\{u_l,\, p_\alpha,\, \theta,\, F_j\right\}(\mathbf{x})\, e^{-i\omega t}\right],$$

$$l = 1,2,3, \qquad \alpha = 1,2,3,4, \qquad j = 1,2,\cdots,8,$$

then from (1.14) we obtain the following system of equations of steady vibrations in the linear theory of thermoelasticity for quadruple porosity solids:

$$\begin{aligned}
&(\mu\,\Delta + \rho\,\omega^2)\mathbf{u} + (\lambda+\mu)\,\nabla \mathrm{div}\mathbf{u} - \nabla\,(\mathbf{ap}) - \varepsilon_0 \nabla\theta = \mathbf{F}^{(1)},\\
&(\mathbf{K}\,\Delta + \mathbf{c})\mathbf{p} + i\omega\,\mathbf{a}\,\mathrm{div}\mathbf{u} + i\omega\boldsymbol{\varepsilon}\,\theta = \mathbf{F}^{(2)}, \qquad (1.15)\\
&(k\Delta + i\omega a_0 T_0)\theta + i\omega T_0\,(\varepsilon_0\,\mathrm{div}\,\mathbf{u} + \boldsymbol{\varepsilon}\,\mathbf{p}) = F_8,
\end{aligned}$$

where ω is the oscillation frequency, $\omega > 0$, $\mathbf{c} = (c_{\alpha\beta})_{4\times 4} = i\omega\mathbf{b} - \mathbf{d}$, $\mathbf{p} = (p_1, p_2, p_3, p_4)$, $\mathbf{ap} = a_\alpha p_\alpha$, $\mathbf{F}^{(1)} = (F_1, F_2, F_3)$, and $\mathbf{F}^{(2)} = (F_4, F_5, F_6, F_7)$.

- *Equations in the Laplace transform space.* We can rewrite the system (1.14) in the Laplace transform space as

$$\begin{aligned}
&(\mu\,\Delta - \rho\,\tau^2)\mathbf{u} + (\lambda+\mu)\,\nabla \mathrm{div}\mathbf{u} - \nabla\,(\mathbf{a}\,\mathbf{p}) - \varepsilon_0 \nabla\theta = \mathbf{F}^{(1)},\\
&(\mathbf{K}\,\Delta + \mathbf{c}')\,\mathbf{p} - \tau\,\mathbf{a}\,\mathrm{div}\mathbf{u} + \tau\boldsymbol{\varepsilon}\,\theta = \mathbf{F}^{(2)}, \qquad (1.16)\\
&(k\Delta - \tau a_0 T_0)\theta - \tau T_0\,(\varepsilon_0\,\mathrm{div}\,\mathbf{u} + \boldsymbol{\varepsilon}\,\mathbf{p}) = F_8,
\end{aligned}$$

where $\mathbf{c}' = (c'_{lj})_{4\times 4} = -(\tau\,\mathbf{b}{+}\mathbf{d})$; τ is a complex number and $\mathrm{Re}\,\tau > 0$. It is easy to verify that the system (1.16) may be obtained from (1.15) by replacing ω by $i\tau$. As in the classical theories of elasticity and thermoelasticity (see, e.g., Kupradze et al. [219]), Eq. (1.16) will be called the equations of *pseudo-oscillations* in the linear theory of thermoelasticity for solid with quadruple porosity. In addition, the system (1.16) plays an important auxiliary role in the study of dynamic problems of the linear theory of thermoelasticity for quadruple porosity solids (for details see Kupradze [216] and Kupradze et al. [219]).
- *Equations of steady vibrations in the quasi-static theory.* Neglecting inertial effect in (1.14), from (1.15) we obtain the system of equations of steady vibrations in the linear quasi-static theory of thermoelasticity for quadruple porosity materials

$$\begin{aligned}
&\mu\,\Delta\mathbf{u} + (\lambda+\mu)\,\nabla \mathrm{div}\mathbf{u} - \nabla\,(\mathbf{ap}) - \varepsilon_0 \nabla\theta = \mathbf{F}^{(1)},\\
&(\mathbf{K}\,\Delta + \mathbf{c})\mathbf{p} + i\omega\mathbf{a}\,\mathrm{div}\mathbf{u} + i\omega\boldsymbol{\varepsilon}\,\theta = \mathbf{F}^{(2)}, \qquad (1.17)\\
&(k\Delta + i\omega a_0 T_0)\theta + i\omega T_0\,(\varepsilon_0\,\mathrm{div}\,\mathbf{u} + \boldsymbol{\varepsilon}\,\mathbf{p}) = F_8,
\end{aligned}$$

- *Equations of equilibrium.* In the equilibrium case ($\omega = 0$), from (1.15) we get the following system of equations in the linear equilibrium theory of thermoelasticity for solids with quadruple porosity

$$\begin{aligned}
&\mu\,\Delta\mathbf{u} + (\lambda+\mu)\,\nabla\mathrm{div}\mathbf{u} - \nabla\,(\mathbf{ap}) - \varepsilon_0\nabla\theta = \mathbf{F}^{(1)},\\
&(\mathbf{K}\,\Delta - \mathbf{d})\mathbf{p} = \mathbf{F}^{(2)},\\
&k\Delta\theta = F_8.
\end{aligned}\tag{1.18}$$

- *Equations of steady vibrations of heat conduction for quadruple porosity rigid body.* Obviously, neglecting the displacement field in (1.14), from (1.15) we get the system of equations of steady vibrations in the linear theory of heat conduction in a quadruple porosity rigid body

$$\begin{aligned}
&(\mathbf{K}\,\Delta + \mathbf{c})\mathbf{p} + i\omega\boldsymbol{\varepsilon}\,\theta = \mathbf{F}^{(2)},\\
&(k\Delta + i\omega a_0 T_0)\theta + i\omega T_0\boldsymbol{\varepsilon}\,\mathbf{p} = F_8.
\end{aligned}\tag{1.19}$$

1.5 Equations of Elasticity

We now consider the isothermal case $\hat{\theta}(\mathbf{x},t) \equiv 0$. Obviously, in this case from (1.14) we obtain the following system of equations of motion in the linear theory of elasticity for materials with quadruple porosity expressed in terms of the displacement vector $\hat{\mathbf{u}}$ and the pressures $\hat{p}_\alpha (\alpha = 1, 2, 3, 4)$ (see Straughan [324]):

$$\begin{aligned}
&\mu\Delta\hat{\mathbf{u}} + (\lambda+\mu)\nabla\mathrm{div}\,\hat{\mathbf{u}} - \nabla(\mathbf{a}\,\hat{\mathbf{p}}) - \rho\,\ddot{\hat{\mathbf{u}}} = \hat{\mathbf{F}}^{(1)},\\
&\mathbf{K}\,\Delta\,\hat{\mathbf{p}} - \mathbf{b}\,\dot{\hat{\mathbf{p}}} - \mathbf{d}\hat{\mathbf{p}} - \mathbf{a}\,\mathrm{div}\dot{\hat{\mathbf{u}}} = \hat{\mathbf{F}}^{(2)}.
\end{aligned}\tag{1.20}$$

It is easy to verify that the following systems of equations in the linear theory of elasticity for materials with quadruple porosity may be obtained from (1.20):

- *Equations of steady vibrations*

$$\begin{aligned}
&(\mu\,\Delta + \rho\,\omega^2)\mathbf{u} + (\lambda+\mu)\,\nabla\mathrm{div}\mathbf{u} - \nabla\,(\mathbf{ap}) = \mathbf{F}^{(1)},\\
&(\mathbf{K}\,\Delta + \mathbf{c})\mathbf{p} + i\omega\mathbf{a}\,\mathrm{div}\mathbf{u} = \mathbf{F}^{(2)}.
\end{aligned}\tag{1.21}$$

- *Equations in the Laplace transform space*

$$(\mu\,\Delta - \rho\,\tau^2)\mathbf{u} + (\lambda+\mu)\,\nabla\mathrm{div}\mathbf{u} - \nabla\,(\mathbf{a}\,\mathbf{p}) = \mathbf{F}^{(1)},$$
$$(\mathbf{K}\,\Delta + \mathbf{c}')\,\mathbf{p} - \tau\,\mathbf{a}\,\mathrm{div}\mathbf{u} = \mathbf{F}^{(2)}. \tag{1.22}$$

- *Equations of steady vibrations in the quasi-static theory*

$$\mu\,\Delta\mathbf{u} + (\lambda+\mu)\,\nabla\mathrm{div}\mathbf{u} - \nabla\,(\mathbf{a}\mathbf{p}) = \mathbf{F}^{(1)},$$
$$(\mathbf{K}\,\Delta + \mathbf{c})\mathbf{p} + i\omega\mathbf{a}\,\mathrm{div}\mathbf{u} = \mathbf{F}^{(2)}. \tag{1.23}$$

- *Equations of equilibrium*

$$\mu\,\Delta\mathbf{u} + (\lambda+\mu)\,\nabla\mathrm{div}\mathbf{u} - \nabla\,(\mathbf{a}\mathbf{p}) = \mathbf{F}^{(1)},$$
$$(\mathbf{K}\,\Delta - \mathbf{d})\mathbf{p} = \mathbf{F}^{(2)}. \tag{1.24}$$

- *Equations of steady vibrations for quadruple porosity rigid body*

$$(\mathbf{K}\,\Delta + \mathbf{c})\mathbf{p} = \mathbf{F}^{(2)}. \tag{1.25}$$

- *Equations of equilibrium for quadruple porosity rigid body*

$$(\mathbf{K}\,\Delta - \mathbf{d})\mathbf{p} = \mathbf{F}^{(2)}. \tag{1.26}$$

Remark 1.1 Clearly, if $\mathbf{a} = \mathbf{0}$, then the system (1.21) is uncoupled for the vectors $\mathbf{u}$ and $\mathbf{p}$, and consequently, from (1.21) one may derive the system of equations of steady vibrations for the classical theory of elasticity (see Kupradze et al. [219]) and the system of equations of steady vibrations for quadruple porosity rigid body (see Chap. 4). Henceforth, throughout this book we adopt that $\mathbf{a} \neq \mathbf{0}$.

1.6 Matrix Representation of the Basic Equations

It is convenient to write the basic systems of Eqs. (1.15)–(1.19) and (1.21)–(1.26) in the matrix form.

We introduce the second order matrix differential operators with constant coefficients:

1.

$$
\begin{aligned}
&\mathbf{A}(\mathbf{D_x},\omega)=\big(A_{lj}(\mathbf{D_x},\omega)\big)_{8\times 8},\\
&A_{lj}(\mathbf{D_x},\omega)=(\mu\Delta+\rho\omega^2)\delta_{lj}+(\lambda+\mu)\frac{\partial^2}{\partial x_l\partial x_j},\\
&A_{l;\alpha+3}(\mathbf{D_x},\omega)=-a_\alpha\frac{\partial}{\partial x_l},\qquad A_{l8}(\mathbf{D_x},\omega)=-\varepsilon_0\frac{\partial}{\partial x_l},\\
&A_{\alpha+3;j}(\mathbf{D_x},\omega)=i\omega a_\alpha\frac{\partial}{\partial x_j},\qquad A_{\alpha+3;\beta+3}(\mathbf{D_x},\omega)=k_{\alpha\beta}\Delta+c_{\alpha\beta},\\
&A_{\alpha+3;8}(\mathbf{D_x},\omega)=i\omega\varepsilon_\alpha,\qquad A_{8j}(\mathbf{D_x},\omega)=i\omega T_0\varepsilon_0\frac{\partial}{\partial x_j},\\
&A_{8;\alpha+3}(\mathbf{D_x},\omega)=i\omega T_0\varepsilon_\alpha,\qquad A_{88}(\mathbf{D_x},\omega)=k\Delta+i\omega a_0T_0,\\
&l,j=1,2,3,\qquad \alpha,\beta=1,2,3,4.
\end{aligned}
\tag{1.27}
$$

2.

$$
\begin{aligned}
&\mathbf{A}^{(q)}(\mathbf{D_x},\omega)=\Big(A^{(q)}_{lj}(\mathbf{D_x},\omega)\Big)_{8\times 8},\\
&A^{(q)}_{lj}(\mathbf{D_x},\omega)=\mu\Delta\delta_{lj}+(\lambda+\mu)\frac{\partial^2}{\partial x_l\partial x_j},\\
&A^{(q)}_{lm}(\mathbf{D_x},\omega)=A_{lm}(\mathbf{D_x},\omega),\qquad A^{(q)}_{ms}(\mathbf{D_x},\omega)=A_{ms}(\mathbf{D_x},\omega),\\
&l,j=1,2,3,\qquad m=4,5,\cdots,8,\qquad s=1,2,\cdots,8.
\end{aligned}
\tag{1.28}
$$

3.

$$
\begin{aligned}
&\mathbf{A}^{(e)}(\mathbf{D_x})=\Big(A^{(e)}_{lj}(\mathbf{D_x})\Big)_{8\times 8},\\
&A^{(e)}_{ls}(\mathbf{D_x})=A^{(q)}_{ls}(\mathbf{D_x}),\\
&A^{(e)}_{\alpha+3;\beta+3}(\mathbf{D_x})=k_{\alpha\beta}\Delta-d_{\alpha\beta},\qquad A^{(e)}_{88}(\mathbf{D_x})=k\Delta,\\
&A^{(e)}_{\alpha+3;l}(\mathbf{D_x})=A^{(e)}_{\alpha+3;8}(\mathbf{D_x})=A^{(e)}_{8m}(\mathbf{D_x})=0,\qquad l=1,2,3,\\
&\alpha,\beta=1,2,3,4,\qquad s=1,2,\cdots,8,\qquad m=1,2,\cdots,7.
\end{aligned}
\tag{1.29}
$$

4.

$$\mathbf{A}^{(r)}(\mathbf{D_x},\omega)=\left(A^{(r)}_{lj}(\mathbf{D_x},\omega)\right)_{5\times 5},$$

$$A^{(r)}_{\alpha\beta}(\mathbf{D_x},\omega)=k_{\alpha\beta}\Delta+c_{\alpha\beta},\qquad A^{(r)}_{\alpha 5}(\mathbf{D_x},\omega)=i\omega\varepsilon_\alpha,$$
$$A^{(r)}_{5\alpha}(\mathbf{D_x},\omega)=i\omega T_0\varepsilon_\alpha,\qquad A^{(r)}_{55}(\mathbf{D_x},\omega)=k\Delta+i\omega a_0T_0, \tag{1.30}$$
$$\alpha,\beta=1,2,3,4.$$

Taking into account (1.27)–(1.30) the systems (1.15)–(1.19) can be rewritten in the following matrix form

$$\mathbf{A}(\mathbf{D_x},\omega)\,\mathbf{U}(\mathbf{x})=\mathbf{F}(\mathbf{x}), \tag{1.31}$$

$$\mathbf{A}(\mathbf{D_x},i\tau)\,\mathbf{U}(\mathbf{x})=\mathbf{F}(\mathbf{x}), \tag{1.32}$$

$$\mathbf{A}^{(q)}(\mathbf{D_x},\omega)\,\mathbf{U}(\mathbf{x})=\mathbf{F}(\mathbf{x}), \tag{1.33}$$

$$\mathbf{A}^{(e)}(\mathbf{D_x})\,\mathbf{U}(\mathbf{x})=\mathbf{F}(\mathbf{x}), \tag{1.34}$$

$$\mathbf{A}^{(r)}(\mathbf{D_x},\omega)\,\boldsymbol{\vartheta}=\mathbf{F}'(\mathbf{x}), \tag{1.35}$$

respectively, where $\mathbf{U}=(\mathbf{u},\mathbf{p},\theta)$ and $\mathbf{F}=(\mathbf{F}^{(1)},\mathbf{F}^{(2)},F_8)$ are eight-component vector functions, $\boldsymbol{\vartheta}=(\mathbf{p},\theta)$ and $\mathbf{F}'=(\mathbf{F}^{(2)},F_8)$ are five-component vector functions, and $\mathbf{x}\in\mathbb{R}^3$.

Now we introduce the matrix differential operators:

$$\boldsymbol{\mathcal{A}}(\mathbf{D_x},\omega)=\left(\mathcal{A}_{lj}(\mathbf{D_x},\omega)\right)_{7\times 7},\qquad \boldsymbol{\mathcal{A}}^{(q)}(\mathbf{D_x},\omega)=\left(\mathcal{A}^{(q)}_{lj}(\mathbf{D_x},\omega)\right)_{7\times 7},$$
$$\boldsymbol{\mathcal{A}}^{(e)}(\mathbf{D_x})=\left(\mathcal{A}^{(e)}_{lj}(\mathbf{D_x})\right)_{7\times 7},\qquad \boldsymbol{\mathcal{A}}^{(r)}(\mathbf{D_x},\omega)=\left(\mathcal{A}^{(r)}_{lj}(\mathbf{D_x},\omega)\right)_{4\times 4},$$
$$\boldsymbol{\mathcal{A}}^{(re)}(\mathbf{D_x})=\left(\mathcal{A}^{(re)}_{lj}(\mathbf{D_x})\right)_{4\times 4}, \tag{1.36}$$

where

$$\mathcal{A}_{lj}(\mathbf{D_x},\omega)=A_{lj}(\mathbf{D_x},\omega),\qquad \mathcal{A}^{(q)}_{lj}(\mathbf{D_x},\omega)=A^{(q)}_{lj}(\mathbf{D_x},\omega),$$
$$\mathcal{A}^{(e)}_{lj}(\mathbf{D_x})=A^{(e)}_{lj}(\mathbf{D_x}),\qquad \mathcal{A}^{(r)}_{\alpha\beta}(\mathbf{D_x},\omega)=k_{\alpha\beta}\Delta+c_{\alpha\beta}, \tag{1.37}$$

$$\mathcal{A}^{(re)}_{\alpha\beta}(\mathbf{D_x}) = k_{\alpha\beta}\Delta - d_{\alpha\beta}, \qquad l, j = 1, 2, \cdots, 7,$$

$$\alpha, \beta = 1, 2, 3, 4.$$

Obviously, employing (1.36) and (1.37), the systems (1.21)–(1.26) can be rewritten in the following matrix form

$$\mathcal{A}(\mathbf{D_x}, \omega)\, \mathbf{V}(\mathbf{x}) = \mathcal{F}(\mathbf{x}), \tag{1.38}$$

$$\mathcal{A}(\mathbf{D_x}, i\tau)\, \mathbf{V}(\mathbf{x}) = \mathcal{F}(\mathbf{x}), \tag{1.39}$$

$$\mathcal{A}^{(q)}(\mathbf{D_x}, \omega)\, \mathbf{V}(\mathbf{x}) = \mathcal{F}(\mathbf{x}), \tag{1.40}$$

$$\mathcal{A}^{(e)}(\mathbf{D_x})\, \mathbf{V}(\mathbf{x}) = \mathcal{F}(\mathbf{x}), \tag{1.41}$$

$$\mathcal{A}^{(r)}(\mathbf{D_x}, \omega)\, \mathbf{p}(\mathbf{x}) = \mathbf{F}^{(2)}(\mathbf{x}), \tag{1.42}$$

$$\mathcal{A}^{(re)}(\mathbf{D_x})\, \mathbf{p}(\mathbf{x}) = \mathbf{F}^{(2)}(\mathbf{x}), \tag{1.43}$$

respectively, where $\mathbf{V} = (\mathbf{u}, \mathbf{p})$ and $\mathcal{F} = (\mathbf{F}^{(1)}, \mathbf{F}^{(2)})$ are seven-component vector functions, $\mathbf{p}$ and $\mathbf{F}^{(2)}$ are four-component vector functions, and $\mathbf{x} \in \mathbb{R}^3$.

Remark 1.2 Throughout this book we assume that $\mathbf{K}$ and $\mathbf{b}$ are positive definite matrices.

1.7 Stress Operators

Let $\mathbf{n}(\mathbf{x}) = (n_1(\mathbf{x}), n_2(\mathbf{x}), n_3(\mathbf{x}))$ be an arbitrary unit vector. In the sequel we use the matrix differential operators:

(i)

$$\mathbf{P}(\mathbf{D_x}, \mathbf{n}) = (P_{lj}(\mathbf{D_x}, \mathbf{n}))_{8\times 8},$$

$$P_{lj}(\mathbf{D_x}, \mathbf{n}) = \mu\delta_{lj}\frac{\partial}{\partial \mathbf{n}} + \lambda n_l \frac{\partial}{\partial x_j} + \mu n_j \frac{\partial}{\partial x_l},$$

$$P_{l;\alpha+3}(\mathbf{D_x}, \mathbf{n}) = -a_\alpha\, n_l, \qquad P_{l8}(\mathbf{D_x}, \mathbf{n}) = -\varepsilon_0\, n_l, \tag{1.44}$$

$$P_{\alpha+3;\beta+3}(\mathbf{D_x}, \mathbf{n}) = k_{\alpha\beta}\frac{\partial}{\partial \mathbf{n}}, \qquad P_{88}(\mathbf{D_x}, \mathbf{n}) = k\frac{\partial}{\partial \mathbf{n}},$$

$$P_{\alpha+3;j}(\mathbf{D_x}, \mathbf{n}) = P_{\alpha+3;8}(\mathbf{D_x}, \mathbf{n}) = P_{8m}(\mathbf{D_x}, \mathbf{n}) = 0;$$

(ii)

$$\mathcal{P}(\mathbf{D_x}, \mathbf{n}) = (\mathcal{P}_{mr}(\mathbf{D_x}, \mathbf{n}))_{7\times 7}, \qquad \mathcal{P}_{mr}(\mathbf{D_x}, \mathbf{n}) = P_{mr}(\mathbf{D_x}, \mathbf{n}), \tag{1.45}$$

where $l, j = 1, 2, 3$, $\alpha, \beta = 1, 2, 3, 4$ and $m, r = 1, 2, \cdots, 7$.

The operators $\mathcal{P}(\mathbf{D_x}, \mathbf{n})$ and $\mathbf{P}(\mathbf{D_x}, \mathbf{n})$ are called the *stress operators* in the linear theories of elasticity and thermoelasticity of quadruple porosity materials, respectively.

In what follows, we also consider the stress operators of a more general form. We introduce the notation

(i)

$$\mathbf{P}^{(\kappa)}(\mathbf{D_x}, \mathbf{n}) = \left(P^{(\kappa)}_{lj}(\mathbf{D_x}, \mathbf{n})\right)_{8\times 8},$$

$$P^{(\kappa)}_{lj}(\mathbf{D_x}, \mathbf{n}) = \mu\delta_{lj}\frac{\partial}{\partial \mathbf{n}} + (\lambda + \mu)n_l\frac{\partial}{\partial x_j} + \kappa\mathcal{M}_{lj}(\mathbf{D_x}, \mathbf{n}),$$

$$P^{(\kappa)}_{lm}(\mathbf{D_x}, \mathbf{n}) = P_{lm}(\mathbf{D_x}, \mathbf{n}), \qquad P^{(\kappa)}_{mr}(\mathbf{D_x}, \mathbf{n}) = P_{mr}(\mathbf{D_x}, \mathbf{n}), \tag{1.46}$$

$$m = 4, 5, \cdots, 8, \qquad r = 1, 2, \cdots, 8;$$

(ii)

$$\mathcal{P}^{(\kappa)}(\mathbf{D_x}, \mathbf{n}) = \left(\mathcal{P}^{(\kappa)}_{ms}(\mathbf{D_x}, \mathbf{n})\right)_{7\times 7},$$

$$\mathcal{P}^{(\kappa)}_{ms}(\mathbf{D_x}, \mathbf{n}) = P^{(\kappa)}_{ms}(\mathbf{D_x}, \mathbf{n}), \tag{1.47}$$

$$m, s = 1, 2, \cdots, 7,$$

where κ is an arbitrary real number and

$$\mathcal{M}_{lj}(\mathbf{D_x}, \mathbf{n}) = n_j\frac{\partial}{\partial x_l} - n_l\frac{\partial}{\partial x_j}, \qquad l, j = 1, 2, 3. \tag{1.48}$$

The operators $\mathcal{P}^{(\kappa)}(\mathbf{D_x}, \mathbf{n})$ and $\mathbf{P}^{(\kappa)}(\mathbf{D_x}, \mathbf{n})$ are called the *generalized stress operators* in the linear theories of elasticity and thermoelasticity of quadruple porosity materials, respectively.

We will use the generalized stress operators $\mathcal{P}^{(\kappa)}(\mathbf{D_x}, \mathbf{n})$ and $\mathbf{P}^{(\kappa)}(\mathbf{D_x}, \mathbf{n})$ for the following two values of κ:

(i) $\kappa = \mu$. Clearly, on the basis of (1.44)–(1.47) the stress operators $\mathcal{P}(\mathbf{D_x}, \mathbf{n})$ and $\mathbf{P}(\mathbf{D_x}, \mathbf{n})$ are obtained from the generalized stress operators $\mathcal{P}^{(\kappa)}(\mathbf{D_x}, \mathbf{n})$ and $\mathbf{P}^{(\kappa)}(\mathbf{D_x}, \mathbf{n})$ by replacing κ by μ, respectively;

(ii) $\kappa = \mu(\lambda + \mu)(\lambda + 3\mu)^{-1}$. This value of κ will use in the proof of the existence theorems in the quasi-static theories of elasticity and thermoelasticity for quadruple porosity materials by using Fredholm's integral equations (see Sects. 7.6 and 11.9).

The operators $\mathcal{P}^{(\kappa)}$ and $\mathbf{P}^{(\kappa)}$ when $\kappa = \mu(\lambda + \mu)(\lambda + 3\mu)^{-1}$ are denoted by $\mathcal{N}$ and $\mathbf{N}$ and called the *pseudostress operators* in the linear theories of elasticity and thermoelasticity for quadruple porosity materials, respectively.

Chapter 2
Fundamental Solutions in Elasticity

This chapter is concerned with the fundamental solutions of the systems of equations in the linear theory of elasticity for materials with quadruple porosity.

In Sect. 2.1, the brief history of the fundamental solutions of partial differential equations of mathematical physics is given.

In Sect. 2.2, the fundamental solutions of steady vibrations and pseudo-oscillations equations of the linear theory of elasticity for materials with quadruple porosity are constructed by means of elementary functions.

In Sects. 2.3 and 2.4, the fundamental solutions of systems of quasi-static and equilibrium equations in the considered theory are presented, respectively.

In Sects. 2.5 and 2.6, the fundamental solutions of steady vibrations and equilibrium equations of quadruple porosity rigid body are obtained by means of elementary functions, respectively.

In Sect. 2.7, the basic properties of these fundamental solutions are established.

Finally, in Sect. 2.8, the singular solutions in the linear theory of elasticity for materials with quadruple porosity are constructed.

2.1 Fundamental Solutions: An Overview

Fundamental solutions play a pivotal role in investigation of various problems of mathematical physics and continuum mechanics. In the eighteenth and nineteenth centuries, the fundamental solutions of special equations of mathematical physics were constructed. Indeed, d'Alembert [119], Volterra [391], and Poisson [276] constructed the fundamental solutions of 1D, 2D, and 3D wave equations, respectively. Laplace [221] used the fundamental solution of the elliptic operator Δ which bears his name.

The fundamental solution (Kelvin's Matrix) of the system of equilibrium equations in the classical theory of elasticity for 3D isotropic homogeneous solid is

M. Svanadze, *Potential Method in Mathematical Theories of Multi-Porosity Media*,
Interdisciplinary Applied Mathematics 51,
https://doi.org/10.1007/978-3-030-28022-2_2

obtained by Thomson (Lord Kelvin) [371] using elementary functions. By virtue of this solution Somigliana [316] presented the integral representation (Somigliana identity) of the displacement vector in the classical theory of elasticity. Fredholm [143] introduced the fundamental solution of equations in the 3D equilibrium theory of elasticity for anisotropic solids. The fundamental solution (Kupradze's matrix) of equation of steady vibrations in the 3D classical theory of elasticity is constructed by Kupradze [215]. Kupradze and Burchuladze [218] obtained the fundamental matrix of the coupled system of equations of steady vibrations in the classical theory of thermoelasticity for 3D isotropic homogeneous solids. Ortner and Wagner [268] presented the fundamental matrix in the dynamical theory of thermoelasticity. These results formed the ground for the revival of the potential method in the classical and modern theories of elasticity and thermoelasticity for materials with microstructures.

Moreover, Ehrenpreis [136] and Malgrange [238] developed the first existence proofs for fundamental solutions of the partial differential operators. The result of these authors states that every not identically vanishing linear partial differential operator with constant coefficients possesses a fundamental solution in the class of generalized function. Hörmander [167] proved that there always exist "regular" fundamental solutions for this kind of operators.

It is well known that fundamental solutions have occupied a special place in the theory of partial differential equations. They encounter in many mathematical, mechanical, physical, and engineering applications. For historical and bibliographical material on the fundamental solutions of linear partial differential operators see Hörmander [168], Kythe [220], and Ortner and Wagner [269].

For investigating BVPs of the classical and nonclassical theories of elasticity and thermoelasticity by potential method it is necessary to construct fundamental solutions of corresponding systems of partial differential equations and to establish their basic properties. In the classical theory of elasticity, the development of the potential method using the fundamental solutions is a significant achievement (see Kupradze [217]).

Several methods are known for constructing fundamental solutions in the classical theories of elasticity and thermoelasticity (for details see Gurtin [162], Hetnarski and Ignaczak [165], Kupradze et al. [219], and Nowacki [263]). The explicit expressions of fundamental solutions in the theory of elasticity, thermoelasticity, and micropolar elasticity are obtained in different manner by Basheleishvili [27], Dragos [134], Kupradze [215], Kupradze and Burchuladze [218], and Sandru [293].

The fundamental solutions in dynamic poroelasticity are constructed by Burridge and Vargas [66], Cleary [101], and Kaynia and Banerjee [194]. The fundamental solution and the representations of Galerkin type solution in the theory of fluid-saturated porous media are established by de Boer and Svanadze [129, 357]. Basic results on the method of fundamental solutions in poroelasticity are given in the book of Augustin [18].

The fundamental solutions in the theories of double and triple porosity materials are constructed explicitly by Svanadze [331, 336, 345, 368, 370] and Svanadze and De Cicco [358].

In this chapter, the fundamental solutions of systems of equations in the linear theories of elasticity for quadruple porosity materials are constructed explicitly and their basic properties are established.

2.2 Equations of Steady Vibrations and Pseudo-Oscillations

In this section the fundamental solutions of the systems of Eqs. (1.21) and (1.22) are constructed explicitly by means of elementary functions.

We introduce the notation

$$\boldsymbol{\mathcal{A}}^{(0)}(\mathbf{D_x}) = \left(\mathcal{A}^{(0)}_{lj}(\mathbf{D_x})\right)_{7\times 7},$$

$$\mathcal{A}^{(0)}_{lj}(\mathbf{D_x}) = \mu\Delta\delta_{lj} + (\lambda+\mu)\frac{\partial^2}{\partial x_l \partial x_j},$$

$$\mathcal{A}^{(0)}_{\alpha+3;\beta+3}(\mathbf{D_x}) = k_{\alpha\beta}\,\Delta,$$

$$\mathcal{A}^{(0)}_{l;\alpha+3}(\mathbf{D_x}) = \mathcal{A}^{(0)}_{\alpha+3;l}(\mathbf{D_x}) = 0,$$

$$l, j = 1, 2, 3, \qquad \alpha, \beta = 1, 2, 3, 4.$$

Definition 2.1 The matrix differential operator $\boldsymbol{\mathcal{A}}^{(0)}(\mathbf{D_x})$ is called the *principal part* of the operators $\boldsymbol{\mathcal{A}}(\mathbf{D_x}, \omega)$, $\boldsymbol{\mathcal{A}}(\mathbf{D_x}, i\tau)$, $\boldsymbol{\mathcal{A}}^{(q)}(\mathbf{D_x}, \omega)$, and $\boldsymbol{\mathcal{A}}^{(e)}(\mathbf{D_x})$.

Definition 2.2 The operator $\boldsymbol{\mathcal{A}}(\mathbf{D_x}, \omega)$ ($\boldsymbol{\mathcal{A}}(\mathbf{D_x}, i\tau)$, $\boldsymbol{\mathcal{A}}^{(q)}(\mathbf{D_x}, \omega)$, $\boldsymbol{\mathcal{A}}^{(e)}(\mathbf{D_x})$) is said to be *elliptic* if

$$\det \boldsymbol{\mathcal{A}}^{(0)}(\boldsymbol{\psi}) \neq 0,$$

where $\boldsymbol{\psi} = (\psi_1, \psi_2, \psi_3) \neq \mathbf{0}$.

Obviously, by direct calculation we have

$$\det \boldsymbol{\mathcal{A}}^{(0)}(\boldsymbol{\psi}) = \mu^2 \mu_0\, k_0\, |\boldsymbol{\psi}|^{14},$$

where $\mu_0 = \lambda + 2\mu,\ k_0 = \det\mathbf{K}$.

Hence, $\boldsymbol{\mathcal{A}}(\mathbf{D_x}, \omega)$, $\boldsymbol{\mathcal{A}}(\mathbf{D_x}, i\tau)$, $\boldsymbol{\mathcal{A}}^{(q)}(\mathbf{D_x})$, $\boldsymbol{\mathcal{A}}^{(e)}(\mathbf{D_x})$ are the elliptic differential operators if and only if

$$\mu\,\mu_0\,k_0 \neq 0. \tag{2.1}$$

In this chapter we will suppose that the assumption (2.1) holds true.

Definition 2.3 The fundamental solution of system (1.21) (the fundamental matrix of operator $\mathcal{A}(\mathbf{D_x}, \omega)$) is the matrix $\mathbf{\Gamma}(\mathbf{x}, \omega) = \left(\Gamma_{lj}(\mathbf{x}, \omega)\right)_{7\times 7}$ satisfying the following equation in the class of generalized functions

$$\mathcal{A}(\mathbf{D_x}, \omega)\mathbf{\Gamma}(\mathbf{x}, \omega) = \delta(\mathbf{x})\mathbf{I}_7, \tag{2.2}$$

where $\mathbf{x} \in \mathbb{R}^3$.

Now we will construct the matrix $\mathbf{\Gamma}(\mathbf{x}, \omega)$ explicitly. We consider the system of nonhomogeneous equations

$$\begin{aligned} &(\mu\,\Delta + \rho\,\omega^2)\,\mathbf{u} + (\lambda + \mu)\,\nabla \mathrm{div}\mathbf{u} + i\omega\nabla\,(\mathbf{a}\,\mathbf{p}) = \mathbf{F}^{(1)}, \\ &(\mathbf{K}\,\Delta + \mathbf{c})\,\mathbf{p} - \mathbf{a}\,\mathrm{div}\mathbf{u} = \mathbf{F}^{(2)}. \end{aligned} \tag{2.3}$$

As one may easily verify, the system (2.3) may be written in the form

$$\mathcal{A}^\top(\mathbf{D_x}, \omega)\mathbf{V}(\mathbf{x}) = \mathcal{F}(\mathbf{x}), \tag{2.4}$$

where $\mathbf{V} = (\mathbf{u}, \mathbf{p})$ and $\mathcal{F} = (\mathbf{F}^{(1)}, \mathbf{F}^{(2)})$ are seven-component vector functions and $\mathbf{x} \in \mathbb{R}^3$.

Applying the operator div to the first equation of (2.3) we obtain the following system

$$\begin{aligned} &\left(\mu_0\Delta + \rho\,\omega^2\right)\mathrm{div}\mathbf{u} + i\omega\Delta\,(\mathbf{a}\,\mathbf{p}) = \mathrm{div}\,\mathbf{F}^{(1)}, \\ &(\mathbf{K}\,\Delta + \mathbf{c})\,\mathbf{p} - \mathbf{a}\,\mathrm{div}\mathbf{u} = \mathbf{F}^{(2)}. \end{aligned} \tag{2.5}$$

From (2.5) we have

$$\mathcal{B}(\Delta, \omega)\mathcal{V}\,(\mathbf{x}) = \boldsymbol{\varphi}\,(\mathbf{x})\,, \tag{2.6}$$

where $\mathcal{V} = (\mathrm{div}\mathbf{u}, \mathbf{p})$ and $\boldsymbol{\varphi} = (\varphi_1, \varphi_2, \cdots, \varphi_5) = (\mathrm{div}\mathbf{F}^{(1)}, \mathbf{F}^{(2)})$ are five-component vector functions,

$$\mathcal{B}(\Delta, \omega) = \left(\mathcal{B}_{lj}(\Delta, \omega)\right)_{5\times 5}, \qquad \mathcal{B}_{11}(\Delta, \omega) = \mu_0\Delta + \rho\omega^2,$$

$$\mathcal{B}_{1;\alpha+1}(\Delta, \omega) = i\omega a_\alpha\,\Delta, \qquad \mathcal{B}_{\alpha+1;1}(\Delta, \omega) = -a_\alpha,$$

$$\mathcal{B}_{\alpha+1;\beta+1}(\Delta, \omega) = k_{\alpha\beta}\Delta + c_{\alpha\beta}, \qquad \alpha, \beta = 1, 2, 3, 4.$$

We introduce the notation

$$\Lambda_1(\Delta,\omega) = \frac{1}{k_0\mu_0}\det \boldsymbol{\mathcal{B}}(\Delta,\omega).$$

It is easily seen that $\Lambda_1(-\xi,\omega) = 0$ is a fifth degree algebraic equation and there exist five roots $\lambda_1^2, \lambda_2^2, \cdots, \lambda_5^2$ (with respect to ξ). Then we have

$$\Lambda_1(\Delta,\omega) = \prod_{j=1}^{5}(\Delta+\lambda_j^2).$$

The system (2.6) implies

$$\Lambda_1(\Delta,\omega)\,\boldsymbol{\mathcal{V}} = \boldsymbol{\Phi}, \tag{2.7}$$

where

$$\boldsymbol{\Phi} = (\Phi_1, \Phi_2, \cdots, \Phi_5), \qquad \Phi_j = \frac{1}{k_0\mu_0}\sum_{l=1}^{5}\mathcal{B}_{lj}^*\,\varphi_l, \qquad j = 1, 2, \cdots, 5, \tag{2.8}$$

where $\mathcal{B}_{lj}^*$ is the cofactor of element $\mathcal{B}_{lj}$ of the matrix $\boldsymbol{\mathcal{B}}$.

Now applying the operator $\Lambda_1(\Delta,\omega)$ to the first equation of (2.3) and taking into account (2.7), we obtain

$$\Lambda_2(\Delta,\omega)\mathbf{u} = \tilde{\mathbf{F}}, \tag{2.9}$$

where $\Lambda_2(\Delta,\omega) = \Lambda_1(\Delta,\omega)(\Delta+\lambda_6^2)$, $\lambda_6^2 = \dfrac{\rho\omega^2}{\mu}$ and

$$\tilde{\mathbf{F}} = \frac{1}{\mu}\left[\Lambda_1(\Delta,\omega)\mathbf{F}^{(1)} - (\lambda+\mu)\nabla\,\Phi_1 - i\omega a_\alpha\nabla\,\Phi_{\alpha+1}\right]. \tag{2.10}$$

On the basis of (2.7) and (2.9) we get

$$\boldsymbol{\Lambda}(\Delta,\omega)\mathbf{V}(\mathbf{x}) = \tilde{\boldsymbol{\Phi}}(\mathbf{x}), \tag{2.11}$$

where $\tilde{\boldsymbol{\Phi}} = \left(\tilde{\mathbf{F}}, \Phi_2, \Phi_3, \Phi_4, \Phi_5\right)$ is a seven-component vector function and

$$\mathbf{\Lambda}(\Delta,\omega) = \left(\Lambda_{lj}(\Delta,\omega)\right)_{7\times 7}, \qquad \Lambda_{11} = \Lambda_{22} = \Lambda_{33} = \Lambda_2,$$

$$\Lambda_{44} = \Lambda_{55} = \Lambda_{66} = \Lambda_{77} = \Lambda_1, \qquad \Lambda_{lj} = 0,$$

$$l, j = 1, 2, \cdots, 7, \qquad l \neq j.$$

We introduce the notation

$$\begin{aligned}
n_{l1}(\Delta,\omega) &= -\frac{1}{k_0\mu\mu_0}\left[(\lambda+\mu)\mathcal{B}^*_{l1}(\Delta,\omega) + i\omega a_\alpha \mathcal{B}^*_{l;\alpha+1}(\Delta,\omega)\right],\\
n_{lm}(\Delta,\omega) &= \frac{1}{k_0\mu_0}\mathcal{B}^*_{lm}(\Delta,\omega),\\
l &= 1, 2, \cdots, 5, \qquad m = 2, 3, 4, 5.
\end{aligned} \tag{2.12}$$

In view of (2.12), from (2.8) and (2.10) we have

$$\begin{aligned}
\tilde{\mathbf{F}} &= \frac{1}{\mu}\Lambda_1(\Delta,\omega)\mathbf{F}^{(1)} + \sum_{l=1}^{5} n_{l1}(\Delta,\omega)\nabla\varphi_l\\
&= \left[\frac{1}{\mu}\Lambda_1(\Delta,\omega)\mathbf{I}_3 + n_{11}(\Delta,\omega)\,\nabla\mathrm{div}\right]\mathbf{F}^{(1)}\\
&\quad + \sum_{l=2}^{5} n_{l1}(\Delta,\omega)\,\nabla F_{l+2},
\end{aligned} \tag{2.13}$$

$$\Phi_m = n_{1m}(\Delta,\omega)\,\mathrm{div}\mathbf{F}^{(1)} + \sum_{l=2}^{5} n_{lm}(\Delta,\omega)\,\nabla F_{l+2},$$

$$m = 2, 3, 4, 5.$$

Thus, from (2.13) we have

$$\tilde{\mathbf{\Phi}}(\mathbf{x}) = \mathcal{L}^\top(\mathbf{D}_\mathbf{x},\omega)\,\mathcal{F}(\mathbf{x}), \tag{2.14}$$

where

$$
\begin{aligned}
&\mathcal{L}\left(\mathbf{D_x},\omega\right)=\left(\mathcal{L}_{lj}\left(\mathbf{D_x},\omega\right)\right)_{7\times 7},\\
&\mathcal{L}_{lj}\left(\mathbf{D_x},\omega\right)=\frac{1}{\mu}\Lambda_1(\Delta,\omega)\,\delta_{lj}+n_{11}(\Delta,\omega)\frac{\partial^2}{\partial x_l\partial x_j},\\
&\mathcal{L}_{l;\beta+3}\left(\mathbf{D_x},\omega\right)=n_{1;\beta+1}(\Delta,\omega)\frac{\partial}{\partial x_l},\\
&\mathcal{L}_{\alpha+3;l}\left(\mathbf{D_x},\omega\right)=n_{\alpha+1;1}(\Delta,\omega)\frac{\partial}{\partial x_l},\\
&\mathcal{L}_{\alpha\beta}\left(\mathbf{D_x},\omega\right)=n_{\alpha+1;\beta+1}(\Delta,\omega),\\
&l,j=1,2,3,\qquad \alpha,\beta=1,2,3,4.
\end{aligned}
\tag{2.15}
$$

By virtue of (2.4) and (2.14) from (2.11) it follows that $\mathbf{\Lambda V}=\mathcal{L}^\top\mathcal{A}^\top\mathbf{V}$. It is obvious that $\mathcal{L}^\top\mathcal{A}^\top=\mathbf{\Lambda}$ and, hence,

$$
\mathcal{A}(\mathbf{D_x},\omega)\mathcal{L}(\mathbf{D_x},\omega)=\mathbf{\Lambda}(\Delta,\omega). \tag{2.16}
$$

We assume that $\lambda_l^2\neq\lambda_j^2$, where $l,j=1,2,\cdots,6$ and $l\neq j$. Let

$$
\begin{aligned}
&\mathbf{Y}(\mathbf{x},\omega)=\left(Y_{lj}(\mathbf{x},\omega)\right)_{7\times 7},\\
&Y_{11}(\mathbf{x},\omega)=Y_{22}(\mathbf{x},\omega)=Y_{33}(\mathbf{x},\omega)=\sum_{j=1}^{6}\eta_{2j}\gamma^{(j)}(\mathbf{x},\omega),\\
&Y_{44}(\mathbf{x},\omega)=Y_{55}(\mathbf{x},\omega)=Y_{66}(\mathbf{x},\omega)=Y_{77}(\mathbf{x},\omega)=\sum_{j=1}^{5}\eta_{1j}\gamma^{(j)}(\mathbf{x},\omega),\\
&Y_{lj}(\mathbf{x},\omega)=0,\qquad l\neq j,\qquad l,j=1,2,\cdots,7,
\end{aligned}
\tag{2.17}
$$

where

$$
\gamma^{(j)}(\mathbf{x},\omega)=-\frac{e^{i\lambda_j|\mathbf{x}|}}{4\pi\,|\mathbf{x}|} \tag{2.18}
$$

is the fundamental solution of Helmholtz' equation, i.e., $(\Delta+\lambda_j^2)\gamma^{(j)}(\mathbf{x},\omega)=\delta\left(\mathbf{x}\right)$ and

$$\eta_{1m} = \prod_{l=1,\, l\neq m}^{5} (\lambda_l^2 - \lambda_m^2)^{-1},$$

$$\eta_{2j} = \prod_{l=1,\, l\neq j}^{6} (\lambda_l^2 - \lambda_j^2)^{-1}, \tag{2.19}$$

$$m = 1, 2, \cdots, 5, \qquad j = 1, 2, \cdots, 6.$$

Lemma 2.1 *The matrix* $\mathbf{Y}(\mathbf{x}, \omega)$ *is the fundamental solution of the operator* $\mathbf{\Lambda}(\Delta, \omega)$, *that is,*

$$\mathbf{\Lambda}(\Delta, \omega)\mathbf{Y}(\mathbf{x}, \omega) = \delta(\mathbf{x})\, \mathbf{I}_7, \tag{2.20}$$

where $\mathbf{x} \in \mathbb{R}^3$.

Proof It suffices to show that Y_{11} and Y_{44} are the fundamental solutions of operators $\Lambda_2(\Delta)$ and $\Lambda_1(\Delta)$, respectively, i.e.,

$$\Lambda_2(\Delta, \omega) Y_{11}(\mathbf{x}, \omega) = \delta(\mathbf{x}) \tag{2.21}$$

and

$$\Lambda_1(\Delta, \omega) Y_{44}(\mathbf{x}, \omega) = \delta(\mathbf{x}).$$

It is easy to verify that from (2.18) and (2.19) it follows that

$$\sum_{j=1}^{5} \eta_{1j} = 0, \qquad \sum_{j=2}^{5} \eta_{1j}(\lambda_1^2 - \lambda_j^2) = 0,$$

$$\sum_{j=3}^{5} \eta_{1j}(\lambda_1^2 - \lambda_j^2)(\lambda_2^2 - \lambda_j^2) = 0,$$

$$\sum_{j=4}^{5} \eta_{1j}(\lambda_1^2 - \lambda_j^2)(\lambda_2^2 - \lambda_j^2)(\lambda_3^2 - \lambda_j^2) = 0,$$

$$\eta_{15} \prod_{j=1}^{4} (\lambda_j^2 - \lambda_5^2) = 1,$$

$$(\Delta + \lambda_l^2)\gamma^{(j)}(\mathbf{x}, \omega) = \delta(\mathbf{x}) + (\lambda_l^2 - \lambda_j^2)\gamma^{(j)}(\mathbf{x}, \omega),$$

$$l, j = 1, 2, \cdots, 5, \qquad \mathbf{x} \in \mathbb{R}^3.$$

Taking into account these equalities we have

$$\Lambda_1(\Delta,\omega)Y_{44}(\mathbf{x},\omega)$$

$$=\prod_{l=2}^{5}(\Delta+\lambda_l^2)\sum_{j=1}^{5}\eta_{1j}\left[\delta(\mathbf{x})+(\lambda_1^2-\lambda_j^2)\gamma^{(j)}(\mathbf{x},\omega)\right]$$

$$=\prod_{l=2}^{5}(\Delta+\lambda_l^2)\sum_{j=2}^{5}\eta_{1j}(\lambda_1^2-\lambda_j^2)\gamma^{(j)}(\mathbf{x},\omega)$$

$$=\prod_{l=3}^{5}(\Delta+\lambda_l^2)\sum_{j=2}^{5}\eta_{1j}(\lambda_1^2-\lambda_j^2)\left[\delta(\mathbf{x})+(\lambda_2^2-\lambda_j^2)\gamma^{(j)}(\mathbf{x},\omega)\right]$$

$$=(\Delta+\lambda_5^2)\gamma^{(5)}(\mathbf{x},\omega)=\delta(\mathbf{x})\,.$$

Equation (2.21) is proved quite similarly. □

We introduce the matrix

$$\mathbf{\Gamma}(\mathbf{x},\omega)=\boldsymbol{\mathcal{L}}(\mathbf{D_x},\omega)\,\mathbf{Y}(\mathbf{x},\omega)\,. \tag{2.22}$$

Using identities (2.16) and (2.20) from (2.22) we get

$$\boldsymbol{\mathcal{A}}(\mathbf{D_x},\omega)\,\mathbf{\Gamma}(\mathbf{x},\omega)=\boldsymbol{\mathcal{A}}(\mathbf{D_x},\omega)\,\boldsymbol{\mathcal{L}}(\mathbf{D_x},\omega)\,\mathbf{Y}(\mathbf{x},\omega)$$

$$=\mathbf{\Lambda}(\Delta,\omega)\,\mathbf{Y}(\mathbf{x},\omega)=\delta(\mathbf{x})\,\mathbf{I}_7.$$

Hence, $\mathbf{\Gamma}(\mathbf{x},\omega)$ is the solution of (2.2). We have thereby proved the following.

Theorem 2.1 *The matrix $\mathbf{\Gamma}(\mathbf{x},\omega)$ defined by (2.22) is the fundamental solution of (1.21), where the matrices $\boldsymbol{\mathcal{L}}(\mathbf{D_x},\omega)$ and $\mathbf{Y}(\mathbf{x},\omega)$ are given by (2.15) and (2.17), respectively.*

Obviously, each element $\Gamma_{lj}(\mathbf{x},\omega)$ of the matrix $\mathbf{\Gamma}(\mathbf{x},\omega)$ is represented in the following form

$$\Gamma_{lj}(\mathbf{x},\omega)=\mathcal{L}_{lj}(\mathbf{D_x},\omega)\,Y_{11}(\mathbf{x},\omega)\,,$$

$$\Gamma_{lm}(\mathbf{x},\omega)=\mathcal{L}_{lm}(\mathbf{D_x},\omega)\,Y_{44}(\mathbf{x},\omega)\,, \tag{2.23}$$

$$l=1,2,\cdots,7,\qquad j=1,2,3,\qquad m=4,5,6,7.$$

It is easy to verify that the fundamental solution of the system of equations of pseudo-oscillations (1.22) (the fundamental matrix of the operator $\boldsymbol{\mathcal{A}}(\mathbf{D_x}, i\tau)$) may be obtained from the matrix $\mathbf{\Gamma}(\mathbf{x}, \omega)$ by replacing ω by $i\tau$. We have the following.

Theorem 2.2 *The matrix* $\mathbf{\Gamma}(\mathbf{x}, i\tau)$ *defined by*

$$\mathbf{\Gamma}(\mathbf{x}, i\tau) = \boldsymbol{\mathcal{L}}(\mathbf{D_x}, i\tau)\, \mathbf{Y}(\mathbf{x}, i\tau)$$

is the fundamental solution of system (1.22), where $\boldsymbol{\mathcal{L}}(\mathbf{D_x}, i\tau)$ *and* $\mathbf{Y}(\mathbf{x}, i\tau)$ *are obtained from the matrices* $\boldsymbol{\mathcal{L}}(\mathbf{x}, \omega)$ *and* $\mathbf{Y}(\mathbf{x}, \omega)$ *by replacing* ω *by* $i\tau$*, respectively.*

Obviously, the matrices $\mathbf{\Gamma}(\mathbf{x}, \omega)$ and $\mathbf{\Gamma}(\mathbf{x}, i\tau)$ are constructed explicitly by means of six metaharmonic functions $\gamma^{(j)}$ $(j = 1, 2, \cdots, 6)$ (see (2.18)).

By the method, developed in this section, we can construct the fundamental solutions of the systems of equations of quasi-static (1.23) and equilibrium (1.24).

2.3 Equations of Quasi-Static

In this section the fundamental solution of the system of Eq. (1.23) is constructed explicitly by means of elementary functions. We introduce the notation:

(i)

$$\boldsymbol{\mathcal{B}}^{(q)}(\Delta, \omega) = \left(\mathcal{B}^{(q)}_{lj}(\Delta, \omega)\right)_{5\times 5}, \qquad \tilde{\boldsymbol{\mathcal{B}}}(\Delta, \omega) = \left(\tilde{\mathcal{B}}_{lj}(\Delta, \omega)\right)_{5\times 5},$$

$$\mathcal{B}^{(q)}_{11}(\Delta, \omega) = \mu_0, \qquad \mathcal{B}^{(q)}_{1m}(\Delta, \omega) = i\omega a_{m-1},$$

$$\tilde{\mathcal{B}}_{1r}(\Delta, \omega) = \Delta\, \mathcal{B}^{(q)}_{1r}(\Delta, \omega),$$

$$\mathcal{B}^{(q)}_{mr}(\Delta, \omega) = \tilde{\mathcal{B}}_{mr}(\Delta, \omega) = \mathcal{B}_{mr}(\Delta, \omega),$$

$$l = 1, 2, 3, \qquad m = 2, 3, 4, 5, \qquad r = 1, 2, \cdots, 5.$$

(ii)

$$\Lambda^{(q)}_1(\Delta, \omega) = \frac{1}{k_0 \mu_0}\, \Delta \det \boldsymbol{\mathcal{B}}^{(q)}(\Delta, \omega).$$

It is easily seen that $\det \boldsymbol{\mathcal{B}}^{(q)}(-\xi, \omega) = 0$ is a fourth degree algebraic equation and there exist four roots $\tilde{\lambda}_1^2$, $\tilde{\lambda}_2^2$, $\tilde{\lambda}_3^2$ and $\tilde{\lambda}_4^2$ (with respect to ξ). Then we have

$$\Lambda^{(q)}_1(\Delta, \omega) = \Delta \prod_{j=1}^{4} (\Delta + \tilde{\lambda}_j^2).$$

(iii)

$$\mathcal{L}^{(q)}(\mathbf{D}_{\mathbf{x}},\omega)=\left(\mathcal{L}^{(q)}_{lj}(\mathbf{D}_{\mathbf{x}},\omega)\right)_{7\times 7},$$

$$\mathcal{L}^{(q)}_{lj}(\mathbf{D}_{\mathbf{x}},\omega)=\frac{1}{\mu}\Lambda^{(q)}_{1}(\Delta,\omega)\,\delta_{lj}$$

$$-\frac{1}{k_0\mu\mu_0}\left[(\lambda+\mu)\tilde{\mathcal{B}}^{*}_{11}(\Delta,\omega)+i\omega a_\alpha\tilde{\mathcal{B}}^{*}_{1;\alpha+1}(\Delta,\omega)\right]\frac{\partial^2}{\partial x_l\partial x_j},$$

$$\mathcal{L}^{(q)}_{l;\beta+3}(\mathbf{D}_{\mathbf{x}},\omega)=\frac{1}{k_0\mu_0}\tilde{\mathcal{B}}^{*}_{1;\beta+1}(\Delta,\omega)\frac{\partial}{\partial x_l}, \tag{2.24}$$

$$\mathcal{L}^{(q)}_{\alpha+3;l}(\mathbf{D}_{\mathbf{x}},\omega)=\frac{1}{k_0\mu_0}\tilde{\mathcal{B}}^{*}_{\alpha+1;1}(\Delta,\omega)\frac{\partial}{\partial x_l},$$

$$\mathcal{L}^{(q)}_{\alpha+3;\beta+3}(\mathbf{D}_{\mathbf{x}},\omega)=\frac{1}{k_0\mu_0}\tilde{\mathcal{B}}^{*}_{\alpha+1;\beta+1}(\Delta,\omega),$$

$$l,j=1,2,3,\qquad \alpha,\beta=1,2,3,4,$$

where $\tilde{\mathcal{B}}^{*}_{lj}$ is the cofactor of element $\tilde{\mathcal{B}}_{lj}$ of the matrix $\tilde{\boldsymbol{\mathcal{B}}}$.

(iv)

$$\mathbf{Y}^{(q)}(\mathbf{x})=\left(Y^{(q)}_{lj}(\mathbf{x})\right)_{7\times 7},$$

$$Y^{(q)}_{11}(\mathbf{x})=Y^{(q)}_{22}(\mathbf{x})=Y^{(q)}_{33}(\mathbf{x})=\sum_{j=1}^{6}\tilde{\eta}_{2j}\tilde{\gamma}^{(j)}(\mathbf{x}),$$

$$Y^{(q)}_{44}(\mathbf{x})=Y^{(q)}_{55}(\mathbf{x})=Y^{(q)}_{66}(\mathbf{x})=Y^{(q)}_{77}(\mathbf{x})=\sum_{j=1}^{5}\tilde{\eta}_{1j}\tilde{\gamma}^{(j)}(\mathbf{x}), \tag{2.25}$$

$$Y^{(q)}_{lj}(\mathbf{x})=0,\qquad l\neq j,\qquad l,j=1,2,\cdots,7,$$

where the functions $\tilde{\gamma}^{(j)}$ are defined by

$$\tilde{\gamma}^{(m)}(\mathbf{x})=-\frac{e^{i\tilde{\lambda}_m|\mathbf{x}|}}{4\pi\,|\mathbf{x}|},\qquad \tilde{\gamma}^{(5)}(\mathbf{x})=-\frac{1}{4\pi\,|\mathbf{x}|}, \tag{2.26}$$

$$\tilde{\gamma}^{(6)}(\mathbf{x})=-\frac{|\mathbf{x}|}{8\pi},\qquad m=1,2,3,4$$

and

$$\tilde{\eta}_{1l} = -\left[\tilde{\lambda}_l^2 \prod_{j=1,\, j\neq l}^{4} (\tilde{\lambda}_j^2 - \tilde{\lambda}_l^2)\right]^{-1}, \qquad \tilde{\eta}_{2l} = \left[\tilde{\lambda}_l^4 \prod_{j=1,\, j\neq l}^{4} (\tilde{\lambda}_j^2 - \tilde{\lambda}_l^2)\right]^{-1},$$

$$\tilde{\eta}_{25} = -\left(\tilde{\lambda}_1\tilde{\lambda}_2\tilde{\lambda}_3\tilde{\lambda}_4\right)^{-4}\left[\tilde{\lambda}_1^2(\tilde{\lambda}_2^2\tilde{\lambda}_3^2 + \tilde{\lambda}_2^2\tilde{\lambda}_4^2 + \tilde{\lambda}_3^2\tilde{\lambda}_4^2) + \tilde{\lambda}_2^2\tilde{\lambda}_3^2\tilde{\lambda}_4^2\right],$$

$$\tilde{\eta}_{15} = \tilde{\eta}_{26} = \left(\tilde{\lambda}_1\tilde{\lambda}_2\tilde{\lambda}_3\tilde{\lambda}_4\right)^{-2}, \qquad l = 1, 2, 3, 4.$$

Obviously,

$$(\Delta + \tilde{\lambda}_l^2)\tilde{\gamma}^{(l)}(\mathbf{x}) = \delta(\mathbf{x}), \qquad \Delta\tilde{\gamma}^{(5)}(\mathbf{x}) = \delta(\mathbf{x}),$$

$$\Delta^2\tilde{\gamma}^{(6)}(\mathbf{x}) = \delta(\mathbf{x}), \qquad l = 1, 2, 3, 4$$

and consequently,

$$\mathbf{\Lambda}^{(q)}(\Delta, \omega)\mathbf{Y}^{(q)}(\mathbf{x}) = \delta(\mathbf{x})\,\mathbf{I}_7, \tag{2.27}$$

where $\mathbf{x} \in \mathbb{R}^3$ and

$$\mathbf{\Lambda}^{(q)}(\Delta, \omega) = \left(\Lambda_{lj}^{(q)}(\Delta, \omega)\right)_{7\times 7}, \qquad \Lambda_{11}^{(q)} = \Lambda_{22}^{(q)} = \Lambda_{33}^{(q)} = \Delta\Lambda_1^{(q)},$$

$$\Lambda_{44}^{(q)} = \Lambda_{55}^{(q)} = \Lambda_{66}^{(q)} = \Lambda_{77}^{(q)} = \Lambda_1^{(q)}, \qquad \Lambda_{lj}^{(q)} = 0,$$

$$l \neq j, \qquad l, j = 1, 2, \cdots, 7.$$

On the basis of identities (2.27) and

$$\mathcal{A}^{(q)}(\mathbf{D}_{\mathbf{x}}, \omega)\,\mathcal{L}^{(q)}(\mathbf{D}_{\mathbf{x}}, \omega) = \mathbf{\Lambda}^{(q)}(\Delta, \omega)$$

we have the following.

Theorem 2.3 *The matrix* $\mathbf{\Gamma}^{(q)}(\mathbf{x}, \omega)$ *defined by*

$$\mathbf{\Gamma}^{(q)}(\mathbf{x}, \omega) = \mathcal{L}^{(q)}(\mathbf{D}_{\mathbf{x}}, \omega)\,\mathbf{Y}^{(q)}(\mathbf{x})$$

is the fundamental solution of system (1.23), where the matrices $\mathcal{L}^{(q)}(\mathbf{D}_{\mathbf{x}}, \omega)$ *and* $\mathbf{Y}^{(q)}(\mathbf{x})$ *are given by (2.24) and (2.25), respectively.*

Obviously, the matrix $\mathbf{\Gamma}^{(q)}(\mathbf{x}, \omega)$ is constructed explicitly by means of six elementary functions $\tilde{\gamma}^{(j)}$ $(j = 1, 2, \cdots, 6)$ (see (2.26)).

2.4 Equations of Equilibrium

In this section the fundamental solution of the system of Eq. (1.24) is constructed explicitly by means of elementary functions. It is easily seen that

$$\frac{1}{k_0}\det\,(\mathbf{K}\xi-\mathbf{d})=\det\mathbf{K}^{-1}\det\,(\mathbf{K}\xi-\mathbf{d})=\det\left[\mathbf{K}^{-1}\,(\mathbf{K}\xi-\mathbf{d})\right]$$
$$=\det\left(\mathbf{I}_4\xi-\mathbf{K}^{-1}\mathbf{d}\right).$$

Clearly, $\mathbf{d}$ is a positive semi-definite matrix (see (1.7)) and $\det\mathbf{d}=0$. By the assumption that $\mathbf{K}$ is a positive definite matrix (see Remark 1.2), the matrix $\mathbf{K}^{-1}$ is the positive definite. Hence, $\mathbf{K}^{-1}\mathbf{d}$ is a positive semi-definite matrix and has nonnegative eigenvalues κ_j^2 $(j=1,2,3,4)$. In addition, on the basis of identity $\det\mathbf{d}=0$, it follows that $\kappa_4=0$. We assume that $\kappa_j>0$ and $\kappa_j\neq\kappa_m$ $(j,m=1,2,3)$. Obviously, we have

$$\frac{1}{k_0}\det\,(\mathbf{K}\Delta-\mathbf{d})=\Delta\prod_{j=1}^{3}\left(\Delta-\kappa_j^2\right).$$

We introduce the notation:

(i)

$$\Lambda_1^{(e)}(\Delta)=\frac{1}{k_0}\Delta\det\boldsymbol{\mathcal{A}}^{(re)}\,(\Delta),$$

where the matrix differential operator $\boldsymbol{\mathcal{A}}^{(re)}(\mathbf{D_x})$ is defined by (1.36), (1.37) and has the form

$$\boldsymbol{\mathcal{A}}^{(re)}(\mathbf{D_x})=\mathbf{K}\Delta-\mathbf{d}.$$

Then we have

$$\Lambda_1^{(e)}(\Delta)=\Delta^2\prod_{j=1}^{3}(\Delta-\kappa_j^2).$$

(ii)

$$\boldsymbol{\mathcal{L}}^{(e)}\,(\mathbf{D_x})=\left(\mathcal{L}_{lj}^{(e)}\,(\mathbf{D_x})\right)_{7\times7},$$

$$\mathcal{L}_{lj}^{(e)}\,(\mathbf{D_x})=\frac{1}{\mu}\Delta\,\delta_{lj}-\frac{\lambda+\mu}{\mu\mu_0}\frac{\partial^2}{\partial x_l\partial x_j},\qquad \mathcal{L}_{\alpha+3;\,j}^{(e)}\,(\mathbf{D_x})=0,$$

$$\mathcal{L}^{(e)}_{l;\beta+3}(\mathbf{D_x}) = \frac{1}{k_0\mu_0}\, a_\alpha \mathcal{A}^{(re)*}_{\alpha\beta}(\mathbf{D_x})\frac{\partial}{\partial x_l},$$

$$\mathcal{L}^{(e)}_{\alpha+3;\beta+3}(\mathbf{D_x}) = \frac{1}{k_0}\Delta \mathcal{A}^{(re)*}_{\alpha\beta}(\mathbf{D_x}), \tag{2.28}$$

$$l, j = 1, 2, 3, \qquad \alpha, \beta = 1, 2, 3, 4.$$

where $\mathcal{A}^{(re)*}_{\alpha\beta}$ is the cofactor of element $\mathcal{A}^{(re)}_{\alpha\beta}$ of the matrix $\boldsymbol{\mathcal{A}}^{(re)}$.

(iii)

$$\mathbf{Y}^{(e)}(\mathbf{x}) = \left(Y^{(e)}_{lj}(\mathbf{x})\right)_{7\times 7},$$

$$Y^{(e)}_{11}(\mathbf{x}) = Y^{(e)}_{22}(\mathbf{x}) = Y^{(e)}_{33}(\mathbf{x}) = \tilde{\gamma}^{(6)}(\mathbf{x}),$$

$$Y^{(e)}_{44}(\mathbf{x}) = Y^{(e)}_{55}(\mathbf{x}) = Y^{(e)}_{66}(\mathbf{x}) = Y^{(e)}_{77}(\mathbf{x}) \tag{2.29}$$

$$= \eta_m \phi^{(m)}(\mathbf{x}) + \eta_4 \tilde{\gamma}^{(5)}(\mathbf{x}) + \eta_5 \tilde{\gamma}^{(6)}(\mathbf{x}),$$

$$Y^{(e)}_{lj}(\mathbf{x}) = 0, \qquad l \neq j, \qquad l, j = 1, 2, \cdots, 7,$$

where the functions $\tilde{\gamma}^{(5)}$ and $\tilde{\gamma}^{(6)}$ are defined by (2.26),

$$\phi^{(m)}(\mathbf{x}) = -\frac{e^{-\kappa_m |\mathbf{x}|}}{4\pi\, |\mathbf{x}|} \tag{2.30}$$

and

$$\eta_m = \left[\kappa_m^4 \prod_{j=1,\, j\neq m}^{3} (\kappa_j^2 - \kappa_m^2)\right]^{-1},$$

$$\eta_4 = -\frac{\kappa_1^2\kappa_2^2 + \kappa_1^2\kappa_3^2 + \kappa_2^2\kappa_3^2}{\kappa_1^4\kappa_2^4\kappa_3^4},$$

$$\eta_5 = -\frac{1}{\kappa_1^2\kappa_2^2\kappa_3^2}, \qquad m = 1, 2, 3.$$

Obviously, $(\Delta - \kappa_m^2)\phi^{(m)}(\mathbf{x}) = \delta(\mathbf{x})$, $m = 1, 2, 3$. Consequently, we obtain

$$\mathbf{\Lambda}^{(e)}(\Delta)\mathbf{Y}^{(e)}(\mathbf{x}) = \delta(\mathbf{x})\, \mathbf{J}_7, \tag{2.31}$$

where $\mathbf{x} \in \mathbb{R}^3$ and

$$\mathbf{\Lambda}^{(e)}(\Delta) = \left(\Lambda_{lj}^{(e)}(\Delta)\right)_{7\times 7},$$

$$\Lambda_{11}^{(e)}(\Delta) = \Lambda_{22}^{(e)}(\Delta) = \Lambda_{33}^{(e)}(\Delta) = \Delta^2,$$

$$\Lambda_{44}^{(e)}(\Delta) = \Lambda_{55}^{(e)}(\Delta) = \Lambda_{66}^{(e)}(\Delta) = \Lambda_{77}^{(e)}(\Delta) = \Lambda_{1}^{(e)}(\Delta),$$

$$\Lambda_{lj}^{(e)}(\Delta) = 0, \qquad l \neq j \qquad l, j = 1, 2, \cdots, 7.$$

On the basis of identities (2.31) and

$$\mathcal{A}^{(e)}(\mathbf{D_x})\, \mathcal{L}^{(e)}(\mathbf{D_x}) = \mathbf{\Lambda}^{(e)}(\Delta)$$

we have the following.

Theorem 2.4 *The matrix* $\mathbf{\Gamma}^{(e)}(\mathbf{x})$ *defined by*

$$\mathbf{\Gamma}^{(e)}(\mathbf{x}) = \mathcal{L}^{(e)}(\mathbf{D_x})\, \mathbf{Y}^{(e)}(\mathbf{x})$$

is the fundamental solution of system (1.24), where the matrices $\mathcal{L}^{(e)}(\mathbf{D_x})$ *and* $\mathbf{Y}^{(e)}(\mathbf{x})$ *are given by (2.28) and (2.29), respectively.*

Obviously, the matrix $\mathbf{\Gamma}^{(e)}(\mathbf{x})$ is constructed explicitly by means of five elementary functions $\phi^{(j)}$ $(j = 1, 2, 3)$, $\tilde{\gamma}^{(5)}$, and $\tilde{\gamma}^{(6)}$ (see (2.26) and (2.30)).

Theorem 2.4 leads to the following.

Theorem 2.5 *The fundamental matrix of the operator* $\mathcal{A}^{(0)}(\mathbf{D_x})$ *is the matrix* $\mathbf{\Gamma}^{(0)}(\mathbf{x}) = \left(\Gamma_{lj}^{(0)}(\mathbf{x})\right)_{7\times 7}$*, where*

$$\Gamma_{lj}^{(0)}(\mathbf{x}) = \frac{1}{\mu}\left(\Delta\, \delta_{lj} - \frac{\lambda+\mu}{\mu_0}\frac{\partial^2}{\partial x_l \partial x_j}\right)\tilde{\gamma}^{(6)}(\mathbf{x}),$$

$$\Gamma_{\alpha+3;\beta+3}^{(0)}(\mathbf{x}) = \frac{1}{k_0} k_{\alpha\beta}^{*}\tilde{\gamma}^{(5)}(\mathbf{x}),$$

$$\Gamma_{l;\alpha+3}^{(0)}(\mathbf{x}) = \Gamma_{\alpha+3;l}^{(0)}(\mathbf{x}) = 0,$$

$$l, j = 1, 2, 3, \qquad \alpha, \beta = 1, 2, 3, 4,$$

$k_{\alpha\beta}^{*}$ *is the cofactor of element* $k_{\alpha\beta}$ *of the matrix* $\mathbf{K}$*;* $\tilde{\gamma}^{(5)}(\mathbf{x})$ *and* $\tilde{\gamma}^{(6)}(\mathbf{x})$ *are given by (2.26).*

2.5 Equations of Steady Vibrations of Rigid Body

We introduce the notation:

(i)

$$\Lambda_1^{(r)}(\Delta, \omega) = \frac{1}{k_0} \det \boldsymbol{\mathcal{A}}^{(r)}(\mathbf{D_x}, \omega),$$

where the matrix differential operator $\boldsymbol{\mathcal{A}}^{(r)}(\mathbf{D_x}, \omega)$ is defined by (1.36) and (1.37). It is easily seen that $\Lambda_1^{(r)}(-\xi, \omega) = 0$ is a fourth degree algebraic equation and there exist four roots ν_1^2, ν_2^2, ν_3^2 and ν_4^2 (with respect to ξ). We assume that $\nu_j \neq \nu_m$ $(j, m = 1, 2, 3, 4)$. Obviously, we have

$$\Lambda_1^{(r)}(\Delta, \omega) = \prod_{j=1}^{4} (\Delta + \nu_j^2).$$

(ii)

$$\mathcal{L}^{(r)}(\Delta, \omega) = \left(\mathcal{L}_{lj}^{(r)}(\Delta, \omega)\right)_{4\times 4},$$

$$\mathcal{L}_{\alpha\beta}^{(r)}(\Delta, \omega) = \frac{1}{k_0} \mathcal{A}_{\alpha\beta}^{(r)*}(\Delta, \omega), \tag{2.32}$$

$$\alpha, \beta = 1, 2, 3, 4,$$

where $\mathcal{A}_{\alpha\beta}^{(r)*}$ is the cofactor of element $\mathcal{A}_{\alpha\beta}^{(r)}$ of the matrix $\boldsymbol{\mathcal{A}}^{(r)}$.

(iii)

$$\mathbf{Y}^{(r)}(\mathbf{x}, \omega) = \left(Y_{lj}^{(r)}(\mathbf{x}, \omega)\right)_{4\times 4},$$

$$Y_{11}^{(r)}(\mathbf{x}, \omega) = Y_{22}^{(r)}(\mathbf{x}, \omega) = Y_{33}^{(r)}(\mathbf{x}, \omega)$$

$$= Y_{44}^{(r)}(\mathbf{x}, \omega) = \sum_{m=1}^{4} \tilde{\eta}_m \tilde{\phi}^{(m)}(\mathbf{x}), \tag{2.33}$$

$$Y_{lj}^{(r)}(\mathbf{x}, \omega) = 0, \qquad l \neq j, \qquad l, j = 1, 2, 3, 4,$$

where the functions $\tilde{\phi}^{(m)}$ are defined by

$$\tilde{\phi}^{(m)}(\mathbf{x}) = -\frac{e^{i\nu_m|\mathbf{x}|}}{4\pi\,|\mathbf{x}|}, \qquad m = 1, 2, 3, 4 \tag{2.34}$$

and

$$\tilde{\eta}_l = \prod_{j=1,\, j \neq l}^{4} (\nu_j^2 - \nu_l^2)^{-1}.$$

Obviously,

$$\mathbf{\Lambda}^{(r)}(\Delta, \omega)\mathbf{Y}^{(r)}(\mathbf{x}, \omega) = \delta(\mathbf{x})\, \mathbf{I}_4, \tag{2.35}$$

where $\mathbf{x} \in \mathbb{R}^3$ and

$$\mathbf{\Lambda}^{(r)}(\Delta, \omega) = \left(\Lambda_{lj}^{(r)}(\Delta, \omega)\right)_{4\times 4},$$

$$\Lambda_{11}^{(r)} = \Lambda_{22}^{(r)} = \Lambda_{33}^{(r)} = \Lambda_{44}^{(r)} = \Lambda_1^{(r)},$$

$$\Lambda_{lj}^{(r)} = 0, \qquad l \neq j, \qquad l, j = 1, 2, 3, 4.$$

On the basis of identities (2.35) and

$$\boldsymbol{\mathcal{A}}^{(r)}(\mathbf{D}_{\mathbf{x}}, \omega)\, \boldsymbol{\mathcal{L}}^{(r)}(\mathbf{D}_{\mathbf{x}}, \omega) = \mathbf{\Lambda}^{(r)}(\Delta, \omega)$$

we have the following.

Theorem 2.6 *The matrix* $\mathbf{\Gamma}^{(r)}(\mathbf{x}, \omega)$ *defined by*

$$\mathbf{\Gamma}^{(r)}(\mathbf{x}, \omega) = \boldsymbol{\mathcal{L}}^{(r)}(\Delta, \omega)\, \mathbf{Y}^{(r)}(\mathbf{x}, \omega)$$

is the fundamental solution of system (1.25), where the matrices $\boldsymbol{\mathcal{L}}^{(r)}(\Delta, \omega)$ *and* $\mathbf{Y}^{(r)}(\mathbf{x}, \omega)$ *are given by (2.32) and (2.33), respectively.*

Obviously, $\mathbf{\Gamma}^{(r)}(\mathbf{x}, \omega)$ is a symmetric matrix, $\mathbf{\Gamma}^{(r)^\top}(\mathbf{x}, \omega) = \mathbf{\Gamma}^{(r)}(\mathbf{x}, \omega)$ and constructed explicitly by means of four metaharmonic functions $\tilde{\phi}^{(j)}$ $(j = 1, 2, 3, 4)$ (see (2.34)).

2.6 Equations of Equilibrium of Rigid Body

We introduce the notation:

(i)

$$\Lambda^{(re)}(\Delta) = \frac{1}{k_0} \det \boldsymbol{\mathcal{A}}^{(re)}(\mathbf{D}_{\mathbf{x}}) = \frac{1}{k_0} \det(\mathbf{K}\Delta - \mathbf{d}).$$

It is easily seen that (see Sect. 2.4)

$$\Lambda^{(re)}(\Delta) = \Delta \prod_{j=1}^{3} \left(\Delta - \kappa_j^2\right).$$

(ii)

$$\mathcal{L}^{(re)}(\Delta) = \left(\mathcal{L}^{(re)}_{\alpha\beta}(\Delta)\right)_{4\times 4},$$

$$\mathcal{L}^{(re)}_{\alpha\beta}(\Delta) = \frac{1}{k_0}\mathcal{A}^{(re)^*}_{\alpha\beta}(\mathbf{D_x}), \tag{2.36}$$

$$\alpha, \beta = 1, 2, 3, 4,$$

where $\mathcal{A}^{(re)^*}_{\alpha\beta}$ is defined in Sect. 2.4.

(iii)

$$Y^{(re)}(\mathbf{x}) = \sum_{m=1}^{3} \eta_m^{(re)} \phi^{(m)}(\mathbf{x}) + \eta_4^{(re)} \tilde{\gamma}^{(5)}(\mathbf{x}), \tag{2.37}$$

where $\phi^{(m)}$ $(m = 1, 2, 3)$ and $\tilde{\gamma}^{(5)}$ are given by (2.30) and (2.26), respectively;

$$\eta_m^{(re)} = \frac{1}{\kappa_m^2} \prod_{l=1,\, l\neq m}^{3} (\kappa_l^2 - \kappa_m^2)^{-1},$$

$$\eta_4^{(re)} = -\frac{1}{(\kappa_1\kappa_2\kappa_3)^2}, \qquad m = 1, 2, 3.$$

Obviously,

$$\Lambda^{(re)}(\Delta) Y^{(re)}(\mathbf{x}) = \delta(\mathbf{x}), \tag{2.38}$$

where $\mathbf{x} \in \mathbb{R}^3$.

On the basis of identities (2.38) and

$$\mathcal{A}^{(re)}(\mathbf{D_x}) \mathcal{L}^{(re)}(\Delta) = \Lambda^{(re)}(\Delta)\, \mathbf{I}_4$$

we have the following.

Theorem 2.7 *The matrix* $\mathbf{\Gamma}^{(re)}(\mathbf{x})$ *defined by*

$$\mathbf{\Gamma}^{(re)}(\mathbf{x}) = \mathcal{L}^{(re)}(\Delta) Y^{(re)}(\mathbf{x}) \tag{2.39}$$

is the fundamental solution of (1.26), where $\mathcal{L}^{(re)}(\Delta)$ *and* $Y^{(re)}(\mathbf{x})$ *are given by (2.36) and (2.37), respectively.*

Obviously, $\boldsymbol{\Gamma}^{(re)}(\mathbf{x})$ is a symmetric matrix, i.e., $\boldsymbol{\Gamma}^{(re)^\top}(\mathbf{x}) = \boldsymbol{\Gamma}^{(re)}(\mathbf{x})$ and constructed explicitly by means of four elementary functions $\phi_m^{(re)}(\mathbf{x})$ $(m = 1, 2, 3)$ and $\tilde{\gamma}^{(5)}$ (see (2.30) and (2.26)).

It is easy to verify that the fundamental solution of the system of equations

$$\mathbf{K}\Delta\mathbf{p}(\mathbf{x}) = \mathbf{0} \tag{2.40}$$

is the matrix

$$\boldsymbol{\Phi}^{(re)}(\mathbf{x}) = \left(\Phi_{\alpha\beta}^{(re)}(\mathbf{x})\right)_{4\times 4}, \tag{2.41}$$

where

$$\Phi_{\alpha\beta}^{(re)}(\mathbf{x}) = \frac{1}{k_0}\, k_{\alpha\beta}^* \tilde{\gamma}^{(5)}(\mathbf{x}), \tag{2.42}$$

where $k_{\alpha\beta}^*$ is the cofactor of element $k_{\alpha\beta}$ of the matrix $\mathbf{K}$.

2.7 Basic Properties of Fundamental Solutions

We now can establish the basic properties of the matrices $\boldsymbol{\Gamma}(\mathbf{x}, \omega)$, $\boldsymbol{\Gamma}(\mathbf{x}, i\tau)$, $\boldsymbol{\Gamma}^{(q)}(\mathbf{x}, \omega)$, $\boldsymbol{\Gamma}^{(e)}(\mathbf{x})$, $\boldsymbol{\Gamma}^{(r)}(\mathbf{x}, \omega)$, and $\boldsymbol{\Gamma}^{(re)}(\mathbf{x})$.

Theorems 2.1–2.4, 2.6, and 2.7 lead to the following results.

Theorem 2.8 *If* $\mathbf{F}^{(1)} = \mathbf{0}$ *and* $\mathbf{F}^{(2)} = \mathbf{0}$*, then the matrices* $\boldsymbol{\Gamma}(\mathbf{x}, \omega)$, $\boldsymbol{\Gamma}(\mathbf{x}, i\tau)$, $\boldsymbol{\Gamma}^{(q)}(\mathbf{x}, \omega)$, $\boldsymbol{\Gamma}^{(e)}(\mathbf{x})$, $\boldsymbol{\Gamma}^{(r)}(\mathbf{x}, \omega)$, *and* $\boldsymbol{\Gamma}^{(re)}(\mathbf{x})$ *are solutions of Eqs. (1.38)–(1.43), respectively, i.e.*

$$\begin{aligned}
&\mathcal{A}(\mathbf{D_x}, \omega)\boldsymbol{\Gamma}(\mathbf{x}, \omega) = \mathbf{0}, && \mathcal{A}(\mathbf{D_x}, i\tau)\boldsymbol{\Gamma}(\mathbf{x}, i\tau) = \mathbf{0},\\
&\mathcal{A}^{(q)}(\mathbf{D_x}, \omega)\boldsymbol{\Gamma}^{(q)}(\mathbf{x}, \omega) = \mathbf{0}, && \mathcal{A}^{(e)}(\mathbf{D_x})\boldsymbol{\Gamma}^{(e)}(\mathbf{x}) = \mathbf{0},\\
&\mathcal{A}^{(r)}(\mathbf{D_x}, \omega)\boldsymbol{\Gamma}^{(r)}(\mathbf{x}, \omega) = \mathbf{0}, && \mathcal{A}^{(re)}(\mathbf{D_x})\boldsymbol{\Gamma}^{(re)}(\mathbf{x}) = \mathbf{0}
\end{aligned} \tag{2.43}$$

at every point $\mathbf{x} \in \mathbb{R}^3$ *except the origin.*

Theorem 2.9 *The relations*

$$\Gamma_{lj}(\mathbf{x},\omega) = O\left(|\mathbf{x}|^{-1}\right), \qquad \Gamma_{\alpha+3;\beta+3}(\mathbf{x},\omega) = O\left(|\mathbf{x}|^{-1}\right),$$

$$\Gamma_{l;\alpha+3}(\mathbf{x},\omega) = O(1), \qquad \Gamma_{\alpha+3;j}(\mathbf{x},\omega) = O(1),$$

$$l, j = 1, 2, 3, \qquad \alpha, \beta = 1, 2, 3, 4$$

hold the neighborhood of the origin.

Theorem 2.10 *The relations*

$$\frac{\partial}{\partial x_m}\Gamma_{lj}(\mathbf{x},\omega) = O\left(|\mathbf{x}|^{-2}\right), \qquad \frac{\partial}{\partial x_m}\Gamma_{\alpha+3;\beta+3}(\mathbf{x},\omega) = O\left(|\mathbf{x}|^{-2}\right),$$

$$\frac{\partial}{\partial x_m}\Gamma_{l;\alpha+3}(\mathbf{x},\omega) = O\left(|\mathbf{x}|^{-1}\right), \qquad \frac{\partial}{\partial x_m}\Gamma_{\alpha+3;j}(\mathbf{x},\omega) = O\left(|\mathbf{x}|^{-1}\right),$$

$$l, j, m = 1, 2, 3, \qquad \alpha, \beta = 1, 2, 3, 4$$

hold the neighborhood of the origin.

We shall use the following.

Lemma 2.2 *If η_{1m} $(m = 1, 2, \cdots, 5)$ is defined by (2.19), then*

$$\sum_{m=1}^{5} \eta_{1m} = 0, \qquad \sum_{m=1}^{5} \lambda_m^2 \eta_{1m} = 0, \qquad \sum_{m=1}^{5} \lambda_m^4 \eta_{1m} = 0,$$

$$\sum_{m=1}^{5} \lambda_m^6 \eta_{1m} = 0, \qquad \sum_{m=1}^{5} \lambda_m^8 \eta_{1m} = 1, \tag{2.44}$$

$$\sum_{m=1}^{5} \frac{\eta_{1m}}{\lambda_m^2} = \frac{1}{(\lambda_1\lambda_2\lambda_3\lambda_4\lambda_5)^2} = \frac{k_0\mu_0}{\rho\omega^2 \mathcal{B}_{11}^*(0,\omega)}$$

and

$$\sum_{m=1}^{5} c_m = -\frac{1}{\rho\omega^2}, \qquad \sum_{m=1}^{5} \lambda_m^2 c_m = -\frac{1}{\mu_0}, \tag{2.45}$$

where

$$c_m = -\frac{1}{k_0\mu_0\lambda_m^2}\eta_{1m}\mathcal{B}_{11}^*(-\lambda_m^2,\omega), \qquad m = 1, 2, \cdots, 5.$$

Proof On the basis of (2.19), the identities (2.44) are proved by direct calculations. Moreover, it is easy to verify that $\mathcal{B}^*_{11}(-\lambda_m^2,\omega)$ can be rewritten in the following form

$$\mathcal{B}^*_{11}(-\lambda_m^2,\omega) = k_0\lambda_m^8 + q_1\lambda_m^6 + q_2\lambda_m^4 + q_3\lambda_m^2 + \mathcal{B}^*_{11}(0,\omega),$$
$$m = 1, 2, \cdots, 5, \tag{2.46}$$

where the values q_1, q_2, and q_3 are independent on λ_m. By virtue of (2.44) and (2.46), we obtain

$$\sum_{m=1}^{5} \frac{1}{\lambda_m^2}\eta_{1m}\mathcal{B}^*_{11}(-\lambda_m^2,\omega)$$
$$= \sum_{m=1}^{5} \eta_{1m}\left[k_0\lambda_m^6 + q_1\lambda_m^4 + q_2\lambda_m^2 + q_3 + \frac{1}{\lambda_m^2}\mathcal{B}^*_{11}(0,\omega)\right]$$
$$= \mathcal{B}^*_{11}(0,\omega)\sum_{m=1}^{5}\frac{\eta_{1m}}{\lambda_m^2} = \frac{k_0\mu_0}{\rho\omega^2}$$

and we have

$$\sum_{m=1}^{5} c_m = -\frac{1}{k_0\mu_0}\sum_{m=1}^{5}\frac{1}{\lambda_m^2}\eta_{1m}\mathcal{B}^*_{11}(-\lambda_m^2,\omega) = -\frac{1}{\rho\omega^2}.$$

Similarly, by virtue of (2.44) and (2.46), we obtain

$$\sum_{m=1}^{5} \eta_{1m}\mathcal{B}^*_{11}(-\lambda_m^2,\omega)$$
$$= \sum_{m=1}^{5} \eta_{1m}\left[k_0\lambda_m^8 + q_1\lambda_m^6 + q_2\lambda_m^4 + q_3\lambda_m^2 + \mathcal{B}^*_{11}(0,\omega)\right] = k_0$$

and we also get

$$\sum_{m=1}^{5} \lambda_m^2 c_m = -\frac{1}{kk_0\mu_0}\sum_{m=1}^{5}\eta_{1m}\mathcal{B}^*_{11}(-\lambda_m^2,\omega) = -\frac{1}{\mu_0}.$$

Hence, we have desired results. □

Now we can establish the singular part of the matrix $\mathbf{\Gamma}(\mathbf{x},\omega)$ in the neighborhood of the origin.

Theorem 2.11 *The relations*

$$\Gamma_{lj}(\mathbf{x},\omega)-\Gamma^{(0)}_{lj}(\mathbf{x})=const+O(|\mathbf{x}|),\qquad l,j=1,2,\cdots,7 \tag{2.47}$$

hold in the neighborhood of the origin.

Proof On the basis of (2.15) and (2.17) from (2.23) for $l,j=1,2,3$ it follows that

$$\begin{aligned}\Gamma_{lj}(\mathbf{x},\omega)&=\left[\frac{1}{\mu}\Lambda_1(\Delta,\omega)\,\delta_{lj}+n_{11}(\Delta,\omega)\frac{\partial^2}{\partial x_l\partial x_j}\right]\sum_{m=1}^{6}\eta_{2m}\gamma^{(m)}(\mathbf{x},\omega)\\&=\sum_{m=1}^{6}\eta_{2m}\left[\frac{1}{\mu}\Lambda_1(-\lambda_m^2,\omega)\,\delta_{lj}+n_{11}(-\lambda_m^2,\omega)\frac{\partial^2}{\partial x_l\partial x_j}\right]\gamma^{(m)}(\mathbf{x},\omega)\\&=\sum_{m=1}^{6}\eta_{2m}\left[n_{11}(-\lambda_m^2,\omega)-\frac{1}{\mu\lambda_m^2}\Lambda_1(-\lambda_m^2,\omega)\right]\frac{\partial^2}{\partial x_l\partial x_j}\gamma^{(m)}(\mathbf{x},\omega)\\&\quad+\sum_{m=1}^{6}\frac{1}{\mu\lambda_m^2}\eta_{2m}\Lambda_1(-\lambda_m^2,\omega)R_{lj}(\mathbf{D_x})\gamma^{(m)}(\mathbf{x},\omega),\end{aligned} \tag{2.48}$$

where

$$R_{lj}(\mathbf{D_x})=\frac{\partial^2}{\partial x_l\partial x_j}-\Delta\delta_{lj}.$$

In view of identity

$$\begin{aligned}\det\mathcal{B}(\Delta,\omega)&=(\mu_0\Delta+\rho\omega^2)\mathcal{B}^*_{11}(\Delta,\omega)\\&\quad+i\omega\beta_j\Delta\mathcal{B}^*_{1;j+1}(\Delta,\omega)+i\omega\varepsilon_0T_0\Delta\mathcal{B}^*_{15}(\Delta,\omega)\end{aligned}$$

from (2.12) we obtain

$$\begin{aligned}\Delta n_{11}(\Delta,\omega)&=-\frac{1}{k_0\mu\mu_0}\left[\det\mathcal{B}(\Delta,\omega)-(\mu\Delta+\rho\omega^2)\mathcal{B}^*_{11}(\Delta,\omega)\right]\\&=-\frac{1}{\mu}\Lambda_1(\Delta,\omega)+\frac{1}{kk_0\mu_0}(\Delta+\lambda_6^2)\mathcal{B}^*_{11}(\Delta,\omega).\end{aligned} \tag{2.49}$$

Obviously, from (2.49) we get

$$\begin{aligned}\left[n_{11}(-\lambda_m^2,\omega)-\frac{1}{\mu\lambda_m^2}\Lambda_1(-\lambda_m^2,\omega)\right]\gamma^{(m)}(\mathbf{x},\omega)\\=-\frac{1}{kk_0\mu_0\lambda_m^2}(\lambda_6^2-\lambda_m^2)\mathcal{B}_{11}^*(-\lambda_m^2,\omega)\gamma^{(m)}(\mathbf{x},\omega),\end{aligned}\tag{2.50}$$

$$m=1,2,\cdots,6.$$

By virtue of (2.50) from (2.48) we obtain

$$\begin{aligned}\Gamma_{lj}(\mathbf{x},\omega)=\sum_{m=1}^{5}\left[-\frac{1}{kk_0\mu_0\lambda_m^2}(\lambda_6^2-\lambda_m^2)\eta_{2m}\mathcal{B}_{11}^*(-\lambda_m^2,\omega)\right]\frac{\partial^2}{\partial x_l\partial x_j}\gamma^{(m)}(\mathbf{x},\omega)\\+\frac{1}{\mu}\sum_{m=1}^{6}\frac{1}{\lambda_m^2}\eta_{2m}\Lambda_1(-\lambda_m^2,\omega)R_{lj}(\mathbf{D_x})\gamma^{(m)}(\mathbf{x},\omega).\end{aligned}\tag{2.51}$$

On the basis of identities

$$(\lambda_6^2-\lambda_j^2)\eta_{2j}=\eta_{1j},\qquad j=1,2,\cdots,5,$$

$$\eta_{2m}\Lambda_1(-\lambda_m^2,\omega)=\begin{cases}0 & \text{for } m=1,2,\cdots,5\\ 1 & \text{for } m=6\end{cases}$$

from (2.51) we have

$$\Gamma_{lj}(\mathbf{x},\omega)=\sum_{m=1}^{5}c_m\frac{\partial^2}{\partial x_l\partial x_j}\gamma^{(m)}(\mathbf{x},\omega)+\frac{1}{\rho\omega^2}R_{lj}(\mathbf{D_x})\gamma^{(6)}(\mathbf{x},\omega).\tag{2.52}$$

Clearly, in the neighborhood of the origin we have

$$\begin{aligned}\gamma^{(m)}(\mathbf{x},\omega)&=-\frac{1}{4\pi|\mathbf{x}|}\sum_{n=0}^{\infty}\frac{(i\lambda_m|\mathbf{x}|)^n}{n!}\\&=\tilde{\gamma}^{(5)}(\mathbf{x})-\frac{i\lambda_m}{4\pi}-\lambda_m^2\tilde{\gamma}^{(6)}(\mathbf{x})+\xi_m(\mathbf{x}),\end{aligned}\tag{2.53}$$

where $m=1,2,\cdots,6$ and

$$\xi_m(\mathbf{x})=-\frac{1}{4\pi|\mathbf{x}|}\sum_{n=3}^{\infty}\frac{(i\lambda_m|\mathbf{x}|)^n}{n!}.$$

Obviously,

$$\xi_m(\mathbf{x}) = O\left(|\mathbf{x}|^2\right), \qquad \xi_{m,j}(\mathbf{x}) = O\left(|\mathbf{x}|\right),$$
$$\xi_{m,lj}(\mathbf{x}) = const + O\left(|\mathbf{x}|\right), \tag{2.54}$$
$$l, j = 1, 2, 3, \qquad m = 1, 2, \cdots, 6.$$

On the basis of (2.52) and Theorem 2.5 it follows that

$$\Gamma_{lj}(\mathbf{x}, \omega) - \Gamma_{lj}^{(0)}(\mathbf{x}) = \frac{\partial^2}{\partial x_l \partial x_j}\left[\sum_{m=1}^{5} c_m \gamma^{(m)}(\mathbf{x}, \omega) - \frac{1}{\mu_0}\tilde{\gamma}^{(6)}(\mathbf{x})\right]$$
$$+ R_{lj}(\mathbf{D}_\mathbf{x})\left[\frac{1}{\rho\omega^2}\gamma^{(6)}(\mathbf{x}, \omega) + \frac{1}{\mu}\tilde{\gamma}^{(6)}(\mathbf{x})\right]. \tag{2.55}$$

On the other hand, using (2.45) and (2.53) we have

$$\sum_{m=1}^{5} c_m \gamma^{(m)}(\mathbf{x}, \omega) - \frac{1}{\mu_0}\tilde{\gamma}^{(6)}(\mathbf{x})$$
$$= \sum_{m=1}^{5} c_m \left[\tilde{\gamma}^{(5)}(\mathbf{x}) - \frac{i\lambda_m}{4\pi} + \xi_m(\mathbf{x})\right] - \left(\sum_{m=1}^{5} c_m \lambda_m^2 + \frac{1}{\mu_0}\right)\tilde{\gamma}^{(6)}(\mathbf{x})$$
$$= -\frac{1}{\rho\omega^2}\tilde{\gamma}^{(5)}(\mathbf{x}) - \frac{i}{4\pi}\sum_{m=1}^{5} \lambda_m c_m + \sum_{m=1}^{5} c_m \xi_m(\mathbf{x}),$$

$$\frac{1}{\rho\omega^2}\gamma^{(6)}(\mathbf{x}, \omega) + \frac{1}{\mu}\tilde{\gamma}^{(6)}(\mathbf{x})$$
$$= \frac{1}{\rho\omega^2}\left(\tilde{\gamma}^{(5)}(\mathbf{x}) - \frac{i\lambda_6}{4\pi} - \lambda_6^2\tilde{\gamma}^{(6)}(\mathbf{x}) + \xi_6(\mathbf{x})\right) + \frac{1}{\mu}\tilde{\gamma}^{(6)}(\mathbf{x})$$
$$= \frac{1}{\rho\omega^2}\tilde{\gamma}^{(5)}(\mathbf{x}) - \frac{i\lambda_6}{4\pi\rho\omega^2} + \frac{1}{\rho\omega^2}\xi_6(\mathbf{x}). \tag{2.56}$$

Taking into account (2.54), (2.56), and $\Delta\gamma^{(5)}(\mathbf{x}) = 0$ for $\mathbf{x} \neq \mathbf{0}$, from (2.55) we finally obtain

$$\Gamma_{lj}(\mathbf{x},\omega)-\Gamma^{(0)}_{lj}(\mathbf{x})$$

$$=-\frac{1}{\rho\omega^2}\left[\frac{\partial^2}{\partial x_l\partial x_j}-R_{lj}(\mathbf{D_x})\right]\tilde{\gamma}^{(5)}(\mathbf{x})+\sum_{m=1}^{5}c_m\xi_{m,lj}(\mathbf{x})$$

$$+\frac{1}{\rho\omega^2}R_{lj}(\mathbf{D_x})\xi_6(\mathbf{x})$$

$$=-\frac{1}{\rho\omega^2}\Delta\tilde{\gamma}^{(5)}(\mathbf{x})+const+O\left(|\mathbf{x}|\right)=const+O\left(|\mathbf{x}|\right),$$

$$l,j=1,2,3.$$

The other formulas of (2.47) can be proved quite similarly. □

Thus, on the basis of Theorems 2.9 and 2.11 the matrix $\mathbf{\Gamma}^{(0)}(\mathbf{x})$ is the singular part of the fundamental solution $\mathbf{\Gamma}(\mathbf{x},\omega)$ in the neighborhood of the origin. Obviously, $\mathbf{\Gamma}^{(0)}(\mathbf{x})$ is the symmetrical matrix, i.e., $\left[\mathbf{\Gamma}^{(0)}\right]^\top=\mathbf{\Gamma}^{(0)}$.

In a similar manner we can prove the following theorems.

Theorem 2.12 *The relations*

$$\begin{aligned}
&\Gamma_{lj}(\mathbf{x},i\tau)-\Gamma^{(0)}_{lj}(\mathbf{x})=const+O\left(|\mathbf{x}|\right),\\
&\Gamma^{(q)}_{lj}(\mathbf{x},\omega)-\Gamma^{(0)}_{lj}(\mathbf{x})=const+O\left(|\mathbf{x}|\right),\\
&\Gamma^{(e)}_{lj}(\mathbf{x})-\Gamma^{(0)}_{lj}(\mathbf{x})=const+O\left(|\mathbf{x}|\right),\qquad l,j=1,2,\cdots,7
\end{aligned}\tag{2.57}$$

hold in the neighborhood of the origin.

Thus, the matrix $\mathbf{\Gamma}^{(0)}(\mathbf{x})$ is the singular part of the fundamental solutions $\mathbf{\Gamma}(\mathbf{x},i\tau)$, $\mathbf{\Gamma}^{(q)}(\mathbf{x},\omega)$, and $\mathbf{\Gamma}^{(e)}(\mathbf{x},\omega)$ in the neighborhood of the origin.

Theorem 2.13 *The relations*

$$\begin{aligned}
&\Gamma^{(r)}_{\alpha\beta}(\mathbf{x},\omega)-\Phi^{(re)}_{\alpha\beta}(\mathbf{x})=const+O\left(|\mathbf{x}|\right),\\
&\Gamma^{(re)}_{\alpha\beta}(\mathbf{x})-\Phi^{(re)}_{\alpha\beta}(\mathbf{x})=const+O\left(|\mathbf{x}|\right),\qquad \alpha,\beta=1,2,3,4
\end{aligned}\tag{2.58}$$

hold in the neighborhood of the origin, where the matrix $\mathbf{\Phi}^{(re)}(\mathbf{x})$ is the fundamental solution of (2.40) and defined by (2.41) and (2.42).

Thus, on the basis of Theorem 2.13 the matrix $\mathbf{\Phi}^{(re)}(\mathbf{x})$ is the singular part of the fundamental solutions $\mathbf{\Gamma}^{(r)}(\mathbf{x},\omega)$ and $\mathbf{\Gamma}^{(re)}(\mathbf{x})$ in the neighborhood of the origin.

2.8 Singular Solutions in Elasticity

In this section we shall construct the solutions of Eqs. (1.38)–(1.43) for $\mathcal{F}(\mathbf{x}) = \mathbf{0}$ playing a major role in the investigation of the BVPs of the linear theory of elasticity for materials with quadruple porosity. The following theorem is valid.

Theorem 2.14 *Every column of the matrices*

$$\left[\mathcal{P}^{(\kappa)}(\mathbf{D_x}, \mathbf{n}(\mathbf{x}))\mathbf{\Gamma}^\top(\mathbf{x}-\mathbf{y}, \omega)\right]^\top,$$

$$\left[\mathcal{P}^{(\kappa)}(\mathbf{D_x}, \mathbf{n}(\mathbf{x}))\mathbf{\Gamma}^\top(\mathbf{x}-\mathbf{y}, i\tau)\right]^\top,$$

$$\left\{\mathcal{P}^{(\kappa)}(\mathbf{D_x}, \mathbf{n}(\mathbf{x}))[\mathbf{\Gamma}^{(q)}(\mathbf{x}-\mathbf{y}, \omega)]^\top\right\}^\top,$$

$$\left\{\mathcal{P}^{(\kappa)}(\mathbf{D_x}, \mathbf{n}(\mathbf{x}))[\mathbf{\Gamma}^{(e)}(\mathbf{x}-\mathbf{y})]^\top\right\}^\top,$$

$$\left\{\mathcal{P}^{(\kappa)}(\mathbf{D_x}, \mathbf{n}(\mathbf{x}))\mathbf{\Gamma}^{(0)}(\mathbf{x}-\mathbf{y})\right\}^\top$$

with respect to the point $\mathbf{x}$ *satisfies the homogeneous equations*

$$\begin{aligned}
&\mathcal{A}(\mathbf{D_x}, \omega)\left[\mathcal{P}^{(\kappa)}(\mathbf{D_x}, \mathbf{n}(\mathbf{x}))\mathbf{\Gamma}^\top(\mathbf{x}-\mathbf{y}, \omega)\right]^\top = \mathbf{0},\\
&\mathcal{A}(\mathbf{D_x}, -i\tau)\left[\mathcal{P}^{(\kappa)}(\mathbf{D_x}, \mathbf{n}(\mathbf{x}))\mathbf{\Gamma}^\top(\mathbf{x}-\mathbf{y}, -i\tau)\right]^\top = \mathbf{0},\\
&\mathcal{A}^{(q)}(\mathbf{D_x}, \omega)\left\{\mathcal{P}^{(\kappa)}(\mathbf{D_x}, \mathbf{n}(\mathbf{x}))[\mathbf{\Gamma}^{(q)}(\mathbf{x}-\mathbf{y}, \omega)]^\top\right\}^\top = \mathbf{0},\\
&\mathcal{A}^{(e)}(\mathbf{D_x})\left\{\mathcal{P}^{(\kappa)}(\mathbf{D_x}, \mathbf{n}(\mathbf{x}))[\mathbf{\Gamma}^{(e)}(\mathbf{x}-\mathbf{y})]^\top\right\}^\top = \mathbf{0},\\
&\mathcal{A}^{(0)}(\mathbf{D_x})\left\{\mathcal{P}^{(\kappa)}(\mathbf{D_x}, \mathbf{n}(\mathbf{x}))\mathbf{\Gamma}^{(0)}(\mathbf{x}-\mathbf{y})\right\}^\top = \mathbf{0}
\end{aligned} \tag{2.59}$$

everywhere in $\mathbb{R}^3$ *except* $\mathbf{x} = \mathbf{y}$*, respectively, where* κ *is an arbitrary real number.*

Proof Keeping in mind (2.43) it follows that

$$\left\{\mathcal{A}(\mathbf{D}_{\mathbf{x}},\omega)\left[\mathcal{P}^{(\kappa)}(\mathbf{D}_{\mathbf{x}},\mathbf{n}(\mathbf{x}))\boldsymbol{\Gamma}^{\top}(\mathbf{x}-\mathbf{y},\omega)\right]^{\top}\right\}_{lj}$$

$$=\sum_{m=1}^{7}\mathcal{A}_{lm}(\mathbf{D}_{\mathbf{x}},\omega)\left[\mathcal{P}^{(\kappa)}(\mathbf{D}_{\mathbf{x}},\mathbf{n}(\mathbf{x}))\boldsymbol{\Gamma}^{\top}(\mathbf{x}-\mathbf{y},\omega)\right]_{jm}$$

$$=\sum_{m,s=1}^{7}\mathcal{A}_{lm}(\mathbf{D}_{\mathbf{x}},\omega)\mathcal{P}^{(\kappa)}_{js}(\mathbf{D}_{\mathbf{x}},\mathbf{n}(\mathbf{x}))\Gamma_{ms}(\mathbf{x}-\mathbf{y},\omega)$$

$$=\sum_{m,s=1}^{7}\mathcal{P}^{(\kappa)}_{js}(\mathbf{D}_{\mathbf{x}},\mathbf{n}(\mathbf{x}))\mathcal{A}_{lm}(\mathbf{D}_{\mathbf{x}},\omega)\Gamma_{ms}(\mathbf{x}-\mathbf{y},\omega)=0$$

which is the first formula involved in Eq. (2.59). The other formulas in (2.59) can be proved quite similarly. □

Theorem 2.10 leads to the following.

Theorem 2.15 *The relations*

$$\left[\mathcal{P}^{(\kappa)}(\mathbf{D}_{\mathbf{x}},\mathbf{n}(\mathbf{x}))\boldsymbol{\Gamma}^{\top}(\mathbf{x},\omega)\right]_{lj}=O\left(|\mathbf{x}|^{-2}\right),$$

$$\left[\mathcal{P}^{(\kappa)}(\mathbf{D}_{\mathbf{x}},\mathbf{n}(\mathbf{x}))\boldsymbol{\Gamma}^{\top}(\mathbf{x},i\tau)\right]_{lj}=O\left(|\mathbf{x}|^{-2}\right),$$

$$\left\{\mathcal{P}^{(\kappa)}(\mathbf{D}_{\mathbf{x}},\mathbf{n}(\mathbf{x}))[\boldsymbol{\Gamma}^{(q)}(\mathbf{x},\omega)]^{\top}\right\}_{lj}=O\left(|\mathbf{x}|^{-2}\right),$$

$$\left\{\mathcal{P}^{(\kappa)}(\mathbf{D}_{\mathbf{x}},\mathbf{n}(\mathbf{x}))[\boldsymbol{\Gamma}^{(e)}(\mathbf{x})]^{\top}\right\}_{lj}=O\left(|\mathbf{x}|^{-2}\right),$$

$$\left\{\mathcal{P}^{(\kappa)}(\mathbf{D}_{\mathbf{x}},\mathbf{n}(\mathbf{x}))\boldsymbol{\Gamma}^{(0)}(\mathbf{x})\right\}_{lj}=O\left(|\mathbf{x}|^{-2}\right),\qquad l,j=1,2,\cdots,7$$

hold the neighborhood of the origin, where κ is an arbitrary real number.

Now we can establish the singular part of the matrices $\mathcal{P}^{(\kappa)}(\mathbf{D}_{\mathbf{x}},\mathbf{n}(\mathbf{x}))\boldsymbol{\Gamma}^{\top}(\mathbf{x},\omega)$, $\mathcal{P}^{(\kappa)}(\mathbf{D}_{\mathbf{x}},\mathbf{n}(\mathbf{x}))\boldsymbol{\Gamma}^{\top}(\mathbf{x},i\tau)$, $\mathcal{P}^{(\kappa)}(\mathbf{D}_{\mathbf{x}},\mathbf{n}(\mathbf{x}))[\boldsymbol{\Gamma}^{(q)}(\mathbf{x},\omega)]^{\top}$ and $\mathcal{P}^{(\kappa)}(\mathbf{D}_{\mathbf{x}},\mathbf{n}(\mathbf{x}))[\boldsymbol{\Gamma}^{(e)}(\mathbf{x})]^{\top}$ in the neighborhood of the origin.

Theorem 2.11 and the relations (2.57) lead to the following.

Theorem 2.16 *The relations*

$$\begin{aligned}
&\left\{\mathcal{P}^{(\kappa)}(\mathbf{D}_\mathbf{x}, \mathbf{n}(\mathbf{x}))\mathbf{\Gamma}^\top(\mathbf{x}, \omega)\right\}_{lj} \\
&\qquad - \left\{\mathcal{P}^{(\kappa)}(\mathbf{D}_\mathbf{x}, \mathbf{n}(\mathbf{x}))\mathbf{\Gamma}^{(0)}(\mathbf{x})\right\}_{lj} = O\left(|\mathbf{x}|^{-1}\right), \\
&\left\{\mathcal{P}^{(\kappa)}(\mathbf{D}_\mathbf{x}, \mathbf{n}(\mathbf{x}))\mathbf{\Gamma}^\top(\mathbf{x}, i\tau)\right\}_{lj} \\
&\qquad - \left\{\mathcal{P}^{(\kappa)}(\mathbf{D}_\mathbf{x}, \mathbf{n}(\mathbf{x}))\mathbf{\Gamma}^{(0)}(\mathbf{x})\right\}_{lj} = O\left(|\mathbf{x}|^{-1}\right), \\
&\left\{\mathcal{P}^{(\kappa)}(\mathbf{D}_\mathbf{x}, \mathbf{n}(\mathbf{x}))\left[\mathbf{\Gamma}^{(q)}(\mathbf{x}, \omega)\right]^\top\right\}_{lj} \\
&\qquad - \left\{\mathcal{P}^{(\kappa)}(\mathbf{D}_\mathbf{x}, \mathbf{n}(\mathbf{x}))\mathbf{\Gamma}^{(0)}(\mathbf{x})\right\}_{lj} = O\left(|\mathbf{x}|^{-1}\right), \\
&\left\{\mathcal{P}^{(\kappa)}(\mathbf{D}_\mathbf{x}, \mathbf{n}(\mathbf{x}))\mathbf{\Gamma}^{(e)}(\mathbf{x})\right\}_{lj} \\
&\qquad - \left\{\mathcal{P}^{(\kappa)}(\mathbf{D}_\mathbf{x}, \mathbf{n}(\mathbf{x}))\mathbf{\Gamma}^{(0)}(\mathbf{x})\right\}_{lj} = O\left(|\mathbf{x}|^{-1}\right), \\
&l, j = 1, 2, \cdots, 7
\end{aligned}$$

hold in the neighborhood of the origin, where κ is an arbitrary real number.

Thus, the matrix $\mathcal{P}^{(\kappa)}(\mathbf{D}_\mathbf{x}, \mathbf{n}(\mathbf{x}))\mathbf{\Gamma}^{(0)}(\mathbf{x})$ is the singular part of the matrices $\mathcal{P}^{(\kappa)}(\mathbf{D}_\mathbf{x}, \mathbf{n}(\mathbf{x}))\mathbf{\Gamma}^\top(\mathbf{x}, \omega)$, $\mathcal{P}^{(\kappa)}(\mathbf{D}_\mathbf{x}, \mathbf{n}(\mathbf{x}))\mathbf{\Gamma}^\top(\mathbf{x}, i\tau)$, $\mathcal{P}^{(\kappa)}(\mathbf{D}_\mathbf{x}, \mathbf{n}(\mathbf{x}))\left[\mathbf{\Gamma}^{(q)}(\mathbf{x}, \omega)\right]^\top$ and $\mathcal{P}^{(\kappa)}(\mathbf{D}_\mathbf{x}, \mathbf{n}(\mathbf{x}))\left[\mathbf{\Gamma}^{(e)}(\mathbf{x})\right]^\top$ in the neighborhood of the origin for arbitrary real number κ.

We now establish an important property of matrix $\mathcal{N}(\mathbf{D}_\mathbf{x}, \mathbf{n}(\mathbf{x}))\mathbf{\Gamma}^{(0)}(\mathbf{x})$, where the matrix differential operator $\mathcal{N}(\mathbf{D}_\mathbf{x}, \mathbf{n}(\mathbf{x}))$ introduced in the end of Sect. 1.7.

Clearly, the matrix $\mathbf{\Gamma}^{(0)}(\mathbf{x})$ can be written as

$$\begin{aligned}
&\mathbf{\Gamma}^{(0)}(\mathbf{x}) = \left(\Gamma^{(0)}_{lj}(\mathbf{x})\right)_{7\times 7}, \qquad \Gamma^{(0)}_{lj}(\mathbf{x}) = \lambda' \frac{\delta_{lj}}{|\mathbf{x}|} + \mu' \frac{x_l x_j}{|\mathbf{x}|^3}, \\
&\Gamma^{(0)}_{\alpha+3;\beta+3}(\mathbf{x}) = \frac{1}{k_0} k^*_{\alpha\beta}\tilde{\gamma}^{(5)}(\mathbf{x}), \qquad \Gamma^{(0)}_{l;\alpha+3}(\mathbf{x}) = \Gamma^{(0)}_{\alpha+3;l}(\mathbf{x}) = 0, \\
&\lambda' = -\frac{\lambda+3\mu}{8\pi\mu\mu_0}, \qquad \mu' = -\frac{\lambda+\mu}{8\pi\mu\mu_0}, \\
&l, j = 1, 2, 3, \qquad \alpha, \beta = 1, 2, 3, 4,
\end{aligned} \tag{2.60}$$

$k^*_{\alpha\beta}$ is the cofactor of element $k_{\alpha\beta}$ of the matrix $\mathbf{K}$, and $\tilde{\gamma}^{(5)}(\mathbf{x})$ is given by (2.26).

On the other hand, we can write the matrix differential operator $\mathcal{N}(\mathbf{D_x}, \mathbf{n}(\mathbf{x}))$ in the following form

$$\begin{aligned}
&\mathcal{N}(\mathbf{D_x}, \mathbf{n}) = (\mathcal{N}_{ms}(\mathbf{D_x}, \mathbf{n}))_{7\times 7}, \\
&\mathcal{N}_{lj}(\mathbf{D_x}, \mathbf{n}) = \mu\delta_{lj}\frac{\partial}{\partial \mathbf{n}} + (\lambda+\mu)n_l\frac{\partial}{\partial x_j} + \kappa_0\mathcal{M}_{lj}(\mathbf{D_x}, \mathbf{n}), \\
&\mathcal{N}_{l;\alpha+3}(\mathbf{D_x}, \mathbf{n}) = -a_\alpha\, n_l, \qquad \mathcal{N}_{\alpha+3;\beta+3}(\mathbf{D_x}, \mathbf{n}) = k_{\alpha\beta}\frac{\partial}{\partial \mathbf{n}}, \\
&\mathcal{N}_{\alpha+3;j}(\mathbf{D_x}, \mathbf{n}) = 0,
\end{aligned} \tag{2.61}$$

where $l, j = 1, 2, 3,\ \alpha, \beta = 1, 2, 3, 4$ and

$$\kappa_0 = \frac{\mu\mu'}{\lambda'} = \frac{\mu(\lambda+\mu)}{\lambda+3\mu}. \tag{2.62}$$

On the basis of (1.47), (1.48), and (2.60) by direct calculation for the matrix

$$\mathcal{P}^{(\kappa)}(\mathbf{D_x}, \mathbf{n}(\mathbf{x}))\mathbf{\Gamma}^{(0)}(\mathbf{x}) = \left(\left[\mathcal{P}^{(\kappa)}(\mathbf{D_x}, \mathbf{n}(\mathbf{x}))\mathbf{\Gamma}^{(0)}(\mathbf{x})\right]_{lj}\right)_{7\times 7}$$

we obtain

$$\begin{aligned}
&\left[\mathcal{P}^{(\kappa)}(\mathbf{D_x}, \mathbf{n}(\mathbf{x}))\mathbf{\Gamma}^{(0)}(\mathbf{x})\right]_{lj} = \frac{x_m n_m(\mathbf{x})\delta_{lj}}{4\pi\,|\mathbf{x}|^3} \\
&\quad +\mathcal{M}_{lm}(\mathbf{D_x}, \mathbf{n})\left[\lambda'(\kappa-\kappa_0)\frac{\delta_{mj}}{|\mathbf{x}|} + \mu'(\kappa+\mu)\frac{x_m x_j}{|\mathbf{x}|^3}\right], \\
&\left[\mathcal{P}^{(\kappa)}(\mathbf{D_x}, \mathbf{n}(\mathbf{x}))\mathbf{\Gamma}^{(0)}(\mathbf{x})\right]_{l;\beta+3} = -\frac{a_\alpha k^*_{\alpha\beta}}{k_0}\, n_l(\mathbf{x})\tilde{\gamma}^{(5)}(\mathbf{x}), \\
&\left[\mathcal{P}^{(\kappa)}(\mathbf{D_x}, \mathbf{n}(\mathbf{x}))\mathbf{\Gamma}^{(0)}(\mathbf{x})\right]_{\beta+3;l} = 0, \\
&\left[\mathcal{P}^{(\kappa)}(\mathbf{D_x}, \mathbf{n}(\mathbf{x}))\mathbf{\Gamma}^{(0)}(\mathbf{x})\right]_{m+3;\beta+3} = \delta_{m\beta}\frac{\partial}{\partial \mathbf{n}(\mathbf{x})}\tilde{\gamma}^{(5)}(\mathbf{x}),
\end{aligned} \tag{2.63}$$

where $l, j = 1, 2, 3,\ m, \beta = 1, 2, 3, 4$ and κ is an arbitrary real number.

Taking into account (2.61) for the matrix

$$\mathcal{N}(\mathbf{D_x}, \mathbf{n}(\mathbf{x}))\mathbf{\Gamma}^{(0)}(\mathbf{x}) = \left(\left[\mathcal{N}(\mathbf{D_x}, \mathbf{n}(\mathbf{x}))\mathbf{\Gamma}^{(0)}(\mathbf{x})\right]_{lj}\right)_{7\times 7}$$

from (2.63) we deduce that

$$\begin{aligned}
&\left[\mathcal{N}(\mathbf{D_x},\mathbf{n}(\mathbf{x}))\mathbf{\Gamma}^{(0)}(\mathbf{x})\right]_{lj}\\
&\quad=\frac{x_m n_m(\mathbf{x})\delta_{lj}}{4\pi\,|\mathbf{x}|^3}+\mu'(\kappa_0+\mu)\mathcal{M}_{lm}(\mathbf{D_x},\mathbf{n})\frac{x_m x_j}{|\mathbf{x}|^3},\\
&\left[\mathcal{N}(\mathbf{D_x},\mathbf{n}(\mathbf{x}))\mathbf{\Gamma}^{(0)}(\mathbf{x})\right]_{l;\beta+3}=-\frac{a_\alpha k^*_{\alpha\beta}}{k_0}\,n_l(\mathbf{x})\tilde{\gamma}^{(5)}(\mathbf{x}),\\
&\left[\mathcal{N}(\mathbf{D_x},\mathbf{n}(\mathbf{x}))\mathbf{\Gamma}^{(0)}(\mathbf{x})\right]_{\beta+3;l}=0,\\
&\left[\mathcal{N}(\mathbf{D_x},\mathbf{n}(\mathbf{x}))\mathbf{\Gamma}^{(0)}(\mathbf{x})\right]_{m+3;\beta+3}=\delta_{m\beta}\frac{\partial}{\partial\mathbf{n}(\mathbf{x})}\tilde{\gamma}^{(5)}(\mathbf{x}),
\end{aligned}\tag{2.64}$$

where $l,\ j=1,2,3,\ m,\beta=1,2,3,4$.

We are now ready to prove the following.

Theorem 2.17 *If S is the Liapunov surface of classes $C^{1,\nu}$ $(0<\nu\le 1)$ and $\mathbf{x},\mathbf{y}\in S$, then the relations*

$$\begin{aligned}
&\left[\mathcal{N}(\mathbf{D_x},\mathbf{n}(\mathbf{x}))\mathbf{\Gamma}^{(0)}(\mathbf{x}-\mathbf{y})\right]_{lj}=O\left(|\mathbf{x}-\mathbf{y}|^{-2+\nu}\right),\\
&\left[\mathcal{N}(\mathbf{D_x},\mathbf{n}(\mathbf{x}))\mathbf{\Gamma}^{(0)}(\mathbf{x}-\mathbf{y})\right]_{l,\beta+3}=O\left(|\mathbf{x}-\mathbf{y}|^{-1}\right),\\
&\left[\mathcal{N}(\mathbf{D_x},\mathbf{n}(\mathbf{x}))\mathbf{\Gamma}^{(0)}(\mathbf{x}-\mathbf{y})\right]_{m+3,\beta+3}=O\left(|\mathbf{x}-\mathbf{y}|^{-2+\nu}\right),
\end{aligned}\tag{2.65}$$

hold in the neighborhood of $\mathbf{x}=\mathbf{y}$, where $l,\ j=1,2,3,\ m,\beta=1,2,3,4$.

Proof Obviously, from (2.64) it follows that

$$\begin{aligned}
&\left[\mathcal{N}(\mathbf{D_x},\mathbf{n}(\mathbf{x}))\mathbf{\Gamma}^{(0)}(\mathbf{x}-\mathbf{y})\right]_{lj}=\frac{(x_m-y_m)n_m(\mathbf{x})\delta_{lj}}{4\pi\,|\mathbf{x}-\mathbf{y}|^3}\\
&\quad+\mu'(\kappa_0+\mu)\mathcal{M}_{lm}(\mathbf{D_x},\mathbf{n})\frac{(x_m-y_m)(x_j-y_j)}{|\mathbf{x}-\mathbf{y}|^3},\\
&\left[\mathcal{N}(\mathbf{D_x},\mathbf{n}(\mathbf{x}))\mathbf{\Gamma}^{(0)}(\mathbf{x}-\mathbf{y})\right]_{l;\beta+3}=-\frac{a_\alpha k^*_{\alpha\beta}}{k_0}\,n_l(\mathbf{x})\tilde{\gamma}^{(5)}(\mathbf{x}-\mathbf{y}),\\
&\left[\mathcal{N}(\mathbf{D_x},\mathbf{n}(\mathbf{x}))\mathbf{\Gamma}^{(0)}(\mathbf{x}-\mathbf{y})\right]_{\beta+3;l}=0,
\end{aligned}\tag{2.66}$$

$$\left[\mathcal{N}(\mathbf{D_x}, \mathbf{n}(\mathbf{x}))\boldsymbol{\Gamma}^{(0)}(\mathbf{x}-\mathbf{y})\right]_{m+3;\beta+3} = \delta_{m\beta}\frac{\partial}{\partial \mathbf{n}(\mathbf{x})}\tilde{\gamma}^{(5)}(\mathbf{x}-\mathbf{y}),$$

where $l, j = 1, 2, 3,\ m, \beta = 1, 2, 3, 4$.

On the basis of relations (see Günther [161])

$$\frac{(x_m - y_m)n_m(\mathbf{x})}{|\mathbf{x}-\mathbf{y}|^3} = O\left(|\mathbf{x}-\mathbf{y}|^{-2+\nu}\right),$$

$$\frac{\partial}{\partial \mathbf{n}(\mathbf{x})}\tilde{\gamma}^{(5)}(\mathbf{x}-\mathbf{y}) = O\left(|\mathbf{x}-\mathbf{y}|^{-2+\nu}\right)$$

and the estimate (see Kupradze et al. [219])

$$\mathcal{M}_{lm}(\mathbf{D_x}, \mathbf{n})\frac{(x_m - y_m)(x_j - y_j)}{|\mathbf{x}-\mathbf{y}|^3} = O\left(|\mathbf{x}-\mathbf{y}|^{-2+\nu}\right)$$

for $\mathbf{x}, \mathbf{y} \in S$ from (2.66) we obtain the relations (2.65). □

Hence, the matrix $\mathcal{N}(\mathbf{D_x}, \mathbf{n}(\mathbf{x}))\boldsymbol{\Gamma}^{(0)}(\mathbf{x}-\mathbf{y})$ is weakly singular in the neighborhood of $\mathbf{x} = \mathbf{y}$ for $\mathbf{x}, \mathbf{y} \in S$.

Clearly, Theorem 2.16 leads to the following.

Corollary 2.1 *The relations*

$$\begin{aligned}
&\left\{\mathcal{N}(\mathbf{D_x}, \mathbf{n}(\mathbf{x}))\boldsymbol{\Gamma}^\top(\mathbf{x}, \omega)\right\}_{lj} \\
&\qquad - \left\{\mathcal{N}(\mathbf{D_x}, \mathbf{n}(\mathbf{x}))\boldsymbol{\Gamma}^{(0)}(\mathbf{x})\right\}_{lj} = O\left(|\mathbf{x}|^{-1}\right), \\
&\left\{\mathcal{N}(\mathbf{D_x}, \mathbf{n}(\mathbf{x}))\boldsymbol{\Gamma}^\top(\mathbf{x}, i\tau)\right\}_{lj} \\
&\qquad - \left\{\mathcal{N}(\mathbf{D_x}, \mathbf{n}(\mathbf{x}))\boldsymbol{\Gamma}^{(0)}(\mathbf{x})\right\}_{lj} = O\left(|\mathbf{x}|^{-1}\right), \\
&\left\{\mathcal{N}(\mathbf{D_x}, \mathbf{n}(\mathbf{x}))\left[\boldsymbol{\Gamma}^{(q)}(\mathbf{x}, \omega)\right]^\top\right\}_{lj} \\
&\qquad - \left\{\mathcal{N}(\mathbf{D_x}, \mathbf{n}(\mathbf{x}))\boldsymbol{\Gamma}^{(0)}(\mathbf{x})\right\}_{lj} = O\left(|\mathbf{x}|^{-1}\right), \\
&\left\{\mathcal{N}(\mathbf{D_x}, \mathbf{n}(\mathbf{x}))\left[\boldsymbol{\Gamma}^{(e)}(\mathbf{x})\right]^\top\right\}_{lj} \\
&\qquad - \left\{\mathcal{N}(\mathbf{D_x}, \mathbf{n}(\mathbf{x}))\boldsymbol{\Gamma}^{(0)}(\mathbf{x})\right\}_{lj} = O\left(|\mathbf{x}|^{-1}\right), \\
&l, j = 1, 2, \cdots, 7
\end{aligned}$$

hold in the neighborhood of the origin.

Thus, the matrix $\mathcal{N}(\mathbf{D_x}, \mathbf{n}(\mathbf{x}))\boldsymbol{\Gamma}^{(0)}(\mathbf{x})$ is the singular part of the matrices $\mathcal{N}(\mathbf{D_x}, \mathbf{n}(\mathbf{x}))\boldsymbol{\Gamma}^\top(\mathbf{x}, \omega)$, $\mathcal{N}(\mathbf{D_x}, \mathbf{n}(\mathbf{x}))\boldsymbol{\Gamma}^\top(\mathbf{x}, i\tau)$, $\mathcal{N}(\mathbf{D_x}, \mathbf{n}(\mathbf{x}))\left[\boldsymbol{\Gamma}^{(q)}(\mathbf{x}, \omega)\right]^\top$, and $\mathcal{N}(\mathbf{D_x}, \mathbf{n}(\mathbf{x}))\left[\boldsymbol{\Gamma}^{(e)}(\mathbf{x})\right]^\top$ in the neighborhood of the origin.

Chapter 3
Galerkin-Type Solutions and Green's Formulas in Elasticity

This chapter is concerned with the Galerkin-type representations of general solutions and Green's formulas in the linear theory of elasticity for materials with quadruple porosity.

In Sect. 3.1, a brief review of the representations of general solutions in the classical theories of elasticity and thermoelasticity is introduced.

In Sect. 3.2, the Galerkin-type representations of general solutions of the systems (1.21) to (1.25) in the linear theory of elasticity for materials with quadruple porosity are established.

In Sect. 3.3, the short history of Green's classical formulas (first, second, and third identities) is introduced.

In Sects. 3.4 and 3.5, Green's identities for equations of steady vibrations and equations in Laplace transform space of the linear theory of quadruple porosity elasticity are obtained, respectively.

In Sects. 3.6 and 3.7, Green's identities for equations of quasi-static and equilibrium of the considered theory are presented, respectively.

Finally, in Sects. 3.8 and 3.9, Green's identities for equations of steady vibrations and equilibrium of quadruple porosity rigid body are given, respectively.

3.1 On the Representation of General Solution

Contemporary treatment of the various BVPs of the classical theories of elasticity and thermoelasticity usually begins with the representation of a solution of field equations in terms of elementary (harmonic, biharmonic, metaharmonic, etc.) functions. In the classical theory of elasticity the Boussinesq–Somigliana–Galerkin, Boussinesq–Papkovitch–Neuber, Green–Lamé, Naghdi–Hsu, and Cauchy–Kovalevski–Somigliana solutions are well known (see Gurtin [162] and

M. Svanadze, *Potential Method in Mathematical Theories of Multi-Porosity Media*, Interdisciplinary Applied Mathematics 51,
https://doi.org/10.1007/978-3-030-28022-2_3

Hetnarski and Ignaczak [165]). A review of the history of these solutions is given in the paper by Wang et al. [393].

Iacovache [170] obtained the Galerkin-type solution (see Galerkin [145]) of equations of classical elastokinetics. Nowacki [264, 265] and Sandru [293] presented the formulas of representations of solutions in the classical theory of thermodynamics and the theory of micropolar elasticity.

Biot [43] obtained the Boussinesq–Papkovitch–Neuber-type solution in the theory of elasticity for a single porosity material and Verruijt [388] established the completeness of this solution. Svanadze and de Boer [357] presented the representations of Galerkin-type solutions of equations of motion and steady vibrations in the linear theory of the liquid-saturated porous medium.

Chandrasekharaiah [74, 75] obtained the Boussinesq–Papkovitch–Neuber, Green–Lamé, and Cauchy–Kovalevski–Somigliana type solutions in the theory elasticity for materials with single voids. Ciarletta [88] proved the representation theorem of Galerkin-type in the theory of thermoelastic materials with single voids.

In addition, Unger and Aifantis [382] established the completeness of the Boussinesq–Papkovitch–Neuber-type solution in the theory of consolidation with double porosity. Svanadze [350, 351] obtained the representations of Galerkin-type solutions in the linear theories of elasticity and thermoelasticity for triple porosity materials and established the completeness of these solutions.

3.2 Galerkin-Type Solutions

3.2.1 Equations of Steady Vibrations

The next two theorems provide a Galerkin-type solution to system (1.21).

Theorem 3.1 *Let*

$$\begin{aligned} u_l(\mathbf{x}) &= \frac{1}{\mu}\Lambda_1(\Delta,\omega)\,w_l(\mathbf{x}) + n_{11}(\Delta,\omega)\,w_{j,lj}(\mathbf{x}) \\ &\quad + n_{1;\alpha+1}(\Delta,\omega)\,\phi_{\alpha,l}(\mathbf{x}), \\ p_\alpha(\mathbf{x}) &= n_{\alpha+1;1}(\Delta,\omega) w_{j,j}(\mathbf{x}) + n_{\alpha+1;\beta+1}(\Delta,\omega)\phi_\beta(\mathbf{x}), \end{aligned} \tag{3.1}$$

where $l = 1,2,3,\quad \alpha = 1,2,3,4,\quad \mathbf{w} = (w_1,w_2,w_3) \in C^{14}(\Omega),\quad \boldsymbol{\phi} = (\phi_1,\phi_2,\phi_3,\phi_4) \in C^{12}(\Omega)$,

$$\Lambda_2(\Delta,\omega)\,\mathbf{w}(\mathbf{x}) = \mathcal{F}^{(1)}(\mathbf{x}), \qquad \Lambda_1(\Delta,\omega)\,\boldsymbol{\phi}(\mathbf{x}) = \mathcal{F}^{(2)}(\mathbf{x}), \tag{3.2}$$

$\mathcal{F}^{(1)}$ *and* $\mathcal{F}^{(2)}$ *are three and four component vector-functions, respectively, then* $\mathbf{V} = (\mathbf{u},\mathbf{p})$ *is a solution of system (1.21).*

Proof By virtue of (2.15) we can rewrite the Eqs. (3.1) and (3.2) in the form

$$\mathbf{V}(\mathbf{x}) = \mathcal{L}(\mathbf{D}_{\mathbf{x}}, \omega)\, \widetilde{\mathbf{w}}(\mathbf{x}) \tag{3.3}$$

and

$$\boldsymbol{\Lambda}(\Delta, \omega)\, \widetilde{\mathbf{w}}(\mathbf{x}) = \mathcal{F}(\mathbf{x}), \tag{3.4}$$

respectively, where $\widetilde{\mathbf{w}} = (\mathbf{w}, \boldsymbol{\phi})$, $\mathcal{F} = (\mathcal{F}^{(1)}, \mathcal{F}^{(2)})$. Clearly, on the basis of (2.16), (3.3), and (3.4) it follows that

$$\mathcal{A}(\mathbf{D}_{\mathbf{x}}, \omega)\mathbf{V} = \mathcal{A}(\mathbf{D}_{\mathbf{x}}, \omega)\mathcal{L}(\mathbf{D}_{\mathbf{x}}, \omega)\widetilde{\mathbf{w}} = \Lambda(\Delta, \omega)\widetilde{\mathbf{w}} = \mathcal{F}.$$

Thus, the vector $\mathbf{V}$ is a solution of the system (1.21). □

Theorem 3.2 *If* $\mathbf{V} = (\mathbf{u}, \mathbf{p})$ *is a solution of system (1.21) in* Ω*, then* $\mathbf{V}$ *is represented by (3.1), where* $\widetilde{\mathbf{w}} = (\mathbf{w}, \boldsymbol{\phi})$ *is a solution of (3.2) and* Ω *is a finite domain in* $\mathbb{R}^3$.

Proof Let $\mathbf{V}$ be a solution of system (1.21). Obviously, if $\boldsymbol{\Gamma}'(\mathbf{x})$ is the fundamental matrix of the operator $\mathcal{L}(\mathbf{D}_{\mathbf{x}}, \omega)$ (see (2.15)), then the vector function (the volume potential)

$$\widetilde{\mathbf{w}}(\mathbf{x}) = \int_{\Omega} \boldsymbol{\Gamma}'(\mathbf{x} - \mathbf{y})\mathbf{V}(\mathbf{y})d\mathbf{y}$$

is a solution of (3.3).

On the other hand, by virtue of (1.37), (2.16), and (3.3) we have

$$\mathcal{F}(\mathbf{x}) = \mathcal{A}(\mathbf{D}_{\mathbf{x}}, \omega)\mathbf{V}(\mathbf{x}) = \mathcal{A}(\mathbf{D}_{\mathbf{x}}, \omega)\mathcal{L}(\mathbf{D}_{\mathbf{x}}, \omega)\, \widetilde{\mathbf{w}}(\mathbf{x}) = \boldsymbol{\Lambda}(\Delta, \omega)\, \widetilde{\mathbf{w}}(\mathbf{x}).$$

Hence, $\widetilde{\mathbf{w}}$ is a solution of (3.4). □

Thus, on the basis of Theorems 3.1 and 3.2 the completeness of the Galerkin-type solution of system (1.21) is proved.

Remark 3.1 Quite similarly as in Theorem 2.1 we can construct explicitly the fundamental matrix $\boldsymbol{\Gamma}'(\mathbf{x})$ of the operator $\mathcal{L}(\mathbf{D}_{\mathbf{x}}, \omega)$ by elementary functions.

3.2.2 Equations in the Laplace Transform Space

Theorems 3.1 and 3.2 lead to the representation of the Galerkin-type solution to system (1.22) in the Laplace transform space. We have the following.

Theorem 3.3 *Let*

$$u_l(\mathbf{x}) = \frac{1}{\mu}\Lambda_1(\Delta, i\tau)\, w_l(\mathbf{x}) + n_{11}(\Delta, i\tau)\, w_{j,lj}(\mathbf{x})$$
$$+ n_{1;\alpha+1}(\Delta, i\tau)\, \phi_{\alpha,l}(\mathbf{x}), \tag{3.5}$$
$$p_\alpha(\mathbf{x}) = n_{\alpha+1;1}(\Delta, i\tau) w_{j,j}(\mathbf{x}) + n_{\alpha+1;\beta+1}(\Delta, i\tau)\phi_\beta(\mathbf{x}),$$

where $l = 1, 2, 3$, $\alpha = 1, 2, 3, 4$, $\mathbf{w} = (w_1, w_2, w_3) \in C^{14}(\Omega)$, $\boldsymbol{\phi} = (\phi_1, \phi_2, \phi_3, \phi_4) \in C^{12}(\Omega)$,

$$\Lambda_2(\Delta, i\tau)\, \mathbf{w}(\mathbf{x}) = \mathcal{F}^{(1)}(\mathbf{x}), \qquad \Lambda_1(\Delta, i\tau)\, \boldsymbol{\phi}(\mathbf{x}) = \mathcal{F}^{(2)}(\mathbf{x}), \tag{3.6}$$

$\mathcal{F}^{(1)}$ *and* $\mathcal{F}^{(2)}$ *are three and four component vector-functions, respectively,* τ *is a complex number with* $\operatorname{Im}\tau > 0$, *then* $\mathbf{V} = (\mathbf{u}, \mathbf{p})$ *is a solution of system (1.22).*

Theorem 3.4 *If* $\mathbf{V} = (\mathbf{u}, \mathbf{p})$ *is a solution of system (1.22) in* Ω, *then* $\mathbf{V}$ *is represented by (3.5), where* $\tilde{\mathbf{w}} = (\mathbf{w}, \boldsymbol{\phi})$ *is a solution of (3.6) and* Ω *is a finite domain in* $\mathbb{R}^3$.

Thus, on the basis of Theorems 3.3 and 3.4 the completeness of the Galerkin-type solution of system (1.22) in the Laplace transform space is proved.

3.2.3 Equations of Quasi-Static

Quite similarly, on the basis of (2.24) we can obtain the representation of the Galerkin-type solution to system of equations of steady vibrations (1.23) in the quasi-static theory of elasticity for materials with quadruple porosity. We have the following.

Theorem 3.5 *Let*

$$u_l(\mathbf{x}) = \frac{1}{\mu}\Lambda_1^{(q)}(\Delta, \omega)\, w_l(\mathbf{x})$$
$$- \frac{1}{k_0\mu\mu_0}\left[(\lambda+\mu)\tilde{\mathcal{B}}^*_{11}(\Delta, \omega) + i\omega a_\alpha \tilde{\mathcal{B}}^*_{1;\alpha+1}(\Delta, \omega)\right] w_{j,lj}(\mathbf{x})$$
$$+ \frac{1}{k_0\mu_0}\tilde{\mathcal{B}}^*_{1;\beta+1}(\Delta, \omega)\, \phi_{\beta,l}(\mathbf{x}), \tag{3.7}$$
$$p_\alpha(\mathbf{x}) = \frac{1}{k_0\mu_0}\tilde{\mathcal{B}}^*_{\alpha+1;1}(\Delta, \omega) w_{j,j}(\mathbf{x}) + \frac{1}{k_0\mu_0}\tilde{\mathcal{B}}^*_{\alpha+1;\beta+1}(\Delta, \omega)\phi_\beta(\mathbf{x}),$$

where $l = 1, 2, 3,\ \alpha = 1, 2, 3, 4,\ \mathbf{w} = (w_1, w_2, w_3) \in C^{14}(\Omega),\ \boldsymbol{\phi} = (\phi_1, \phi_2, \phi_3, \phi_4) \in C^{12}(\Omega)$,

$$\Delta\Lambda_1^{(q)}(\Delta, \omega)\,\mathbf{w}(\mathbf{x}) = \mathcal{F}^{(1)}(\mathbf{x}), \qquad \Lambda_1^{(q)}(\Delta, \omega)\,\boldsymbol{\phi}(\mathbf{x}) = \mathcal{F}^{(2)}(\mathbf{x}), \tag{3.8}$$

$\mathcal{F}^{(1)}$ *and* $\mathcal{F}^{(2)}$ *are three and four component vector-functions, respectively, then* $\mathbf{V} = (\mathbf{u}, \mathbf{p})$ *is a solution of system (1.23).*

Theorem 3.6 *If* $\mathbf{V} = (\mathbf{u}, \mathbf{p})$ *is a solution of system (1.23) in* Ω*, then* $\mathbf{V}$ *is represented by (3.7), where* $\widetilde{\mathbf{w}} = (\mathbf{w}, \boldsymbol{\phi})$ *is a solution of (3.8) and* Ω *is a finite domain in* $\mathbb{R}^3$.

Thus, on the basis of Theorems 3.5 and 3.6 the completeness of the Galerkin-type solution of system (1.23) is proved in the quasi-static theory of elasticity for materials with quadruple porosity.

3.2.4 Equations of Equilibrium

By virtue of (2.28) we prove the completeness of the Galerkin-type solution of system (1.24) in the equilibrium theory of elasticity for materials with quadruple porosity. Namely, the following theorems are valid.

Theorem 3.7 *Let*

$$\begin{aligned} u_l(\mathbf{x}) &= \frac{1}{\mu}\Delta\, w_l(\mathbf{x}) - \frac{\lambda+\mu}{\mu\mu_0}\, w_{j,lj}(\mathbf{x}) + \frac{1}{k_0\mu_0}\mathcal{A}_{\alpha\beta}^{(re)^*}(\mathbf{D_x})a_\alpha\, \phi_{\beta,l}(\mathbf{x}), \\ p_\alpha(\mathbf{x}) &= \frac{1}{k_0}\Delta\mathcal{A}_{\alpha\beta}^{(re)^*}(\mathbf{D_x})\phi_\beta(\mathbf{x}), \end{aligned} \tag{3.9}$$

where $l = 1, 2, 3,\ \alpha = 1, 2, 3, 4,\ \mathbf{w} = (w_1, w_2, w_3) \in C^{4}(\Omega),\ \boldsymbol{\phi} = (\phi_1, \phi_2, \phi_3, \phi_4) \in C^{10}(\Omega)$,

$$\Delta^2\,\mathbf{w}(\mathbf{x}) = \mathcal{F}^{(1)}(\mathbf{x}), \qquad \Lambda_1^{(e)}(\Delta)\,\boldsymbol{\phi}(\mathbf{x}) = \mathcal{F}^{(2)}(\mathbf{x}), \tag{3.10}$$

$\mathcal{F}^{(1)}$ *and* $\mathcal{F}^{(2)}$ *are three and four component vector-functions, respectively, then* $\mathbf{V} = (\mathbf{u}, \mathbf{p})$ *is a solution of system (1.24).*

Theorem 3.8 *If* $\mathbf{V} = (\mathbf{u}, \mathbf{p})$ *is a solution of system (1.24) in* Ω*, then* $\mathbf{V}$ *is represented by (3.9), where* $\widetilde{\mathbf{w}} = (\mathbf{w}, \boldsymbol{\phi})$ *is a solution of (3.10) and* Ω *is a finite domain in* $\mathbb{R}^3$.

3.2.5 Equations of Steady Vibrations for Rigid Body with Quadruple Porosity

By virtue of (2.32) we prove the completeness of the Galerkin-type solution of system (1.25) in the theory of rigid body with quadruple porosity. Namely, the following theorems are valid.

Theorem 3.9 *Let*

$$p_\alpha(\mathbf{x}) = \frac{1}{k_0}\mathcal{A}^{(r)*}_{\alpha\beta}(\mathbf{D_x}, \omega)\phi_\beta(\mathbf{x}), \tag{3.11}$$

where $\phi_\alpha \in C^8(\Omega)$, $\alpha = 1, 2, 3, 4$,

$$\Lambda_1^{(r)}(\Delta, \omega)\,\boldsymbol{\phi}(\mathbf{x}) = \mathcal{F}^{(2)}(\mathbf{x}), \tag{3.12}$$

$\mathcal{F}^{(2)}$ *is a four component vector-function, then* $\mathbf{p}$ *is a solution of system (1.25).*

Theorem 3.10 *If* $\mathbf{p}$ *is a solution of system (1.25) in* Ω, *then* $\mathbf{p}$ *is represented by (3.11), where* $\boldsymbol{\phi}$ *is a solution of (3.12) and* Ω *is a finite domain in* $\mathbb{R}^3$.

3.3 Green's Formulas in Mathematical Physics

One of the most celebrated techniques of mathematical physics is the divergence theorem, which transforms a volume integral into a surface integral. In 1828, G. Green [157] presented the following three elegant identities (Green's formulas), where the relationship between volume and surface integrals is established:

(i) *Green's first identity*

$$\int_{\Omega^+} [u(\mathbf{x})\Delta v(\mathbf{x}) + \nabla u(\mathbf{x}) \cdot \nabla v(\mathbf{x})]\, d\mathbf{x} = \int_S u(\mathbf{z})\frac{\partial v(\mathbf{z})}{\partial \mathbf{n}(\mathbf{z})} d_\mathbf{z}S;$$

(ii) *Green's second identity*

$$\int_{\Omega^+} [u(\mathbf{x})\Delta v(\mathbf{x}) - v(\mathbf{x})\Delta u(\mathbf{x})]\, d\mathbf{x} = \int_S \left[u(\mathbf{z})\frac{\partial v(\mathbf{z})}{\partial \mathbf{n}(\mathbf{z})} - v(\mathbf{z})\frac{\partial u(\mathbf{z})}{\partial \mathbf{n}(\mathbf{z})}\right] d_\mathbf{z}S;$$

(iii) *Green's third identity*

$$u(\mathbf{x}) = \int_S \left[\frac{\partial \tilde{\gamma}(\mathbf{x}-\mathbf{z})}{\partial \mathbf{n}(\mathbf{z})} u(\mathbf{z}) - \tilde{\gamma}(\mathbf{x}-\mathbf{z})\frac{\partial u(\mathbf{z})}{\partial \mathbf{n}(\mathbf{z})}\right] d_\mathbf{z}S$$

$$+ \int_{\Omega^+} \tilde{\gamma}(\mathbf{x}-\mathbf{y})\Delta u(\mathbf{y})d\mathbf{y} \quad \text{for} \quad \mathbf{x} \in \Omega^+,$$

where u and v are regular functions in Ω^+ (i.e., $u, v \in C^2(\Omega^+) \cap C^1(\overline{\Omega^+})$), $\tilde{\gamma}(\mathbf{x})$ is the fundamental solution of the Laplace equation and given by

$$\tilde{\gamma}(\mathbf{x}) = -\frac{1}{4\pi |\mathbf{x}|}.$$

Obviously, Green's third identity is the integral representation of a regular function in Ω^+.

This representation formula contains the convolution type three integrals (potentials): two surface potentials (named as the single-layer and double-layer potentials) and the volume potential (named as the Newtonian potential). For a harmonic function such a representation formula immediately yields its analyticity. Besides, the formula for representation of solutions in the form of potentials initiated the introduction of the Green function that had played an outstanding role in the development of mathematical physics (for details, see Günther [161], Hsiao and Wendland [169], and Kellogg [195]).

Note that the integral representation formula is not very useful for solving Dirichlet or Neumann BVPs for the Laplace equation, since they require data of both u and $\frac{\partial u}{\partial \mathbf{n}}$ on the boundary.

In the next sections of this chapter, we establish Green's identities in the linear theory of elasticity for materials with quadruple porosity. These identities play an important role in the study of BVPs of the considered theory.

3.4 Green's Formulas for Equations of Steady Vibrations

In this section Green's identities for equations of steady vibrations in the linear theory of elasticity for materials with quadruple porosity are established.

Definition 3.1 A vector function $\mathbf{V} = (\mathbf{u}, \mathbf{p}) = (V_1, V_2, \cdots, V_7)$ is called *regular* of the class $\mathfrak{B}^{(s)}$ in Ω^- (or Ω^+) if

(i)

$$V_l \in C^2(\Omega^-) \cap C^1(\overline{\Omega^-}) \qquad (\text{or } V_l \in C^2(\Omega^+) \cap C^1(\overline{\Omega^+})),$$

(ii)

$$\mathbf{V} = \sum_{j=1}^{6} \mathbf{V}^{(j)}, \quad \mathbf{V}^{(j)} = (V_1^{(j)}, V_2^{(j)}, \cdots, V_7^{(j)}),$$

$$V_l^{(j)} \in C^2(\Omega^-) \cap C^1(\bar{\Omega}^-),$$

(iii) $(\Delta+\lambda_j^2)V_l^{(j)}(\mathbf{x})=0$ and

$$\left(\frac{\partial}{\partial|\mathbf{x}|}-i\lambda_j\right)V_l^{(j)}(\mathbf{x})=e^{i\lambda_j|\mathbf{x}|}o(|\mathbf{x}|^{-1})\qquad\text{for}\qquad|\mathbf{x}|\gg 1,\tag{3.13}$$

where $V_r^{(6)}=0,\ j=1,2,\cdots,6,\ l=1,2,\cdots,7$ and $r=4,5,6,7$.

Vekua [386] proved that the relation (3.13) implies

$$U_l^{(j)}(\mathbf{x})=e^{i\lambda_j|\mathbf{x}|}O(|\mathbf{x}|^{-1})\qquad\text{for}\qquad|\mathbf{x}|\gg 1,\tag{3.14}$$

where $j=1,2,\cdots,6,\quad l=1,2,\cdots,7$.

Relations (3.13) and (3.14) are Sommerfeld–Kupradze type radiation conditions in the linear theory of elasticity for solids with quadruple porosity.

3.4.1 Green's First Identity

In the sequel we use the matrix differential operators:

1.

$$\mathcal{A}^{(c)}(\mathbf{D_x})=\left(\mathcal{A}_{lj}^{(c)}(\mathbf{D_x})\right)_{3\times3},\qquad \mathcal{A}_{lj}^{(c)}(\mathbf{D_x})=\mu\Delta\delta_{lj}+(\lambda+\mu)\frac{\partial^2}{\partial x_l\partial x_j},$$

$$\mathcal{A}^{(1)}(\mathbf{D_x},\omega)=\left(\mathcal{A}_{lr}^{(1)}(\mathbf{D_x},\omega)\right)_{3\times7},\qquad \mathcal{A}_{lr}^{(1)}(\mathbf{D_x},\omega)=\mathcal{A}_{lr}(\mathbf{D_x},\omega),$$

$$\mathcal{A}^{(2)}(\mathbf{D_x},\omega)=\left(\mathcal{A}_{\alpha r}^{(2)}(\mathbf{D_x},\omega)\right)_{4\times7},\qquad \mathcal{A}_{\alpha r}^{(2)}(\mathbf{D_x},\omega)=\mathcal{A}_{\alpha+3;r}(\mathbf{D_x},\omega);$$

2.

$$\mathcal{P}^{(c)}(\mathbf{D_x},\mathbf{n})=(\mathcal{P}_{lj}^{(c)}(\mathbf{D_x},\mathbf{n}))_{3\times3},\qquad \mathcal{P}_{lj}^{(c)}(\mathbf{D_x},\mathbf{n})=\mathcal{P}_{lj}(\mathbf{D_x},\mathbf{n}),$$

$$\mathcal{P}^{(1)}(\mathbf{D_x},\mathbf{n})=(\mathcal{P}_{lr}^{(1)}(\mathbf{D_x},\mathbf{n}))_{3\times7},\qquad \mathcal{P}_{lr}^{(0)}(\mathbf{D_x},\mathbf{n})=\mathcal{P}_{lr}(\mathbf{D_x},\mathbf{n}),$$

$$\mathcal{P}^{(2)}(\mathbf{D_x},\mathbf{n})=(\mathcal{P}_{\alpha\beta}^{(2)}(\mathbf{D_x},\mathbf{n}))_{4\times4},\qquad \mathcal{P}_{\alpha\beta}^{(2)}(\mathbf{D_x},\mathbf{n})=\mathcal{P}_{\alpha+3;\beta+3}(\mathbf{D_x},\mathbf{n}),$$

where $l, j = 1, 2, 3,\ \alpha, \beta = 1, 2, 3, 4,\ r = 1, 2, \cdots, 7$; the matrix differential operators $\boldsymbol{\mathcal{A}}$ and $\boldsymbol{\mathcal{P}}$ are defined by (1.36) and (1.45), respectively. Obviously, the matrix $\boldsymbol{\mathcal{P}}^{(c)}$ is the stress operator in the classical theory of elasticity (see, e.g., Kupradze et al. [219]).

Let $\mathbf{u}' = (u_1', u_2', u_3')$ and $\mathbf{p}' = (p_1', p_2', p_3', p_4')$ be three- and four-component vector functions, respectively; $\mathbf{V}' = (\mathbf{u}', \mathbf{p}')$. We introduce the notation

$$W_0(\mathbf{u}, \mathbf{u}') = \frac{1}{3}(3\lambda + 2\mu)\operatorname{div}\mathbf{u}\operatorname{div}\overline{\mathbf{u}'}$$
$$+\frac{\mu}{2}\sum_{l,j=1;\,l\neq j}^{3}\left(\frac{\partial u_j}{\partial x_l} + \frac{\partial u_l}{\partial x_j}\right)\left(\frac{\partial \overline{u_j'}}{\partial x_l} + \frac{\partial \overline{u_l'}}{\partial x_j}\right)$$
$$+\frac{\mu}{3}\sum_{l,j=1}^{3}\left(\frac{\partial u_l}{\partial x_l} - \frac{\partial u_j}{\partial x_j}\right)\left(\frac{\partial \overline{u_l'}}{\partial x_l} - \frac{\partial \overline{u_j'}}{\partial x_j}\right), \tag{3.15}$$
$$W_1(\mathbf{V}, \mathbf{u}') = W_0(\mathbf{u}, \mathbf{u}') - \rho\omega^2\mathbf{u}\cdot\mathbf{u}' - a_\alpha\, p_\alpha \operatorname{div}\overline{\mathbf{u}'},$$
$$W_2(\mathbf{V}, \mathbf{p}') = k_{\alpha\beta}\nabla p_\alpha \cdot \nabla \overline{p_\beta'} - c_{\alpha\beta}\, p_\beta\, \overline{p_\alpha'} - i\omega\, a_\alpha\, \overline{p_\alpha'}\operatorname{div}\mathbf{u},$$
$$W(\mathbf{V}, \mathbf{V}') = W_1(\mathbf{V}, \mathbf{u}') + W_2(\mathbf{V}, \mathbf{p}').$$

We have the following.

Theorem 3.11 *If* $\mathbf{V} = (\mathbf{u}, \mathbf{p})$ *is a regular vector of the class* $\mathfrak{B}^{(s)}$ *in* Ω^+, $u_j', p_\alpha' \in C^1(\Omega^+) \cap C(\overline{\Omega^+})$, $j = 1, 2, 3$, $\alpha = 1, 2, 3, 4$, *then*

$$\int_{\Omega^+}\left[\boldsymbol{\mathcal{A}}(\mathbf{D_x}, \omega)\,\mathbf{V}(\mathbf{x})\cdot\mathbf{V}'(\mathbf{x}) + W(\mathbf{V}, \mathbf{V}')\right]d\mathbf{x}$$
$$= \int_S \boldsymbol{\mathcal{P}}(\mathbf{D_z}, \mathbf{n})\mathbf{V}(\mathbf{z})\cdot\mathbf{V}'(\mathbf{z})\, d_{\mathbf{z}}S, \tag{3.16}$$

where $\boldsymbol{\mathcal{A}}(\mathbf{D_x}, \omega)$, $\boldsymbol{\mathcal{P}}(\mathbf{D_z}, \mathbf{n})$ *and* $W(\mathbf{V}, \mathbf{V}')$ *are defined by (1.36), (1.45), and (3.15), respectively.*

Proof In the classical theory of elasticity (see, e.g., Kupradze et al. [219]), on the basis of the divergence theorem the following identities are proved

$$\int_{\Omega^+} \Big[\mathcal{A}^{(c)}(\mathbf{D_x})\,\mathbf{u}(\mathbf{x})\cdot\mathbf{u}'(\mathbf{x}) + W_0(\mathbf{u},\mathbf{u}')\Big]\,d\mathbf{x}$$

$$= \int_S \mathcal{P}^{(c)}(\mathbf{D_z},\mathbf{n})\mathbf{u}(\mathbf{z})\cdot\mathbf{u}'(\mathbf{z})\,d_{\mathbf{z}}S, \tag{3.17}$$

$$\int_{\Omega^+} \Big[\Delta p_\beta(\mathbf{x})\,\overline{p'_\alpha(\mathbf{x})} + \nabla p_\beta(\mathbf{x})\cdot\nabla p'_\alpha(\mathbf{x})\Big]\,d\mathbf{x} = \int_S \frac{\partial p_\beta(\mathbf{z})}{\partial\mathbf{n}(\mathbf{z})}\overline{p'_\alpha(\mathbf{z})}\,d_{\mathbf{z}}S,$$

$$\alpha,\beta = 1,2,3,4.$$

By virtue of (3.17) we have

$$\int_{\Omega^+} \Big[\mathcal{A}^{(1)}(\mathbf{D_x},\omega)\,\mathbf{V}(\mathbf{x})\cdot\mathbf{u}'(\mathbf{x}) + W_1(\mathbf{V},\mathbf{u}')\Big]\,d\mathbf{x}$$

$$= \int_S \mathcal{P}^{(1)}(\mathbf{D_z},\mathbf{n})\mathbf{V}(\mathbf{z})\cdot\mathbf{u}'(\mathbf{z})\,d_{\mathbf{z}}S, \tag{3.18}$$

$$\int_{\Omega^+} \Big[\mathcal{A}^{(2)}(\mathbf{D_x},\omega)\,\mathbf{V}(\mathbf{x})\cdot\mathbf{p}'(\mathbf{x}) + W_2(\mathbf{V},\mathbf{p}')\Big]\,d\mathbf{x}$$

$$= \int_S \mathcal{P}^{(2)}(\mathbf{D_z},\mathbf{n})\mathbf{p}(\mathbf{z})\cdot\mathbf{p}'(\mathbf{z})\,d_{\mathbf{z}}S.$$

It is easy to verify that the identity (3.16) may be obtained from (3.18). □

Theorem 3.11 and the radiation conditions (3.13) and (3.14) lead to the following.

Theorem 3.12 *If* $\mathbf{V} = (\mathbf{u},\mathbf{p})$ *and* $\mathbf{V}' = (\mathbf{u}',\mathbf{p}')$ *are regular vectors of the class* $\mathfrak{B}^{(s)}$ *in* Ω^-, *then*

$$\int_{\Omega^-} \big[\mathcal{A}(\mathbf{D_x})\,\mathbf{V}(\mathbf{x})\cdot\mathbf{V}'(\mathbf{x}) + W(\mathbf{V},\mathbf{V}')\big]\,d\mathbf{x}$$

$$= -\int_S \mathcal{P}(\mathbf{D_z},\mathbf{n})\mathbf{V}(\mathbf{z})\cdot\mathbf{V}'(\mathbf{z})\,d_{\mathbf{z}}S, \tag{3.19}$$

where $\mathcal{A}(\mathbf{D_x})$, $\mathcal{P}(\mathbf{D_z},\mathbf{n})$ *and* $W(\mathbf{V},\mathbf{V}')$ *are defined by (1.36), (1.45), and (3.15), respectively.*

The formulas (3.16) and (3.19) are Green's first identities of the steady vibrations equations in the linear theory of quadruple porosity elasticity for domains Ω^+ and Ω^-, respectively.

3.4.2 Green's Second Identity

Clearly, the matrix differential operator $\tilde{\mathcal{A}}(\mathbf{D_x},\omega) = \left(\tilde{A}_{lj}(\mathbf{D_x},\omega)\right)_{7\times 7}$ is the associate operator of $\mathcal{A}(\mathbf{D_x},\omega)$, where $\tilde{\mathcal{A}}(\mathbf{D_x},\omega) = \mathcal{A}^\top(-\mathbf{D_x},\omega)$. It is easy to verify that the operator $\tilde{\mathcal{A}}(\mathbf{D_x},\omega)$ may be obtained from the operator $\mathcal{A}(\mathbf{D_x},\omega)$ by replacing a_α by $i\omega a_\alpha$ $(\alpha = 1,2,3,4)$ and vice versa. It is easy to see that the associated system of homogeneous equations of (1.21) is

$$\begin{aligned} &(\mu\,\Delta + \rho\,\omega^2)\tilde{\mathbf{u}} + (\lambda+\mu)\,\nabla\mathrm{div}\tilde{\mathbf{u}} - i\omega\nabla\,(\mathbf{a}\,\tilde{\mathbf{p}}) = \mathbf{0},\\ &(\mathbf{K}\,\Delta + \mathbf{c})\,\tilde{\mathbf{p}} + \mathbf{a}\,\mathrm{div}\tilde{\mathbf{u}} = \mathbf{0}, \end{aligned} \tag{3.20}$$

where $\tilde{\mathbf{u}}$ and $\tilde{\mathbf{p}}$ are three- and four-component vector functions on $\Omega^\pm$, respectively.

Let $\mathbf{V} = (\mathbf{u},\mathbf{p}) = (V_1, V_2, \cdots, V_7)$, the vector $\tilde{\mathbf{V}}_j$ is the j-th column of the matrix $\tilde{\mathbf{V}} = (\tilde{V}_{lj})_{7\times 7}$, $\tilde{\mathbf{u}}_j = (\tilde{V}_{1j}, \tilde{V}_{2j}, \tilde{V}_{3j})^\top$, $\tilde{\mathbf{p}}_j = (\tilde{p}_{1j}, \tilde{p}_{2j}, \tilde{p}_{3j}, \tilde{p}_{4j})^\top = (\tilde{V}_{4j}, \tilde{V}_{5j}, \tilde{V}_{6j}, \tilde{V}_{7j})^\top$, $j = 1, 2, \cdots, 7$.

Theorem 3.13 *If* $\mathbf{V}$ *and* $\tilde{\mathbf{V}}_j$ $(j = 1, 2, \cdots, 7)$ *are regular vectors of the class* $\mathfrak{B}^{(s)}$ *in* Ω^+, *then*

$$\begin{aligned} &\int\limits_{\Omega^+} \left\{[\tilde{\mathcal{A}}(\mathbf{D_y},\omega)\tilde{\mathbf{V}}(\mathbf{y})]^\top \mathbf{V}(\mathbf{y}) - [\tilde{\mathbf{V}}(\mathbf{y})]^\top \mathcal{A}(\mathbf{D_y},\omega)\mathbf{V}(\mathbf{y})\right\} d\mathbf{y}\\ &\quad = \int\limits_S \left\{[\tilde{\mathcal{P}}(\mathbf{D_z},\mathbf{n})\tilde{\mathbf{V}}(\mathbf{z})]^\top \mathbf{V}(\mathbf{z}) - [\tilde{\mathbf{V}}(\mathbf{z})]^\top \mathcal{P}(\mathbf{D_z},\mathbf{n})\mathbf{V}(\mathbf{z})\right\} d_{\mathbf{z}}S, \end{aligned} \tag{3.21}$$

where the matrices $\mathcal{A}(\mathbf{D_x},\omega)$ *and* $\mathcal{P}(\mathbf{D_z},\mathbf{n})$ *are given by (1.36) and (1.45), respectively; the operator* $\tilde{\mathcal{P}}(\mathbf{D_z},\mathbf{n})$ *is defined by*

$$\begin{aligned} &\tilde{\mathcal{P}}(\mathbf{D_x},\mathbf{n}) = \left(\tilde{\mathcal{P}}_{ms}(\mathbf{D_x},\mathbf{n})\right)_{7\times 7},\\ &\tilde{\mathcal{P}}_{lj}(\mathbf{D_x},\mathbf{n}) = \mu\delta_{lj}\frac{\partial}{\partial\mathbf{n}} + \lambda n_l\frac{\partial}{\partial x_j} + \mu n_j\frac{\partial}{\partial x_l},\\ &\tilde{\mathcal{P}}_{l;\alpha+3}(\mathbf{D_x},\mathbf{n}) = -i\omega a_\alpha\, n_l, \qquad \tilde{\mathcal{P}}_{\alpha+3;j}(\mathbf{D_x},\mathbf{n}) = 0,\\ &\tilde{\mathcal{P}}_{\alpha+3;\beta+3}(\mathbf{D_x},\mathbf{n}) = k_{lj}\frac{\partial}{\partial\mathbf{n}}, \qquad l,j = 1,2,3, \qquad \alpha,\beta = 1,2,3,4. \end{aligned} \tag{3.22}$$

Proof We consider the difference $(\tilde{\mathcal{A}}\tilde{\mathbf{V}})^\top \mathbf{V} - (\tilde{\mathbf{V}})^\top \mathcal{A}\mathbf{V}$. The m-th $(m = 1, 2, \cdots, 7)$ component of this vector may be written in the form

$$
\begin{aligned}
[(\tilde{\mathcal{A}}\tilde{\mathbf{V}})^\top \mathbf{V} - \tilde{\mathbf{V}}^\top \mathcal{A}\mathbf{V}]_m &= \sum_{r=1}^{7} \left[(\tilde{\mathcal{A}}\tilde{\mathbf{V}})_{rm} V_r - \tilde{V}_{rm} (\mathcal{A}\mathbf{V})_r \right] \\
&= \left[\mathcal{A}^{(c)}\tilde{\mathbf{u}}_m + \rho\omega^2 \tilde{\mathbf{u}}_m - i\omega\nabla\left(\mathbf{a}\tilde{\mathbf{p}}_m\right)\right]\mathbf{u} + \left(\mathbf{K}\,\Delta\,\tilde{\mathbf{p}}_m + \mathbf{c}\,\tilde{\mathbf{p}}_m + \mathbf{a}\,\mathrm{div}\tilde{\mathbf{u}}_m\right)\mathbf{p} \\
&\quad - \tilde{\mathbf{u}}_m \left[\mathcal{A}^{(c)}\mathbf{u} + \rho\omega^2 \mathbf{u} - \nabla\left(\mathbf{a}\,\mathbf{p}\right)\right] - \tilde{\mathbf{p}}_m \left(\mathbf{K}\,\Delta\,\mathbf{p} + \mathbf{c}\,\mathbf{p} + i\omega\mathbf{a}\,\mathrm{div}\mathbf{u}\right) \\
&= \left(\mathcal{A}^{(c)}\tilde{\mathbf{u}}_m\,\mathbf{u} - \tilde{\mathbf{u}}_m \mathcal{A}^{(c)}\mathbf{u}\right) + \left(\mathbf{p}\mathbf{K}\Delta\,\tilde{\mathbf{p}}_m - \tilde{\mathbf{p}}\mathbf{K}\Delta\,\mathbf{p}\right) \\
&\quad + a_\alpha\left(p_\alpha \mathrm{div}\tilde{\mathbf{u}}_m + \tilde{\mathbf{u}}_m \nabla p_j\right) - i\omega a_\alpha\left(\mathbf{u}\nabla\tilde{p}_{\alpha m} + \tilde{p}_{\alpha m}\,\mathrm{div}\mathbf{u}\right) \\
&= \left(\mathcal{A}^{(c)}\tilde{\mathbf{u}}_m\,\mathbf{u} - \tilde{\mathbf{u}}_m \mathcal{A}^{(c)}\mathbf{u}\right) + \mathbf{K}\left(\mathbf{p}\Delta\,\tilde{\mathbf{p}}_m - \tilde{\mathbf{p}}_m\,\Delta\,\mathbf{p}\right) \\
&\quad + a_\alpha\left(p_\alpha\,\tilde{u}_{lm}\right)_{,l} - i\omega a_\alpha\left(u_l\,\tilde{p}_{\alpha m}\right)_{,l}\,.
\end{aligned}
\tag{3.23}
$$

Integrating over Ω^+ and using the identities (see, e.g., Kupradze et al. [219])

$$
\int_{\Omega^+} \left(\mathcal{A}^{(c)}\tilde{\mathbf{u}}_j\,\mathbf{u} - \tilde{\mathbf{u}}_j\,\mathcal{A}^{(c)}\mathbf{u}\right) d\mathbf{y} = \int_S \left(\mathcal{P}^{(c)}\tilde{\mathbf{u}}_j\,\mathbf{u} - \tilde{\mathbf{u}}_j \mathcal{P}^{(c)}\mathbf{u}\right) d_{\mathbf{z}}S,
$$

$$
\int_{\Omega^+} \left(\mathbf{p}\Delta\,\tilde{\mathbf{p}}_m - \tilde{\mathbf{p}}_m\,\Delta\,\mathbf{p}\right) d\mathbf{y} = \int_S \left(\mathbf{p}\frac{\partial\tilde{\mathbf{p}}_m}{\partial\mathbf{n}} - \tilde{\mathbf{p}}_m\frac{\partial\mathbf{p}}{\partial\mathbf{n}}\right) d_{\mathbf{z}}S,
$$

$$
\int_{\Omega^+} \left(p_\alpha\,\tilde{u}_{lm}\right)_{,l} d\mathbf{y} = \int_S p_\alpha\,\tilde{\mathbf{u}}_m\,\mathbf{n}\,d_{\mathbf{z}}S,
$$

$$
\int_{\Omega^+} \left(u_l\,\tilde{p}_{\alpha m}\right)_{,l} d\mathbf{y} = \int_S \tilde{p}_{\alpha m}\mathbf{u}\,\mathbf{n}\,d_{\mathbf{z}}S,
$$

from (3.23) it follows that

$$
\begin{aligned}
&\int_{\Omega^+} \left[(\tilde{\mathcal{A}}\tilde{\mathbf{V}})^\top \mathbf{V} - \tilde{\mathbf{V}}^\top \mathcal{A}\mathbf{V}\right]_m d\mathbf{y} \\
&\quad = \int_S \left[\left(\mathcal{P}^{(c)}\tilde{\mathbf{u}}_m\,\mathbf{u} - \tilde{\mathbf{u}}_m \mathcal{P}^{(c)}\mathbf{u}\right) + \mathbf{K}\left(\mathbf{p}\frac{\partial\tilde{\mathbf{p}}_m}{\partial\mathbf{n}} - \tilde{\mathbf{p}}_m\frac{\partial\mathbf{p}}{\partial\mathbf{n}}\right)\right] d_{\mathbf{z}}S \\
&\qquad + \int_S \left(\mathbf{a}\mathbf{p}\tilde{\mathbf{u}}_m - i\omega\mathbf{a}\tilde{\mathbf{p}}_m\,\mathbf{u}\right)\mathbf{n}d_{\mathbf{z}}S.
\end{aligned}
$$

Hence, we can write

$$\int_{\Omega^+} \left[(\tilde{\mathcal{A}}\tilde{\mathbf{V}})^\top \mathbf{V} - \tilde{\mathbf{V}}^\top \mathcal{A}\mathbf{V} \right]_m d\mathbf{y}$$

$$= \int_S \left[\left(\mathcal{P}^{(c)} \tilde{\mathbf{u}}_m - i\omega \mathbf{a}\tilde{\mathbf{p}}_m \mathbf{n} \right) \mathbf{u} + \mathbf{K} \frac{\partial \tilde{\mathbf{p}}_m}{\partial \mathbf{n}} \mathbf{p} \right] d_{\mathbf{z}} S$$

$$- \int_S \left[\tilde{\mathbf{u}}_m \left(\mathcal{P}^{(c)} \mathbf{u} - \mathbf{a}\mathbf{p} \right) + \tilde{\mathbf{p}}_m \, \mathbf{K} \frac{\partial \mathbf{p}}{\partial \mathbf{n}} \right] d_{\mathbf{z}} S$$

$$= \int_S \left\{ [\tilde{\mathcal{P}}(\mathbf{D}_{\mathbf{z}}, \mathbf{n}) \tilde{\mathbf{V}}(\mathbf{z})]^\top \mathbf{V}(\mathbf{z}) - [\tilde{\mathbf{V}}(\mathbf{z})]^\top \mathcal{P}(\mathbf{D}_{\mathbf{z}}, \mathbf{n}) \mathbf{V}(\mathbf{z}) \right\}_m d_{\mathbf{z}} S$$

and formula (3.21) is thereby proved. □

Theorem 3.13 and the radiation conditions (3.13) and (3.14) lead to the following.

Theorem 3.14 *If* $\mathbf{V}$ *and* $\tilde{\mathbf{V}}_j$ $(j = 1, 2, \cdots, 7)$ *are regular vectors of the class* $\mathfrak{B}^{(s)}$ *in* Ω^-*, then*

$$\begin{aligned} &\int_{\Omega^-} \left\{ [\tilde{\mathcal{A}}(\mathbf{D}_{\mathbf{y}}, \omega) \tilde{\mathbf{V}}(\mathbf{y})]^\top \mathbf{V}(\mathbf{y}) - [\tilde{\mathbf{V}}(\mathbf{y})]^\top \mathcal{A}(\mathbf{D}_{\mathbf{y}}, \omega) \mathbf{V}(\mathbf{y}) \right\} d\mathbf{y} \\ &\quad = - \int_S \left\{ [\tilde{\mathcal{P}}(\mathbf{D}_{\mathbf{z}}, \mathbf{n}) \tilde{\mathbf{V}}(\mathbf{z})]^\top \mathbf{V}(\mathbf{z}) - [\tilde{\mathbf{V}}(\mathbf{z})]^\top \mathcal{P}(\mathbf{D}_{\mathbf{z}}, \mathbf{n}) \mathbf{V}(\mathbf{z}) \right\} d_{\mathbf{z}} S, \end{aligned} \tag{3.24}$$

where the operator $\tilde{\mathcal{P}}(\mathbf{D}_{\mathbf{z}}, \mathbf{n})$ *is defined by (3.22).*

The formulas (3.21) and (3.24) are Green's second identities of the steady vibrations equations in the linear theory of quadruple porosity elasticity for domains Ω^+ and Ω^-, respectively.

3.4.3 Green's Third Identity

Let $\tilde{\mathbf{\Gamma}}(\mathbf{x}, \omega)$ be the fundamental solution of the system (3.20) (the fundamental matrix of operator $\tilde{\mathcal{A}}(\mathbf{D}_{\mathbf{y}}, \omega)$). Obviously, the matrix $\tilde{\mathbf{\Gamma}}(\mathbf{x}, \omega)$ satisfies the following condition

$$\tilde{\mathbf{\Gamma}}(\mathbf{x}, \omega) = \mathbf{\Gamma}^\top(-\mathbf{x}, \omega), \tag{3.25}$$

where $\mathbf{\Gamma}(\mathbf{x}, \omega)$ is the fundamental solution of the system (1.21) (see Theorem 2.1).

We denote by $\mathbb{B}(\mathbf{x},\epsilon)$ the (open) ball of radius ϵ and center $\mathbf{x}$. Let $\mathbf{x}\in\Omega^+$. Pick $\epsilon>0$ such that $\overline{\mathbb{B}(\mathbf{x},\epsilon)}\subset\Omega^+$. Applying the formula (3.21) in the domain $\Omega^+\backslash\overline{\mathbb{B}(\mathbf{x},\epsilon)}$ with $\tilde{\mathbf{V}}(\mathbf{y})=\tilde{\mathbf{\Gamma}}(\mathbf{y}-\mathbf{x},\omega)$, and letting $\epsilon\to 0$. By virtue of (3.25) from (3.21) and (3.24) we obtain the following results.

Theorem 3.15 *If* $\mathbf{V}$ *is a regular vector of the class* $\mathfrak{B}^{(s)}$ *in* Ω^+*, then*

$$\mathbf{V}(\mathbf{x})=\int\limits_S\Big\{[\tilde{\mathcal{P}}(\mathbf{D_z},\mathbf{n})\mathbf{\Gamma}^\top(\mathbf{x}-\mathbf{z},\omega)]^\top\mathbf{V}(\mathbf{z})-\mathbf{\Gamma}(\mathbf{x}-\mathbf{z},\omega)\,\mathcal{P}(\mathbf{D_z},\mathbf{n})\mathbf{V}(\mathbf{z})\Big\}\,d_{\mathbf{z}}S$$
$$+\int\limits_{\Omega^+}\mathbf{\Gamma}(\mathbf{x}-\mathbf{y},\omega)\,\mathcal{A}(\mathbf{D_y},\omega)\mathbf{V}(\mathbf{y})d\mathbf{y}\qquad \textit{for}\quad \mathbf{x}\in\Omega^+. \tag{3.26}$$

Theorem 3.16 *If* $\mathbf{V}$ *is a regular vector of the class* $\mathfrak{B}^{(s)}$ *in* Ω^-*, then*

$$\mathbf{V}(\mathbf{x})=-\int\limits_S\Big\{[\tilde{\mathcal{P}}(\mathbf{D_z},\mathbf{n})\mathbf{\Gamma}^\top(\mathbf{x}-\mathbf{z},\omega)]^\top\mathbf{V}(\mathbf{z})-\mathbf{\Gamma}(\mathbf{x}-\mathbf{z},\omega)\,\mathcal{P}(\mathbf{D_z},\mathbf{n})\mathbf{V}(\mathbf{z})\Big\}\,d_{\mathbf{z}}S$$
$$+\int\limits_{\Omega^-}\mathbf{\Gamma}(\mathbf{x}-\mathbf{y},\omega)\,\mathcal{A}(\mathbf{D_y},\omega)\mathbf{V}(\mathbf{y})d\mathbf{y}\qquad \textit{for}\quad \mathbf{x}\in\Omega^-. \tag{3.27}$$

The formulas (3.26) and (3.27) are Green's third identities of the steady vibrations equations in the linear theory of quadruple porosity elasticity for domains Ω^+ and Ω^-, respectively.

Theorems 3.15 and 3.16 lead to the following.

Corollary 3.1 *If* $\mathbf{V}$ *is a regular solution of the class* $\mathfrak{B}^{(s)}$ *of the homogeneous equation*

$$\mathcal{A}(\mathbf{D_x},\omega)\,\mathbf{V}(\mathbf{x})=\mathbf{0} \tag{3.28}$$

in Ω^+*, then*

$$\mathbf{V}(\mathbf{x})=\int\limits_S\Big\{[\tilde{\mathcal{P}}(\mathbf{D_z},\mathbf{n})\mathbf{\Gamma}^\top(\mathbf{x}-\mathbf{z},\omega)]^\top\mathbf{V}(\mathbf{z})-\mathbf{\Gamma}(\mathbf{x}-\mathbf{z},\omega)\,\mathcal{P}(\mathbf{D_z},\mathbf{n})\mathbf{V}(\mathbf{z})\Big\}\,d_{\mathbf{z}}S \tag{3.29}$$

for $\mathbf{x}\in\Omega^+$.

Corollary 3.2 *If* $\mathbf{V}$ *is a regular solution of the class* $\mathfrak{B}^{(s)}$ *of (3.28) in* Ω^-*, then*

$$\mathbf{V}(\mathbf{x}) =$$

$$-\int\limits_S \left\{[\tilde{\mathcal{P}}(\mathbf{D_z},\mathbf{n})\mathbf{\Gamma}^\top(\mathbf{x}-\mathbf{z},\omega)]^\top \mathbf{V}(\mathbf{z}) - \mathbf{\Gamma}(\mathbf{x}-\mathbf{z},\omega)\,\mathcal{P}(\mathbf{D_z},\mathbf{n})\mathbf{V}(\mathbf{z})\right\} d_\mathbf{z}S \tag{3.30}$$

for $\mathbf{x} \in \Omega^-$.

The formulas (3.29) and (3.30) are the Somigliana-type integral representations of a regular solution of the steady vibrations homogeneous equation (3.28) in the linear theory of quadruple porosity elasticity for domains Ω^+ and Ω^-, respectively.

3.5 Green's Formulas for Equations in the Laplace Transform Space

Obviously, from Green's formulas of the steady vibrations equations we obtain Green's formulas of the equations of pseudo-oscillations by replacing ω by $i\tau$, where τ is a complex number with $\mathrm{Im}\tau > 0$ (see Sect. 3.4).

We introduce the notation

$$\mathcal{B}(\Delta, i\tau) = \left(\mathcal{B}_{lj}(\Delta, i\tau)\right)_{5\times 5}, \qquad \mathcal{B}_{11}(\Delta, i\tau) = \mu_0\Delta - \rho\tau^2,$$

$$\mathcal{B}_{1;\alpha+1}(\Delta, i\tau) = -\tau a_\alpha\,\Delta, \qquad \mathcal{B}_{\alpha+1;1}(\Delta, i\tau) = -a_\alpha,$$

$$\mathcal{B}_{\alpha+1;\beta+1}(\Delta, i\tau) = k_{\alpha\beta}\Delta + c'_{\alpha\beta}, \qquad \alpha, \beta = 1, 2, 3, 4$$

and

$$\Lambda_1(\Delta, i\tau) = \frac{1}{k_0\mu_0}\det \mathcal{B}\,(\Delta, i\tau),$$

where $c'_{\alpha\beta} = -(\tau b_{\alpha\beta} + d_{\alpha\beta})$.

It is easily seen that $\Lambda_1(-\xi, i\tau) = 0$ is a fifth degree algebraic equation and there exist five roots $\check{\lambda}_1^2, \check{\lambda}_2^2, \cdots, \check{\lambda}_5^2$ (with respect to ξ). In the following we assume that $\mathrm{Im}\check{\lambda}_j > 0$ for $j = 1, 2, \cdots, 6$, where $\check{\lambda}_6^2 = -\dfrac{\rho\tau^2}{\mu}$.

Definition 3.2 A vector function $\mathbf{V} = (\mathbf{u}, \mathbf{p}) = (V_1, V_2, \cdots, V_7)$ is called *regular* of the class $\mathfrak{B}$ in Ω^- (or Ω^+) if

(i)

$$V_l \in C^2(\Omega^-) \cap C^1(\overline{\Omega^-}) \qquad (\text{or } V_l \in C^2(\Omega^+) \cap C^1(\overline{\Omega^+})),$$

(ii)

$$V_l(\mathbf{x}) = O(|\mathbf{x}|^{-1}) \qquad V_{l,j}(\mathbf{x}) = o(|\mathbf{x}|^{-1}) \qquad \text{for} \qquad |\mathbf{x}| \gg 1, \tag{3.31}$$

where $j = 1, 2, 3$ and $l = 1, 2, \cdots, 7$.

3.5.1 Green's First Identity

We introduce the notation

$$\begin{aligned} W_1^{(l)}(\mathbf{V}, \mathbf{u}') &= W_0(\mathbf{u}, \mathbf{u}') + \rho\tau^2 \mathbf{u} \cdot \mathbf{u}' - a_\alpha\, p_\alpha \operatorname{div} \overline{\mathbf{u}'}, \\ W_2^{(l)}(\mathbf{V}, \mathbf{p}') &= k_{\alpha\beta} \nabla p_\alpha \cdot \nabla p'_\beta - c'_{\alpha\beta}\, p_\beta \overline{p'_\alpha} - \tau\, a_\alpha \overline{p'_\alpha} \operatorname{div} \mathbf{u}, \\ W^{(l)}(\mathbf{V}, \mathbf{V}') &= W_1^{(l)}(\mathbf{V}, \mathbf{u}') + W_2^{(l)}(\mathbf{V}, \mathbf{p}'), \end{aligned} \tag{3.32}$$

where $W_0(\mathbf{u}, \mathbf{u}')$ is given by (3.15); $\mathbf{u}' = (u'_1, u'_2, u'_3)$ and $\mathbf{p}' = (p'_1, p'_2, p'_3, p'_4)$ are three- and four-component vector functions, respectively; $\mathbf{V}' = (\mathbf{u}', \mathbf{p}')$.

Keeping in mind Theorem 3.11 we have the following.

Theorem 3.17 *If* $\mathbf{V} = (\mathbf{u}, \mathbf{p})$ *is a regular vector of the class* $\mathfrak{B}$ *in* Ω^+, $u'_j, p'_\alpha \in C^1(\Omega^+) \cap C(\overline{\Omega^+})$, $j = 1, 2, 3$, $\alpha = 1, 2, 3, 4$, *then*

$$\begin{aligned} &\int\limits_{\Omega^+} \Big[\mathcal{A}(\mathbf{D_x}, i\tau)\, \mathbf{V}(\mathbf{x}) \cdot \mathbf{V}'(\mathbf{x}) + W^{(l)}(\mathbf{V}, \mathbf{V}')\Big]\, d\mathbf{x} \\ &\qquad = \int\limits_S \mathcal{P}(\mathbf{D_z}, \mathbf{n}) \mathbf{V}(\mathbf{z}) \cdot \mathbf{V}'(\mathbf{z})\, d_{\mathbf{z}}S, \end{aligned} \tag{3.33}$$

where $\mathcal{A}(\mathbf{D_x}, i\tau)$, $\mathcal{P}(\mathbf{D_z}, \mathbf{n})$ *and* $W^{(l)}(\mathbf{V}, \mathbf{V}')$ *are defined by (1.36), (1.45), and (3.32), respectively.*

Theorem 3.17 and the conditions at infinity (3.31) lead to the following.

Theorem 3.18 *If* $\mathbf{V} = (\mathbf{u}, \mathbf{p})$ *and* $\mathbf{V}' = (\mathbf{u}', \mathbf{p}')$ *are regular vectors of the class* $\mathfrak{B}$ *in* Ω^-, *then*

$$\int_{\Omega^+} \left[\mathcal{A}(\mathbf{D_x}, i\tau)\, \mathbf{V}(\mathbf{x}) \cdot \mathbf{V}'(\mathbf{x}) + W^{(l)}(\mathbf{V}, \mathbf{V}')\right] d\mathbf{x}$$
$$= -\int_S \mathcal{P}(\mathbf{D_z}, \mathbf{n})\mathbf{V}(\mathbf{z}) \cdot \mathbf{V}'(\mathbf{z})\, d_{\mathbf{z}}S, \tag{3.34}$$

The formulas (3.33) and (3.34) are Green's first identities in the Laplace transform space of the linear theory of elasticity of quadruple porosity materials for domains Ω^+ and Ω^-, respectively.

3.5.2 Green's Second Identity

Let

$$\tilde{\mathcal{P}}'(\mathbf{D_x}, \mathbf{n}) = (\tilde{\mathcal{P}}'_{lj}(\mathbf{D_x}, \mathbf{n}))_{7\times 7},$$

$$\tilde{\mathcal{P}}'_{lj}(\mathbf{D_x}, \mathbf{n}) = \mu\delta_{lj}\frac{\partial}{\partial \mathbf{n}} + \lambda n_l \frac{\partial}{\partial x_j} + \mu n_j \frac{\partial}{\partial x_l},$$

$$\tilde{\mathcal{P}}'_{l;\alpha+3}(\mathbf{D_x}, \mathbf{n}) = \tau a_\alpha n_l, \qquad \tilde{\mathcal{P}}'_{\alpha+3;j}(\mathbf{D_x}, \mathbf{n}) = 0, \tag{3.35}$$

$$\tilde{\mathcal{P}}'_{\alpha+3;\beta+3}(\mathbf{D_x}, \mathbf{n}) = k_{lj}\frac{\partial}{\partial \mathbf{n}}, \qquad l, j = 1, 2, 3, \qquad \alpha, \beta = 1, 2, 3, 4;$$

It is easy to verify that the matrix differential operator $\tilde{\mathcal{P}}'(\mathbf{D_x}, \mathbf{n})$ may be obtained from the operator $\tilde{\mathcal{P}}(\mathbf{D_x}, \mathbf{n})$ by replacing ω by $i\tau$.

Theorems 3.13 and 3.14 lead to the following results.

Theorem 3.19 *If* $\mathbf{V}$ *and* $\tilde{\mathbf{V}}_j$ $(j = 1, 2, \cdots, 7)$ *are regular vectors of the class* $\mathfrak{B}$ *in* Ω^+, *then*

$$\int_{\Omega^+} \left\{[\tilde{\mathcal{A}}(\mathbf{D_y}, i\tau)\tilde{\mathbf{V}}(\mathbf{y})]^\top \mathbf{V}(\mathbf{y}) - [\tilde{\mathbf{V}}(\mathbf{y})]^\top \mathcal{A}(\mathbf{D_y}, i\tau)\mathbf{V}(\mathbf{y})\right\} d\mathbf{y}$$
$$= \int_S \left\{[\tilde{\mathcal{P}}'(\mathbf{D_z}, \mathbf{n})\tilde{\mathbf{V}}(\mathbf{z})]^\top \mathbf{V}(\mathbf{z}) - [\tilde{\mathbf{V}}(\mathbf{z})]^\top \mathcal{P}(\mathbf{D_z}, \mathbf{n})\mathbf{V}(\mathbf{z})\right\} d_{\mathbf{z}}S, \tag{3.36}$$

where the operators $\mathcal{P}$ *and* $\tilde{\mathcal{P}}'$ *are given by (1.45) and (3.35), respectively.*

Theorem 3.19 and the radiation conditions (3.31) lead to the following.

Theorem 3.20 *If* $\mathbf{V}$ *and* $\tilde{\mathbf{V}}_j$ $(j = 1, 2, \cdots, 5)$ *are regular vectors of the class* $\mathfrak{B}$ *in* Ω^-, *then*

$$\int\limits_{\Omega^-} \left\{ [\tilde{\mathcal{A}}(\mathbf{D_y}, i\tau)\tilde{\mathbf{V}}(\mathbf{y})]^\top \mathbf{V}(\mathbf{y}) - [\tilde{\mathbf{V}}(\mathbf{y})]^\top \mathcal{A}(\mathbf{D_y}, i\tau)\mathbf{V}(\mathbf{y}) \right\} d\mathbf{y}$$
$$= - \int\limits_S \left\{ [\tilde{\mathcal{P}}'(\mathbf{D_z}, \mathbf{n})\tilde{\mathbf{V}}(\mathbf{z})]^\top \mathbf{V}(\mathbf{z}) - [\tilde{\mathbf{V}}(\mathbf{z})]^\top \mathcal{P}(\mathbf{D_z}, \mathbf{n})\mathbf{V}(\mathbf{z}) \right\} d_{\mathbf{z}}S, \tag{3.37}$$

where the operators $\mathcal{P}$ *and* $\tilde{\mathcal{P}}'$ *are given by (1.45) and (3.35), respectively.*

The formulas (3.36) and (3.37) are Green's second identities in the Laplace transform space of the linear theory of quadruple porosity elasticity for domains Ω^+ and Ω^-, respectively.

3.5.3 Green's Third Identity

By virtue of (3.36) and (3.37) we have the following theorems.

Theorem 3.21 *If* $\mathbf{V}$ *is a regular vector of the class* $\mathfrak{B}$ *in* Ω^+, *then*

$$\mathbf{V}(\mathbf{x}) = \int\limits_S \left\{ [\tilde{\mathcal{P}}'(\mathbf{D_z}, \mathbf{n})\mathbf{\Gamma}^\top(\mathbf{x}-\mathbf{z}, i\tau)]^\top \mathbf{V}(\mathbf{z}) - \mathbf{\Gamma}(\mathbf{x}-\mathbf{z}, i\tau)\,\mathcal{P}(\mathbf{D_z}, \mathbf{n})\mathbf{V}(\mathbf{z}) \right\} d_{\mathbf{z}}S$$
$$+ \int\limits_{\Omega^+} \mathbf{\Gamma}(\mathbf{x}-\mathbf{y}, i\tau)\,\mathcal{A}(\mathbf{D_y}, i\tau)\mathbf{V}(\mathbf{y})d\mathbf{y} \qquad \text{for} \quad \mathbf{x} \in \Omega^+. \tag{3.38}$$

Theorem 3.22 *If* $\mathbf{V}$ *is a regular vector of the class* $\mathfrak{B}$ *in* Ω^-, *then*

$$\mathbf{V}(\mathbf{x}) = - \int\limits_S \left\{ [\tilde{\mathcal{P}}'(\mathbf{D_z}, \mathbf{n})\mathbf{\Gamma}^\top(\mathbf{x}-\mathbf{z}, i\tau)]^\top \mathbf{V}(\mathbf{z}) - \mathbf{\Gamma}(\mathbf{x}-\mathbf{z}, i\tau)\,\mathcal{P}(\mathbf{D_z}, \mathbf{n})\mathbf{V}(\mathbf{z}) \right\} d_{\mathbf{z}}S$$
$$+ \int\limits_{\Omega^-} \mathbf{\Gamma}(\mathbf{x}-\mathbf{y}, i\tau)\,\mathcal{A}(\mathbf{D_y}, i\tau)\mathbf{V}(\mathbf{y})d\mathbf{y} \qquad \text{for} \quad \mathbf{x} \in \Omega^-. \tag{3.39}$$

The formulas (3.38) and (3.39) are Green's third identities in the Laplace transform space of the linear theory of quadruple porosity elasticity for domains Ω^+ and Ω^-, respectively.

3.6 Green's Formulas for Quasi-Static Equations

Let $\mathbf{u}' = (u_1', u_2', u_3')$ and $\mathbf{p}' = (p_1', p_2', p_3', p_4')$ are three- and four-component vector functions, respectively; $\mathbf{V}' = (\mathbf{u}', \mathbf{p}')$. We introduce the notation

$$W^{(q)}(\mathbf{V}, \mathbf{V}') = W_0(\mathbf{u}, \mathbf{u}') + W_2(\mathbf{V}, \mathbf{p}') - a_\alpha\, p_\alpha \operatorname{div} \overline{\mathbf{u}'}, \tag{3.40}$$

where $W_0(\mathbf{u}, \mathbf{u}')$ and $W_2(\mathbf{V}, \mathbf{p}')$ are given by (3.15).

We have the following.

Theorem 3.23 *If* $u_j, p_\alpha \in C^2(\Omega^+) \cap C^1(\overline{\Omega^+})$ *and* $u_j', p_\alpha' \in C^1(\Omega^+) \cap C(\overline{\Omega^+})$, $j = 1, 2, 3$, $\alpha = 1, 2, 3, 4$, *then*

$$\int\limits_{\Omega^+} \left[\boldsymbol{\mathcal{A}}^{(q)}(\mathbf{D_x}, \omega)\, \mathbf{V}(\mathbf{x}) \cdot \mathbf{V}'(\mathbf{x}) + W^{(q)}(\mathbf{V}, \mathbf{V}')\right] d\mathbf{x} = \int\limits_S \boldsymbol{\mathcal{P}}(\mathbf{D_z}, \mathbf{n})\mathbf{V}(\mathbf{z}) \cdot \mathbf{V}'(\mathbf{z})\, d_{\mathbf{z}}S, \tag{3.41}$$

where $\boldsymbol{\mathcal{A}}^{(q)}(\mathbf{D_x}, \omega)$, $\boldsymbol{\mathcal{P}}(\mathbf{D_z}, \mathbf{n})$ *and* $W^{(q)}(\mathbf{V}, \mathbf{V}')$ *are defined by (1.35), (1.45), and (3.40), respectively.*

Clearly, the matrix differential operator $\tilde{\boldsymbol{\mathcal{A}}}^{(q)}(\mathbf{D_x}, \omega) = \left(\tilde{A}_{lj}^{(q)}(\mathbf{D_x}, \omega)\right)_{7\times 7}$ is the associate operator of $\boldsymbol{\mathcal{A}}^{(q)}(\mathbf{D_x}, \omega)$, where $\tilde{\boldsymbol{\mathcal{A}}}^{(q)}(\mathbf{D_x}, \omega) = \left[\boldsymbol{\mathcal{A}}^{(q)}(-\mathbf{D_x}, \omega)\right]^\top$. It is easy to verify that the operator $\tilde{\boldsymbol{\mathcal{A}}}^{(q)}(\mathbf{D_x}, \omega)$ may be obtained from the operator $\boldsymbol{\mathcal{A}}^{(q)}(\mathbf{D_x}, \omega)$ by replacing a_α by $i\omega a_\alpha$ $(\alpha = 1, 2, 3, 4)$ and vice versa. It is easy to see that the associated system of homogeneous equations of (1.23) is

$$\begin{aligned} &\mu\, \Delta \tilde{\mathbf{u}} + (\lambda + \mu)\, \nabla \operatorname{div} \tilde{\mathbf{u}} - i\omega \nabla\, (\mathbf{a}\, \tilde{\mathbf{p}}) = \mathbf{0}, \\ &(\mathbf{K}\, \Delta + \mathbf{c})\, \tilde{\mathbf{p}} + \mathbf{a} \operatorname{div} \tilde{\mathbf{u}} = \mathbf{0}, \end{aligned} \tag{3.42}$$

where $\tilde{\mathbf{u}}$ and $\tilde{\mathbf{p}}$ are three- and four-component vector functions on $\Omega^\pm$, respectively.

Let $\mathbf{V} = (\mathbf{u}, \mathbf{p}) = (V_1, V_2, \cdots, V_7)$ and $\tilde{\mathbf{V}} = (\tilde{V}_{lj})_{7\times 7}$. By virtue of (3.21) and (3.26) we get the following results.

Theorem 3.24 *If $V_l, \tilde{V}_{lj} \in C^2(\Omega^+) \cap C^1(\overline{\Omega^+})$ $(l, j = 1, 2, \cdots, 7)$, then*

$$\int\limits_{\Omega^+} \Big\{[\tilde{\mathcal{A}}^{(q)}(\mathbf{D_y}, \omega)\tilde{\mathbf{V}}(\mathbf{y})]^\top \mathbf{V}(\mathbf{y}) - [\tilde{\mathbf{V}}(\mathbf{y})]^\top \mathcal{A}^{(q)}(\mathbf{D_y}, \omega)\mathbf{V}(\mathbf{y})\Big\}\, d\mathbf{y}$$
$$= \int\limits_S \Big\{[\tilde{\mathcal{P}}(\mathbf{D_z}, \mathbf{n})\tilde{\mathbf{V}}(\mathbf{z})]^\top \mathbf{V}(\mathbf{z}) - [\tilde{\mathbf{V}}(\mathbf{z})]^\top \mathcal{P}(\mathbf{D_z}, \mathbf{n})\mathbf{V}(\mathbf{z})\Big\}\, d_{\mathbf{z}}S, \tag{3.43}$$

where the matrices $\mathcal{A}^{(q)}(\mathbf{D_x}, \omega)$, $\mathcal{P}(\mathbf{D_z}, \mathbf{n})$, and $\tilde{\mathcal{P}}(\mathbf{D_z}, \mathbf{n})$ are given by (1.37), (1.45), and (3.22), respectively.

Let $\tilde{\mathbf{\Gamma}}^{(q)}(\mathbf{x}, \omega)$ is the fundamental solution of the system (3.42) (the fundamental matrix of operator $\tilde{\mathcal{A}}^{(q)}(\mathbf{D_y}, \omega)$). Obviously, the matrix $\tilde{\mathbf{\Gamma}}^{(q)}(\mathbf{x}, \omega)$ satisfies the following condition

$$\tilde{\mathbf{\Gamma}}^{(q)}(\mathbf{x}, \omega) = \mathbf{\Gamma}^{(q)^\top}(-\mathbf{x}, \omega), \tag{3.44}$$

where $\mathbf{\Gamma}^{(q)}(\mathbf{x}, \omega)$ is the fundamental solution of the system (1.23) (see Theorem 2.3).

Theorem 3.24 and Eq. (3.44) lead to the following results.

Theorem 3.25 *If $u_l, p_\alpha \in C^2(\Omega^+) \cap C^1(\overline{\Omega^+})$ $(l = 1, 2, 3,\ \alpha = 1, 2, 3, 4)$, then*

$$\mathbf{V}(\mathbf{x}) =$$
$$\int\limits_S \Big\{[\tilde{\mathcal{P}}(\mathbf{D_z}, \mathbf{n})\mathbf{\Gamma}^{(q)^\top}(\mathbf{x}-\mathbf{z}, \omega)]^\top \mathbf{V}(\mathbf{z}) - \mathbf{\Gamma}^{(q)}(\mathbf{x}-\mathbf{z}, \omega)\,\mathcal{P}(\mathbf{D_z}, \mathbf{n})\mathbf{V}(\mathbf{z})\Big\}\, d_{\mathbf{z}}S$$
$$+ \int\limits_{\Omega^+} \mathbf{\Gamma}^{(q)}(\mathbf{x}-\mathbf{y}, \omega)\,\mathcal{A}^{(q)}(\mathbf{D_y}, \omega)\mathbf{V}(\mathbf{y})d\mathbf{y} \qquad \textit{for} \quad \mathbf{x} \in \Omega^+. \tag{3.45}$$

The formulas (3.41), (3.43), and (3.45) are Green's first, second, and third identities for quasi-static equations in the linear theory of elasticity of quadruple porosity materials for domains Ω^+, respectively.

Obviously, quite similar manner, (3.41), (3.43), and (3.45) imply Green's identities for quasi-static equations in the linear theory of elasticity of quadruple porosity materials for domains Ω^-.

3.7 Green's Formulas for Equilibrium Equations

We introduce the notation

$$W^{(e)}(\mathbf{V},\mathbf{V}') = W_0(\mathbf{u},\mathbf{u}') - a_\alpha\, p_\alpha\, \mathrm{div}\,\overline{\mathbf{u}'} + k_{\alpha\beta}\nabla p_\alpha\cdot\nabla p'_\beta + d_{\alpha\beta}\, p_\beta\, \overline{p'_\alpha}, \tag{3.46}$$

where $W_0(\mathbf{u},\mathbf{u}')$ is given by (3.15), $\mathbf{u}' = (u'_1, u'_2, u'_3)$ and $\mathbf{p}' = (p'_1, p'_2, p'_3, p'_4)$ are three- and four-component vector functions, respectively; $\mathbf{V}' = (\mathbf{u}',\mathbf{p}')$ and $\mathbf{V} = (\mathbf{u},\mathbf{p})$.

We have the following.

Theorem 3.26 *If* $u_j, p_\alpha \in C^2(\Omega^+)\cap C^1(\overline{\Omega^+})$ *and* $u'_j, p'_\alpha \in C^1(\Omega^+)\cap C(\overline{\Omega^+})$, $j = 1,2,3$, $\alpha = 1,2,3,4$, *then*

$$\int_{\Omega^+}\left[\mathcal{A}^{(e)}(\mathbf{D_x})\,\mathbf{V}(\mathbf{x})\cdot\mathbf{V}'(\mathbf{x}) + W^{(e)}(\mathbf{V},\mathbf{V}')\right]d\mathbf{x} = \int_S \mathcal{P}(\mathbf{D_z},\mathbf{n})\mathbf{V}(\mathbf{z})\cdot\mathbf{V}'(\mathbf{z})\, d_{\mathbf{z}}S, \tag{3.47}$$

where $\mathcal{A}^{(e)}(\mathbf{D_x})$, $\mathcal{P}(\mathbf{D_z},\mathbf{n})$ *and* $W^{(e)}(\mathbf{V},\mathbf{V}')$ *are defined by (1.37), (1.45), and (3.46), respectively.*

Denote the matrix differential operator $\tilde{\mathcal{A}}^{(e)}(\mathbf{D_x}) = \left(\tilde{A}^{(e)}_{lj}(\mathbf{D_x})\right)_{7\times 7}$ by

$$\tilde{\mathcal{A}}^{(e)}(\mathbf{D_x}) = \left[\mathcal{A}^{(e)}(-\mathbf{D_x})\right]^{\top}.$$

We introduce the notation

$$\begin{aligned}
&\tilde{\mathcal{P}}^{(e)}(\mathbf{D_x},\mathbf{n}) = (\tilde{\mathcal{P}}^{(e)}_{lj}(\mathbf{D_x},\mathbf{n}))_{7\times 7},\\
&\tilde{\mathcal{P}}^{(e)}_{lj}(\mathbf{D_x},\mathbf{n}) = \mu\delta_{lj}\frac{\partial}{\partial\mathbf{n}} + \lambda n_l\frac{\partial}{\partial x_j} + \mu n_j\frac{\partial}{\partial x_l},\\
&\tilde{\mathcal{P}}^{(e)}_{\alpha+3;j}(\mathbf{D_x},\mathbf{n}) = \tilde{\mathcal{P}}^{(e)}_{l;\alpha+3}(\mathbf{D_x},\mathbf{n}) = 0,\\
&\tilde{\mathcal{P}}^{(e)}_{\alpha+3;\beta+3}(\mathbf{D_x},\mathbf{n}) = k_{lj}\tfrac{\partial}{\partial\mathbf{n}},\\
&l, j = 1,2,3, \qquad \alpha,\beta = 1,2,3,4.
\end{aligned} \tag{3.48}$$

Let $\mathbf{V} = (\mathbf{u}, \mathbf{p}) = (V_1, V_2, \cdots, V_7)$, $\tilde{\mathbf{V}} = (\tilde{V}_{lj})_{7\times 7}$. We get the following.

Theorem 3.27 *If* $V_l, \tilde{V}_{lj} \in C^2(\Omega^+) \cap C^1(\overline{\Omega^+})$ $(l, j = 1, 2, \cdots, 7)$, *then*

$$\int\limits_{\Omega^+} \left\{ [\tilde{\mathcal{A}}^{(e)}(\mathbf{D_y})\tilde{\mathbf{V}}(\mathbf{y})]^\top \mathbf{V}(\mathbf{y}) - [\tilde{\mathbf{V}}(\mathbf{y})]^\top \mathcal{A}^{(e)}(\mathbf{D_y})\mathbf{V}(\mathbf{y}) \right\} dy$$
$$= \int\limits_S \left\{ [\tilde{\mathcal{P}}^{(e)}(\mathbf{D_z}, \mathbf{n})\tilde{\mathbf{V}}(\mathbf{z})]^\top \mathbf{V}(\mathbf{z}) - [\tilde{\mathbf{V}}(\mathbf{z})]^\top \mathcal{P}(\mathbf{D_z}, \mathbf{n})\mathbf{V}(\mathbf{z}) \right\} d_{\mathbf{z}}S, \tag{3.49}$$

where the matrices $\mathcal{A}^{(e)}(\mathbf{D_x})$, $\mathcal{P}(\mathbf{D_z}, \mathbf{n})$, *and* $\tilde{\mathcal{P}}^{(e)}(\mathbf{D_z}, \mathbf{n})$ *are given by (1.37), (1.45), and (3.48), respectively.*

Let $\tilde{\mathbf{\Gamma}}^{(e)}(\mathbf{x})$ be the fundamental matrix of operator $\tilde{\mathcal{A}}^{(e)}(\mathbf{D_y})$. Obviously, the matrix $\tilde{\mathbf{\Gamma}}^{(e)}(\mathbf{x})$ satisfies the following condition

$$\tilde{\mathbf{\Gamma}}^{(e)}(\mathbf{x}) = \mathbf{\Gamma}^{(e)^\top}(-\mathbf{x}), \tag{3.50}$$

where $\mathbf{\Gamma}^{(e)}(\mathbf{x})$ is the fundamental solution of the system (1.24) (see Theorem 2.4).

Theorem 3.27 and Eq. (3.50) lead to the following results.

Theorem 3.28 *If* $u_l, p_\alpha \in C^2(\Omega^+) \cap C^1(\overline{\Omega^+})$ $(l = 1, 2, 3,\ \alpha = 1, 2, 3, 4)$, *then*

$$\mathbf{V}(\mathbf{x}) =$$
$$\int\limits_S \left\{ [\tilde{\mathcal{P}}^{(e)}(\mathbf{D_z}, \mathbf{n})\mathbf{\Gamma}^{(e)^\top}(\mathbf{x}-\mathbf{z})]^\top \mathbf{V}(\mathbf{z}) - \mathbf{\Gamma}^{(e)}(\mathbf{x}-\mathbf{z})\, \mathcal{P}(\mathbf{D_z}, \mathbf{n})\mathbf{V}(\mathbf{z}) \right\} d_{\mathbf{z}}S$$
$$+ \int\limits_{\Omega^+} \mathbf{\Gamma}^{(e)}(\mathbf{x}-\mathbf{y})\, \mathcal{A}^{(e)}(\mathbf{D_y})\mathbf{V}(\mathbf{y})dy \qquad for \quad \mathbf{x} \in \Omega^+. \tag{3.51}$$

The formulas (3.47), (3.49), and (3.51) are Green's first, second, and third identities of the equilibrium equations in the linear theory of quadruple porosity elasticity for domain Ω^+, respectively. Obviously, quite similar manner, these formulas imply Green's identities of the equilibrium equations in the same theory for domain Ω^-.

3.8 Green's Formulas for Steady Vibrations Equations of Rigid Body

In this section Green's identities for equations of steady vibrations of rigid body are established.

Definition 3.3 A vector function $\mathbf{p} = (p_1, p_2, p_3, p_4)$ is called regular of the class $\mathfrak{B}^{(r)}$ in Ω^- (or Ω^+) if

(i)

$$p_\alpha \in C^2(\Omega^-) \cap C^1(\overline{\Omega^-}) \qquad (\text{or } p_\alpha \in C^2(\Omega^+) \cap C^1(\overline{\Omega^+})),$$

(ii)

$$p_\alpha(\mathbf{x}) = O(|\mathbf{x}|^{-1}) \qquad p_{\alpha,j}(\mathbf{x}) = o(|\mathbf{x}|^{-1}) \qquad \text{for} \qquad |\mathbf{x}| \gg 1, \tag{3.52}$$

where $j = 1, 2, 3$ and $\alpha = 1, 2, 3, 4$.

We introduce the notation

$$W^{(r)}(\mathbf{p}, \mathbf{p}') = k_{\alpha\beta} \nabla p_\alpha \cdot \nabla p'_\beta - c_{\alpha\beta} p_\beta \overline{p'_\alpha}, \tag{3.53}$$

where $\mathbf{p}' = (p'_1, p'_2, p'_3, p'_4)$ is four-component vector function.

We have the following.

Theorem 3.29 *If* $\mathbf{p}$ *is a regular vector of the class* $\mathfrak{B}^{(r)}$ *in* Ω^+ *and* $p'_\alpha \in C^1(\Omega^+) \cap C(\overline{\Omega^+})$, $\alpha = 1, 2, 3, 4$, *then*

$$\int_{\Omega^+} \left[\mathcal{A}^{(r)}(\mathbf{D_x}, \omega)\, \mathbf{p}(\mathbf{x}) \cdot \mathbf{p}'(\mathbf{x}) + W^{(r)}(\mathbf{p}, \mathbf{p}') \right] d\mathbf{x} = \int_S \mathcal{P}^{(r)}(\mathbf{D_z}, \mathbf{n}) \mathbf{p}(\mathbf{z}) \cdot \mathbf{p}'(\mathbf{z})\, d_{\mathbf{z}} S, \tag{3.54}$$

where $\mathcal{A}^{(r)}(\mathbf{D_x}, \omega)$ *and* $W^{(r)}(\mathbf{p}, \mathbf{p}')$ *are defined by (1.37) and (3.53), respectively, and*

$$\mathcal{P}^{(r)}(\mathbf{D_z}, \mathbf{n}) = \left(\mathcal{P}^{(r)}_{\alpha\beta}(\mathbf{D_x}, \mathbf{n})\right)_{4\times 4}, \qquad \mathcal{P}^{(r)}_{\alpha\beta}(\mathbf{D_x}, \mathbf{n}) = k_{\alpha\beta} \frac{\partial}{\partial \mathbf{n}}, \tag{3.55}$$

$\alpha, \beta = 1, 2, 3, 4.$

The following theorems are proved in a similar way.

Theorem 3.30 *If* $p_\alpha, \tilde{p}_{\alpha\beta} \in C^2(\Omega^+) \cap C^1(\overline{\Omega^+})$ $(\alpha, \beta = 1, 2, 3, 4)$, *then*

$$\int\limits_{\Omega^+} \Big\{[\mathcal{A}^{(r)}(\mathbf{D_y}, \omega)\tilde{\mathbf{p}}(\mathbf{y})]^\top \mathbf{p}(\mathbf{y}) - [\tilde{\mathbf{p}}(\mathbf{y})]^\top \mathcal{A}^{(r)}(\mathbf{D_y}, \omega)\mathbf{p}(\mathbf{y})\Big\}\, d\mathbf{y}$$
$$= \int\limits_S \Big\{[\mathcal{P}^{(r)}(\mathbf{D_z}, \mathbf{n})\tilde{\mathbf{p}}(\mathbf{z})]^\top \mathbf{p}(\mathbf{z}) - [\tilde{\mathbf{p}}(\mathbf{z})]^\top \mathcal{P}^{(r)}(\mathbf{D_z}, \mathbf{n})\mathbf{p}(\mathbf{z})\Big\}\, d_\mathbf{z}S, \tag{3.56}$$

where the matrices $\mathcal{A}^{(r)}(\mathbf{D_x}, \omega)$ *and* $\mathcal{P}^{(r)}(\mathbf{D_z}, \mathbf{n})$ *are given by (1.36) and (3.55), respectively.*

Theorem 3.31 *If* $\mathbf{p}$ *is regular vector of the class* $\mathfrak{B}^{(r)}$ *in* Ω^+, *then*

$$\mathbf{p}(\mathbf{x}) =$$
$$\int\limits_S \Big\{[\mathcal{P}^{(r)}(\mathbf{D_z}, \mathbf{n})\boldsymbol{\Gamma}^{(r)}(\mathbf{x}-\mathbf{z}, \omega)]^\top \mathbf{p}(\mathbf{z}) - \boldsymbol{\Gamma}^{(r)}(\mathbf{x}-\mathbf{z}, \omega)\, \mathcal{P}^{(r)}(\mathbf{D_z}, \mathbf{n})\mathbf{p}(\mathbf{z})\Big\}\, d_\mathbf{z}S$$
$$+ \int\limits_{\Omega^+} \boldsymbol{\Gamma}^{(r)}(\mathbf{x}-\mathbf{y}, \omega)\, \mathcal{A}^{(r)}(\mathbf{D_y}, \omega)\mathbf{p}(\mathbf{y})d\mathbf{y} \qquad \text{for} \quad \mathbf{x} \in \Omega^+, \tag{3.57}$$

where $\boldsymbol{\Gamma}^{(r)}(\mathbf{x}, \omega)$ *is the fundamental matrix of the operator* $\mathcal{A}^{(r)}(\mathbf{D_y}, \omega)$ *and defined in Theorem 2.6.*

The formulas (3.54), (3.56), and (3.57) are Green's first, second, and third identities for the steady vibrations equations of quadruple porosity rigid body for domain Ω^+, respectively. Quite similar manner, the identities (3.54), (3.56), (3.57), and condition (3.52) at infinity imply Green's identities for the steady vibrations equations of quadruple porosity rigid body for domain Ω^-. Namely, the following theorems are valid.

Theorem 3.32 *If* $\mathbf{p}$ *regular vector of the class* $\mathfrak{B}^{(r)}$ *in* Ω^-, $p'_\alpha \in C^1(\Omega^-) \cap C(\overline{\Omega^-})$ *and* p'_α $(\alpha = 1, 2, 3, 4)$ *satisfies the conditions (3.52) at infinity, then*

$$\int\limits_{\Omega^-} \Big[\mathcal{A}^{(r)}(\mathbf{D_x}, \omega)\, \mathbf{p}(\mathbf{x}) \cdot \mathbf{p}'(\mathbf{x}) + W^{(r)}(\mathbf{p}, \mathbf{p}')\Big]\, d\mathbf{x}$$
$$= -\int\limits_S \mathcal{P}^{(r)}(\mathbf{D_z}, \mathbf{n})\mathbf{p}(\mathbf{z}) \cdot \mathbf{p}'(\mathbf{z})\, d_\mathbf{z}S. \tag{3.58}$$

Theorem 3.33 *If* $p_\alpha, \tilde{p}_{\alpha\beta} \in C^2(\Omega^-) \cap C^1(\overline{\Omega^-})$, p_α, *and* $\tilde{p}_{\alpha\beta}$ $(\alpha, \beta = 1, 2, 3, 4)$ *satisfy the conditions (3.52) at infinity, then*

$$\int_{\Omega^-} \left\{ [\mathcal{A}^{(r)}(\mathbf{D}_\mathbf{y}, \omega)\tilde{\mathbf{p}}(\mathbf{y})]^\top \mathbf{p}(\mathbf{y}) - [\tilde{\mathbf{p}}(\mathbf{y})]^\top \mathcal{A}^{(r)}(\mathbf{D}_\mathbf{y}, \omega)\mathbf{p}(\mathbf{y}) \right\} d\mathbf{y}$$

$$= -\int_S \left\{ [\mathcal{P}^{(r)}(\mathbf{D}_\mathbf{z}, \mathbf{n})\tilde{\mathbf{p}}(\mathbf{z})]^\top \mathbf{p}(\mathbf{z}) - [\tilde{\mathbf{p}}(\mathbf{z})]^\top \mathcal{P}^{(r)}(\mathbf{D}_\mathbf{z}, \mathbf{n})\mathbf{p}(\mathbf{z}) \right\} d_\mathbf{z} S.$$

Theorem 3.34 *If* $\mathbf{p}$ *is regular vector of the class* $\mathfrak{B}^{(r)}$ *in* Ω^-, *then*

$$\mathbf{p}(\mathbf{x}) =$$

$$-\int_S \left\{ [\mathcal{P}^{(r)}(\mathbf{D}_\mathbf{z}, \mathbf{n})\mathbf{\Gamma}^{(r)}(\mathbf{x}-\mathbf{z}, \omega)]^\top \mathbf{p}(\mathbf{z}) - \mathbf{\Gamma}^{(r)}(\mathbf{x}-\mathbf{z}, \omega)\, \mathcal{P}^{(r)}(\mathbf{D}_\mathbf{z}, \mathbf{n})\mathbf{p}(\mathbf{z}) \right\} d_\mathbf{z} S$$

$$+\int_{\Omega^-} \mathbf{\Gamma}^{(r)}(\mathbf{x}-\mathbf{y}, \omega)\, \mathcal{A}^{(r)}(\mathbf{D}_\mathbf{y}, \omega)\mathbf{p}(\mathbf{y}) d\mathbf{y} \qquad \textit{for} \quad \mathbf{x} \in \Omega^-.$$

3.9 Green's Formulas for Equilibrium Equations of Rigid Body

We introduce the notation

$$W^{(re)}(\mathbf{p}, \mathbf{p}') = k_{\alpha\beta} \nabla p_\alpha \cdot \nabla p'_\beta + d_{\alpha\beta} p_\beta \overline{p'_\alpha}, \tag{3.59}$$

where $\mathbf{p}' = (p'_1, p'_2, p'_3, p'_4)$ is four-component vector function.

We have the following.

Theorem 3.35 *If* $\mathbf{p}$ *is a regular vector of the class* $\mathfrak{B}^{(r)}$ *in* $\Omega^\pm$ *and* $p'_\alpha \in C^1(\Omega^\pm) \cap C(\overline{\Omega^\pm})$, $\alpha = 1, 2, 3, 4$, *then*

$$\int_{\Omega^\pm} \left[\mathcal{A}^{(re)}(\mathbf{D}_\mathbf{x})\, \mathbf{p}(\mathbf{x}) \cdot \mathbf{p}'(\mathbf{x}) + W^{(re)}(\mathbf{p}, \mathbf{p}') \right] d\mathbf{x}$$

$$= \pm \int_S \mathcal{P}^{(r)}(\mathbf{D}_\mathbf{z}, \mathbf{n})\mathbf{p}(\mathbf{z}) \cdot \mathbf{p}'(\mathbf{z})\, d_\mathbf{z} S, \tag{3.60}$$

where $\mathcal{A}^{(re)}(\mathbf{D}_\mathbf{x}, \omega)$, $\mathcal{P}^{(r)}(\mathbf{D}_\mathbf{z}, \mathbf{n})$, *and* $W^{(re)}(\mathbf{p}, \mathbf{p}')$ *are defined by (1.37), (3.55), and (3.59), respectively.*

Theorem 3.36 *If* $p_\alpha, \tilde{p}_{\alpha\beta} \in C^2(\Omega^\pm) \cap C^1(\overline{\Omega^\pm})$ $(\alpha, \beta = 1, 2, 3, 4)$, *then*

$$\int_{\Omega^\pm} \left\{[\boldsymbol{\mathcal{A}}^{(re)}(\mathbf{D_y})\tilde{\mathbf{p}}(\mathbf{y})]^\top \mathbf{p}(\mathbf{y}) - [\tilde{\mathbf{p}}(\mathbf{y})]^\top \boldsymbol{\mathcal{A}}^{(re)}(\mathbf{D_y})\mathbf{p}(\mathbf{y})\right\} d\mathbf{y}$$
$$= \pm \int_S \left\{[\boldsymbol{\mathcal{P}}^{(r)}(\mathbf{D_z}, \mathbf{n})\tilde{\mathbf{p}}(\mathbf{z})]^\top \mathbf{p}(\mathbf{z}) - [\tilde{\mathbf{p}}(\mathbf{z})]^\top \boldsymbol{\mathcal{P}}^{(r)}(\mathbf{D_z}, \mathbf{n})\mathbf{p}(\mathbf{z})\right\} d_{\mathbf{z}}S. \tag{3.61}$$

Theorem 3.37 *If* $\mathbf{p}$ *is regular vector of the class* $\mathfrak{B}^{(r)}$ *in* $\Omega^\pm$, *then*

$$\mathbf{p}(\mathbf{x}) =$$
$$\pm \int_S \left\{[\boldsymbol{\mathcal{P}}^{(r)}(\mathbf{D_z}, \mathbf{n})\boldsymbol{\Gamma}^{(re)}(\mathbf{x}-\mathbf{z})]^\top \mathbf{p}(\mathbf{z}) - \boldsymbol{\Gamma}^{(re)}(\mathbf{x}-\mathbf{z})\, \boldsymbol{\mathcal{P}}^{(r)}(\mathbf{D_z}, \mathbf{n})\mathbf{p}(\mathbf{z})\right\} d_{\mathbf{z}}S$$
$$+ \int_{\Omega^+} \boldsymbol{\Gamma}^{(re)}(\mathbf{x}-\mathbf{y})\, \boldsymbol{\mathcal{A}}^{(re)}(\mathbf{D_y})\mathbf{p}(\mathbf{y})d\mathbf{y} \qquad \textit{for} \quad \mathbf{x} \in \Omega^\pm, \tag{3.62}$$

where $\boldsymbol{\Gamma}^{(re)}(\mathbf{x})$ is the fundamental matrix of the operator $\boldsymbol{\mathcal{A}}^{(re)}(\mathbf{D_y})$ and defined by (2.39).

The formulas (3.60), (3.61), and (3.62) are Green's first, second, and third identities for the equilibrium equations of quadruple porosity rigid body for domain $\Omega^\pm$, respectively.

Chapter 4
Problems of Steady Vibrations of Rigid Body

We begin our investigation of the BVPs in the linear theory of elasticity for quadruple porosity materials by means of the potential method and the theory of integral equations. Namely, in this chapter, the BVPs of steady vibrations of the linear theory for quadruple porosity rigid body are investigated by means of the potential method and the theory of Fredholm integral equations.

In Sect. 4.1, the basic internal and external BVPs of steady vibrations of this theory are formulated.

In Sect. 4.2, the uniqueness theorems for classical solutions of these BVPs are proved by employing techniques based on Green's identities.

Then, in Sect. 4.3, the surface and volume potentials of the considered theory are constructed and their basic properties are established.

Furthermore, in Sect. 4.4, some useful Fredholm integral operators are introduced and Fredholm's classical theorems are formulated.

Finally, in Sect. 4.5, the BVPs of steady vibrations are reduced to the equivalent integral equations for which the Fredholm's theorems are valid. The existence theorems for classical solutions of the BVPs of steady vibrations are proved by means of the potential method.

4.1 Basic Boundary Value Problems

In this chapter we assume that the values ν_1, ν_2, ν_3, and ν_4 are complex numbers with $\mathrm{Im}\nu_\alpha > 0$, where ν_α $(\alpha = 1, 2, 3, 4)$ is given in Sect. 2.5.

The basic internal and external BVPs of steady vibration in the linear theory of quadruple porosity rigid body are formulated as follows.

Find a regular (classical) solution of the class $\mathfrak{B}^{(r)}$ in Ω^+ to system

$$\mathcal{A}^{(r)}(\mathbf{D_x}, \omega)\mathbf{p}(\mathbf{x}) = \mathbf{F}(\mathbf{x}) \tag{4.1}$$

M. Svanadze, *Potential Method in Mathematical Theories of Multi-Porosity Media*, Interdisciplinary Applied Mathematics 51,
https://doi.org/10.1007/978-3-030-28022-2_4

for $\mathbf{x} \in \Omega^+$ satisfying the boundary condition

$$\lim_{\Omega^+ \ni \mathbf{x} \to \mathbf{z} \in S} \mathbf{p}(\mathbf{x}) \equiv \{\mathbf{p}(\mathbf{z})\}^+ = \mathbf{f}(\mathbf{z}) \tag{4.2}$$

in the *Problem* $(I_r)^+_{\mathbf{F},\mathbf{f}}$, and

$$\{\mathcal{P}^{(r)}(\mathbf{D}_\mathbf{z}, \mathbf{n})\mathbf{p}(\mathbf{z})\}^+ = \mathbf{f}(\mathbf{z}) \tag{4.3}$$

in the *Problem* $(II_r)^+_{\mathbf{F},\mathbf{f}}$, where $\mathcal{A}^{(r)}(\mathbf{D}_\mathbf{x}, \omega) = \mathbf{K}\Delta + \mathbf{c}$, the matrix differential operator $\mathcal{P}^{(r)}(\mathbf{D}_\mathbf{z}, \mathbf{n})$ is defined by (3.55), $\mathbf{F}$ and $\mathbf{f}$ are prescribed four-component vector functions, $\mathbf{c} = (c_{\alpha\beta})_{4\times4} = i\omega\mathbf{b} - \mathbf{d}$. Here $\mathbf{b}$ is the cross-coupling compressibility tensor (see (1.4)) and $\mathbf{d}$ is the internal transport tensor (see (1.7)).

Find a regular (classical) solution of the class $\mathfrak{B}^{(r)}$ to system (4.1) for $\mathbf{x} \in \Omega^-$ satisfying the boundary condition

$$\lim_{\Omega^- \ni \mathbf{x} \to \mathbf{z} \in S} \mathbf{p}(\mathbf{x}) \equiv \{\mathbf{p}(\mathbf{z})\}^- = \mathbf{f}(\mathbf{z}) \tag{4.4}$$

in the *Problem* $(I_r)^-_{\mathbf{F},\mathbf{f}}$, and

$$\{\mathcal{P}^{(r)}(\mathbf{D}_\mathbf{z}, \mathbf{n})\mathbf{p}(\mathbf{z})\}^- = \mathbf{f}(\mathbf{z}) \tag{4.5}$$

in the *Problem* $(II_r)^-_{\mathbf{F},\mathbf{f}}$. Here $\mathbf{F}$ and $\mathbf{f}$ are prescribed four-component vector functions, and supp $\mathbf{F}$ is a finite domain in Ω^-. Note that the class $\mathfrak{B}^{(r)}$ of vector functions is given by Definition 3.3.

Furthermore, the BVPs $(I_r)^\pm_{\mathbf{F},\mathbf{f}}$ and $(II_r)^\pm_{\mathbf{F},\mathbf{f}}$ are Dirichlet and Neumann type BVPs in the linear theory of quadruple porosity rigid body.

Throughout this book we assume that second order tensors $k_{\alpha\beta}$ and $b_{\alpha\beta}$ are positive definite.

4.2 Uniqueness Theorems

In this section the uniqueness theorems for classical solutions of the BVPs $(I_r)^\pm_{\mathbf{F},\mathbf{f}}$ and $(II_r)^\pm_{\mathbf{F},\mathbf{f}}$ are proved by using Green's identities.

Theorem 4.1 *The internal BVP* $(\mathbb{K}_r)^+_{\mathbf{F},\mathbf{f}}$ *admits at most one regular (classical) solution of the class* $\mathfrak{B}^{(r)}$ *in* Ω^+, *where* $\mathbb{K} = I, II$.

Proof Suppose that there are two regular solutions of problem $(\mathbb{K}_r)^+_{\mathbf{F},\mathbf{f}}$. Then their difference $\mathbf{p}$ corresponds to zero data ($\mathbf{F} = \mathbf{f} = \mathbf{0}$), i.e., on the basis of (4.1)–(4.3) the vector $\mathbf{p}$ is a regular solution of the system of homogeneous equations

$$(\mathbf{K}\Delta + \mathbf{c})\mathbf{p}(\mathbf{x}) = \mathbf{0} \tag{4.6}$$

for $\mathbf{x} \in \Omega^+$ satisfying the boundary condition

$$\{\mathcal{P}^{(r)}(\mathbf{D_z}, \mathbf{n})\mathbf{p}(\mathbf{z}) \cdot \mathbf{p}(\mathbf{z})\}^+ = 0 \tag{4.7}$$

for $\mathbf{z} \in S$.

On account of (4.6) and (4.7) from Green's formulas (3.54) for $\mathbf{p}' = \mathbf{p}$ we may deduce that

$$\int_{\Omega^+} W^{(r)}(\mathbf{p}, \mathbf{p}) d\mathbf{x} = 0, \tag{4.8}$$

where

$$W^{(r)}(\mathbf{p}, \mathbf{p}) = k_{\alpha\beta} \nabla p_\alpha \cdot \nabla p_\beta - c_{\alpha\beta} p_\alpha \overline{p_\beta}. \tag{4.9}$$

Then, employing the inequalities

$$k_{\alpha\beta} \nabla p_\alpha \cdot \nabla p_\beta \geq 0, \qquad b_{\alpha\beta} p_\alpha \overline{p_\beta} \geq 0,$$

$$d_{\alpha\beta} p_\alpha \overline{p_\beta} = \gamma_{12}|p_1 - p_2|^2 + \gamma_{13}|p_1 - p_3|^2 + \gamma_{14}|p_1 - p_4|^2 \tag{4.10}$$

$$+\gamma_{23}|p_2 - p_3|^2 + \gamma_{24}|p_2 - p_4|^2 + \gamma_{34}|p_3 - p_4|^2 \geq 0$$

we may derive from (4.9)

$$\mathrm{Im} W^{(r)}(\mathbf{p}, \mathbf{p}) = -\omega b_{\alpha\beta} p_\alpha \overline{p_\beta} \leq 0. \tag{4.11}$$

Now use inequality (4.11) in (4.8) to derive

$$b_{\alpha\beta} p_\alpha \overline{p_\beta} = 0. \tag{4.12}$$

Obviously, the last equation implies $\mathbf{p}(\mathbf{x}) \equiv \mathbf{0}$ for $\mathbf{x} \in \Omega^+$. Hence, uniqueness of a solution of BVP $(\mathbb{K}_r)^+_{\mathbf{F},\mathbf{f}}$ follows, where $\mathbb{K} = I, II$. □

Theorem 4.2 *The external BVP $(\mathbb{K}_r)^-_{\mathbf{F},\mathbf{f}}$ admits at most one regular (classical) solution of the class $\mathfrak{B}^{(r)}$ in Ω^-, where $\mathbb{K} = I, II$.*

Proof Suppose that there are two regular solutions of problem $(\mathbb{K}_r)^-_{\mathbf{F},\mathbf{f}}$. Then their difference $\mathbf{p}$ corresponds to zero data ($\mathbf{F} = \mathbf{f} = \mathbf{0}$), i.e., on the basis of (4.1), (4.4), and (4.5) the vector $\mathbf{p}$ is a regular solution of homogeneous system of Eq. (4.6) for $\mathbf{x} \in \Omega^-$ satisfying the homogeneous boundary condition

$$\{\mathcal{P}^{(r)}(\mathbf{D_z}, \mathbf{n})\mathbf{p}(\mathbf{z}) \cdot \mathbf{p}(\mathbf{z})\}^- = 0 \tag{4.13}$$

for $\mathbf{z} \in S$.

On account of (4.6) and (4.13) from Green's formula (3.58) for $\mathbf{p}' = \mathbf{p}$ we may deduce that

$$\int\limits_{\Omega^-} W^{(r)}(\mathbf{p}, \mathbf{p}) d\mathbf{x} = 0, \tag{4.14}$$

where $W^{(r)}(\mathbf{p}, \mathbf{p})$ is defined by (4.9). Then, employing the inequalities (4.10) and (4.11) from (4.14) we may derive Eq. (4.12). Hence $\mathbf{p}(\mathbf{x}) \equiv \mathbf{0}$ for $\mathbf{x} \in \Omega^-$ and the uniqueness of a solution of BVP $(\mathbb{K}_r)^-_{\mathbf{F},\mathbf{f}}$ follows, where $\mathbb{K} = I, II$. □

4.3 Basic Properties of Potentials

On the basis of integral representation of regular vector (see (3.57)) we introduce the following notation

$$\mathbf{Z}^{(r,1)}(\mathbf{x}, \mathbf{g}) = \int\limits_S \mathbf{\Gamma}^{(r)}(\mathbf{x} - \mathbf{y}, \omega)\, \mathbf{g}(\mathbf{y}) d_{\mathbf{y}} S,$$

$$\mathbf{Z}^{(r,2)}(\mathbf{x}, \mathbf{g}) = \int\limits_S [\mathcal{P}^{(r)}(\mathbf{D}_{\mathbf{y}}, \mathbf{n}) \mathbf{\Gamma}^{(r)}(\mathbf{x} - \mathbf{y}, \omega)]^\top \mathbf{g}(\mathbf{y}) d_{\mathbf{y}} S,$$

$$\mathbf{Z}^{(r,3)}(\mathbf{x}, \mathbf{h}, \Omega^\pm) = \int\limits_{\Omega^\pm} \mathbf{\Gamma}^{(r)}(\mathbf{x} - \mathbf{y}, \omega)\, \mathbf{h}(\mathbf{y}) d\mathbf{y},$$

where $\mathbf{g}$ and $\mathbf{h}$ are four-component vectors.

As in mathematical physics (see, e.g., Günther [161], Hsiao and Wendland [169], Kellogg [195]), the vector functions $\mathbf{Z}^{(r,1)}(\mathbf{x}, \mathbf{g})$, $\mathbf{Z}^{(r,2)}(\mathbf{x}, \mathbf{g})$, and $\mathbf{Z}^{(r,3)}(\mathbf{x}, \mathbf{h}, \Omega^\pm)$ are called *the single-layer, double-layer, and volume potentials* in the steady vibrations problems of the linear theory of quadruple porosity rigid body, respectively.

Obviously, on the basis of Theorems 3.31 and 3.34, the regular vector in Ω^+ and Ω^- is represented by sum of the surface (single-layer and double-layer) and volume potentials as follows

$$\mathbf{p}(\mathbf{x}) = \mathbf{Z}^{(r,2)}\left(\mathbf{x}, \{\mathbf{p}\}^+\right) - \mathbf{Z}^{(r,1)}\left(\mathbf{x}, \{\mathcal{P}^{(r)}\mathbf{p}\}^+\right) + \mathbf{Z}^{(r,3)}\left(\mathbf{x}, \mathcal{A}^{(r)}\mathbf{p}, \Omega^+\right)$$

for $\mathbf{x} \in \Omega^+$ and

$$\mathbf{p}(\mathbf{x}) = -\mathbf{Z}^{(r,2)}\left(\mathbf{x}, \{\mathbf{p}\}^-\right) + \mathbf{Z}^{(r,1)}\left(\mathbf{x}, \{\mathcal{P}^{(r)}\mathbf{p}\}^-\right) + \mathbf{Z}^{(r,3)}\left(\mathbf{x}, \mathcal{A}^{(r)}\mathbf{p}, \Omega^-\right)$$

for $\mathbf{x} \in \Omega^-$, respectively.

We now establish the basic properties of these potentials.

4.3.1 Single-Layer Potential

We have the following results.

Theorem 4.3 *If S is the Liapunov surface of the class $C^{1,\nu}$, $0 < \nu \leq 1$, $\mathbf{g}$ is an integrable vector function on S, then*

(i)

$$\mathbf{Z}^{(r,1)}(\cdot, \mathbf{g}) \in C^{\infty}(\Omega^{\pm}),$$

(ii) $\mathbf{Z}^{(r,1)}(\cdot, \mathbf{g})$ *is a solution of the homogeneous equation (4.6) in Ω^+ and Ω^-, i.e.*

$$\mathcal{A}^{(r)}(\mathbf{D_x}, \omega)\, \mathbf{Z}^{(r,1)}(\mathbf{x}, \mathbf{g}) = \mathbf{0} \tag{4.15}$$

for $\mathbf{x} \in \Omega^{\pm}$.

Theorem 4.4 *If $S \in C^{1,\nu}$, $0 < \nu \leq 1$, $\mathbf{g}$ is an integrable and bounded vector function on S, then*

$$Z_{\alpha}^{(r,1)}(\mathbf{x}, \mathbf{g}) = O\left(\frac{e^{-\nu_0|\mathbf{x}|}}{|\mathbf{x}|}\right), \qquad Z_{\alpha,j}^{(r,1)}(\mathbf{x}, \mathbf{g}) = O\left(\frac{e^{-\nu_0|\mathbf{x}|}}{|\mathbf{x}|^2}\right) \tag{4.16}$$

for $|\mathbf{x}| \to +\infty$, *where* $\mathbf{Z}^{(r,1)} = \left(Z_1^{(r,1)}, Z_2^{(r,1)}, Z_3^{(r,1)}, Z_4^{(r,1)}\right)$, $j = 1, 2, 3$, $\alpha = 1, 2, 3, 4$ *and* $\nu_0 = \min\{\mathrm{Im}\nu_1, \mathrm{Im}\nu_2, \mathrm{Im}\nu_3, \mathrm{Im}\nu_4\} > 0$.

Theorem 4.5 *If $S \in C^{1,\nu}$, $0 < \nu \leq 1$, $\mathbf{g}$ is an integrable and bounded vector function on S, then*

(i) the integrals $\mathbf{Z}^{(r,1)}(\mathbf{z}, \mathbf{g})$, $\left\{\mathbf{Z}^{(r,1)}(\mathbf{z}, \mathbf{g})\right\}^+$, *and* $\left\{\mathbf{Z}^{(r,1)}(\mathbf{z}, \mathbf{g})\right\}^-$ *exist for* $\mathbf{z} \in S$ *and*

$$\mathbf{Z}^{(r,1)}(\mathbf{z}, \mathbf{g}) = \left\{\mathbf{Z}^{(r,1)}(\mathbf{z}, \mathbf{g})\right\}^+ = \left\{\mathbf{Z}^{(r,1)}(\mathbf{z}, \mathbf{g})\right\}^-, \tag{4.17}$$

(ii)

$$\mathbf{Z}^{(r,1)}(\cdot, \mathbf{g}) \in C(\mathbb{R}^3),$$

(iii)

$$\mathcal{P}^{(r)}(\mathbf{D_z}, \mathbf{n})\, \mathbf{Z}^{(r,1)}(\cdot, \mathbf{g}) \in C^{0,\nu'}(S),$$

where $0 < \nu' < \nu \le 1$ *and*

$$\left\{\mathbf{Z}^{(r,1)}(\mathbf{z},\mathbf{g})\right\}^{\pm} \equiv \lim_{\Omega^{\pm}\ni\mathbf{x}\to\mathbf{z}\in S} \mathbf{Z}^{(r,1)}(\mathbf{x},\mathbf{g}).$$

Theorem 4.6 *If* $S \in C^{1,\nu}$, $0 < \nu \le 1$, $\mathbf{g}$ *is a continuous vector function on* S, *then the vector functions* $\mathcal{P}^{(r)}(\mathbf{D_z},\mathbf{n})\,\mathbf{Z}^{(r,1)}(\mathbf{z},\mathbf{g})$, $\left\{\mathcal{P}^{(r)}(\mathbf{D_z},\mathbf{n})\,\mathbf{Z}^{(r,1)}(\mathbf{z},\mathbf{g})\right\}^{+}$, *and* $\left\{\mathcal{P}^{(r)}(\mathbf{D_z},\mathbf{n})\,\mathbf{Z}^{(r,1)}(\mathbf{z},\mathbf{g})\right\}^{-}$ *exist for* $\mathbf{z} \in S$ *and*

$$\left\{\mathcal{P}^{(r)}(\mathbf{D_z},\mathbf{n})\,\mathbf{Z}^{(r,1)}(\mathbf{z},\mathbf{g})\right\}^{\pm} = \mp\frac{1}{2}\mathbf{g}(\mathbf{z}) + \mathcal{P}^{(r)}(\mathbf{D_z},\mathbf{n})\,\mathbf{Z}^{(r,1)}(\mathbf{z},\mathbf{g}), \tag{4.18}$$

where

$$\left\{\mathcal{P}^{(r)}(\mathbf{D_z},\mathbf{n})\,\mathbf{Z}^{(r,1)}(\mathbf{z},\mathbf{g})\right\}^{\pm} \equiv \lim_{\Omega^{\pm}\ni\mathbf{x}\to\mathbf{z}\in S} \mathcal{P}^{(r)}(\mathbf{D_x},\mathbf{n})\,\mathbf{Z}^{(r,1)}(\mathbf{x},\mathbf{g}).$$

We now prove the following useful property of the single-layer potential.

Theorem 4.7 *The single-layer potential* $\mathbf{Z}^{(r,1)}(\mathbf{x},\mathbf{g}) \equiv \mathbf{0}$ *for* $\mathbf{x} \in \Omega^{+}$ *(or* $\mathbf{x} \in \Omega^{-}$*) if and only if* $\mathbf{g}(\mathbf{z}) \equiv \mathbf{0}$, *where* $\mathbf{z} \in S$.

Proof Obviously, if $\mathbf{g}(\mathbf{z}) \equiv \mathbf{0}$ for $\mathbf{z} \in S$, then $\mathbf{Z}^{(r,1)}(\mathbf{x},\mathbf{g}) \equiv \mathbf{0}$ for $\mathbf{x} \in \Omega^{\pm}$. Let $\mathbf{Z}^{(r,1)}(\mathbf{x},\mathbf{g}) \equiv \mathbf{0}$ for $\mathbf{x} \in \Omega^{+}$. Then we have

$$\left\{\mathbf{Z}^{(r,1)}(\mathbf{z},\mathbf{g})\right\}^{+} = \left\{\mathcal{P}^{(r)}(\mathbf{D_z},\mathbf{n})\,\mathbf{Z}^{(r,1)}(\mathbf{z},\mathbf{g})\right\}^{+} = \mathbf{0}. \tag{4.19}$$

On the basis of (4.19) from (4.17) it follows that

$$\left\{\mathbf{Z}^{(r,1)}(\mathbf{z},\mathbf{g})\right\}^{-} = \mathbf{0} \tag{4.20}$$

for $\mathbf{z} \in S$. Furthermore, the vector function $\mathbf{Z}^{(r,1)}(\mathbf{x},\mathbf{g})$ is a solution of the homogeneous equation (4.15) for $\mathbf{x} \in \Omega^{-}$ and satisfies the homogeneous boundary condition (4.20), i.e., $\mathbf{Z}^{(r,1)}(\mathbf{x},\mathbf{g})$ is a regular solution of the external BVP $(I_r)^{-}_{\mathbf{0},\mathbf{0}}$. Keeping in mind Theorem 4.2 we may see that $\mathbf{Z}^{(r,1)}(\mathbf{x},\mathbf{g}) \equiv \mathbf{0}$ for $\mathbf{x} \in \Omega^{-}$. Then we have

$$\left\{\mathcal{P}^{(r)}(\mathbf{D_z},\mathbf{n})\,\mathbf{Z}^{(r,1)}(\mathbf{z},\mathbf{g})\right\}^{-} = \mathbf{0}. \tag{4.21}$$

Combining (4.18) with (4.19) and (4.21) we may further conclude that $\mathbf{g}(\mathbf{z}) \equiv \mathbf{0}$ for $\mathbf{z} \in S$.

Quite similarly, if $\mathbf{Z}^{(r,1)}(\mathbf{x},\mathbf{g}) \equiv \mathbf{0}$ for $\mathbf{x} \in \Omega^{-}$, then the vector function $\mathbf{Z}^{(r,1)}(\mathbf{x},\mathbf{g})$ is a regular solution of the internal BVP $(I_r)^{+}_{\mathbf{0},\mathbf{0}}$. Keeping in mind

Theorem 4.1 we may see that $\mathbf{Z}^{(r,1)}(\mathbf{x}, \mathbf{g}) \equiv \mathbf{0}$ for $\mathbf{x} \in \Omega^+$. These equations allow us to show $\mathbf{g}(\mathbf{z}) \equiv \mathbf{0}$ for $\mathbf{z} \in S$. □

4.3.2 Double-Layer Potential

On the basis of (2.58) we have the following consequences.

Theorem 4.8 *If $S \in C^{1,\nu}$, $0 < \nu \le 1$, $\mathbf{g}$ is an integrable vector function on S, then*

(i)

$$\mathbf{Z}^{(r,2)}(\cdot, \mathbf{g}) \in C^\infty(\Omega^\pm),$$

(ii) $\mathbf{Z}^{(r,2)}(\cdot, \mathbf{g})$ is a solution of the homogeneous equation (4.6) in Ω^+ and Ω^-, i.e.

$$\mathcal{A}^{(r)}(\mathbf{D_x}, \omega)\, \mathbf{Z}^{(r,2)}(\mathbf{x}, \mathbf{g}) = \mathbf{0} \tag{4.22}$$

for $\mathbf{x} \in \Omega^\pm$.

Theorem 4.9 *If $S \in C^{1,\nu}$, $0 < \nu \le 1$, $\mathbf{g}$ is an integrable and bounded vector function on S, $\mathbf{Z}^{(r,2)} = \left(Z_1^{(r,2)}, Z_2^{(r,2)}, Z_3^{(r,2)}, Z_4^{(r,2)}\right)$, then*

$$Z_\alpha^{(r,2)}(\mathbf{x}, \mathbf{g}) = O\left(\frac{e^{-\nu_0|\mathbf{x}|}}{|\mathbf{x}|^2}\right), \qquad Z_{\alpha,j}^{(r,2)}(\mathbf{x}, \mathbf{g}) = O\left(\frac{e^{-\nu_0|\mathbf{x}|}}{|\mathbf{x}|^3}\right) \tag{4.23}$$

for $|\mathbf{x}| \to +\infty$, where $j = 1, 2, 3$ and $\alpha = 1, 2, 3, 4$.

Theorem 4.10 *If $S \in C^{1,\nu}$, $0 < \nu \le 1$, then the matrices $\mathcal{P}^{(r)}(\mathbf{D_z}, \mathbf{n})\mathbf{\Gamma}^{(r)}(\mathbf{z} - \mathbf{y}, \omega)$ and $[\mathcal{P}^{(r)}(\mathbf{D_y}, \mathbf{n})\mathbf{\Gamma}^{(r)}(\mathbf{z} - \mathbf{y}, \omega)]^\top$ are weakly singular kernels of $\mathcal{P}^{(r)}(\mathbf{D_z}, \mathbf{n})\mathbf{Z}^{(r,1)}(\mathbf{z}, \mathbf{g})$ and $\mathbf{Z}^{(r,2)}(\mathbf{z}, \mathbf{g})$, respectively, i.e.*

$$\begin{aligned} \left|[\mathcal{P}^{(r)}(\mathbf{D_z}, \mathbf{n})\mathbf{\Gamma}^{(r)}(\mathbf{z} - \mathbf{y}, \omega)]_{\alpha\beta}\right| &< \frac{\text{const}}{|\mathbf{z} - \mathbf{y}|^{2-\nu}}, \\ \left|[\mathcal{P}^{(r)}(\mathbf{D_y}, \mathbf{n})\mathbf{\Gamma}^{(r)}(\mathbf{z} - \mathbf{y}, \omega)]_{\alpha\beta}\right| &< \frac{\text{const}}{|\mathbf{z} - \mathbf{y}|^{2-\nu}} \end{aligned} \tag{4.24}$$

for $\mathbf{z}, \mathbf{y} \in S$, $\alpha, \beta = 1, 2, 3, 4$, and const *is a positive number independent on $\mathbf{z}$ and $\mathbf{y}$.*

Theorem 4.11 *If $S \in C^{1,\nu}$, $0 < \nu \le 1$, $\mathbf{g}$ is a continuous vector function on S, then*

(i) the integrals $\mathbf{Z}^{(r,2)}(\mathbf{z}, \mathbf{g})$, $\left\{\mathbf{Z}^{(r,2)}(\mathbf{z}, \mathbf{g})\right\}^+$ and $\left\{\mathbf{Z}^{(r,2)}(\mathbf{z}, \mathbf{g})\right\}^-$ exist,

(ii)

$$\left\{\mathbf{Z}^{(r,2)}(\mathbf{z},\mathbf{g})\right\}^{\pm} = \pm\frac{1}{2}\mathbf{g}(\mathbf{z}) + \mathbf{Z}^{(r,2)}(\mathbf{z},\mathbf{g}) \tag{4.25}$$

for $\mathbf{z} \in S$.

Theorem 4.12 *If* $S \in C^{1,\nu}$, $\mathbf{g} \in C^{0,\nu'}(S)$, $0 < \nu' < \nu \le 1$, *then*

$$\mathbf{Z}^{(r,2)}(\cdot,\mathbf{g}) \in C^{0,\nu'}(S).$$

We now formulate the Liapunov–Tauber type theorem in the linear theory of quadruple porosity rigid body.

Theorem 4.13 *If* $S \in C^{1,\nu}$, $\mathbf{g} \in C^{1,\nu}(S)$, $0 < \nu \le 1$, *then*

(i) the vector functions $\left\{\mathcal{P}^{(r)}(\mathbf{D}_\mathbf{z},\mathbf{n})\mathbf{Z}^{(r,2)}(\mathbf{z},\mathbf{g})\right\}^{+}$ *and* $\left\{\mathcal{P}^{(r)}(\mathbf{D}_\mathbf{z},\mathbf{n})\mathbf{Z}^{(r,2)}(\mathbf{z},\mathbf{g})\right\}^{-}$ *exist,*

(ii) they are elements of the class $C^{1,\nu}(S)$, *and*

(iii)

$$\left\{\mathcal{P}^{(r)}(\mathbf{D}_\mathbf{z},\mathbf{n})\mathbf{Z}^{(r,2)}(\mathbf{z},\mathbf{g})\right\}^{+} = \left\{\mathcal{P}^{(r)}(\mathbf{D}_\mathbf{z},\mathbf{n})\mathbf{Z}^{(r,2)}(\mathbf{z},\mathbf{g})\right\}^{-}$$

for $\mathbf{z} \in S$.

4.3.3 Volume Potential

We have the following results.

Theorem 4.14 *If* $S \in C^{1,\nu}$, $\mathbf{h}$ *is an integrable and bounded vector function in* Ω^+, *then*

(i)

$$\mathbf{Z}^{(r,3)}(\cdot,\mathbf{h},\Omega^+) \in C^{\infty}(\Omega^-),$$

(ii)

$$\mathbf{Z}^{(r,3)}(\cdot,\mathbf{h},\Omega^+) \in C^{1,\nu}(\mathbb{R}^3),$$

(iii) $\mathbf{Z}^{(r,3)}(\cdot,\mathbf{h},\Omega^+)$ *is a solution of the homogeneous equation*

$$\mathcal{A}^{(r)}(\mathbf{D}_\mathbf{x},\omega)\,\mathbf{Z}^{(r,3)}(\mathbf{x},\mathbf{h},\Omega^+) = \mathbf{0}$$

for $\mathbf{x} \in \Omega^-$.

Theorem 4.15 *If* $S \in C^{1,\nu}$, $0 < \nu \leq 1$, $\mathbf{h} \in C^1(\Omega^+)$, *then*

$$\mathbf{Z}^{(r,3)}(\cdot, \mathbf{h}, \Omega^+) \in C^2(\Omega^+).$$

Theorem 4.16 *If* $S \in C^{1,\nu}$, $\mathbf{h}$ *is an integrable and bounded vector function in* Ω^- *and* supp $\mathbf{h}$ *is a finite domain in* Ω^-, *then*

(i)

$$\mathbf{Z}^{(r,3)}(\cdot, \mathbf{h}, \Omega^-) \in C^\infty(\Omega^+),$$

(ii)

$$\mathbf{Z}^{(r,3)}(\cdot, \mathbf{h}, \Omega^-) \in C^{1,\nu}(\mathbb{R}^3),$$

(iii) $\mathbf{Z}^{(r,3)}(\cdot, \mathbf{h}, \Omega^-)$ *is a solution of the homogeneous equation (4.27), i.e.*

$$\mathcal{A}^{(r)}(\mathbf{D}_\mathbf{x}, \omega)\, \mathbf{Z}^{(r,3)}(\mathbf{x}, \mathbf{h}, \Omega^-) = \mathbf{0}$$

for $\mathbf{x} \in \Omega^+$.

Theorem 4.17 *If* $S \in C^{1,\nu}$, $0 < \nu \leq 1$, $\mathbf{h} \in C^1(\Omega^-)$, *and* supp $\mathbf{h}$ *is a finite domain in* Ω^-, *then*

$$\mathbf{Z}^{(r,3)}(\cdot, \mathbf{h}, \Omega^-) \in C^2(\Omega^-).$$

We now formulate Poisson-type theorems in the linear theory of quadruple porosity rigid body for domains Ω^+ and Ω^-.

Theorem 4.18 *If* $S \in C^{1,\nu}$, $0 < \nu \leq 1$, $\mathbf{h} \in C^{0,\nu}(\Omega^+)$, *then*

$$\mathcal{A}^{(r)}(\mathbf{D}_\mathbf{x}, \omega)\, \mathbf{Z}^{(r,3)}(\mathbf{x}, \mathbf{h}, \Omega^+) = \mathbf{h}(\mathbf{x})$$

for $\mathbf{x} \in \Omega^+$.

Theorem 4.19 *If* $S \in C^{1,\nu}$, $0 < \nu \leq 1$, $\mathbf{h} \in C^{0,\nu}(\Omega^+)$, *and* supp $\mathbf{h}$ *is a finite domain in* Ω^-, *then*

$$\mathcal{A}^{(r)}(\mathbf{D}_\mathbf{x}, \omega)\, \mathbf{Z}^{(r,3)}(\mathbf{x}, \mathbf{h}, \Omega^-) = \mathbf{h}(\mathbf{x})$$

for $\mathbf{x} \in \Omega^-$.

4.4 Fredholm Operators

In the theory of integral equations, a Fredholm operator is an operator that arises in the Fredholm theory of integral equations. We introduce the integral operators

$$\begin{aligned}
\mathcal{K}^{(r,1)}\mathbf{g}(\mathbf{z}) &\equiv \frac{1}{2}\,\mathbf{g}(\mathbf{z}) + \mathbf{Z}^{(r,2)}(\mathbf{z},\mathbf{g}),\\
\mathcal{K}^{(r,2)}\mathbf{h}(\mathbf{z}) &\equiv \frac{1}{2}\,\mathbf{h}(\mathbf{z}) + \mathcal{P}^{(r)}(\mathbf{D}_{\mathbf{z}},\mathbf{n})\mathbf{Z}^{(r,1)}(\mathbf{z},\mathbf{h}),\\
\mathcal{K}^{(r,3)}\mathbf{g}(\mathbf{z}) &\equiv -\frac{1}{2}\,\mathbf{g}(\mathbf{z}) + \mathbf{Z}^{(r,2)}(\mathbf{z},\mathbf{g}),\\
\mathcal{K}^{(r,4)}\mathbf{h}(\mathbf{z}) &\equiv -\frac{1}{2}\,\mathbf{h}(\mathbf{z}) + \mathcal{P}^{(r)}(\mathbf{D}_{\mathbf{z}},\mathbf{n})\mathbf{Z}^{(r,1)}(\mathbf{z},\mathbf{h}),
\end{aligned} \tag{4.26}$$

where $\mathbf{g}$ and $\mathbf{h}$ are four-component vector functions on S and $\mathbf{z} \in S$. On the basis of (4.24), the operator $\mathcal{K}^{(r,\alpha)}$ $(\alpha = 1,2,3,4)$ has a weakly singular kernel and consequently, $\mathcal{K}^{(r,\alpha)}$ is a Fredholm operator. Moreover, operators $\mathcal{K}^{(r,j)}$ and $\mathcal{K}^{(r,j+1)}$ are adjoint with respect to each other, where $j = 1, 3$.

We can formulate Fredholm's classical theorems for the integral operators $\mathcal{K}^{(r,\alpha)}$ $(\alpha = 1,2,3,4)$ as follows (see, e.g., Courant and Hilbert [105] and Vladimirov [390]):

(i) **Fredholm's first theorem (alternative)**: *Either the homogeneous integral equation*

$$\mathcal{K}^{(r,j)}\mathbf{g}(\mathbf{z}) = \mathbf{0} \tag{4.27}$$

has a nontrivial solution, or the nonhomogeneous integral equation

$$\mathcal{K}^{(r,j)}\mathbf{g}(\mathbf{z}) = \mathbf{f}(\mathbf{z}) \tag{4.28}$$

can be solved uniquely for an arbitrary vector function $\mathbf{f}$*, where* $j = 1, 2, 3, 4$.

(ii) **Fredholm's second theorem**: *The homogeneous integral equation (4.27) and*

$$\mathcal{K}^{(r,j+1)}\mathbf{h}(\mathbf{z}) = \mathbf{0} \tag{4.29}$$

have the same finite number of linearly independent solutions for $j = 1, 3$.

(iii) **Fredholm's third theorem**: *The nonhomogeneous integral equation (4.28) is solvable if and only if the right-hand side* $\mathbf{f}$ *satisfies the condition*

$$\int_S \mathbf{f}(\mathbf{z}) \cdot \mathbf{h}(\mathbf{z})\, d_{\mathbf{z}}S = 0 \tag{4.30}$$

for all solutions $\mathbf{h}(\mathbf{z})$ *to the homogeneous adjoint equation (4.29), where* $j = 1, 3$.

Clearly, Eqs. (4.28) and (4.29) are the Fredholm integral equations of the second kind.

4.5 Existence Theorems

We are now in a position to study the existence of regular (classical) solutions of the BVPs $(\mathbb{K}_r)^+_{\mathbf{F},\mathbf{f}}$ and $(\mathbb{K}_r)^-_{\mathbf{F},\mathbf{f}}$ by means of the potential method and the theory of Fredholm integral equations, where $\mathbb{K} = I, II$.

Obviously, by Theorems 4.18 and 4.19 the volume potential $\mathbf{Z}^{(r,3)}(\mathbf{x}, \mathbf{F}, \Omega^{\pm})$ is a particular regular solution of the nonhomogeneous equation (4.1), where $\mathbf{F} \in C^{0,\nu'}(\Omega^{\pm})$, $0 < \nu' \leq 1$; $\operatorname{supp}\mathbf{F}$ is a finite domain in Ω^-. Therefore, in this section we will consider problems $(\mathbb{K}_r)^+_{\mathbf{0},\mathbf{f}}$ and $(\mathbb{K}_r)^-_{\mathbf{0},\mathbf{f}}$, where $\mathbb{K} = I, II$.

Problem $(I_r)^+_{\mathbf{0},\mathbf{f}}$ We seek a regular solution of the class $\mathfrak{B}^{(r)}$ in Ω^+ to this problem in the form

$$\mathbf{p}(\mathbf{x}) = \mathbf{Z}^{(r,2)}(\mathbf{x}, \mathbf{g}) \qquad \text{for} \qquad \mathbf{x} \in \Omega^+, \tag{4.31}$$

where $\mathbf{g}$ is the required four-component vector function.

By Eq. (4.22) the vector function $\mathbf{p}$ is a solution of (4.6) for $\mathbf{x} \in \Omega^+$. Keeping in mind the boundary condition (4.2) and using (4.25) and (4.26), from (4.31) we obtain, for determining the unknown vector $\mathbf{g}$, a Fredholm integral equation

$$\mathcal{K}^{(r,1)}\mathbf{g}(\mathbf{z}) = \mathbf{f}(\mathbf{z}) \qquad \text{for} \quad \mathbf{z} \in S. \tag{4.32}$$

We prove that (4.32) is always solvable for an arbitrary vector $\mathbf{f}$. Let us consider the adjoint homogeneous equation

$$\mathcal{K}^{(r,2)}\mathbf{h}_0(\mathbf{z}) = \mathbf{0} \qquad \text{for} \quad \mathbf{z} \in S, \tag{4.33}$$

where $\mathbf{h}_0$ is the required four-component vector function. Now we prove that (4.33) has only the trivial solution.

Indeed, let $\mathbf{h}_0$ be a solution of the homogeneous equation (4.33). On the basis of (4.15), (4.16), (4.18), and (4.33) the vector function $\mathbf{Z}^{(r,1)}(\mathbf{x}, \mathbf{h}_0)$ is a regular solution of problem $(II_r)^-_{\mathbf{0},\mathbf{0}}$. Using Theorem 4.2, the problem $(II_r)^-_{\mathbf{0},\mathbf{0}}$ has only the trivial solution, that is

$$\mathbf{Z}^{(r,1)}(\mathbf{x}, \mathbf{h}_0) \equiv \mathbf{0} \qquad \text{for} \qquad \mathbf{x} \in \Omega^-. \tag{4.34}$$

On the other hand, by (4.17) and (4.34) we get

$$\{\mathbf{Z}^{(r,1)}(\mathbf{z},\mathbf{h}_0)\}^+ = \{\mathbf{Z}^{(r,1)}(\mathbf{z},\mathbf{h}_0)\}^- = \mathbf{0} \qquad \text{for} \qquad \mathbf{z} \in S,$$

i.e., the vector $\mathbf{Z}^{(r,1)}(\mathbf{x},\mathbf{h}_0)$ is a regular solution of problem $(I_r)^+_{\mathbf{0},\mathbf{0}}$. Using Theorem 4.1, the problem $(I_r)^+_{\mathbf{0},\mathbf{0}}$ has only the trivial solution, that is

$$\mathbf{Z}^{(r,1)}(\mathbf{x},\mathbf{h}_0) \equiv \mathbf{0} \qquad \text{for} \qquad \mathbf{x} \in \Omega^+. \tag{4.35}$$

By virtue of (4.34), (4.35), and identity (4.18) we obtain

$$\mathbf{h}_0(\mathbf{z}) = \{\mathcal{P}^{(r)}(\mathbf{D}_\mathbf{z},\mathbf{n})\mathbf{Z}^{(r,1)}(\mathbf{z},\mathbf{h}_0)\}^- - \{\mathcal{P}^{(r)}(\mathbf{D}_\mathbf{z},\mathbf{n})\mathbf{Z}^{(r,1)}(\mathbf{z},\mathbf{h}_0)\}^+ = \mathbf{0}$$

for $\mathbf{z} \in S$. Thus, the homogeneous equation (4.33) has only the trivial solution and therefore (4.32) is always solvable for an arbitrary vector $\mathbf{f}$.

We have thereby proved the following.

Theorem 4.20 *If $S \in C^{1,\nu}$, $\mathbf{f} \in C^{1,\nu'}(S)$, $0 < \nu' < \nu \le 1$, then a regular (classical) solution of the class $\mathfrak{B}^{(r)}$ of problem $(I_r)^+_{\mathbf{0},\mathbf{f}}$ exists, is unique, and is represented by double-layer potential (4.31), where $\mathbf{g}$ is a solution of the Fredholm integral equation (4.32) which is always solvable for an arbitrary vector $\mathbf{f}$.*

Problem $(II_r)^-_{\mathbf{0},\mathbf{f}}$ We seek a regular solution of the class $\mathfrak{B}^{(r)}$ in Ω^- to this problem in the form

$$\mathbf{p}(\mathbf{x}) = \mathbf{Z}^{(r,1)}(\mathbf{x},\mathbf{h}) \qquad \text{for} \qquad \mathbf{x} \in \Omega^-, \tag{4.36}$$

where $\mathbf{h}$ is the required four-component vector function.

Obviously, by Theorem 4.3 the vector function $\mathbf{p}$ is a solution of (4.6) for $\mathbf{x} \in \Omega^-$. Keeping in mind the boundary condition (4.5) and using (4.18), from (4.36) we obtain, for determining the unknown vector $\mathbf{h}$, a Fredholm integral equation

$$\mathcal{K}^{(r,2)}\mathbf{h}(\mathbf{z}) = \mathbf{f}(\mathbf{z}) \qquad \text{for} \quad \mathbf{z} \in S. \tag{4.37}$$

It has been proved above that the corresponding homogeneous equation (4.34) has only the trivial solution. Hence, it follows that (4.37) is always solvable.

We have thereby proved the following.

Theorem 4.21 *If $S \in C^{1,\nu}$, $\mathbf{f} \in C^{0,\nu'}(S)$, $0 < \nu' < \nu \le 1$, then a regular (classical) solution of the class $\mathfrak{B}^{(r)}$ of problem $(II_r)^-_{\mathbf{0},\mathbf{f}}$ exists, is unique, and is represented by single-layer potential (4.36), where $\mathbf{h}$ is a solution of the Fredholm integral equation (4.37) which is always solvable for an arbitrary vector $\mathbf{f}$.*

Problem $(II_r)^+_{\mathbf{0},\mathbf{f}}$ We seek a regular solution of the class $\mathfrak{B}^{(r)}$ in Ω^+ to this problem in the form

$$\mathbf{p}(\mathbf{x}) = \mathbf{Z}^{(r,1)}(\mathbf{x},\mathbf{g}) \qquad \text{for} \qquad \mathbf{x} \in \Omega^+, \tag{4.38}$$

where $\mathbf{g}$ is the required four-component vector function.

Obviously, by Theorem 4.3 the vector function $\mathbf{p}$ is a solution of (4.6) for $\mathbf{x} \in \Omega^+$. Keeping in mind the boundary condition (4.3) and using (4.18), from (4.38) we obtain, for determining the unknown vector $\mathbf{g}$, the following integral equation

$$\mathcal{K}^{(r,4)}\mathbf{g}(\mathbf{z}) = \mathbf{f}(\mathbf{z}) \qquad \text{for} \qquad \mathbf{z} \in S. \tag{4.39}$$

We prove that (4.39) is always solvable for an arbitrary vector $\mathbf{f}$. Let us consider the corresponding homogeneous equation

$$\mathcal{K}^{(r,4)}\mathbf{g}_0(\mathbf{z}) = \mathbf{0} \qquad \text{for} \quad \mathbf{z} \in S, \tag{4.40}$$

where $\mathbf{g}_0$ is the required four-component vector function. Now we prove that (4.40) has only the trivial solution.

Indeed, let $\mathbf{g}_0$ be a solution of the homogeneous equation (4.40). On the basis of Theorem 3.3 and (3.35) the vector $\mathbf{Z}^{(r,1)}(\mathbf{x}, \mathbf{g}_0)$ is a regular solution of problem $(II_r)^+_{\mathbf{0},\mathbf{0}}$. Using Theorem 4.1, the problem $(II_r)^+_{\mathbf{0},\mathbf{0}}$ has only the trivial solution, that is

$$\mathbf{Z}^{(r,1)}(\mathbf{x}, \mathbf{g}_0) = \mathbf{0} \qquad \text{for} \qquad \mathbf{x} \in \Omega^+. \tag{4.41}$$

On the other hand, by (4.17) and (4.41) we get $\{\mathbf{Z}^{(r,1)}(\mathbf{z}, \mathbf{g}_0)\}^- = \mathbf{0}$ for $\mathbf{z} \in S$, i.e., the vector $\mathbf{Z}^{(r,1)}(\mathbf{x}, \mathbf{g}_0)$ is a regular solution of problem $(I_r)^-_{\mathbf{0},\mathbf{0}}$. On the basis of Theorem 4.2, the problem $(I_r)^-_{\mathbf{0},\mathbf{0}}$ has only the trivial solution, that is

$$\mathbf{Z}^{(r,1)}(\mathbf{x}, \mathbf{g}_0) = \mathbf{0} \qquad \text{for} \qquad \mathbf{x} \in \Omega^-. \tag{4.42}$$

By virtue of (4.41), (4.42), and identity (4.18) we obtain

$$\mathbf{g}_0(\mathbf{z}) = \{\mathcal{P}^{(r)}(\mathbf{D}_\mathbf{z}, \mathbf{n})\mathbf{Z}^{(r,1)}(\mathbf{z}, \mathbf{g}_0)\}^- - \{\mathcal{P}^{(r)}(\mathbf{D}_\mathbf{z}, \mathbf{n})\mathbf{Z}^{(r,1)}(\mathbf{z}, \mathbf{g}_0)\}^+ = \mathbf{0}.$$

Thus, the homogeneous equation (4.40) has only a trivial solution and therefore (4.39) is always solvable for an arbitrary vector $\mathbf{f}$.

We have thereby proved

Theorem 4.22 *If $S \in C^{1,\nu}$, $\mathbf{f} \in C^{0,\nu'}(S)$, $0 < \nu' < \nu \leq 1$, then a regular (classical) solution of the class $\mathfrak{B}^{(r)}$ of problem $(II_r)^+_{\mathbf{0},\mathbf{f}}$ exists, is unique, and is represented by single-layer potential (4.38), where $\mathbf{g}$ is a solution of the Fredholm integral equation (4.39) which is always solvable for an arbitrary vector* $\mathbf{f}$.

Problem $(I_r)^-_{\mathbf{0},\mathbf{f}}$ We seek a regular (classical) solution of the class $\mathfrak{B}^{(r)}$ in Ω^- to this problem in the form

$$\mathbf{p}(\mathbf{x}) = \mathbf{Z}^{(r,2)}(\mathbf{x}, \mathbf{h}) \qquad \text{for} \qquad \mathbf{x} \in \Omega^-, \tag{4.43}$$

where $\mathbf{h}$ is the required four-component vector function.

Obviously, by Theorem 4.8 and (4.23) the vector function **p** is a solution of (4.6) for $\mathbf{x} \in \Omega^-$. Keeping in mind the boundary condition (4.4) and using (4.25), from (4.43) we obtain, for determining the unknown vector **h**, the following integral equation

$$\mathcal{K}^{(r,3)}\mathbf{h}(\mathbf{z}) = \mathbf{f}(\mathbf{z}) \qquad \text{for} \qquad \mathbf{z} \in S. \tag{4.44}$$

It has been proved above that the adjoint homogeneous equation (4.40) has only the trivial solution. Hence, it follows that (4.44) is always solvable.

We have thereby proved the following.

Theorem 4.23 *If $S \in C^{1,\nu}$, $\mathbf{f} \in C^{1,\nu'}(S)$, $0 < \nu' < \nu \leq 1$, then a regular (classical) solution of the class $\mathfrak{B}^{(r)}$ of problem $(I_r)^-_{\mathbf{0},\mathbf{f}}$ exists, is unique, and is represented by double-layer potential (4.43), where* **h** *is a solution of the Fredholm integral equation (4.44) which is always solvable for an arbitrary vector* **f**.

Chapter 5
Problems of Equilibrium of Rigid Body

In this chapter, the BVPs of equilibrium of the linear theory for quadruple porosity rigid body are investigated by means of the potential method and the theory of Fredholm integral equations.

In Sect. 5.1, the basic internal and external BVPs of equilibrium of this theory are formulated.

In Sect. 5.2, the uniqueness theorems for classical solutions of these BVPs are proved.

In Sect. 5.3, the surface and volume potentials of the considered theory are constructed and their basic properties are established.

Finally, in Sect. 5.4, the BVPs of equilibrium are reduced to the equivalent integral equations for which Fredholm's theorems are valid and the existence theorems for classical solutions of these BVPs are proved by means of the potential method.

5.1 Basic Boundary Value Problems

The basic internal and external BVPs of equilibrium in the linear theory of quadruple porosity rigid body are formulated as follows.

Find a regular (classical) solution of the class $\mathfrak{B}^{(r)}$ in Ω^+ to system

$$(\mathbf{K}\Delta - \mathbf{d})\mathbf{p}(\mathbf{x}) = \mathbf{F}(\mathbf{x}) \tag{5.1}$$

for $\mathbf{x} \in \Omega^+$ satisfying the boundary condition (4.2) in the *Problem* $(I_{re})^+_{\mathbf{F},\mathbf{f}}$, and the boundary condition (4.3) in the *Problem* $(II_{re})^+_{\mathbf{F},\mathbf{f}}$, where the matrix differential operator $\mathcal{P}^{(r)}(\mathbf{D_z}, \mathbf{n})$ is defined by (3.55), $\mathbf{F}$ and $\mathbf{f}$ are prescribed four-component vector functions, and $\mathbf{d}$ is the internal transport tensor (see (1.7)).

M. Svanadze, *Potential Method in Mathematical Theories of Multi-Porosity Media*, Interdisciplinary Applied Mathematics 51,
https://doi.org/10.1007/978-3-030-28022-2_5

Find a regular (classical) solution to system (5.1) for $\mathbf{x} \in \Omega^-$ satisfying the boundary condition (4.3) in the *Problem* $(I_{re})^-_{\mathbf{F},\mathbf{f}}$, and the boundary condition (4.4) in the *Problem* $(II_{re})^-_{\mathbf{F},\mathbf{f}}$. Here $\mathbf{F}$ and $\mathbf{f}$ are prescribed four-component vector functions, supp $\mathbf{F}$ is a finite domain in Ω^-. Note that the class $\mathfrak{B}^{(r)}$ of vector functions is given by Definition 3.3.

Furthermore, the BVPs $(I_{re})^{\pm}_{\mathbf{F},\mathbf{f}}$ and $(II_{re})^{\pm}_{\mathbf{F},\mathbf{f}}$ are Dirichlet and Neumann type BVPs in the linear theory of quadruple porosity rigid body.

5.2 Uniqueness Theorems

In this section the uniqueness theorems for classical solutions of the BVPs $(I_{re})^{\pm}_{\mathbf{F},\mathbf{f}}$ and $(II_{re})^{\pm}_{\mathbf{F},\mathbf{f}}$ are proved by using Green's identities.

Theorem 5.1 *The internal BVP* $(I_{re})^{+}_{\mathbf{F},\mathbf{f}}$ *admits at most one regular (classical) solution of the class* $\mathfrak{B}^{(r)}$.

Proof Suppose that there are two regular solutions of problem $(I_{re})^{+}_{\mathbf{F},\mathbf{f}}$. Then their difference $\mathbf{p}$ corresponds to zero data ($\mathbf{F} = \mathbf{f} = \mathbf{0}$), i.e., on the basis of (5.1) and (4.2) the vector $\mathbf{p}$ is a regular solution of system of homogeneous equations

$$(\mathbf{K}\Delta - \mathbf{d})\mathbf{p}(\mathbf{x}) = \mathbf{0} \tag{5.2}$$

for $\mathbf{x} \in \Omega^+$ satisfying the boundary condition

$$\{\mathbf{p}(\mathbf{z})\}^+ = \mathbf{0} \tag{5.3}$$

for $\mathbf{z} \in S$.

On account of (5.2) and (5.3) from Green's formula (3.60) for Ω^+ and $\mathbf{p}' = \mathbf{p}$ we may deduce that

$$\int_{\Omega^+} W^{(re)}(\mathbf{p}, \mathbf{p}) d\mathbf{x} = 0, \tag{5.4}$$

where $W^{(re)}(\mathbf{p}, \mathbf{p})$ defined by (3.59). Then, employing the inequalities (4.10) we may derive from (5.4)

$$k_{\alpha\beta} \nabla p_\alpha \cdot \nabla p_\beta = 0. \tag{5.5}$$

Obviously, (5.5) implies

$$p_\alpha(\mathbf{x}) = c_\alpha = \text{const}, \qquad \alpha = 1, 2, 3, 4 \tag{5.6}$$

for $\mathbf{x} \in \Omega^+$. On the basis of the homogeneous boundary condition (5.3) from (5.6) we have $\mathbf{p}(\mathbf{x}) = \mathbf{0}$ for $\mathbf{x} \in \Omega^+$. Hence, uniqueness of a solution of BVP $(I_{re})^+_{\mathbf{F},\mathbf{f}}$ follows. □

Theorem 5.2 *Any two regular solutions of the internal BVP* $(II_{re})^+_{\mathbf{F},\mathbf{f}}$ *may differ only for an additive constant vector* $\mathbf{p} = (p_1, p_2, p_3, p_4)$, *where* $p_\alpha(\mathbf{x})$ *is given by (5.6) and* $\alpha = 1, 2, 3, 4$.

Proof Suppose that there are two regular solutions of problem $(II_{re})^+_{\mathbf{F},\mathbf{f}}$. Then their difference $\mathbf{p}$ corresponds to zero data ($\mathbf{F} = \mathbf{f} = \mathbf{0}$), i.e., on the basis of (5.1) and (4.3) the vector $\mathbf{p}$ is a regular solution of system of homogeneous equations (5.2) for $\mathbf{x} \in \Omega^+$ satisfying the homogeneous boundary condition

$$\{\mathcal{P}^{(r)}(\mathbf{D}_{\mathbf{z}}, \mathbf{n})\mathbf{p}(\mathbf{z})\}^+ = \mathbf{0} \tag{5.7}$$

for $\mathbf{z} \in S$. Quite similarly as in previous theorem, by virtue of (5.2) and (5.7) from (3.60) we obtain the desired equation (5.6). □

Theorem 5.3 *The external BVP* $(\mathbb{K}_{re})^-_{\mathbf{F},\mathbf{f}}$ *admits at most one regular (classical) solution of the class* $\mathfrak{B}^{(r)}$, *where* $\mathbb{K} = I, II$.

Proof Suppose that there are two regular solutions of problem $(\mathbb{K}_{re})^-_{\mathbf{F},\mathbf{f}}$. Then their difference $\mathbf{p}$ corresponds to zero data ($\mathbf{F} = \mathbf{f} = \mathbf{0}$), i.e., the vector $\mathbf{p}$ is a regular solution of homogeneous system of Eq. (5.1) for $\mathbf{x} \in \Omega^-$ satisfying the boundary condition

$$\{\mathcal{P}^{(r)}(\mathbf{D}_{\mathbf{z}}, \mathbf{n})\mathbf{p}(\mathbf{z}) \cdot \mathbf{p}(\mathbf{z})\}^- = 0 \tag{5.8}$$

for $\mathbf{z} \in S$.

On account of (5.1) and (5.8) from Green's formula (3.60) for Ω^- and $\mathbf{p}' = \mathbf{p}$ we may deduce that

$$\int_{\Omega^-} W^{(re)}(\mathbf{p}, \mathbf{p}) d\mathbf{x} = 0, \tag{5.9}$$

where $W^{(re)}(\mathbf{p}, \mathbf{p})$ is defined by (3.59). Then, similarly, we may derive (5.6) from (5.9). On the basis of condition at infinity (3.52) we get $\mathbf{p}(\mathbf{x}) = \mathbf{0}$ for $\mathbf{x} \in \Omega^-$ and the uniqueness of a solution of BVP $(\mathbb{K}_{re})^-_{\mathbf{F},\mathbf{f}}$ follows, where $\mathbb{K} = I, II$. □

5.3 Basic Properties of Potentials

On the basis of integral representation of regular vector (see (3.62)) we introduce the following notations

$$\mathbf{Z}^{(re,1)}(\mathbf{x},\mathbf{g}) = \int\limits_S \mathbf{\Gamma}^{(re)}(\mathbf{x}-\mathbf{y})\,\mathbf{g}(\mathbf{y})d_{\mathbf{y}}S,$$

$$\mathbf{Z}^{(re,2)}(\mathbf{x},\mathbf{g}) = \int\limits_S [\boldsymbol{\mathcal{P}}^{(r)}(\mathbf{D}_{\mathbf{y}},\mathbf{n})\mathbf{\Gamma}^{(re)}(\mathbf{x}-\mathbf{y})]^\top\,\mathbf{g}(\mathbf{y})d_{\mathbf{y}}S,$$

$$\mathbf{Z}^{(re,3)}(\mathbf{x},\boldsymbol{\phi},\Omega^\pm) = \int\limits_{\Omega^\pm} \mathbf{\Gamma}^{(re)}(\mathbf{x}-\mathbf{y})\,\mathbf{h}(\mathbf{y})d\mathbf{y},$$

where $\mathbf{g}$ and $\mathbf{h}$ are four-component vectors.

As in mathematical physics (see, e.g., Günther [83]), the vector functions $\mathbf{Z}^{(re,1)}(\mathbf{x},\mathbf{g})$, $\mathbf{Z}^{(re,2)}(\mathbf{x},\mathbf{g})$, and $\mathbf{Z}^{(re,3)}(\mathbf{x},\mathbf{h},\Omega^\pm)$ are called *single-layer, double-layer, and volume potentials* in the linear equilibrium theory of quadruple porosity rigid body, respectively.

Obviously, on the basis of Theorem 3.37, the regular vector in $\Omega^\pm$ is represented by sum of the surface (single-layer and double-layer) and volume potentials as follows

$$\mathbf{p}(\mathbf{x}) = \mathbf{Z}^{(re,2)}\left(\mathbf{x},\{\mathbf{p}\}^+\right) - \mathbf{Z}^{(re,1)}\left(\mathbf{x},\{\boldsymbol{\mathcal{P}}^{(r)}\mathbf{p}\}^+\right) + \mathbf{Z}^{(re,3)}\left(\mathbf{x},(\mathbf{K}\Delta-\mathbf{d})\mathbf{p},\Omega^+\right)$$

for $\mathbf{x}\in\Omega^+$ and

$$\mathbf{p}(\mathbf{x}) = -\mathbf{Z}^{(re,2)}\left(\mathbf{x},\{\mathbf{p}\}^-\right) + \mathbf{Z}^{(re,1)}\left(\mathbf{x},\{\boldsymbol{\mathcal{P}}^{(r)}\mathbf{p}\}^-\right) + \mathbf{Z}^{(re,3)}\left(\mathbf{x},(\mathbf{K}\Delta-\mathbf{d})\mathbf{p},\Omega^-\right)$$

for $\mathbf{x}\in\Omega^-$, respectively.

We now establish the basic properties of potentials $\mathbf{Z}^{(re,1)}(\mathbf{x},\mathbf{g})$, $\mathbf{Z}^{(re,2)}(\mathbf{x},\mathbf{g})$ and $\mathbf{Z}^{(re,3)}(\mathbf{x},\boldsymbol{\phi},\Omega^\pm)$.

5.3.1 Single-Layer Potential

We have the following results.

Theorem 5.4 *If S is the Liapunov surface of the class $C^{1,\nu}$, $0<\nu\le 1$, $\mathbf{g}$ is an integrable vector function on S, then*

(i)

$$\mathbf{Z}^{(re,1)}(\cdot,\mathbf{g}) \in C^\infty(\Omega^\pm),$$

(ii) $\mathbf{Z}^{(re,1)}(\cdot,\mathbf{g})$ *is a solution of the homogeneous equation (5.2) in Ω^+ and Ω^-, i.e.*

$$(\mathbf{K}\Delta - \mathbf{d})\,\mathbf{Z}^{(re,1)}(\mathbf{x},\mathbf{g}) = \mathbf{0} \tag{5.10}$$

for $\mathbf{x} \in \Omega^{\pm}$.

Theorem 5.5 *If* $S \in C^{1,\nu}$, $0 < \nu \le 1$, $\mathbf{g}$ *is an integrable and bounded vector function on S, then*

$$Z_{\alpha}^{(re,1)}(\mathbf{x},\mathbf{g}) = O\left(\frac{1}{|\mathbf{x}|}\right), \qquad Z_{\alpha,j}^{(re,1)}(\mathbf{x},\mathbf{g}) = O\left(\frac{1}{|\mathbf{x}|^2}\right) \tag{5.11}$$

for $|\mathbf{x}| \to +\infty$, *where* $\mathbf{Z}^{(re,1)} = \left(Z_1^{(re,1)}, Z_2^{(re,1)}, Z_3^{(re,1)}, Z_4^{(re,1)}\right)$, $j = 1, 2, 3$ *and* $\alpha = 1, 2, 3, 4$.

Theorem 5.6 *If* $S \in C^{1,\nu}$, $0 < \nu \le 1$, $\mathbf{g}$ *is an integrable and bounded vector function on S, then*

(i) the integrals $\mathbf{Z}^{(re,1)}(\mathbf{z},\mathbf{g})$, $\{\mathbf{Z}^{(re,1)}(\mathbf{z},\mathbf{g})\}^{+}$ *and* $\{\mathbf{Z}^{(re,1)}(\mathbf{z},\mathbf{g})\}^{-}$ *exist for* $\mathbf{z} \in S$ *and*

$$\mathbf{Z}^{(re,1)}(\mathbf{z},\mathbf{g}) = \left\{\mathbf{Z}^{(re,1)}(\mathbf{z},\mathbf{g})\right\}^{+} = \left\{\mathbf{Z}^{(re,1)}(\mathbf{z},\mathbf{g})\right\}^{-}, \tag{5.12}$$

(ii)

$$\mathbf{Z}^{(re,1)}(\cdot,\mathbf{g}) \in C(\mathbb{R}^3),$$

(iii)

$$\mathcal{P}^{(r)}(\mathbf{D}_{\mathbf{z}},\mathbf{n})\,\mathbf{Z}^{(re,1)}(\cdot,\mathbf{g}) \in C^{0,\nu'}(S),$$

where $0 < \nu' < \nu \le 1$ *and*

$$\left\{\mathbf{Z}^{(re,1)}(\mathbf{z},\mathbf{g})\right\}^{\pm} \equiv \lim_{\Omega^{\pm} \ni \mathbf{x} \to \mathbf{z} \in S} \mathbf{Z}^{(re,1)}(\mathbf{x},\mathbf{g}).$$

Theorem 5.7 *If* $S \in C^{1,\nu}$, $0 < \nu \le 1$, $\mathbf{g}$ *is a continuous vector function on S, then the vector functions* $\mathcal{P}^{(r)}(\mathbf{D}_{\mathbf{z}},\mathbf{n})\,\mathbf{Z}^{(re,1)}(\mathbf{z},\mathbf{g})$, $\{\mathcal{P}^{(r)}(\mathbf{D}_{\mathbf{z}},\mathbf{n})\,\mathbf{Z}^{(re,1)}(\mathbf{z},\mathbf{g})\}^{+}$ *and* $\{\mathcal{P}^{(r)}(\mathbf{D}_{\mathbf{z}},\mathbf{n})\,\mathbf{Z}^{(re,1)}(\mathbf{z},\mathbf{g})\}^{-}$ *exist for* $\mathbf{z} \in S$ *and*

$$\left\{\mathcal{P}^{(r)}(\mathbf{D}_{\mathbf{z}},\mathbf{n})\,\mathbf{Z}^{(re,1)}(\mathbf{z},\mathbf{g})\right\}^{\pm} = \mp\frac{1}{2}\mathbf{g}(\mathbf{z}) + \mathcal{P}^{(r)}(\mathbf{D}_{\mathbf{z}},\mathbf{n})\,\mathbf{Z}^{(re,1)}(\mathbf{z},\mathbf{g}), \tag{5.13}$$

where

$$\left\{\mathcal{P}^{(r)}(\mathbf{D}_{\mathbf{z}},\mathbf{n})\,\mathbf{Z}^{(re,1)}(\mathbf{z},\mathbf{g})\right\}^{\pm} \equiv \lim_{\Omega^{\pm} \ni \mathbf{x} \to \mathbf{z} \in S} \mathcal{P}^{(r)}(\mathbf{D}_{\mathbf{z}},\mathbf{n})\,\mathbf{Z}^{(re,1)}(\mathbf{x},\mathbf{g}).$$

Theorem 5.8 *The single-layer potential* $\mathbf{Z}^{(re,1)}(\mathbf{x},\mathbf{g}) \equiv \mathbf{0}$ *for* $\mathbf{x} \in \Omega^+$ *(or* $\mathbf{x} \in \Omega^-$*) if and only if* $\mathbf{g}(\mathbf{z}) \equiv \mathbf{0}$*, where* $\mathbf{z} \in S$.

5.3.2 Double-Layer Potential

We have the following results.

Theorem 5.9 *If* $S \in C^{1,\nu}$, $0 < \nu \le 1$, $\mathbf{g}$ *is an integrable vector function on* S*, then*

(i)

$$\mathbf{Z}^{(re,2)}(\cdot,\mathbf{g}) \in C^{\infty}(\Omega^{\pm}),$$

(ii) $\mathbf{Z}^{(re,2)}(\cdot,\mathbf{g})$ *is a solution of the homogeneous equation (5.2) in* Ω^+ *and* Ω^-*, i.e.*

$$(\mathbf{K}\Delta - \mathbf{d})\; \mathbf{Z}^{(re,2)}(\mathbf{x},\mathbf{g}) = \mathbf{0} \tag{5.14}$$

for $\mathbf{x} \in \Omega^{\pm}$.

Theorem 5.10 *If* $S \in C^{1,\nu}$, $0 < \nu \le 1$, $\mathbf{g}$ *is an integrable and bounded vector function on* S*, then*

$$Z^{(re,2)}_{\alpha}(\mathbf{x},\mathbf{g}) = O\left(\frac{1}{|\mathbf{x}|^2}\right), \qquad Z^{(re,2)}_{\alpha,j}(\mathbf{x},\mathbf{g}) = O\left(\frac{1}{|\mathbf{x}|^3}\right) \tag{5.15}$$

for $|\mathbf{x}| \to +\infty$*, where* $\mathbf{Z}^{(re,2)} = \left(Z^{(re,2)}_1, Z^{(re,2)}_2, Z^{(re,2)}_3, Z^{(re,2)}_4\right)$, $j = 1,2,3$*, and* $\alpha = 1,2,3,4$.

Theorem 5.11 *If* $S \in C^{1,\nu}$, $0 < \nu \le 1$*, then the matrices* $\mathcal{P}^{(r)}(\mathbf{D}_{\mathbf{z}},\mathbf{n})\mathbf{\Gamma}^{(re)}(\mathbf{z}-\mathbf{y})$ *and* $[\mathcal{P}^{(r)}(\mathbf{D}_{\mathbf{y}},\mathbf{n})\mathbf{\Gamma}^{(re)}(\mathbf{z}-\mathbf{y})]^{\top}$ *are the weakly singular kernels of* $\mathcal{P}^{(r)}(\mathbf{D}_{\mathbf{z}},\mathbf{n})\mathbf{Z}^{(re,1)}(\mathbf{z},\mathbf{g})$ *and the double-layer potential* $\mathbf{Z}^{(re,2)}(\mathbf{z},\mathbf{g})$*, respectively, i.e.*

$$\begin{aligned} \left|[\mathcal{P}^{(r)}(\mathbf{D}_{\mathbf{z}},\mathbf{n})\mathbf{\Gamma}^{(re)}(\mathbf{z}-\mathbf{y})]_{\alpha\beta}\right| &< \frac{\text{const}}{|\mathbf{z}-\mathbf{y}|^{2-\nu}}, \\ \left|[\mathcal{P}^{(r)}(\mathbf{D}_{\mathbf{y}},\mathbf{n})\mathbf{\Gamma}^{(re)}(\mathbf{z}-\mathbf{y})]_{\alpha\beta}\right| &< \frac{\text{const}}{|\mathbf{z}-\mathbf{y}|^{2-\nu}} \end{aligned} \tag{5.16}$$

for $\mathbf{z},\mathbf{y} \in S$, $\alpha,\beta = 1,2,3,4$*, and* const *is a positive number independent on* $\mathbf{z}$ *and* $\mathbf{y}$.

Theorem 5.12 *If* $S \in C^{1,\nu}$, $0 < \nu \le 1$, $\mathbf{g}$ *is a continuous vector function on* S*, then*

(i) the integrals $\mathbf{Z}^{(re,2)}(\mathbf{z},\mathbf{g})$, $\{\mathbf{Z}^{(re,2)}(\mathbf{z},\mathbf{g})\}^{+}$ *and* $\{\mathbf{Z}^{(re,2)}(\mathbf{z},\mathbf{g})\}^{-}$ *exist,*

(ii)

$$\left\{\mathbf{Z}^{(re,2)}(\mathbf{z},\mathbf{g})\right\}^{\pm} = \pm\frac{1}{2}\mathbf{g}(\mathbf{z}) + \mathbf{Z}^{(re,2)}(\mathbf{z},\mathbf{g}) \tag{5.17}$$

for $\mathbf{z} \in S$.

Theorem 5.13 *If* $S \in C^{1,\nu}$, $\mathbf{g} \in C^{0,\nu'}(S)$, $0 < \nu' < \nu \le 1$, *then*

$$\mathbf{Z}^{(re,2)}(\cdot,\mathbf{g}) \in C^{0,\nu'}(S).$$

Theorem 5.14 *If* $S \in C^{1,\nu}$, $\mathbf{g} \in C^{1,\nu}(S)$, $0 < \nu \le 1$, *then*

(i) the vector functions $\{\mathcal{P}^{(r)}(\mathbf{D_z},\mathbf{n})\mathbf{Z}^{(re,2)}(\mathbf{z},\mathbf{g})\}^{+}$ *and* $\{\mathcal{P}^{(r)}(\mathbf{D_z},\mathbf{n})\mathbf{Z}^{(re,2)}(\mathbf{z},\mathbf{g})\}^{-}$ *exist,*

(ii) they are elements of the class $C^{1,\nu}(S)$, *and*

(iii)

$$\left\{\mathcal{P}^{(r)}(\mathbf{D_z},\mathbf{n})\mathbf{Z}^{(re,2)}(\mathbf{z},\mathbf{g})\right\}^{+} = \left\{\mathcal{P}^{(r)}(\mathbf{D_z},\mathbf{n})\mathbf{Z}^{(re,2)}(\mathbf{z},\mathbf{g})\right\}^{-} \tag{5.18}$$

for $\mathbf{z} \in S$.

Theorem 5.14 is the Liapunov–Tauber type theorem in the linear equilibrium theory of quadruple porosity rigid body.

5.3.3 Volume Potential

We have the following results.

Theorem 5.15 *If* $S \in C^{1,\nu}$, $\mathbf{h}$ *is an integrable and bounded vector function in* Ω^{+}, *then*

(i)

$$\mathbf{Z}^{(re,3)}(\cdot,\mathbf{h},\Omega^{+}) \in C^{\infty}(\Omega^{-}),$$

(ii)

$$\mathbf{Z}^{(re,3)}(\cdot,\mathbf{h},\Omega^{+}) \in C^{1,\nu}(\mathbb{R}^3),$$

(iii) $\mathbf{Z}^{(re,3)}(\cdot,\mathbf{h})$ *is a solution of the homogeneous equation (5.2), i.e.*

$$(\mathbf{K}\Delta - \mathbf{d})\,\mathbf{Z}^{(re,3)}(\mathbf{x},\mathbf{h},\Omega^{+}) = \mathbf{0} \tag{5.19}$$

for $\mathbf{x} \in \Omega^{-}$.

Theorem 5.16 *If* $S \in C^{1,\nu}$, $0 < \nu \le 1$, $\mathbf{h} \in C^1(\Omega^+)$, *then*

$$\mathbf{Z}^{(re,3)}(\cdot, \mathbf{h}, \Omega^+) \in C^2(\Omega^+).$$

Theorem 5.17 *If* $S \in C^{1,\nu}$, $\mathbf{h}$ *is an integrable and bounded vector function in* Ω^- *and* supp $\mathbf{h}$ *is a finite domain in* Ω^-, *then*

(i)

$$\mathbf{Z}^{(re,3)}(\cdot, \mathbf{h}, \Omega^-) \in C^\infty(\Omega^-),$$

(ii)

$$\mathbf{Z}^{(re,3)}(\cdot, \mathbf{h}, \Omega^-) \in C^{1,\nu}(\mathbb{R}^3),$$

(iii) $\mathbf{Z}^{(re,3)}(\cdot, \mathbf{h}, \Omega^-)$ *is a solution of the homogeneous equation (5.19), i.e.*

$$(\mathbf{K}\Delta - \mathbf{d})\ \mathbf{Z}^{(re,3)}(\mathbf{x}, \mathbf{h}, \Omega^-) = \mathbf{0}$$

for $\mathbf{x} \in \Omega^+$.

Theorem 5.18 *If* $S \in C^{1,\nu}$, $0 < \nu \le 1$, $\mathbf{h} \in C^1(\Omega^-)$, and supp $\mathbf{h}$ *is a finite domain in* Ω^-, *then*

$$\mathbf{Z}^{(re,3)}(\cdot, \mathbf{h}, \Omega^-) \in C^2(\Omega^-).$$

Theorem 5.19 *If* $S \in C^{1,\nu}$, $0 < \nu \le 1$, $\mathbf{h} \in C^{0,\nu}(\Omega^+)$, *then*

$$(\mathbf{K}\Delta - \mathbf{d})\ \mathbf{Z}^{(re,3)}(\mathbf{x}, \mathbf{h}, \Omega^+) = \mathbf{h}(\mathbf{x}) \tag{5.20}$$

for $\mathbf{x} \in \Omega^+$.

Theorem 5.20 *If* $S \in C^{1,\nu}$, $0 < \nu \le 1$, $\mathbf{h} \in C^{0,\nu}(\Omega^+)$, and supp $\mathbf{h}$ *is a finite domain in* Ω^-, *then*

$$(\mathbf{K}\Delta - \mathbf{d})\ \mathbf{Z}^{(re,3)}(\mathbf{x}, \mathbf{h}, \Omega^-) = \mathbf{h}(\mathbf{x}) \tag{5.21}$$

for $\mathbf{x} \in \Omega^-$.

Theorems 5.19 and 5.20 are Poisson type theorems in the linear equilibrium theory of quadruple porosity rigid body for domains Ω^+ and Ω^-, respectively.

5.4 Existence Theorems

We introduce the integral operators

$$
\begin{aligned}
\mathcal{K}^{(re,1)}\mathbf{g}(\mathbf{z}) &\equiv \frac{1}{2}\,\mathbf{g}(\mathbf{z}) + \mathbf{Z}^{(re,2)}(\mathbf{z},\mathbf{g}),\\
\mathcal{K}^{(re,2)}\mathbf{h}(\mathbf{z}) &\equiv \frac{1}{2}\,\mathbf{h}(\mathbf{z}) + \mathcal{P}^{(r)}(\mathbf{D}_{\mathbf{z}},\mathbf{n})\mathbf{Z}^{(re,1)}(\mathbf{z},\mathbf{h}),\\
\mathcal{K}^{(re,3)}\mathbf{g}(\mathbf{z}) &\equiv -\frac{1}{2}\,\mathbf{g}(\mathbf{z}) + \mathbf{Z}^{(re,2)}(\mathbf{z},\mathbf{g}),\\
\mathcal{K}^{(re,4)}\mathbf{h}(\mathbf{z}) &\equiv -\frac{1}{2}\,\mathbf{h}(\mathbf{z}) + \mathcal{P}^{(r)}(\mathbf{D}_{\mathbf{z}},\mathbf{n})\mathbf{Z}^{(re,1)}(\mathbf{z},\mathbf{h}),
\end{aligned}
\tag{5.22}
$$

where $\mathbf{g}$ and $\mathbf{h}$ are four-component vector functions on S and $\mathbf{z} \in S$. On the basis of (5.16), the operator $\mathcal{K}^{(re,\alpha)}$ has a weakly singular kernel and consequently, it is a Fredholm operator. Moreover, operators $\mathcal{K}^{(re,j)}$ and $\mathcal{K}^{(re,j+1)}$ are adjoint with respect to each other, where $j = 1, 3$.

We are now in a position to study the existence of regular solutions of the BVPs $(\mathbb{K}_{re})^{+}_{\mathbf{F},\mathbf{f}}$ and $(\mathbb{K}_{re})^{-}_{\mathbf{F},\mathbf{f}}$, by means of the potential method and the theory of Fredholm integral equations, where $\mathbb{K} = I, II$.

Obviously, on the basis of (5.20) and (5.21) the potential $\mathbf{Z}^{(re,3)}(\mathbf{x},\mathbf{F},\Omega^{\pm})$ is a particular regular solution of the nonhomogeneous equation (5.1), where $\mathbf{F} \in C^{0,\nu'}(\Omega^{\pm})$, $0 < \nu' \le 1$; $\operatorname{supp}\mathbf{F}$ is a finite domain in Ω^{-}. Therefore, in this section we will consider problems $(\mathbb{K}_{re})^{+}_{\mathbf{0},\mathbf{f}}$ and $(\mathbb{K}_{re})^{-}_{\mathbf{0},\mathbf{f}}$, where $\mathbb{K} = I, II$.

Problem $(I_{re})^{+}_{\mathbf{0},\mathbf{f}}$ We seek a regular solution to this problem by means of a double-layer potential

$$
\mathbf{p}(\mathbf{x}) = \mathbf{Z}^{(re,2)}(\mathbf{x},\mathbf{g}) \qquad \text{for} \qquad \mathbf{x} \in \Omega^{+}, \tag{5.23}
$$

where $\mathbf{g}$ is the required four-component vector function.

By virtue of (5.14) the vector function $\mathbf{p}$ is a solution of (5.2) for $\mathbf{x} \in \Omega^{+}$. Keeping in mind the conditions (4.2), (5.17), and using (5.22), from (5.23) we obtain, for determining the unknown vector $\mathbf{g}$, a Fredholm integral equation

$$
\mathcal{K}^{(re,1)}\mathbf{g}(\mathbf{z}) = \mathbf{f}(\mathbf{z}) \qquad \text{for} \quad \mathbf{z} \in S. \tag{5.24}
$$

We prove that (5.24) is always solvable for an arbitrary vector $\mathbf{f}$. Let us consider the adjoint homogeneous equation

$$
\mathcal{K}^{(re,2)}\mathbf{h}_0(\mathbf{z}) = \mathbf{0} \qquad \text{for} \quad \mathbf{z} \in S, \tag{5.25}
$$

where $\mathbf{h}_0$ is the required four-component vector function. Now we prove that (5.25) has only the trivial solution.

Indeed, let $\mathbf{h}_0$ be a solution of the homogeneous equation (5.25). On the basis of (5.10), (5.11), and (5.13) the vector function $\mathbf{Z}^{(re,1)}(\mathbf{x},\mathbf{h}_0)$ is a regular solution of problem $(II_{re})^{-}_{\mathbf{0},\mathbf{0}}$. Using Theorem 5.3, the problem $(II_{re})^{-}_{\mathbf{0},\mathbf{0}}$ has only the trivial solution, that is

$$\mathbf{Z}^{(re,1)}(\mathbf{x}, \mathbf{h}_0) \equiv \mathbf{0} \qquad \text{for} \qquad \mathbf{x} \in \Omega^-. \tag{5.26}$$

On the other hand, by (5.12) and (5.26) we get

$$\{\mathbf{Z}^{(re,1)}(\mathbf{z}, \mathbf{h}_0)\}^+ = \{\mathbf{Z}^{(re,1)}(\mathbf{z}, \mathbf{h}_0)\}^- = \mathbf{0} \qquad \text{for} \qquad \mathbf{z} \in S,$$

i.e., the vector $\mathbf{Z}^{(re,1)}(\mathbf{x}, \mathbf{h}_0)$ is a regular solution of problem $(I_{re})^+_{\mathbf{0},\mathbf{0}}$. Using Theorem 5.1, the problem $(I_{re})^+_{\mathbf{0},\mathbf{0}}$ has only the trivial solution, that is

$$\mathbf{Z}^{(re,1)}(\mathbf{x}, \mathbf{h}_0) \equiv \mathbf{0} \qquad \text{for} \qquad \mathbf{x} \in \Omega^+. \tag{5.27}$$

By virtue of (5.26), (5.27), and identity (5.13) we obtain

$$\mathbf{h}_0(\mathbf{z}) = \{\mathcal{P}^{(r)}(\mathbf{D_z}, \mathbf{n})\mathbf{Z}^{(re,1)}(\mathbf{z}, \mathbf{h}_0)\}^- - \{\mathcal{P}^{(r)}(\mathbf{D_z}, \mathbf{n})\mathbf{Z}^{(re,1)}(\mathbf{z}, \mathbf{h}_0)\}^+ = \mathbf{0}$$

for $\mathbf{z} \in S$. Thus, the homogeneous equation (5.25) has only the trivial solution and therefore (5.24) is always solvable for an arbitrary vector $\mathbf{f}$.

We have thereby proved the following.

Theorem 5.21 *If $S \in C^{1,\nu}$, $\mathbf{f} \in C^{1,\nu'}(S)$, $0 < \nu' < \nu \le 1$, then a regular (classical) solution of the class $\mathfrak{B}^{(r)}$ of problem $(I_{re})^+_{\mathbf{0},\mathbf{f}}$ exists, is unique, and is represented by double-layer potential (5.23), where $\mathbf{g}$ is a solution of the Fredholm integral equation (5.24) which is always solvable for an arbitrary vector $\mathbf{f}$.*

Problem $(II_{re})^-_{\mathbf{0},\mathbf{f}}$ We seek a regular solution of the class $\mathfrak{B}^{(r)}$ to this problem in the form

$$\mathbf{p}(\mathbf{x}) = \mathbf{Z}^{(re,1)}(\mathbf{x}, \mathbf{h}) \qquad \text{for} \qquad \mathbf{x} \in \Omega^-, \tag{5.28}$$

where $\mathbf{h}$ is the required four-component vector function.

Obviously, by virtue of (5.10) and (5.11) the vector function $\mathbf{p}$ is a regular solution of (5.2) for $\mathbf{x} \in \Omega^-$. Keeping in mind the boundary condition (4.5) and using (5.13), from (5.28) we obtain, for determining the unknown vector $\mathbf{h}$, a Fredholm integral equation

$$\mathcal{K}^{(re,2)}\mathbf{h}(\mathbf{z}) = \mathbf{f}(\mathbf{z}) \qquad \text{for} \quad \mathbf{z} \in S. \tag{5.29}$$

It has been proved above that the corresponding homogeneous equation (5.25) has only the trivial solution. Hence, it follows that (5.29) is always solvable.

We have thereby proved the following.

Theorem 5.22 *If $S \in C^{1,\nu}$, $\mathbf{f} \in C^{0,\nu'}(S)$, $0 < \nu' < \nu \le 1$, then a regular (classical) solution of the class $\mathfrak{B}^{(r)}$ of problem $(II_{re})^-_{\mathbf{0},\mathbf{f}}$ exists, is unique, and is represented by single-layer potential (5.28), where $\mathbf{h}$ is a solution of the Fredholm integral equation (5.29) which is always solvable for an arbitrary vector $\mathbf{f}$.*

Problem $(I_{re})^-_{\mathbf{0},\mathbf{f}}$ We seek a regular (classical) solution of the class $\mathfrak{B}^{(r)}$ to this problem in the form

$$\mathbf{p}(\mathbf{x}) = \mathbf{Z}^{(re,2)}(\mathbf{x},\mathbf{h}) + \mathbf{Z}^{(re,1)}(\mathbf{x},\mathbf{h}) \qquad \text{for} \qquad \mathbf{x} \in \Omega^-, \tag{5.30}$$

where $\mathbf{h}$ is the required four-component vector function.

Obviously, by (5.10), (5.11), (5.14), and (5.15) the vector function $\mathbf{p}$ is a solution of (5.2) for $\mathbf{x} \in \Omega^-$. Keeping in mind the boundary condition (4.4) and using (5.12) and (5.17), from (5.30) we obtain, for determining the unknown vector $\mathbf{h}$, the following Fredholm integral equation

$$\mathcal{K}^{(re,3)}\mathbf{h}(\mathbf{z}) + \mathbf{Z}^{(re,1)}(\mathbf{z},\mathbf{h}) = \mathbf{f}(\mathbf{z}) \qquad \text{for} \qquad \mathbf{z} \in S. \tag{5.31}$$

We prove that (5.31) is always solvable for an arbitrary vector $\mathbf{f}$. Let us consider the homogeneous equation

$$\mathcal{K}^{(re,3)}\mathbf{h}_0(\mathbf{z}) + \mathbf{Z}^{(re,1)}(\mathbf{z},\mathbf{h}_0) = \mathbf{0} \qquad \text{for} \quad \mathbf{z} \in S, \tag{5.32}$$

where $\mathbf{h}_0$ is the required four-component vector function. Now we prove that (5.32) has only the trivial solution.

Indeed, let $\mathbf{h}_0$ be a solution of the homogeneous equation (5.32). On the basis of (5.10), (5.12), (5.14), and (5.17) the vector function

$$\mathbf{p}_0(\mathbf{x}) = \mathbf{Z}^{(re,2)}(\mathbf{x},\mathbf{h}_0) + \mathbf{Z}^{(re,1)}(\mathbf{x},\mathbf{h}_0)$$

is a regular solution of problem $(I_{re})^-_{\mathbf{0},\mathbf{0}}$. Using Theorem 5.3, the problem $(I_{re})^-_{\mathbf{0},\mathbf{0}}$ has only the trivial solution, that is

$$\mathbf{p}_0(\mathbf{x}) \equiv \mathbf{0} \qquad \text{for} \qquad \mathbf{x} \in \Omega^-. \tag{5.33}$$

Clearly, from (5.33) we have

$$\{\mathbf{p}_0(\mathbf{z})\}^- = \mathbf{0}, \qquad \left\{\mathcal{P}^{(r)}(\mathbf{D}_\mathbf{z},\mathbf{n})\mathbf{p}_0(\mathbf{z})\right\}^- = \mathbf{0} \qquad \text{for} \quad \mathbf{z} \in S. \tag{5.34}$$

By virtue of (5.12) and (5.17) we may show

$$\begin{aligned} &\{\mathbf{p}_0(\mathbf{z})\}^+ - \{\mathbf{p}_0(\mathbf{z})\}^- \\ &\quad = \frac{1}{2}\mathbf{h}_0(\mathbf{z}) + \mathbf{Z}^{(re,2)}(\mathbf{x},\mathbf{h}_0) + \mathbf{Z}^{(re,1)}(\mathbf{x},\mathbf{h}_0) \\ &\qquad - \left[-\frac{1}{2}\mathbf{h}_0(\mathbf{z}) + \mathbf{Z}^{(re,2)}(\mathbf{x},\mathbf{h}_0) + \mathbf{Z}^{(re,1)}(\mathbf{x},\mathbf{h}_0)\right] = \mathbf{h}_0(\mathbf{z}). \end{aligned} \tag{5.35}$$

Similarly, on the basis of (5.13) and (5.18) we obtain

$$\begin{aligned}
&\Big\{\mathcal{P}^{(r)}(\mathbf{D_z},\mathbf{n})\mathbf{p}_0(\mathbf{z})\Big\}^+ - \Big\{\mathcal{P}^{(r)}(\mathbf{D_z},\mathbf{n})\mathbf{p}_0(\mathbf{z})\Big\}^- \\
&\quad = -\frac{1}{2}\,\mathbf{h}_0(\mathbf{z}) + \mathbf{Z}^{(re,1)}(\mathbf{x},\mathbf{h}_0) + \Big\{\mathcal{P}^{(r)}(\mathbf{D_z},\mathbf{n})\mathbf{Z}^{(re,2)}(\mathbf{x},\mathbf{h}_0)\Big\}^+ \\
&\qquad - \left[\frac{1}{2}\,\mathbf{h}_0(\mathbf{z}) + \mathbf{Z}^{(re,1)}(\mathbf{x},\mathbf{h}_0) + \Big\{\mathcal{P}^{(r)}(\mathbf{D_z},\mathbf{n})\mathbf{Z}^{(re,2)}(\mathbf{x},\mathbf{h}_0)\Big\}^-\right] \\
&\quad = -\mathbf{h}_0(\mathbf{z}).
\end{aligned} \tag{5.36}$$

Combining (5.34) with (5.35) and (5.36) we may further conclude that

$$\Big\{\mathcal{P}^{(r)}(\mathbf{D_z},\mathbf{n})\mathbf{p}_0(\mathbf{z}) + \mathbf{p}_0(\mathbf{z})\Big\}^+ = \mathbf{0}, \tag{5.37}$$

i.e., the vector $\mathbf{p}_0(\mathbf{z})$ is a regular solution of (5.2) for $\mathbf{x} \in \Omega^+$ satisfies the Robin type homogeneous boundary condition (5.37).

On account of (5.2) and (5.37) from (3.60) for Ω^+ and $\mathbf{p}' = \mathbf{p}$ we may deduce that

$$\int_{\Omega^+} W^{(re)}(\mathbf{p}_0,\mathbf{p}_0)d\mathbf{x} = -\int_S |\mathbf{p}_0(\mathbf{z})|^2 d_{\mathbf{z}}S, \tag{5.38}$$

where

$$W^{(re)}(\mathbf{p}_0,\mathbf{p}_0) = \mathbf{K}\mathbf{p}_0 \cdot \mathbf{p}_0 + \mathbf{d}\mathbf{p}_0 \cdot \mathbf{p}_0.$$

Then, employing the inequalities (4.10) we may derive from (5.38) the following boundary condition

$$\{\mathbf{p}_0(\mathbf{z})\}^+ = \mathbf{0} \qquad \text{for} \quad \mathbf{z} \in S. \tag{5.39}$$

Combining (5.39) with (5.34) and (5.35) we may see that $\mathbf{h}_0(\mathbf{z}) = \mathbf{0}$ for $\mathbf{z} \in S$. Finally, using the Fredholm first theorem (see Sect. 4.4) the integral equation (5.31) is always solvable for an arbitrary vector $\mathbf{f}$.

We have thereby proved the following.

Theorem 5.23 *If $S \in C^{1,\nu}$, $\mathbf{f} \in C^{1,\nu'}(S)$, $0 < \nu' < \nu \le 1$, then a regular (classical) solution of the class $\mathfrak{B}^{(r)}$ of problem $(I_{re})^-_{\mathbf{0},\mathbf{f}}$ exists, is unique, and is represented by sum of the single-layer and double-layer potentials (5.30), where $\mathbf{h}$ is a solution of the Fredholm integral equation (5.31) which is always solvable for an arbitrary vector $\mathbf{f}$.*

Problem $(II_{re})^+_{\mathbf{0},\mathbf{f}}$ We seek a regular solution of the class $\mathfrak{B}^{(r)}$ to this problem in the form

$$\mathbf{p}(\mathbf{x}) = \mathbf{Z}^{(re,1)}(\mathbf{x}, \mathbf{g}) \qquad \text{for} \qquad \mathbf{x} \in \Omega^+, \tag{5.40}$$

where $\mathbf{g}$ is the required four-component vector function.

Obviously, by virtue of (5.10) the vector function $\mathbf{p}$ is a solution of (5.2) for $\mathbf{x} \in \Omega^+$. Keeping in mind the boundary condition (4.3) and using (5.13), from (5.40) we obtain, for determining the unknown vector $\mathbf{g}$, the following Fredholm integral equation

$$\mathcal{K}^{(r,4)}\mathbf{g}(\mathbf{z}) = \mathbf{f}(\mathbf{z}) \qquad \text{for} \qquad \mathbf{z} \in S. \tag{5.41}$$

To investigate the solvability of (5.41) we consider the homogeneous equation

$$\mathcal{K}^{(r,4)}\mathbf{g}(\mathbf{z}) = \mathbf{0} \qquad \text{for} \qquad \mathbf{z} \in S. \tag{5.42}$$

The homogeneous adjoint integral equation of (5.42) has the following form

$$\mathcal{K}^{(r,3)}\mathbf{h}(\mathbf{z}) = \mathbf{0} \qquad \text{for} \qquad \mathbf{z} \in S, \tag{5.43}$$

where $\mathbf{h}$ is a four-component vector function.

It is important to note that the number of the linearly independent solutions of (5.42) and (5.43) depend on the nonnegative six values $\gamma_{12}, \gamma_{13}, \cdots, \gamma_{34}$. In the beginning we consider the case $\mathbf{d} = \mathbf{0}$, i.e., all $\gamma_{\alpha\beta} = 0$.

(a) Let $\mathbf{d} = \mathbf{0}$. We introduce the notation

$$\begin{aligned} \boldsymbol{\chi}^{(11)} &= (1, 0, 0, 0), \qquad & \boldsymbol{\chi}^{(12)} &= (0, 1, 0, 0), \\ \boldsymbol{\chi}^{(13)} &= (0, 0, 1, 0), \qquad & \boldsymbol{\chi}^{(14)} &= (0, 0, 0, 1). \end{aligned} \tag{5.44}$$

Obviously, the system of vectors $\{\boldsymbol{\chi}^{(1j)}\}_{j=1}^4$ is linearly independent. We use the following.

Lemma 5.1 *If* $\mathbf{d} = \mathbf{0}$*, then the homogeneous equations (5.42) and (5.43) have four linearly independent solutions each and they constitute complete system of solutions.*

Proof By Theorem 5.2 each vector $\boldsymbol{\chi}^{(1j)}$ $(j = 1, 2, 3, 4)$ is a regular solution of problem $(II_{re})^+_{\mathbf{0},\mathbf{0}}$, i.e.,

$$\begin{aligned} \mathbf{K}\Delta\boldsymbol{\chi}^{(1j)} &= \mathbf{0} \qquad \text{in} \quad \Omega^+, \\ \left\{\mathcal{P}^{(r)}(\mathbf{D}_\mathbf{z}, \mathbf{n})\boldsymbol{\chi}^{(1j)}\right\}^+ &= \mathbf{0} \qquad \text{on} \quad S. \end{aligned} \tag{5.45}$$

On the other hand, by virtue of (5.45) from formula of integral representation of regular vector (3.62) we obtain

$$\boldsymbol{\chi}^{(1j)} = \mathbf{Z}^{(re,2)}(\mathbf{x}, \boldsymbol{\chi}^{(1j)}) \qquad \text{for} \quad \mathbf{x} \in \Omega^+. \tag{5.46}$$

On the basis of boundary property of double-layer potential (5.17) from (5.46) we have

$$\boldsymbol{\chi}^{(1j)} = \frac{1}{2}\boldsymbol{\chi}^{(1j)} + \mathbf{Z}^{(re,2)}(\mathbf{z}, \boldsymbol{\chi}^{(1j)}) \qquad \text{for} \quad \mathbf{z} \in S, \ j = 1, 2, 3, 4$$

and therefore, $\boldsymbol{\chi}^{(1j)}$ is a solution of Eq. (5.43). Using the Fredholm second theorem (see Sect. 4.4) the integral equations (5.42) and (5.43) have four linearly independent solutions each.

It will now be shown that $\{\boldsymbol{\chi}^{(1j)}\}_{j=1}^{4}$ is a complete system of linearly independent solutions of (5.43).

Let $\{\mathbf{h}^{(j)}(\mathbf{z})\}_{j=1}^{m}$ is the complete system of linearly independent solutions of homogeneous equation (5.43), where $m \geq 4$. We construct single-layer potentials $\mathbf{Z}^{(re,1)}(\mathbf{x}, \mathbf{h}^{(j)})$ $(j = 1, 2, \cdots, m)$. By Theorem 5.8 from identity

$$\mathbf{0} = \sum_{j=1}^{m} r_j \mathbf{Z}^{(re,1)}(\mathbf{x}, \mathbf{h}^{(j)}) = \mathbf{Z}^{(re,1)}(\mathbf{x}, \sum_{j=1}^{m} r_j \mathbf{h}^{(j)}) \qquad \text{for} \quad \mathbf{x} \in \Omega^+$$

it follows that

$$\sum_{j=1}^{m} r_j \mathbf{h}^{(j)}(\mathbf{z}) = \mathbf{0}, \tag{5.47}$$

where $r_1, r_2, \cdots, r_m$ are arbitrary constants, and from (5.47) we have $r_1 = r_2 = \cdots = r_m$. Hence, $\left\{\mathbf{Z}^{(re,1)}(\mathbf{x}, \mathbf{h}^{(j)})\right\}_{j=1}^{m}$ is the system of linearly independent vectors.

On the other hand, it is easy to see that each vector $\mathbf{Z}^{(re,1)}(\mathbf{x}, \mathbf{h}^{(j)})$ $(j = 1, 2, \cdots, m)$ is a regular solution of problem $(II_{re})^+_{\mathbf{0},\mathbf{0}}$. On the basis of Theorem 5.2, vector $\mathbf{Z}^{(re,1)}(\mathbf{x}, \mathbf{h}^{(j)})$ can be written as follows

$$\mathbf{Z}^{(re,1)}(\mathbf{x}, \mathbf{h}^{(j)}) = \sum_{l=1}^{4} r_{jl} \boldsymbol{\chi}^{(1l)} \qquad \text{for} \quad \mathbf{x} \in \Omega^+,$$

where r_{jl} $(j = 1, 2, \cdots, m, \ l = 1, 2, 3, 4)$ are constants. Hence, each vector of system $\left\{\mathbf{Z}^{(re,1)}(\mathbf{x}, \mathbf{h}^{(j)})\right\}_{j=1}^{m}$ is represented by four linearly independent vectors $\boldsymbol{\chi}^{(11)}$, $\boldsymbol{\chi}^{(12)}$, $\boldsymbol{\chi}^{(13)}$, and $\boldsymbol{\chi}^{(14)}$ (see (5.44)). Thus, $m = 4$. □

By virtue of Fredholm's third theorem (see Sect. 4.4), for Eq. (5.41) to be solvable it is necessary and sufficient that (see the condition (4.30))

$$\int_S \mathbf{f}(\mathbf{z})\boldsymbol{\chi}^{(1j)} d_{\mathbf{z}}S = \mathbf{0} \qquad \text{for} \quad j = 1, 2, 3, 4. \tag{5.48}$$

Then, employing (5.44), we may derive from (5.48)

$$\int_S \mathbf{f}(\mathbf{z}) d_{\mathbf{z}}S = \mathbf{0}. \tag{5.49}$$

We have thereby proved the following.

Theorem 5.24 *If $S \in C^{1,\nu}$, $\mathbf{f} \in C^{0,\nu'}(S)$, $0 < \nu' < \nu \le 1$, and $\mathbf{d} = \mathbf{0}$, then the problem $(II_{re})^+_{\mathbf{0},\mathbf{f}}$ is solvable only when condition (5.49) is fulfilled. The solution of this problem is represented by a single-layer potential (5.40) and is determined within an additive constant vector $\tilde{\mathbf{p}} = (c_1, c_2, c_3, c_4)$, where $\mathbf{g}$ is a solution of the solvable Fredholm integral equation (5.41) and $c_\alpha = \text{const}$, $\alpha = 1, 2, 3, 4$.*

(b) We now consider the case where there is only one positive value in the nonnegative internal transport coefficients $\gamma_{\alpha\beta}$ $(\alpha, \beta = 1, 2, 3, 4$ and $\alpha < \beta)$. Clearly, without loss of generality we can assume that $\gamma_{12} > 0$.

We have the following.

Lemma 5.2 *If $\gamma_{12} > 0$ and $\gamma_{13} = \gamma_{14} = \gamma_{23} = \gamma_{24} = \gamma_{34} = 0$, then the homogeneous equation (5.43) has three linearly independent solutions $\boldsymbol{\chi}^{(21)}, \boldsymbol{\chi}^{(22)}, \boldsymbol{\chi}^{(23)}$ and they constitute complete system of solutions, where*

$$\boldsymbol{\chi}^{(21)} = (1, 1, 0, 0), \qquad \boldsymbol{\chi}^{(22)} = (0, 0, 1, 0), \qquad \boldsymbol{\chi}^{(23)} = (0, 0, 0, 1).$$

Quite similarly, as in the case a), on the basis of Lemma 5.2 we obtain the following

Theorem 5.25 *If $S \in C^{1,\nu}$, $\mathbf{f} \in C^{0,\nu'}(S)$, $0 < \nu' < \nu \le 1$, $\gamma_{12} > 0$, and $\gamma_{13} = \gamma_{14} = \gamma_{23} = \gamma_{24} = \gamma_{34} = 0$, then the problem $(II_{re})^+_{\mathbf{0},\mathbf{f}}$ is solvable only when the conditions*

$$\int_S [f_1(\mathbf{z}) + f_2(\mathbf{z})]\, d_{\mathbf{z}}S = 0, \qquad \int_S f_j(\mathbf{z}) d_{\mathbf{z}}S = 0, \qquad j = 3, 4$$

are fulfilled. The solution of this problem is represented by a single-layer potential (5.40) and is determined to within an additive constant vector $\tilde{\mathbf{p}} = (c_1, c_1, c_3, c_4)$, where $\mathbf{g}$ is a solution of the solvable Fredholm integral equation (5.41), $c_\alpha = \text{const}$, $\alpha = 1, 3, 4$ and $\mathbf{f} = (f_1, f_2, f_3, f_4)$.

(c) Similarly we can consider the cases where the homogeneous equation (5.43) has two or one linearly independent solutions. For example, if

$$\gamma_{12} > 0, \qquad \gamma_{13} > 0, \qquad \gamma_{14} = \gamma_{23} = \gamma_{24} = \gamma_{34} = 0, \tag{5.50}$$

then the homogeneous integral equation (5.43) has two linearly independent solutions $\boldsymbol{\chi}^{(31)} = (1,1,1,0)$ and $\boldsymbol{\chi}^{(32)} = (0,0,0,1)$. Moreover, if

$$\gamma_{1j} > 0, \qquad \gamma_{23} = \gamma_{24} = \gamma_{34} = 0, \qquad j = 2,3,4, \tag{5.51}$$

then the homogeneous integral equation (5.43) has one linearly independent solution $\boldsymbol{\chi}^{(41)} = (1,1,1,1)$.

Consequently, the next results are proved in a similar manner.

Theorem 5.26 *If $S \in C^{1,\nu}$, $\mathbf{f} \in C^{0,\nu'}(S)$, $0 < \nu' < \nu \le 1$, and the nonnegative values $\gamma_{\alpha\beta}$ satisfy the assumption (5.50), then the problem $(II_{re})^+_{\mathbf{0},\mathbf{f}}$ is solvable only when the conditions*

$$\int_S [f_1(\mathbf{z}) + f_2(\mathbf{z}) + f_3(\mathbf{z})]\, d_{\mathbf{z}}S = 0, \qquad \int_S f_4(\mathbf{z}) d_{\mathbf{z}}S = 0$$

are fulfilled. The solution of this problem is represented by a single-layer potential (5.40) and is determined to within an additive constant vector $\tilde{\mathbf{p}} = (c_1, c_1, c_1, c_4)$, where $\mathbf{g}$ is a solution of the solvable Fredholm integral equation (5.41), $c_1 = \text{const}$, $c_4 = \text{const}$ and $\mathbf{f} = (f_1, f_2, f_3, f_4)$.

Theorem 5.27 *If $S \in C^{1,\nu}$, $\mathbf{f} \in C^{0,\nu'}(S)$, $0 < \nu' < \nu \le 1$, and the nonnegative values $\gamma_{\alpha\beta}$ satisfy the assumption (5.51), then the problem $(II_{re})^+_{\mathbf{0},\mathbf{f}}$ is solvable only when the condition*

$$\int_S [f_1(\mathbf{z}) + f_2(\mathbf{z}) + f_3(\mathbf{z}) + f_4(\mathbf{z})]\, d_{\mathbf{z}}S = 0$$

is fulfilled. The solution of this problem is represented by a single-layer potential (5.40) and is determined to within an additive constant vector $\tilde{\mathbf{p}} = (c_1, c_1, c_1, c_1)$, where $\mathbf{g}$ is a solution of the solvable Fredholm integral equation (5.41), $c_1 = \text{const}$ and $\mathbf{f} = (f_1, f_2, f_3, f_4)$.

Chapter 6
Problems of Steady Vibrations in Elasticity

In this chapter, the basic internal and external BVPs of steady vibrations in the linear theory of elasticity for quadruple porosity materials are investigated by means of the potential method and the theory of singular integral equations.

In Sect. 6.1, the basic internal and external BVPs of steady vibrations of this theory are formulated.

In Sect. 6.2, the uniqueness theorems for classical solutions of these BVPs are proved by employing techniques based on Green's identities.

Then, in Sect. 6.3, the surface and volume potentials of the considered theory are constructed and their basic properties are established.

Furthermore, in Sect. 6.4, some useful singular integral operators are introduced.

Finally, in Sect. 6.5, the BVPs of steady vibrations are reduced to the equivalent singular integral equations for which Fredholm's theorems are valid. The existence theorems for classical solutions of the BVPs of steady vibrations are proved by means of the potential method.

6.1 Basic Boundary Value Problems

The basic internal and external BVPs of steady vibrations in the linear theory of elasticity for quadruple porosity materials are formulated as follows.

Find a regular (classical) solution of the class $\mathfrak{B}^{(s)}$ in Ω^+ to system

$$\mathcal{A}(\mathbf{D_x}, \omega)\, \mathbf{V}(\mathbf{x}) = \mathbf{F}(\mathbf{x}) \tag{6.1}$$

for $\mathbf{x} \in \Omega^+$ satisfying the boundary condition

$$\lim_{\Omega^+ \ni \mathbf{x} \to \mathbf{z} \in S} \mathbf{V}(\mathbf{x}) \equiv \{\mathbf{V}(\mathbf{z})\}^+ = \mathbf{f}(\mathbf{z}) \tag{6.2}$$

M. Svanadze, *Potential Method in Mathematical Theories of Multi-Porosity Media*, Interdisciplinary Applied Mathematics 51,
https://doi.org/10.1007/978-3-030-28022-2_6

in the *Problem* $(I_s)^+_{\mathbf{F},\mathbf{f}}$, and

$$\lim_{\Omega^+\ni\mathbf{x}\to\mathbf{z}\in S} \mathcal{P}(\mathbf{D_x},\mathbf{n}(\mathbf{z}))\mathbf{V}(\mathbf{x}) \equiv \{\mathcal{P}(\mathbf{D_z},\mathbf{n})\mathbf{V}(\mathbf{z})\}^+ = \mathbf{f}(\mathbf{z}) \tag{6.3}$$

in the *Problem* $(II_s)^+_{\mathbf{F},\mathbf{f}}$, where the matrix differential operators $\mathcal{A}(\mathbf{D_x})$ and $\mathcal{P}(\mathbf{D_z},\mathbf{n})$ are defined by (1.36) and (1.45), respectively; $\mathbf{F}$ and $\mathbf{f}$ are prescribed seven-component vector functions.

Find a regular (classical) solution of the class $\mathfrak{B}^{(s)}$ to system (6.1) for $\mathbf{x}\in\Omega^-$ satisfying the boundary condition

$$\lim_{\Omega^-\ni\mathbf{x}\to\mathbf{z}\in S} \mathbf{V}(\mathbf{x}) \equiv \{\mathbf{V}(\mathbf{z})\}^- = \mathbf{f}(\mathbf{z}) \tag{6.4}$$

in the *Problem* $(I_s)^-_{\mathbf{F},\mathbf{f}}$, and

$$\lim_{\Omega^-\ni\mathbf{x}\to\mathbf{z}\in S} \mathcal{P}(\mathbf{D_x},\mathbf{n}(\mathbf{z}))\mathbf{V}(\mathbf{x}) \equiv \{\mathcal{P}(\mathbf{D_z},\mathbf{n})\mathbf{V}(\mathbf{z})\}^- = \mathbf{f}(\mathbf{z}) \tag{6.5}$$

in the *Problem* $(II_s)^-_{\mathbf{F},\mathbf{f}}$. Here $\mathbf{F}$ and $\mathbf{f}$ are prescribed seven-component vector functions, and $\operatorname{supp}\mathbf{F}$ is a finite domain in Ω^-. Note that the class $\mathfrak{B}^{(s)}$ of vector functions is given by Definition 3.1 (see Sect. 3.4).

6.2 Uniqueness Theorems

One of the oldest problems of the classical theory of elasticity consists in establishing the best conditions on the domain and the Lamé constants λ and μ assuring that the BVPs have at most one solution.

Henceforth, the sufficient conditions will be established for the uniqueness of regular (classical) solutions of BVPs in the theories of elasticity and thermoelasticity for quadruple porosity materials occupying arbitrary 3D domains with a smooth surface.

Moreover, establishing the necessary condition for the uniqueness of solutions represents an open problem in the classical theories of elasticity and thermoelasticity, and generally, in the theory of partial differential equations. The necessary condition for the uniqueness of solutions is established only in the classical theory of elasticity for domains with special geometry. The basic results and historical information on the uniqueness theorems of classical elasticity can be found in the book by Knops and Payne [204] and in the papers by Fosdick et al. [141] and Russo and Starita [292]).

In this section we study the uniqueness of regular solutions of BVPs $(\mathbb{K}_s)^+_{\mathbf{F},\mathbf{f}}$ and $(\mathbb{K}_s)^-_{\mathbf{F},\mathbf{f}}$, where $\mathbb{K}=I, II$. In the sequel we use the following.

Lemma 6.1 *If* $d_{\alpha\beta}p_\beta\overline{p_\alpha}=0$, *then* $d_{\alpha\beta}p_\beta=0$ $(\alpha=1,2,3,4)$ *and vice versa, where the matrix* $\mathbf{d}=(d_{\alpha\beta})_{4\times4}$ *is given by (1.7).*

Proof We now assume that $d_{\alpha\beta} p_\beta \overline{p_\alpha} = 0$. Then from (4.10) one may derive

$$\gamma_{12}|p_1 - p_2|^2 = 0, \qquad \gamma_{13}|p_1 - p_3|^2 = 0, \qquad \gamma_{14}|p_1 - p_4|^2 = 0,$$

$$\gamma_{23}|p_2 - p_3|^2 = 0, \qquad \gamma_{24}|p_2 - p_4|^2 = 0, \qquad \gamma_{34}|p_3 - p_4|^2 = 0.$$

These equations allow us to show the following.

$$\gamma_{12}(p_1 - p_2) = 0, \qquad \gamma_{13}(p_1 - p_3) = 0, \qquad \gamma_{14}(p_1 - p_4) = 0,$$

$$\gamma_{23}(p_2 - p_3) = 0, \qquad \gamma_{24}(p_2 - p_4) = 0, \qquad \gamma_{34}(p_3 - p_4) = 0$$

and finally, combining these relations we obtain the desired result $d_{\alpha\beta} p_\beta = 0$ for $\alpha = 1, 2, 3, 4$.

Clearly, the relations $d_{\alpha\beta} p_\beta = 0$ for $\alpha = 1, 2, 3, 4$ imply $d_{\alpha\beta} p_\beta \overline{p_\alpha} = 0$. □

We have the following consequences.

Theorem 6.1 *If*

$$\mu > 0, \tag{6.6}$$

then any two solutions of the BVP $(I_s)^+_{\mathbf{F},\mathbf{f}}$ *of the class* $\mathfrak{B}^{(s)}$ *may differ only for an additive vector* $\mathbf{V} = (\mathbf{u}, \mathbf{p})$, *where*

$$\mathbf{p}(\mathbf{x}) = \mathbf{0} \qquad for \quad \mathbf{x} \in \Omega^+, \tag{6.7}$$

and vector $\mathbf{u}$ *is a solution of the following BVP*

$$\left(\Delta + \lambda_6^2\right) \mathbf{u}(\mathbf{x}) = \mathbf{0}, \qquad \operatorname{div} \mathbf{u}(\mathbf{x}) = \mathbf{0} \qquad for \quad \mathbf{x} \in \Omega^+, \tag{6.8}$$

$$\{\mathbf{u}(\mathbf{z})\}^+ = 0 \qquad for \quad \mathbf{z} \in S, \tag{6.9}$$

where $\lambda_6^2 = \dfrac{\rho\omega^2}{\mu} > 0$. *In addition, the problems* $(I_s)^+_{\mathbf{0},\mathbf{0}}$ *and (6.8), (6.9) have the same eigenfrequencies.*

Proof Suppose that there are two regular solutions of problem $(I_s)^+_{\mathbf{F},\mathbf{f}}$ in the class $\mathfrak{B}^{(s)}$. Then their difference $\mathbf{V}$ is a regular solution of the internal homogeneous BVP $(I_s)^+_{\mathbf{0},\mathbf{0}}$. Hence, $\mathbf{V}$ is a regular solution of the homogeneous system of equations

$$\begin{aligned} &\left(\mu\,\Delta + \rho\,\omega^2\right)\mathbf{u} + (\lambda + \mu)\,\nabla \mathrm{div}\mathbf{u} - \nabla\,(\mathbf{a}\,\mathbf{p}) = \mathbf{0}, \\ &\mathbf{K}\,\Delta\,\mathbf{p} + \mathbf{c}\,\mathbf{p} + i\omega\mathbf{a}\,\mathrm{div}\mathbf{u} = \mathbf{0} \end{aligned} \tag{6.10}$$

in Ω^+ satisfying the homogeneous boundary condition

$$\{\mathbf{V}(\mathbf{z})\}^+ = 0 \qquad \text{for} \quad \mathbf{z} \in S. \tag{6.11}$$

On the basis of (6.10) and (6.11), from (3.18) we obtain for $\mathbf{u}' = \mathbf{u}$ and $\mathbf{p}' = \mathbf{p}$

$$\int\limits_{\Omega^+} W_1(\mathbf{V}, \mathbf{u}) d\mathbf{x} = 0, \qquad \int\limits_{\Omega^+} W_2(\mathbf{V}, \mathbf{p}) d\mathbf{x} = 0, \tag{6.12}$$

where

$$W_1(\mathbf{V}, \mathbf{u}) = W_0(\mathbf{u}, \mathbf{u}) - \rho\omega^2 |\mathbf{u}|^2 - a_\alpha\, p_\alpha \operatorname{div} \bar{\mathbf{u}},$$

$$W_2(\mathbf{V}, \mathbf{p}) = k_{\alpha\beta} \nabla p_\alpha \cdot \nabla p_\beta - c_{\alpha\beta}\, p_\beta \overline{p_\alpha} - i\omega\, a_\alpha \overline{p_\alpha} \operatorname{div} \mathbf{u},$$

$$W_0(\mathbf{u}, \mathbf{u}) = \frac{1}{3}(3\lambda + 2\mu)\, |\operatorname{div} \mathbf{u}|^2 \tag{6.13}$$

$$+\frac{\mu}{2} \sum_{l,j=1;\, l\neq j}^{3} \left| \frac{\partial u_j}{\partial x_l} + \frac{\partial u_l}{\partial x_j} \right|^2 + \frac{\mu}{3} \sum_{l,j=1}^{3} \left| \frac{\partial u_l}{\partial x_l} - \frac{\partial u_j}{\partial x_j} \right|^2 .$$

Then, we may derive from (6.13)

$$\operatorname{Im} W_1(\mathbf{V}, \mathbf{u}) = -\operatorname{Im} [\mathbf{a}\mathbf{p} \operatorname{div} \bar{\mathbf{u}}] \tag{6.14}$$

and by virtue of (6.14) from the first equation of (6.12) it follows that

$$\int\limits_{\Omega^+} \operatorname{Im} [\mathbf{a}\mathbf{p} \operatorname{div} \bar{\mathbf{u}}]\, d\mathbf{x} = 0. \tag{6.15}$$

On the other hand, by virtue of relations (4.10) we obtain from (6.13)

$$\operatorname{Re} W_2(\mathbf{V}, \mathbf{p}) = k_{\alpha\beta} \nabla p_\beta \cdot \nabla \overline{p_\alpha} + d_{\alpha\beta}\, p_\beta \overline{p_\alpha} - \omega \operatorname{Im} [\mathbf{a}\mathbf{p} \operatorname{div} \bar{\mathbf{u}}] . \tag{6.16}$$

Taking into account (6.15) and (6.16) from the second equation of (6.12) we have

$$\int\limits_{\Omega^+} \left[k_{\alpha\beta} \nabla p_\beta \cdot \nabla \overline{p_\alpha} + d_{\alpha\beta}\, p_\beta \overline{p_\alpha} \right] d\mathbf{x} = 0.$$

On the basis of (4.10) from the last equation it follows that

$$k_{\alpha\beta} \nabla p_\beta \cdot \nabla \overline{p_\alpha} = 0, \qquad d_{\alpha\beta}\, p_\beta \overline{p_\alpha} = 0, \tag{6.17}$$

and hence, from the first equation of (6.17) we get $\nabla p_\alpha = 0$ in Ω^+ and consequently,

$$p_\alpha(\mathbf{x}) = c_\alpha = \text{const}, \qquad \alpha = 1, 2, 3, 4 \tag{6.18}$$

for $\mathbf{x} \in \Omega^+$. By virtue of the homogeneous boundary condition (6.11) from (6.18) we obtain the Eq. (6.7). Then, by using (6.7) from the second equation of (6.10) we have the second equality of (6.8). Taking into account (6.6) the first equation of (6.10) is reduced to the first equality of (6.8). Obviously, from the boundary condition (6.11) we may also obtain (6.9).

Finally, it is easy to see that problems $(I_s)^+_{\mathbf{0},\mathbf{0}}$ and (6.8), (6.9) have the same eigenfrequencies. □

Theorem 6.2 *If the condition (6.6) is satisfied, then any two solutions of the BVP $(II_s)^+_{\mathbf{F},\mathbf{f}}$ of the class $\mathfrak{B}^{(s)}$ may differ only for an additive vector $\mathbf{V} = (\mathbf{u}, \mathbf{p})$, where $\mathbf{p}$ satisfies the condition (6.7), and the vector $\mathbf{u}$ is a solution of (6.8) satisfying the boundary condition*

$$\left\{2\frac{\partial \mathbf{u}(\mathbf{z})}{\partial \mathbf{n}(\mathbf{z})} + [\mathbf{n}(\mathbf{z}) \times \operatorname{curl} \mathbf{u}(\mathbf{z})]\right\}^+ = 0 \qquad for \quad \mathbf{z} \in S. \tag{6.19}$$

In addition, the problems $(II_s)^+_{\mathbf{0},\mathbf{0}}$ and (6.8), (6.23) have the same eigenfrequencies.

Proof Suppose that there are two regular solutions of problem $(II_s)^+_{\mathbf{F},\mathbf{f}}$ in the class $\mathfrak{B}^{(s)}$. Then their difference $\mathbf{V}$ is a regular solution of the homogeneous BVP $(II_s)^+_{\mathbf{0},\mathbf{0}}$. Hence, $\mathbf{V}$ is a regular solution of the homogeneous system (6.10) in Ω^+ satisfying the homogeneous boundary condition

$$\{\mathcal{P}(\mathbf{D_z}, \mathbf{n})\mathbf{V}(\mathbf{z})\}^+ = \mathbf{0}. \tag{6.20}$$

Quite similarly as in Theorem 6.1, on the basis of (6.10) and (6.20), from (3.18) it follows Eqs. (6.17) and (6.18). Taking into account these equations and Lemma 6.1, from system (6.10) we may deduce that

$$\mu\, \Delta\mathbf{u} + (\lambda + \mu)\, \nabla \operatorname{div} \mathbf{u} + \rho\omega^2\, \mathbf{u} = \mathbf{0} \tag{6.21}$$

and

$$b_{\alpha\beta} c_\beta = -a_\alpha \operatorname{div} \mathbf{u}, \qquad \alpha = 1, 2, 3, 4. \tag{6.22}$$

Applying the operator div to (6.21) we obtain

$$\left(\mu_0\, \Delta + \rho\, \omega^2\right) \operatorname{div} \mathbf{u} = 0. \tag{6.23}$$

If $\mu_0 = 0$, then from (6.23) we may derive the second equation of (6.8). On the other hand, if $\mu_0 \neq 0$, then applying the operator $\mu_0\,\Delta + \rho\,\omega^2$ to the equations of system (6.22) and using Eqs. (6.18) and (6.23) we get

$$b_{\alpha\beta} c_\beta = 0, \qquad \alpha = 1, 2, 3, 4. \tag{6.24}$$

Furthermore, if we employ the positive definite assumption of the matrix **b** in (6.24) (see Remark 1.2), then we may obtain Eq. (6.7). Keeping in mind the relation (6.7), from (6.22) we get the second equation of (6.8).

Now taking into accounts (6.6) and (6.7) the first equation of (6.10) is reduced to the first equality of (6.8). Clearly, on the basis of the second equation of (6.8) the boundary condition (6.20) is reduced to (6.19). Moreover, it is easy to see that problems $(II_s)^+_{\mathbf{0},\mathbf{0}}$ and (6.8), (6.19) have the same eigenfrequencies. □

Theorem 6.3 *If the condition (6.6) is satisfied, then the external BVP* $(\mathbb{K}_s)^-_{\mathbf{F},\mathbf{f}}$ *has one regular solution of the class* $\mathfrak{B}^{(s)}$, *where* $\mathbb{K} = I, II$.

Proof Suppose that there are two regular solutions of problem $(\mathbb{K}_s)^-_{\mathbf{F},\mathbf{f}}$ of the class $\mathfrak{B}^{(s)}$. Then their difference **V** is a regular solution of the homogeneous BVP $(\mathbb{K}_s)^-_{\mathbf{0},\mathbf{0}}$. Hence, **V** is a regular solution of the homogeneous system (6.10) in Ω^- satisfying the homogeneous boundary condition

$$\{\mathbf{V}(\mathbf{z})\}^- = 0 \tag{6.25}$$

in the problem $(I_s)^-_{\mathbf{0},\mathbf{0}}$ and

$$\{\mathcal{P}(\mathbf{D_z}, \mathbf{n})\mathbf{V}(\mathbf{z})\}^- = 0 \tag{6.26}$$

in the problem $(II_s)^-_{\mathbf{0},\mathbf{0}}$.

Employing (6.10), (6.25), and (6.26) in the following identities (see (3.18))

$$\int_{\Omega^-} \left[\mathcal{A}^{(1)}(\mathbf{D_x}, \omega)\,\mathbf{V}(\mathbf{x}) \cdot \mathbf{u}(\mathbf{x}) + W_1(\mathbf{V}, \mathbf{u})\right] d\mathbf{x}$$

$$= -\int_S \mathcal{P}^{(1)}(\mathbf{D_z}, \mathbf{n})\mathbf{V}(\mathbf{z}) \cdot \mathbf{u}(\mathbf{z})\, d_{\mathbf{z}}S,$$

$$\int_{\Omega^-} \left[\mathcal{A}^{(2)}(\mathbf{D_x}, \omega)\,\mathbf{V}(\mathbf{x}) \cdot \mathbf{p}(\mathbf{x}) + W_2(\mathbf{V}, \mathbf{p})\right] d\mathbf{x}$$

$$= -\int_S \mathcal{P}^{(2)}(\mathbf{D_z}, \mathbf{n})\mathbf{p}(\mathbf{z}) \cdot \mathbf{p}(\mathbf{z})\, d_{\mathbf{z}}S,$$

we may derive

$$\int_{\Omega^-} W_1(\mathbf{V}, \mathbf{u}) d\mathbf{x} = 0, \qquad \int_{\Omega^-} W_2(\mathbf{V}, \mathbf{p}) d\mathbf{x} = 0, \tag{6.27}$$

where $W_1(\mathbf{V}, \mathbf{u})$ and $W_2(\mathbf{V}, \mathbf{p})$ are defined by (6.13), the operators $\mathcal{P}^{(1)}(\mathbf{D_z}, \mathbf{n})$ and $\mathcal{P}^{(2)}(\mathbf{D_z}, \mathbf{n})$ are given in Sect. 3.4.1.

In a similar manner as in Theorem 6.1, from (6.27) we obtain the relation (6.18). Using the radiation conditions (3.14) in (6.18) we may deduce that

$$\mathbf{p}(\mathbf{x}) = \mathbf{0} \qquad \text{for} \quad \mathbf{x} \in \Omega^-. \tag{6.28}$$

In addition, combining (6.6) and (6.28) with (6.10) we may further conclude that

$$\left(\Delta + \lambda_6^2\right) \mathbf{u}(\mathbf{x}) = \mathbf{0}, \qquad \operatorname{div} \mathbf{u}(\mathbf{x}) = \mathbf{0} \qquad \text{for} \quad \mathbf{x} \in \Omega^-, \tag{6.29}$$

i.e., the vector $\mathbf{u}$ is a solution of Helmholtz equation in Ω^- (see the first equation of (6.29)), satisfying the boundary condition (6.25) or (6.26), and the radiation conditions (3.13) and (3.14). Clearly, we have $\mathbf{u}(\mathbf{x}) \equiv \mathbf{0}$ for $\mathbf{x} \in \Omega^-$. Hence, $\mathbf{V}(\mathbf{x}) \equiv \mathbf{0}$ for $\mathbf{x} \in \Omega^-$ and consequently, a solution to the external BVP $(\mathbb{K}_s)^-_{\mathbf{F},\mathbf{f}}$ is unique in the class $\mathfrak{B}^{(s)}$, where $\mathbb{K} = I, II$. □

Remark 6.1 We note that Theorems 6.1, 6.2, and 6.3 are proved under condition (6.6). Here the Lamé constant λ is an arbitrary real number.

6.3 Basic Properties of Potentials

We assume that the Lamé constants satisfy the inequalities

$$\mu > 0, \qquad \lambda + \mu > 0. \tag{6.30}$$

On the basis of integral representation of regular vectors (see (3.26) and (3.27)) we introduce the following notation:

(i) $\mathbf{Z}^{(s,1)}(\mathbf{x}, \mathbf{g}) = \int_S \mathbf{\Gamma}(\mathbf{x} - \mathbf{y}, \omega) \mathbf{g}(\mathbf{y}) d_{\mathbf{y}} S,$

(ii) $\mathbf{Z}^{(s,2)}(\mathbf{x}, \mathbf{g}) = \int_S [\tilde{\mathcal{P}}(\mathbf{D_y}, \mathbf{n}(\mathbf{y})) \mathbf{\Gamma}^\top(\mathbf{x} - \mathbf{y}, \omega)]^\top \mathbf{g}(\mathbf{y}) d_{\mathbf{y}} S,$

(iii) $\mathbf{Z}^{(s,3)}(\mathbf{x}, \boldsymbol{\phi}, \Omega^\pm) = \int_{\Omega^\pm} \mathbf{\Gamma}(\mathbf{x} - \mathbf{y}, \omega) \boldsymbol{\phi}(\mathbf{y}) d\mathbf{y},$

where $\mathbf{\Gamma}(\mathbf{x},\omega)$ is the fundamental matrix of the operator $\mathcal{A}(\mathbf{D_x},\omega)$ and given by (2.22), the matrix differential operator $\tilde{\mathcal{P}}(\mathbf{D_y},\mathbf{n}(\mathbf{y}))$ is defined by (3.22), $\mathbf{g}$ and $\boldsymbol{\phi}$ are seven-component vector functions.

As in the classical theory of elasticity (see, e.g., Kupradze et al. [219]), the vector functions $\mathbf{Z}^{(s,1)}(\mathbf{x},\mathbf{g})$, $\mathbf{Z}^{(s,2)}(\mathbf{x},\mathbf{g})$ and $\mathbf{Z}^{(s,3)}(\mathbf{x},\boldsymbol{\phi},\Omega^{\pm})$ are called *single-layer, double-layer*, and *volume potentials* in the steady vibration problems of the linear theory of elasticity for quadruple porosity materials, respectively.

Definition 6.1 The integral defined in the sense of the principal value is called the *singular integral.*

On the basis of the properties of fundamental solution $\Gamma(\mathbf{x},\omega)$ (see Sect. 2.5) we have the following results.

Theorem 6.4 *If* $S\in C^{m+1,\nu}$, $\mathbf{g}\in C^{m,\nu'}(S)$, $0<\nu'<\nu\le 1$, *and* m *is a nonnegative integer, then:*

(a)

$$\mathbf{Z}^{(s,1)}(\cdot,\mathbf{g})\in C^{0,\nu'}(\mathbb{R}^3)\cap C^{m+1,\nu'}(\overline{\Omega^{\pm}})\cap C^{\infty}(\Omega^{\pm}),$$

(b)

$$\mathcal{A}(\mathbf{D_x},\omega)\,\mathbf{Z}^{(s,1)}(\mathbf{x},\mathbf{g})=\mathbf{0},$$

(c)

$$\left\{\mathcal{P}(\mathbf{D_z},\mathbf{n}(\mathbf{z}))\,\mathbf{Z}^{(s,1)}(\mathbf{z},\mathbf{g})\right\}^{\pm}=\mp\frac{1}{2}\,\mathbf{g}(\mathbf{z})+\mathcal{P}(\mathbf{D_z},\mathbf{n}(\mathbf{z}))\,\mathbf{Z}^{(s,1)}(\mathbf{z},\mathbf{g}),$$

(d)

$$\mathcal{P}(\mathbf{D_z},\mathbf{n}(\mathbf{z}))\,\mathbf{Z}^{(s,1)}(\mathbf{z},\mathbf{g})$$

is a singular integral, where $\mathbf{z}\in S$, $\mathbf{x}\in\Omega^{\pm}$ *and*

$$\left\{\mathcal{P}(\mathbf{D_z},\mathbf{n}(\mathbf{z}))\,\mathbf{Z}^{(s,1)}(\mathbf{z},\mathbf{g})\right\}^{\pm}\equiv\lim_{\Omega^{\pm}\ni\mathbf{x}\to\,\mathbf{z}\in S}\mathcal{P}(\mathbf{D_x},\mathbf{n}(\mathbf{z}))\,\mathbf{Z}^{(s,1)}(\mathbf{x},\mathbf{g}),$$

(e) the single-layer potential $\mathbf{Z}^{(s,1)}(\mathbf{x},\mathbf{g})\equiv\mathbf{0}$ *for* $\mathbf{x}\in\Omega^{+}$ *(or* $\mathbf{x}\in\Omega^{-}$*) if and only if* $\mathbf{g}(\mathbf{z})\equiv\mathbf{0}$*, where* $\mathbf{z}\in S$.

Theorem 6.5 *If* $S\in C^{m+1,\nu}$, $\mathbf{g}\in C^{m,\nu'}(S)$, $0<\nu'<\nu\le 1$, *then:*

(a)

$$\mathbf{Z}^{(s,2)}(\cdot,\mathbf{g})\in C^{m,\nu'}(\overline{\Omega^{\pm}})\cap C^{\infty}(\Omega^{\pm}),$$

(b)

$$\mathcal{A}(\mathbf{D_x}, \omega)\, \mathbf{Z}^{(s,2)}(\mathbf{x}, \mathbf{g}) = \mathbf{0},$$

(c)

$$\left\{\mathbf{Z}^{(s,2)}(\mathbf{z}, \mathbf{g})\right\}^{\pm} = \pm\frac{1}{2}\mathbf{g}(\mathbf{z}) + \mathbf{Z}^{(s,2)}(\mathbf{z}, \mathbf{g}), \tag{6.31}$$

for the nonnegative integer m,

(d) $\mathbf{Z}^{(s,2)}(\mathbf{z}, \mathbf{g})$ *is a singular integral, where* $\mathbf{z} \in S$,

(e)

$$\left\{\mathcal{P}(\mathbf{D_z}, \mathbf{n}(\mathbf{z}))\, \mathbf{Z}^{(s,2)}(\mathbf{z}, \mathbf{g})\right\}^{+} = \left\{\mathcal{P}(\mathbf{D_z}, \mathbf{n}(\mathbf{z}))\, \mathbf{Z}^{(s,2)}(\mathbf{z}, \mathbf{g})\right\}^{-},$$

for the natural number m, where $\mathbf{z} \in S$, $\mathbf{x} \in \Omega^{\pm}$ *and*

$$\left\{\mathbf{Z}^{(s,2)}(\mathbf{z}, \mathbf{g})\right\}^{\pm} \equiv \lim_{\Omega^{\pm} \ni \mathbf{x} \to \mathbf{z} \in S} \mathbf{Z}^{(s,2)}(\mathbf{x}, \mathbf{g}).$$

Theorem 6.6 *If* $S \in C^{1,\nu}$, $\boldsymbol{\phi} \in C^{0,\nu'}(\Omega^+)$, $0 < \nu' < \nu \le 1$, *then:*

(a)

$$\mathbf{Z}^{(s,3)}(\cdot, \boldsymbol{\phi}, \Omega^+) \in C^{1,\nu'}(\mathbb{R}^3) \cap C^2(\Omega^+) \cap C^{2,\nu'}\left(\overline{\Omega_0^+}\right),$$

(b)

$$\mathcal{A}(\mathbf{D_x}, \omega)\, \mathbf{Z}^{(s,3)}(\mathbf{x}, \boldsymbol{\phi}, \Omega^+) = \boldsymbol{\phi}(\mathbf{x}),$$

where $\mathbf{x} \in \Omega^+$, Ω_0^+ *is a domain in* $\mathbb{R}^3$ *and* $\overline{\Omega_0^+} \subset \Omega^+$.

Theorem 6.7 *If* $S \in C^{1,\nu}$, $\text{supp}\boldsymbol{\phi} = \Omega \subset \Omega^-$, $\boldsymbol{\phi} \in C^{0,\nu'}(\Omega^-)$, $0 < \nu' < \nu \le 1$, *then:*

(a)

$$\mathbf{Z}^{(s,3)}(\cdot, \boldsymbol{\phi}, \Omega^-) \in C^{1,\nu'}(\mathbb{R}^3) \cap C^2(\Omega^-) \cap C^{2,\nu'}(\overline{\Omega_0^-}),$$

(b)

$$\mathcal{A}(\mathbf{D_x}, \omega)\, \mathbf{Z}^{(s,3)}(\mathbf{x}, \boldsymbol{\phi}, \Omega^-) = \boldsymbol{\phi}(\mathbf{x}),$$

where $\mathbf{x} \in \Omega^-$, Ω *is a finite domain in* $\mathbb{R}^3$ *and* $\overline{\Omega_0^-} \subset \Omega^-$.

6.4 Singular Integral Operators

In this and next five chapters the BVPs of the theories of elasticity and thermoelasticity for quadruple porosity materials are reduced to the equivalent singular integral equations. The basic theory of singular integral equations are given in the books Kupradze et al. [219], Mikhlin [252], and Mikhlin and Prössdorf [253].

In 1936, the concept of the symbol was first introduced by Mikhlin [249–251] for singular integral operators on a plane and two-dimensional manifolds, and by Giraud [155] for a multi-dimensional singular integral operator. This concept is extended to pseudo-differential operators by Kohn and Nirenberg [205]. The general concept of symbol is presented by Mikhlin and Prössdorf [253].

The definitions of a normal type singular integral operator, the symbol and the index of operator, and Fredholm's theorems for the singular integral equations are given in the books of Kupradze et al. [219] and Mikhlin [252].

Let $\mathbf{K}(\mathbf{x}, \mathbf{y})$ be a singular kernel of the integral operator $\mathcal{K}$, where

$$\mathcal{K}\mathbf{g}(\mathbf{x}) \equiv \mathbf{g}(\mathbf{x}) + \int_S \mathbf{K}(\mathbf{x}, \mathbf{y})\mathbf{g}(\mathbf{y})d_{\mathbf{y}}S$$

and $\mathbf{g}$ is a vector function on S. It is well known that if $\mathcal{K}$ is of the normal type with an index equal to zero, then Fredholm's theorems are valid for the singular integral equation $\mathcal{K}\mathbf{g}(\mathbf{x}) = \mathbf{f}(\mathbf{x})$, where $\mathbf{g}$ and $\mathbf{f}$ are required and prescribed vector functions, respectively and $\mathbf{x} \in S$ (for details see Kupradze et al. [219] and Mikhlin [252]).

We introduce the notation

$$\begin{aligned}
\mathcal{K}^{(s,1)}\,\mathbf{g}(\mathbf{z}) &\equiv \frac{1}{2}\,\mathbf{g}(\mathbf{z}) + \mathbf{Z}^{(s,2)}(\mathbf{z}, \mathbf{g}),\\
\mathcal{K}^{(s,2)}\,\mathbf{g}(\mathbf{z}) &\equiv -\frac{1}{2}\,\mathbf{g}(\mathbf{z}) + \boldsymbol{\mathcal{P}}(\mathbf{D}_{\mathbf{z}}, \mathbf{n}(\mathbf{z}))\mathbf{Z}^{(s,1)}(\mathbf{z}, \mathbf{g}),\\
\mathcal{K}^{(s,3)}\mathbf{g}(\mathbf{z}) &\equiv -\frac{1}{2}\,\mathbf{g}(\mathbf{z}) + \mathbf{Z}^{(s,2)}(\mathbf{z}, \mathbf{g}),\\
\mathcal{K}^{(s,4)}\mathbf{g}(\mathbf{z}) &\equiv \frac{1}{2}\,\mathbf{g}(\mathbf{z}) + \boldsymbol{\mathcal{P}}(\mathbf{D}_{\mathbf{z}}, \mathbf{n}(\mathbf{z}))\mathbf{Z}^{(s,1)}(\mathbf{z}, \mathbf{g}),\\
\mathcal{K}^{(s)}_{\chi}\mathbf{g}(\mathbf{z}) &\equiv \frac{1}{2}\,\mathbf{g}(\mathbf{z}) + \chi\,\mathbf{Z}^{(s,2)}(\mathbf{z}, \mathbf{g})
\end{aligned} \tag{6.32}$$

for $\mathbf{z} \in S$, where χ is a complex number. Obviously, on the basis of Theorems 6.4 and 6.5, $\mathcal{K}^{(s,j)}$ and $\mathcal{K}^{(s)}_{\chi}$ are the singular integral operators $(j = 1, 2, 3, 4)$.

Let $\boldsymbol{\sigma}^{(s,j)} = (\sigma^{(s,j)}_{lm})_{7\times 7}$ be the symbol of the singular integral operator $\mathcal{K}^{(s,j)}$ $(j = 1, 2, 3, 4)$. Taking into account (6.30) and (6.32) we may see that

$$\det \boldsymbol{\sigma}^{(s,1)} = -\det \boldsymbol{\sigma}^{(s,2)} = -\det \boldsymbol{\sigma}^{(s,3)} = \det \boldsymbol{\sigma}^{(s,4)}$$
$$= -\frac{1}{128}\left[1 - \frac{\mu^2}{(\lambda+2\mu)^2}\right] = -\frac{(\lambda+\mu)(\lambda+3\mu)}{128(\lambda+2\mu)^2} < 0. \tag{6.33}$$

Hence, the operator $\mathcal{K}^{(s,j)}$ is of the normal type, where $j = 1, 2, 3, 4$.

Let $\boldsymbol{\sigma}_\chi^{(s)}$ and $\text{ind}\, \mathcal{K}_\chi^{(s)}$ be the symbol and the index of the operator $\mathcal{K}_\chi^{(s)}$, respectively. It may be easily shown that

$$\det \boldsymbol{\sigma}_\chi^{(s)} = \frac{\mu^2\chi^2 - (\lambda+2\mu)^2}{128(\lambda+2\mu)^2}$$

and $\det \boldsymbol{\sigma}_\chi^{(s)}$ vanishes only at two points χ_1 and χ_2 of the complex plane. By virtue of (6.33) and $\det \boldsymbol{\sigma}_1^{(s)} = \det \boldsymbol{\sigma}^{(s,1)}$ we get $\chi_j \neq 1$ $(j = 1, 2)$ and

$$\text{ind}\, \mathcal{K}_1^{(s)} = \text{ind}\, \mathcal{K}^{(s,1)} = \text{ind}\, \mathcal{K}_0^{(s)} = 0.$$

Quite similarly we obtain

$$\text{ind}\, \mathcal{K}^{(s,2)} = -\text{ind}\, \mathcal{K}^{(s,3)} = 0, \qquad \text{ind}\, \mathcal{K}^{(s,4)} = -\text{ind}\, \mathcal{K}^{(s,1)} = 0.$$

Thus, the singular integral operator $\mathcal{K}^{(s,j)}$ $(j = 1, 2, 3, 4)$ is of the normal type with an index equal to zero. Consequently, Fredholm's theorems are valid for the singular integral operator $\mathcal{K}^{(s,j)}$.

6.5 Existence Theorems

In this section we assume that the condition (6.30) is satisfied. In addition, by Theorems 6.6 and 6.7 the volume potential $\mathbf{Z}^{(s,3)}(\mathbf{x}, \mathbf{F}, \Omega^\pm)$ is a regular solution of the nonhomogeneous equation (6.1), where $\mathbf{F} \in C^{0,\nu'}(\Omega^\pm)$, $0 < \nu' \leq 1$ and $\text{supp}\,\mathbf{F}$ is a finite domain in Ω^-. Therefore, further we will consider problems $(\mathbb{K}_s)^+_{\mathbf{0},\mathbf{f}}$ and $(\mathbb{K}_s)^-_{\mathbf{0},\mathbf{f}}$, where $\mathbb{K} = I, II$.

Now we prove the existence theorems of a regular (classical) solution of the class $\mathfrak{B}^{(s)}$ for these BVPs.

Problem $(I_s)^+_{\mathbf{0},\mathbf{f}}$ We assume that ω is not an eigenfrequency of the BVP $(I_s)^+_{\mathbf{0},\mathbf{0}}$ (see Theorem 6.1). We seek a regular solution of the class $\mathfrak{B}^{(s)}$ to this problem in the form of the double-layer potential

$$\mathbf{V}(\mathbf{x}) = \mathbf{Z}^{(s,2)}(\mathbf{x}, \mathbf{g}) \qquad \text{for} \qquad \mathbf{x} \in \Omega^+, \tag{6.34}$$

where $\mathbf{g}$ is the required seven-component vector function.

Obviously, by Theorem 6.5 the vector function $\mathbf{V}$ is a solution of the homogeneous equation

$$\mathcal{A}(\mathbf{D}_{\mathbf{x}}, \omega)\, \mathbf{V}(\mathbf{x}) = \mathbf{0} \tag{6.35}$$

for $\mathbf{x} \in \Omega^+$. Keeping in mind the boundary condition (6.2) and using (6.31), from (6.34) we obtain, for determining the unknown vector $\mathbf{g}$, a singular integral equation

$$\mathcal{K}^{(s,1)}\, \mathbf{g}(\mathbf{z}) = \mathbf{f}(\mathbf{z}) \qquad \text{for} \quad \mathbf{z} \in S. \tag{6.36}$$

We prove that Eq. (6.36) is always solvable for an arbitrary vector $\mathbf{f}$.

Let us consider the associate homogeneous equation

$$\mathcal{K}^{(s,4)}\, \mathbf{h}(\mathbf{z}) = \mathbf{0} \qquad \text{for} \quad \mathbf{z} \in S, \tag{6.37}$$

where $\mathbf{h}$ is the required seven-component vector function. Now we prove that (6.37) has only the trivial solution.

Indeed, let $\mathbf{h}_0$ be a solution of the homogeneous equation (6.37). On the basis of Theorem 6.4 and Eq. (6.37) the vector function $\mathbf{V}_0(\mathbf{x}) = \mathbf{Z}^{(s,1)}(\mathbf{x}, \mathbf{h}_0)$ is a regular solution of the external homogeneous BVP $(II_s)^-_{\mathbf{0},\mathbf{0}}$. Using Theorem 6.3, the problem $(II_s)^-_{\mathbf{0},\mathbf{0}}$ has only the trivial solution, that is

$$\mathbf{V}_0(\mathbf{x}) \equiv \mathbf{0} \qquad \text{for} \qquad \mathbf{x} \in \Omega^-. \tag{6.38}$$

On the other hand, by Theorem 6.4 and (6.38) we get

$$\{\mathbf{V}_0(\mathbf{z})\}^+ = \{\mathbf{V}_0(\mathbf{z})\}^- = \mathbf{0} \qquad \text{for} \qquad \mathbf{z} \in S,$$

i.e., the vector $\mathbf{V}_0(\mathbf{x})$ is a regular solution of problem $(I_s)^+_{\mathbf{0},\mathbf{0}}$. By virtue of Theorem 6.1 and the assumption that ω is not an eigenfrequency of the BVP $(I_s)^+_{\mathbf{0},\mathbf{0}}$, the problem $(I_s)^+_{\mathbf{0},\mathbf{0}}$ has only the trivial solution, that is

$$\mathbf{V}_0(\mathbf{x}) \equiv \mathbf{0} \qquad \text{for} \qquad \mathbf{x} \in \Omega^+. \tag{6.39}$$

Now by virtue of Theorem 6.4 and the identities (6.38) and (6.39) we obtain

$$\mathbf{h}_0(\mathbf{z}) = \{\mathcal{P}(\mathbf{D}_{\mathbf{z}}, \mathbf{n})\mathbf{V}_0(\mathbf{z})\}^- - \{\mathcal{P}(\mathbf{D}_{\mathbf{z}}, \mathbf{n})\mathbf{V}_0(\mathbf{z})\}^+ = \mathbf{0} \qquad \text{for} \qquad \mathbf{z} \in S.$$

Thus, the homogeneous equation (6.37) has only the trivial solution and therefore on the basis of Fredholm's theorem the integral equation (6.36) is always solvable for an arbitrary vector $\mathbf{f}$. We have thereby proved the following.

Theorem 6.8 *If* $S \in C^{2,\nu}$, $\mathbf{f} \in C^{1,\nu'}(S)$, $0 < \nu' < \nu \le 1$, *the condition (6.30) is satisfied and* ω *is not an eigenfrequency of the BVP* $(I_s)^+_{\mathbf{0},\mathbf{0}}$, *then a regular solution of the class* $\mathfrak{B}^{(s)}$ *of the internal BVP* $(I_s)^+_{\mathbf{0},\mathbf{f}}$ *exists, is unique, and is represented by double-layer potential (6.34), where* $\mathbf{g}$ *is a solution of the singular integral equation (6.36) which is always solvable for an arbitrary vector* $\mathbf{f}$.

Problem $(II_s)^+_{\mathbf{0},\mathbf{f}}$ We assume that ω is not an eigenfrequency of the BVP $(II_s)^+_{\mathbf{0},\mathbf{0}}$ (see Theorem 6.2). We seek a regular solution of the class $\mathfrak{B}^{(s)}$ to this problem in the form of the single-layer potential

$$\mathbf{V}(\mathbf{x}) = \mathbf{Z}^{(s,1)}(\mathbf{x}, \mathbf{g}) \qquad \text{for} \qquad \mathbf{x} \in \Omega^+, \tag{6.40}$$

where $\mathbf{g}$ is the required seven-component vector function.

Obviously, by Theorem 6.4 the vector function $\mathbf{V}$ is a solution of (6.35) for $\mathbf{x} \in \Omega^+$. Keeping in mind the boundary condition (6.3) and using Theorem 6.4, from (6.40) we obtain, for determining the unknown vector $\mathbf{g}$, a singular integral equation

$$\mathcal{K}^{(s,2)}\, \mathbf{g}(\mathbf{z}) = \mathbf{f}(\mathbf{z}) \qquad \text{for} \quad \mathbf{z} \in S. \tag{6.41}$$

We prove that Eq. (6.41) is always solvable for an arbitrary vector $\mathbf{f}$.

Let us consider the homogeneous equation

$$\mathcal{K}^{(s,2)}\, \mathbf{g}_0(\mathbf{z}) = \mathbf{0} \qquad \text{for} \quad \mathbf{z} \in S, \tag{6.42}$$

where $\mathbf{g}_0$ is the required seven-component vector function. Now we prove that (6.42) has only the trivial solution. On the basis of Theorem 6.4 and Eq. (6.42) the vector function $\mathbf{V}_0(\mathbf{x}) = \mathbf{Z}^{(1)}(\mathbf{x}, \mathbf{g}_0)$ is a regular solution of the internal homogeneous BVP $(II_s)^+_{\mathbf{0},\mathbf{0}}$. Using Theorem 6.2 and the assumption that ω is not an eigenfrequency of the problem $(II_s)^+_{\mathbf{0},\mathbf{0}}$, this problem has only the trivial solution, that is

$$\mathbf{V}_0(\mathbf{x}) \equiv \mathbf{0} \qquad \text{for} \qquad \mathbf{x} \in \Omega^+. \tag{6.43}$$

On the other hand, by Theorem 6.4 and (6.43) we get

$$\{\mathbf{V}_0(\mathbf{z})\}^- = \{\mathbf{V}_0(\mathbf{z})\}^+ = \mathbf{0} \qquad \text{for} \qquad \mathbf{z} \in S,$$

i.e., the vector $\mathbf{V}_0(\mathbf{x})$ is a regular solution of problem $(I_s)^-_{\mathbf{0},\mathbf{0}}$. Using Theorem 6.3 the problem $(I_s)^-_{\mathbf{0},\mathbf{0}}$ has only the trivial solution, that is

$$\mathbf{V}_0(\mathbf{x}) \equiv \mathbf{0} \qquad \text{for} \qquad \mathbf{x} \in \Omega^-. \tag{6.44}$$

Now by virtue of Theorem 6.4 and the identities (6.43) and (6.44) we obtain

$$\mathbf{g}_0(\mathbf{z}) = \{\mathcal{P}(\mathbf{D_z}, \mathbf{n})\mathbf{V}_0(\mathbf{z})\}^- - \{\mathcal{P}(\mathbf{D_z}, \mathbf{n})\mathbf{V}_0(\mathbf{z})\}^+ = \mathbf{0} \qquad \text{for} \qquad \mathbf{z} \in S.$$

Thus, the homogeneous equation (6.42) has only the trivial solution and therefore on the basis of Fredholm's theorem the singular integral equation (6.41) is always solvable for an arbitrary vector **f**. We have thereby proved

Theorem 6.9 *If* $S \in C^{2,\nu}$, $\mathbf{f} \in C^{0,\nu'}(S)$, $0 < \nu' < \nu \le 1$, *the condition (6.30) is satisfied and* ω *is not an eigenfrequency of the BVP* $(II_s)^+_{\mathbf{0},\mathbf{0}}$, *then a regular solution of the class* $\mathfrak{B}^{(s)}$ *of the internal BVP* $(II_s)^+_{\mathbf{0},\mathbf{f}}$ *exists, is unique, and is represented by single-layer potential (6.40), where* **g** *is a solution of the singular integral equation (6.41) which is always solvable for an arbitrary vector* **f**.

Problem $(I_s)^-_{\mathbf{0},\mathbf{f}}$ We seek a regular solution of the class $\mathfrak{B}^{(s)}$ to this problem in the sum of the double-layer and single-layer potentials

$$\mathbf{V}(\mathbf{x}) = \mathbf{Z}^{(s,2)}(\mathbf{x},\mathbf{g}) + (1-i)\mathbf{Z}^{(s,1)}(\mathbf{x},\mathbf{g}) \qquad \text{for} \qquad \mathbf{x} \in \Omega^-, \tag{6.45}$$

where **g** is the required seven-component vector function.

Obviously, by Theorems 6.4 and 6.5 the vector function **V** is a solution of (6.35) for $\mathbf{x} \in \Omega^-$. Keeping in mind the boundary condition (6.4) and using (6.31), from (6.45) we obtain, for determining the unknown vector **g**, a singular integral equation

$$\mathcal{K}^{(s,5)}\mathbf{g}(\mathbf{z}) \equiv \mathcal{K}^{(s,3)}\mathbf{g}(\mathbf{z}) + (1-i)\mathbf{Z}^{(s,1)}(\mathbf{z},\mathbf{g}) = \mathbf{f}(\mathbf{z}) \qquad \text{for} \quad \mathbf{z} \in S. \tag{6.46}$$

We prove that Eq. (6.46) is always solvable for an arbitrary vector **f**. Clearly, the singular integral operator $\mathcal{K}^{(s,5)}$ is of the normal type and $\text{ind}\,\mathcal{K}^{(s,5)} = \text{ind}\,\mathcal{K}^{(s,3)} = 0$.

Now we prove that the homogeneous equation

$$\mathcal{K}^{(s,5)}\mathbf{g}_0(\mathbf{z}) = \mathbf{0} \qquad \text{for} \quad \mathbf{z} \in S \tag{6.47}$$

has only a trivial solution. Indeed, let $\mathbf{g}_0$ be a solution of the homogeneous equation (6.47). Then the vector

$$\mathbf{V}_0(\mathbf{x}) \equiv \mathbf{Z}^{(s,2)}(\mathbf{x},\mathbf{g}_0) + (1-i)\mathbf{Z}^{(s,1)}(\mathbf{x},\mathbf{g}_0) \qquad \text{for} \qquad \mathbf{x} \in \Omega^- \tag{6.48}$$

is a regular solution of problem $(I_s)^-_{\mathbf{0},\mathbf{0}}$. Using Theorem 6.3 we have (6.44).

On the other hand, by Theorems 6.4 and 6.5 from (6.48) we get

$$\begin{aligned} &\{\mathbf{V}_0(\mathbf{z})\}^- - \{\mathbf{V}_0(\mathbf{z})\}^+ = -\mathbf{g}_0(\mathbf{z}), \\ &\{\mathcal{P}(\mathbf{D}_\mathbf{z},\mathbf{n})\mathbf{V}_0(\mathbf{z})\}^- - \{\mathcal{P}(\mathbf{D}_\mathbf{z},\mathbf{n})\mathbf{V}_0(\mathbf{z})\}^+ = (1-i)\mathbf{g}_0(\mathbf{z}), \qquad \text{for} \quad \mathbf{z} \in S. \end{aligned} \tag{6.49}$$

On the basis of (6.44) from (6.49) it follows that

$$\{\mathcal{P}(\mathbf{D}_\mathbf{z},\mathbf{n})\mathbf{V}_0(\mathbf{z}) + (1-i)\mathbf{V}_0(\mathbf{z})\}^+ = \mathbf{0} \qquad \text{for} \quad \mathbf{z} \in S. \tag{6.50}$$

Obviously, the vector $\mathbf{V}_0$ is a solution of Eq. (6.35) in Ω^+ satisfying the boundary condition (6.50). It is easy to see that the relation (6.50) can be written as

$$\begin{aligned} &\left\{\mathcal{P}^{(1)}(\mathbf{D_z}, \mathbf{n})\mathbf{V}_0(\mathbf{z}) + (1-i)\mathbf{u}_0(\mathbf{z})\right\}^+ = \mathbf{0}, \\ &\left\{\mathcal{P}^{(2)}(\mathbf{D_z}, \mathbf{n})\mathbf{p}_0(\mathbf{z}) + (1-i)\mathbf{p}_0(\mathbf{z})\right\}^+ = \mathbf{0}, \end{aligned} \tag{6.51}$$

where $\mathbf{u}_0$ and $\mathbf{p}_0$ are three- and four-component vector functions, respectively, $\mathbf{V}_0 = (\mathbf{u}_0, \mathbf{p}_0)$, the operators $\mathcal{P}^{(1)}(\mathbf{D_z}, \mathbf{n})$ and $\mathcal{P}^{(2)}(\mathbf{D_z}, \mathbf{n})$ are given in Sect. 3.4.1.

On the other hand, by virtue of (6.51) from (3.18) it follows that

$$\begin{aligned} &\int_{\Omega^+} W_1(\mathbf{V}_0, \mathbf{u}_0)d\mathbf{x} + (1-i)\int_S |\mathbf{u}_0(\mathbf{z})|^2 d_\mathbf{z}S = 0, \\ &\int_{\Omega^+} W_2(\mathbf{V}_0, \mathbf{p}_0)d\mathbf{x} + (1-i)\int_S |\mathbf{p}_0(\mathbf{z})|^2 d_\mathbf{z}S = 0, \end{aligned} \tag{6.52}$$

where W_1 and W_2 are given by (6.13). The relations (6.52) imply the following.

$$\begin{aligned} &\int_{\Omega^+} \operatorname{Im} W_1(\mathbf{V}_0, \mathbf{u}_0)d\mathbf{x} - \int_S |\mathbf{u}_0(\mathbf{z})|^2 d_\mathbf{z}S = 0, \\ &\int_{\Omega^+} \operatorname{Re} W_2(\mathbf{V}_0, \mathbf{p}_0)d\mathbf{x} + \int_S |\mathbf{p}_0(\mathbf{z})|^2 d_\mathbf{z}S = 0. \end{aligned}$$

Clearly, we may further conclude that

$$\begin{aligned} &\int_{\Omega^+} \left[\operatorname{Re} W_2(\mathbf{V}_0, \mathbf{p}_0) - \omega \operatorname{Im} W_1(\mathbf{V}_0, \mathbf{u}_0)\right] d\mathbf{x} \\ &\qquad + \int_S \left[\omega|\mathbf{u}_0(\mathbf{z})|^2 + |\mathbf{p}_0(\mathbf{z})|^2\right] d_\mathbf{z}S = 0. \end{aligned} \tag{6.53}$$

Then, employing (6.14) and (6.16), we have

$$\operatorname{Re} W_2(\mathbf{V}_0, \mathbf{p}_0) - \omega \operatorname{Im} W_1(\mathbf{V}_0, \mathbf{u}_0) = \mathbf{K}\nabla\mathbf{p}_0 \cdot \nabla\mathbf{p}_0 + \mathbf{d}\mathbf{p}_0 \cdot \mathbf{p}_0 \geq 0 \tag{6.54}$$

and by virtue of (6.54) from (6.53) we may deduce that

$$\int\limits_S \left[\omega|\mathbf{u}_0(\mathbf{z})|^2 + |\mathbf{p}_0(\mathbf{z})|^2\right] d_\mathbf{z}S = 0.$$

Clearly, from the last equation we obtain

$$\{\mathbf{V}_0(\mathbf{z})\}^+ = \mathbf{0} \qquad \text{for} \qquad \mathbf{z} \in S. \tag{6.55}$$

Finally, using (6.44) and (6.55) from the first equation of (6.49) we get $\mathbf{g}_0(\mathbf{z}) \equiv \mathbf{0}$ for $\mathbf{z} \in S$.

Thus, the homogeneous equation (6.47) has only the trivial solution and therefore on the basis of Fredholm's theorem the integral equation (5.46) is always solvable for an arbitrary vector $\mathbf{f}$. We have thereby proved the following.

Theorem 6.10 *If $S \in C^{2,\nu}$, $\mathbf{f} \in C^{1,\nu'}(S)$, $0 < \nu' < \nu \le 1$ and the condition (6.30) is satisfied, then a regular solution of the class $\mathfrak{B}^{(s)}$ of the external BVP $(I_s)^-_{\mathbf{0},\mathbf{f}}$ exists, is unique, and is represented by sum of double-layer and single-layer potentials (6.45), where $\mathbf{g}$ is a solution of the singular integral equation (6.46) which is always solvable for an arbitrary vector $\mathbf{f}$.*

Problem $(II_s)^-_{\mathbf{0},\mathbf{f}}$ We seek a regular solution of the class $\mathfrak{B}^{(s)}$ to this problem in the form

$$\mathbf{V}(\mathbf{x}) = \mathbf{Z}^{(s,1)}(\mathbf{x}, \mathbf{h}) + \hat{\mathbf{V}}(\mathbf{x}) \qquad \text{for} \qquad \mathbf{x} \in \Omega^-, \tag{6.56}$$

where $\mathbf{h}$ is the required seven-component vector function and $\hat{\mathbf{V}}$ is a regular solution (seven-component vector function) of the equation

$$\mathcal{A}(\mathbf{D}_\mathbf{x}, \omega)\, \hat{\mathbf{V}}(\mathbf{x}) = \mathbf{0} \qquad \text{for} \quad \mathbf{x} \in \Omega^-. \tag{6.57}$$

Keeping in mind the boundary condition (6.5) and using Theorem 6.4, from (6.56) we obtain the following singular integral equation for determining the unknown vector $\mathbf{h}$

$$\mathcal{K}^{(s,4)}\, \mathbf{h}(\mathbf{z}) = \hat{\mathbf{f}}(\mathbf{z}) \qquad \text{for} \quad \mathbf{z} \in S, \tag{6.58}$$

where

$$\hat{\mathbf{f}}(\mathbf{z}) = \mathbf{f}(\mathbf{z}) - \left\{\mathcal{P}(\mathbf{D}_\mathbf{z}, \mathbf{n})\hat{\mathbf{V}}(\mathbf{z})\right\}^-. \tag{6.59}$$

We prove that Eq. (6.58) is always solvable for an arbitrary vector $\mathbf{f}$. We assume that the homogeneous integral equation

$$\mathcal{K}^{(s,4)}\, \mathbf{h}(\mathbf{z}) = \mathbf{0} \tag{6.60}$$

has m linearly independent solutions $\{\mathbf{h}^{(l)}(\mathbf{z})\}_{l=1}^{m}$ that are assumed to be orthonormal.

By Fredholm's theorem the solvability condition of Eq. (6.58) can be written as (see the condition (4.30))

$$\int_S \left\{\mathcal{P}(\mathbf{D_z},\mathbf{n})\hat{\mathbf{V}}(\mathbf{z})\right\}^- \cdot \boldsymbol{\psi}^{(l)}(\mathbf{z})d_{\mathbf{z}}S = N_l^{(s)}, \tag{6.61}$$

where

$$N_l^{(s)} = \int_S \mathbf{f}(\mathbf{z}) \cdot \boldsymbol{\psi}^{(l)}(\mathbf{z})d_{\mathbf{z}}S$$

and $\{\boldsymbol{\psi}^{(l)}(\mathbf{z})\}_{l=1}^{m}$ is a complete system of solutions of the homogeneous associated equation of (6.60), i.e.

$$\mathcal{K}^{(s,1)}\,\boldsymbol{\psi}^{(l)}(\mathbf{z}) = \mathbf{0}, \qquad l = 1, 2, \cdots, m.$$

It is easy to see that the condition (6.61) takes the form

$$\int_S \mathbf{h}^{(l)}(\mathbf{z}) \cdot \left\{\hat{\mathbf{V}}(\mathbf{z})\right\}^- d_{\mathbf{z}}S = -N_l^{(s)}, \qquad l = 1, 2, \cdots, m. \tag{6.62}$$

Let the vector $\hat{\mathbf{V}}$ be a solution of (6.57) and satisfies the following boundary condition

$$\left\{\hat{\mathbf{V}}(\mathbf{z})\right\}^- = \tilde{\mathbf{f}}(\mathbf{z}), \tag{6.63}$$

where

$$\tilde{\mathbf{f}}(\mathbf{z}) = \sum_{l=1}^{m} N_l^{(s)}\mathbf{h}^{(l)}(\mathbf{z}) \tag{6.64}$$

which is solvable by virtue of Theorem 6.10. Because of the orthonormalization of $\{\mathbf{h}^{(l)}(\mathbf{z})\}_{l=1}^{m}$, the condition (6.62) is fulfilled automatically and the solvability of Eq. (6.58) is proved. Consequently, the existence of regular solution of problem $(II_s)^-_{\mathbf{0},\mathbf{f}}$ is proved too.

Thus, the following theorem has been proved.

Theorem 6.11 *If $S \in C^{2,\nu}$, $\mathbf{f} \in C^{0,\nu'}(S)$, $0 < \nu' < \nu \le 1$ and the condition (6.30) is satisfied, then a regular solution of the class $\mathfrak{B}^{(s)}$ of the external BVP $(II_s)^-_{\mathbf{0},\mathbf{f}}$ exists, is unique, and is represented by sum (6.56), where* **h** *is a solution of the singular integral equation (6.58) which is always solvable,* $\hat{\mathbf{V}}$ *is the solution of*

BVP (6.57), (6.63) which is always solvable, and the vector functions $\hat{\mathbf{f}}$ *and* $\tilde{\mathbf{f}}$ *are determined by (6.59) and (6.64), respectively.*

It is well known that the existence of classical solutions of the external basic BVPs for Helmholtz' equation is proved by several methods. Indeed, these methods reduce the exterior Neumann problem to the solvable integral equations. The main difference between the above-mentioned methods is the following: some of the methods require knowledge of all resonant eigenvalues and the corresponding null-spaces of the boundary integral operator and its adjoint one, but other methods reduce the exterior Neumann problem to the equivalent uniquely solvable system of boundary integral equations for arbitrary frequency parameter.

A comprehensive review and analysis of these methods for the boundary integral formulations of acoustic scattering problems are presented by Zaman [399]. An extensive useful list of references of this subject is also given in the books of Colton and Kress [102] and Hsiao and Wendland [169]. Moreover, Theorem 6.11 is proved by the same method used to establish the existence of regular solution of the second external BVP of steady vibrations in the classical theory of thermoelasticity (see Kupradze et al. [219]).

Chapter 7
Problems of Quasi-Static in Elasticity

In this chapter, the basic internal and external BVPs of steady vibrations in the quasi-static linear theory of elasticity for quadruple porosity materials are investigated by means of the potential method and the theory of singular integral equations.

In Sects. 7.1 and 7.2, the basic internal and external BVPs of steady vibrations of this theory are formulated and the uniqueness theorems are proved by Green's identities, respectively.

In Sect. 7.3, the surface and volume potentials of the considered theory are constructed and their basic properties are established.

In Sect. 7.4, some useful singular integral operators are introduced.

Then, in Sect. 7.5, the BVPs of steady vibrations in the quasi-static linear theory of elasticity for quadruple porosity materials are reduced to the equivalent singular integral equations for which Fredholm's theorems are valid. The existence theorems for classical solutions of the BVPs of steady vibrations are proved by means of the potential method.

Furthermore, in Sect. 7.6, the first BVPs $(I_q)^+_{\mathbf{0},\mathbf{f}}$ and $(I_q)^-_{\mathbf{0},\mathbf{f}}$ are investigated by means of Fredholm's integral equations.

Finally, in Sect. 7.7, the BVPs of equilibrium in the linear theory of elasticity for quadruple porosity materials are considered.

7.1 Basic Boundary Value Problems

The basic internal and external BVPs of steady vibrations in the quasi-static linear theory of elasticity for quadruple porosity materials are formulated as follows.

Find a regular (classical) solution of the class $\mathfrak{B}$ in Ω^+ to system

$$\mathcal{A}^{(q)}(\mathbf{D}_{\mathbf{x}}, \omega)\, \mathbf{V}(\mathbf{x}) = \mathbf{F}(\mathbf{x}) \tag{7.1}$$

M. Svanadze, *Potential Method in Mathematical Theories of Multi-Porosity Media*, Interdisciplinary Applied Mathematics 51,
https://doi.org/10.1007/978-3-030-28022-2_7

for $\mathbf{x} \in \Omega^+$ satisfying the boundary condition (6.2) in the *Problem* $(I_q)^+_{\mathbf{F},\mathbf{f}}$, and the boundary condition (6.3) in the *Problem* $(II_q)^+_{\mathbf{F},\mathbf{f}}$, where the matrix differential operators $\mathcal{A}^{(q)}(\mathbf{D_x})$ and $\mathcal{P}(\mathbf{D_z}, \mathbf{n})$ are defined by (1.36) and (1.45), respectively; $\mathbf{F}$ and $\mathbf{f}$ are prescribed seven-component vector functions.

Find a regular (classical) solution of the class $\mathfrak{B}$ to system (7.1) for $\mathbf{x} \in \Omega^-$ satisfying the boundary condition (6.4) in the *Problem* $(I_q)^-_{\mathbf{F},\mathbf{f}}$, and the boundary condition (6.5) in the *Problem* $(II_q)^-_{\mathbf{F},\mathbf{f}}$. Here $\mathbf{F}$ and $\mathbf{f}$ are prescribed seven-component vector functions, supp $\mathbf{F}$ is a finite domain in Ω^-. Note that the class $\mathfrak{B}$ of vector functions is given by Definition 3.2 (see Sect. 3.5).

7.2 Uniqueness Theorems

In this section we prove uniqueness of regular solutions of BVPs $(\mathbb{K}_q)^+_{\mathbf{F},\mathbf{f}}$ and $(\mathbb{K}_q)^-_{\mathbf{F},\mathbf{f}}$, where $\mathbb{K} = I, II$. We have the following consequences.

Theorem 7.1 *If the condition (6.6) is satisfied, then the internal BVP* $(I_q)^+_{\mathbf{F},\mathbf{f}}$ *admits at most one regular (classical) solution of the class* $\mathfrak{B}$.

Proof Suppose that there are two regular solutions of problem $(I_q)^+_{\mathbf{F},\mathbf{f}}$. Then their difference $\mathbf{V}$ is a regular solution of the internal homogeneous BVP $(I_q)^+_{\mathbf{0},\mathbf{0}}$. Hence, $\mathbf{V}$ is a regular solution of the system of homogeneous equations

$$\begin{aligned} &\mu\,\Delta\mathbf{u} + (\lambda+\mu)\,\nabla\mathrm{div}\mathbf{u} - \nabla\,(\mathbf{a}\,\mathbf{p}) = \mathbf{0},\\ &\mathbf{K}\,\Delta\,\mathbf{p} + \mathbf{c}\,\mathbf{p} + i\omega\mathbf{a}\,\mathrm{div}\mathbf{u} = \mathbf{0} \end{aligned} \tag{7.2}$$

in Ω^+ satisfying the homogeneous boundary condition (6.11).

On the basis of (6.11) and (7.2), from (3.18) for $\mathbf{u}' = \mathbf{u}$, $\mathbf{p}' = \mathbf{p}$ and $\rho = 0$ we obtain

$$\int_{\Omega^+} W_3(\mathbf{V},\mathbf{u})d\mathbf{x} = 0, \qquad \int_{\Omega^+} W_2(\mathbf{V},\mathbf{p})d\mathbf{x} = 0, \tag{7.3}$$

where

$$W_3(\mathbf{V},\mathbf{u}) = W_0(\mathbf{u},\mathbf{u}) - a_\alpha\, p_\alpha\, \mathrm{div}\,\overline{\mathbf{u}}, \tag{7.4}$$

$W_0(\mathbf{u},\mathbf{u})$ and $W_2(\mathbf{V},\mathbf{p})$ are defined by (6.13).

In view of (6.6) and (7.4) the functions $\mathrm{Re}\,W_2(\mathbf{V},\mathbf{p})$ and $\mathrm{Im}\,W_3(\mathbf{V},\mathbf{u})$ satisfy the relations (6.16) and

$$\mathrm{Im}\,W_3(\mathbf{V},\mathbf{u}) = -\mathrm{Im}\,[\mathbf{ap}\,\mathrm{div}\bar{\mathbf{u}}]\,,$$

respectively. We can easily verify that

$$\mathrm{Re}W_2(\mathbf{V},\mathbf{p}) - \omega\,\mathrm{Im}\,W_3(\mathbf{V},\mathbf{u}) = k_{\alpha\beta}\nabla p_\beta \cdot \nabla p_\alpha + d_{\alpha\beta} p_\beta \overline{p_\alpha}. \tag{7.5}$$

Then, employing (7.5), we may derive from (7.3)

$$\int\limits_{\Omega^+} \left[k_{\alpha\beta}\nabla p_\beta \cdot \nabla p_\alpha + d_{\alpha\beta} p_\beta \overline{p_\alpha}\right] d\mathbf{x} = 0.$$

As in Theorem 6.1 from the last equation we may deduce that

$$p_j(\mathbf{x}) = c_j = \text{const}, \qquad d_{\alpha\beta} p_\beta \overline{p_\alpha} = 0 \qquad \text{for } \mathbf{x} \in \Omega^+, \\ j = 1, 2, 3, 4. \tag{7.6}$$

By virtue of the homogeneous boundary condition (6.11) from (7.6) we have

$$\mathbf{p}(\mathbf{x}) = \mathbf{0} \qquad \text{for} \quad \mathbf{x} \in \Omega^+. \tag{7.7}$$

Furthermore, if we employ (6.6) and (7.7) in (7.2), then we may see that

$$\Delta\mathbf{u}(\mathbf{x}) = \mathbf{0}, \qquad \mathrm{div}\mathbf{u}(\mathbf{x}) = \mathbf{0} \qquad \text{for} \quad \mathbf{x} \in \Omega^+. \tag{7.8}$$

Clearly, if we now return to the homogeneous boundary condition (6.11), then from (7.8) we obtain

$$\mathbf{u}(\mathbf{x}) = \mathbf{0} \qquad \text{for} \quad \mathbf{x} \in \Omega^+.$$

Thus, in view of the last equation and (7.7) a regular solution to the problem $(I_q)^+_{\mathbf{F},\mathbf{f}}$ is unique. □

Theorem 7.2 *If the condition*

$$\mu > 0, \qquad 3\lambda + 2\mu > 0 \tag{7.9}$$

is satisfied, then any two solutions of the BVP $(II_q)^+_{\mathbf{F},\mathbf{f}}$ *of the class* $\mathfrak{B}$ *may differ only for an additive vector* $\mathbf{V} = (\mathbf{u}, \mathbf{p})$*, where* $\mathbf{p}$ *satisfies the condition (7.7),* $\mathbf{u}$ *is a rigid displacement vector*

$$\mathbf{u}(\mathbf{x}) = \mathbf{a}' + [\mathbf{b}' \times \mathbf{x}] \qquad \text{for} \quad \mathbf{x} \in \Omega^+, \tag{7.10}$$

$\mathbf{a}' = (a_1', a_2', a_3')$ *and* $\mathbf{b}' = (b_1', b_2', b_3')$ *are arbitrary complex constant three-component vectors.*

Proof Suppose that there are two regular solutions of problem $(II_q)^+_{\mathbf{F},\mathbf{f}}$. Then their difference $\mathbf{V}$ is a regular solution of the homogeneous BVP $(II_q)^+_{\mathbf{0},\mathbf{0}}$. Hence, $\mathbf{V}$ is a regular solution of the homogeneous system (7.2) in Ω^+ satisfying the homogeneous boundary condition (6.24).

Quite similarly as in Theorem 7.1, on the basis of (6.24) and (7.2), from (3.18) for $\mathbf{u}' = \mathbf{u}$, $\mathbf{p}' = \mathbf{p}$ and $\rho = 0$ we obtain Eq. (7.3) and consequently, we have the relations (7.6). Then by virtue of (7.6) and Lemma 6.1 from the second equation of (7.2) we may deduce that

$$a_\alpha \operatorname{div}\mathbf{u} = b_{\alpha\beta}\, c_\beta$$

for $\mathbf{x} \in \Omega^+$ and $\alpha = 1, 2, 3, 4$. Now employing the last equation in (7.4) we obtain

$$W_3(\mathbf{V}, \mathbf{u}) = W_0(\mathbf{u}, \mathbf{u}) + b_{\alpha\beta}\, c_\alpha\, \overline{c_\beta}. \tag{7.11}$$

In view of the relations (7.9) and (7.11) from (7.3) it follows that

$$W_0(\mathbf{u}, \mathbf{u}) = 0, \qquad b_{\alpha\beta}\, c_\alpha\, \overline{c_\beta} = 0. \tag{7.12}$$

Moreover, from the second equation of (7.12) we may also obtain (7.7). On the other hand, it is well known from the classical theory of elasticity (see Kupradze et al. [219], p. 114) that the first equation of (7.12) implies the relation (7.10) under the condition (7.9). Thus, we have the desired result. □

In a similar manner as in Theorems 7.1 and 7.2, on the basis of the conditions at infinity (3.31) we obtain the following.

Theorem 7.3 *If the condition (6.6) is satisfied, then the external BVP $(I_q)^-_{\mathbf{F},\mathbf{f}}$ has one regular solution of the class $\mathfrak{B}$, and if the condition (7.9) is satisfied, then the BVP $(II_q)^-_{\mathbf{F},\mathbf{f}}$ has one regular solution of the same class.*

7.3 Basic Properties of Potentials

Henceforth, throughout this chapter we adopt that the condition (7.9) is satisfied. On the basis of integral representation of a regular vector (3.45) we introduce the following notation:

(i) $\mathbf{Z}^{(q,1)}(\mathbf{x}, \mathbf{g}) = \displaystyle\int_S \mathbf{\Gamma}^{(q)}(\mathbf{x} - \mathbf{y}, \omega)\mathbf{g}(\mathbf{y}) d_{\mathbf{y}} S,$

(ii) $\mathbf{Z}^{(q,2)}(\mathbf{x}, \mathbf{g}) = \displaystyle\int_S [\tilde{\mathcal{P}}(\mathbf{D}_{\mathbf{y}}, \mathbf{n}(\mathbf{y}))\mathbf{\Gamma}^{(q)^\top}(\mathbf{x} - \mathbf{y}, \omega)]^\top \mathbf{g}(\mathbf{y}) d_{\mathbf{y}} S,$

(iii) $\mathbf{Z}^{(q,3)}(\mathbf{x}, \boldsymbol{\phi}, \Omega^{\pm}) = \int\limits_{\Omega^{\pm}} \boldsymbol{\Gamma}^{(q)}(\mathbf{x} - \mathbf{y}, \omega)\boldsymbol{\phi}(\mathbf{y})d\mathbf{y},$

where $\boldsymbol{\Gamma}^{(q)}(\mathbf{x}, \omega)$ is the fundamental matrix of the operator $\boldsymbol{\mathcal{A}}^{(q)}(\mathbf{D_x}, \omega)$ and defined in Theorem 2.3; the matrix differential operator $\tilde{\boldsymbol{\mathcal{P}}}(\mathbf{D_y}, \mathbf{n}(\mathbf{y}))$ is given by (3.22); $\mathbf{g}$ and $\boldsymbol{\phi}$ are seven-component vector functions.

As in the classical theory of elasticity (see, e.g., Kupradze et al. [219]), the vector functions $\mathbf{Z}^{(q,1)}(\mathbf{x}, \mathbf{g})$, $\mathbf{Z}^{(q,2)}(\mathbf{x}, \mathbf{g})$ and $\mathbf{Z}^{(q,3)}(\mathbf{x}, \boldsymbol{\phi}, \Omega^{\pm})$ are called *single-layer*, *double-layer*, and *volume potentials* in the steady vibration problems of the quasi-static linear theory of elasticity for quadruple porosity materials, respectively.

On the basis of the properties of fundamental solution $\Gamma^{(q)}(\mathbf{x}, \omega)$ (see Sect. 2.5) we have the following results.

Theorem 7.4 *If* $S \in C^{m+1,\nu}$, $\mathbf{g} \in C^{m,\nu'}(S)$, $0 < \nu' < \nu \le 1$, *and* m *is a nonnegative integer, then:*

(a)

$$\mathbf{Z}^{(q,1)}(\cdot, \mathbf{g}) \in C^{0,\nu'}(\mathbb{R}^3) \cap C^{m+1,\nu'}(\overline{\Omega^{\pm}}) \cap C^{\infty}(\Omega^{\pm}),$$

(b)

$$\boldsymbol{\mathcal{A}}^{(q)}(\mathbf{D_x}, \omega)\, \mathbf{Z}^{(q,1)}(\mathbf{x}, \mathbf{g}) = \mathbf{0},$$

(c)

$$\left\{\boldsymbol{\mathcal{P}}(\mathbf{D_z}, \mathbf{n}(\mathbf{z}))\, \mathbf{Z}^{(q,1)}(\mathbf{z}, \mathbf{g})\right\}^{\pm} = \mp\frac{1}{2}\,\mathbf{g}(\mathbf{z}) + \boldsymbol{\mathcal{P}}(\mathbf{D_z}, \mathbf{n}(\mathbf{z}))\, \mathbf{Z}^{(q,1)}(\mathbf{z}, \mathbf{g}), \tag{7.13}$$

(d)

$$\boldsymbol{\mathcal{P}}(\mathbf{D_z}, \mathbf{n}(\mathbf{z}))\, \mathbf{Z}^{(q,1)}(\mathbf{z}, \mathbf{g})$$

is a singular integral, where $\mathbf{z} \in S$, $\mathbf{x} \in \Omega^{\pm}$ *and*

$$\left\{\boldsymbol{\mathcal{P}}(\mathbf{D_z}, \mathbf{n}(\mathbf{z}))\, \mathbf{Z}^{(q,1)}(\mathbf{z}, \mathbf{g})\right\}^{\pm} \equiv \lim_{\Omega^{\pm} \ni \mathbf{x} \to\, \mathbf{z} \in S} \boldsymbol{\mathcal{P}}(\mathbf{D_x}, \mathbf{n}(\mathbf{z}))\, \mathbf{Z}^{(q,1)}(\mathbf{x}, \mathbf{g}),$$

(e) the single-layer potential $\mathbf{Z}^{(q,1)}(\mathbf{x}, \mathbf{g}) \equiv \mathbf{0}$ *for* $\mathbf{x} \in \Omega^{+}$ *(or* $\mathbf{x} \in \Omega^{-}$*) if and only if* $\mathbf{g}(\mathbf{z}) \equiv \mathbf{0}$, *where* $\mathbf{z} \in S$.

Theorem 7.5 *If* $S \in C^{m+1,\nu}$, $\mathbf{g} \in C^{m,\nu'}(S)$, $0 < \nu' < \nu \le 1$, *then:*

(a)

$$\mathbf{Z}^{(q,2)}(\cdot, \mathbf{g}) \in C^{m,\nu'}(\overline{\Omega^{\pm}}) \cap C^{\infty}(\Omega^{\pm}),$$

(b)

$$\mathcal{A}^{(q)}(\mathbf{D_x}, \omega)\, \mathbf{Z}^{(q,2)}(\mathbf{x}, \mathbf{g}) = \mathbf{0},$$

(c)

$$\left\{\mathbf{Z}^{(q,2)}(\mathbf{z}, \mathbf{g})\right\}^{\pm} = \pm\frac{1}{2}\,\mathbf{g}(\mathbf{z}) + \mathbf{Z}^{(q,2)}(\mathbf{z}, \mathbf{g}), \tag{7.14}$$

for the nonnegative integer m,

(d) $\mathbf{Z}^{(q,2)}(\mathbf{z}, \mathbf{g})$ *is a singular integral, where* $\mathbf{z} \in S$,

(e)

$$\left\{\mathcal{P}(\mathbf{D_z}, \mathbf{n}(\mathbf{z}))\, \mathbf{Z}^{(q,2)}(\mathbf{z}, \mathbf{g})\right\}^{+} = \{\mathcal{P}(\mathbf{D_z}, \mathbf{n}(\mathbf{z}))\, \mathbf{Z}^{(q,2)}(\mathbf{z}, \mathbf{g})\}^{-},$$

for the natural number m, *where* $\mathbf{z} \in S$, $\mathbf{x} \in \Omega^{\pm}$ *and*

$$\left\{\mathbf{Z}^{(q,2)}(\mathbf{z}, \mathbf{g})\right\}^{\pm} \equiv \lim_{\Omega^{\pm} \ni \mathbf{x} \to \mathbf{z} \in S} \mathbf{Z}^{(q,2)}(\mathbf{x}, \mathbf{g}).$$

Theorem 7.6 *If* $S \in C^{1,\nu}$, $\boldsymbol{\phi} \in C^{0,\nu'}(\Omega^+)$, $0 < \nu' < \nu \leq 1$, *then:*

(a)

$$\mathbf{Z}^{(q,3)}(\cdot, \boldsymbol{\phi}, \Omega^+) \in C^{1,\nu'}(\mathbb{R}^3) \cap C^2(\Omega^+) \cap C^{2,\nu'}\left(\overline{\Omega_0^+}\right),$$

(b)

$$\mathcal{A}^{(q)}(\mathbf{D_x}, \omega)\, \mathbf{Z}^{(q,3)}(\mathbf{x}, \boldsymbol{\phi}, \Omega^+) = \boldsymbol{\phi}(\mathbf{x}),$$

where $\mathbf{x} \in \Omega^+$, Ω_0^+ *is a domain in* $\mathbb{R}^3$ *and* $\overline{\Omega_0^+} \subset \Omega^+$.

Theorem 7.7 *If* $S \in C^{1,\nu}$, $\text{supp}\boldsymbol{\phi} = \Omega \subset \Omega^-$, $\boldsymbol{\phi} \in C^{0,\nu'}(\Omega^-)$, $0 < \nu' < \nu \leq 1$, *then:*

(a)

$$\mathbf{Z}^{(q,3)}(\cdot, \boldsymbol{\phi}, \Omega^-) \in C^{1,\nu'}(\mathbb{R}^3) \cap C^2(\Omega^-) \cap C^{2,\nu'}(\overline{\Omega_0^-}),$$

(b)

$$\mathcal{A}^{(q)}(\mathbf{D_x}, \omega)\, \mathbf{Z}^{(q,3)}(\mathbf{x}, \boldsymbol{\phi}, \Omega^-) = \boldsymbol{\phi}(\mathbf{x}),$$

where $\mathbf{x} \in \Omega^-$, Ω *is a finite domain in* $\mathbb{R}^3$ *and* $\overline{\Omega_0^-} \subset \Omega^-$.

7.4 Singular Integral Operators

We introduce the notation

$$\begin{aligned}
\mathcal{K}^{(q,1)}\mathbf{g}(\mathbf{z}) &\equiv \frac{1}{2}\mathbf{g}(\mathbf{z}) + \mathbf{Z}^{(q,2)}(\mathbf{z},\mathbf{g}),\\
\mathcal{K}^{(q,2)}\mathbf{g}(\mathbf{z}) &\equiv -\frac{1}{2}\mathbf{g}(\mathbf{z}) + \mathcal{P}(\mathbf{D}_\mathbf{z},\mathbf{n}(\mathbf{z}))\mathbf{Z}^{(q,1)}(\mathbf{z},\mathbf{g}),\\
\mathcal{K}^{(q,3)}\mathbf{g}(\mathbf{z}) &\equiv -\frac{1}{2}\mathbf{g}(\mathbf{z}) + \mathbf{Z}^{(q,2)}(\mathbf{z},\mathbf{g}),\\
\mathcal{K}^{(q,4)}\mathbf{g}(\mathbf{z}) &\equiv \frac{1}{2}\mathbf{g}(\mathbf{z}) + \mathcal{P}(\mathbf{D}_\mathbf{z},\mathbf{n}(\mathbf{z}))\mathbf{Z}^{(q,1)}(\mathbf{z},\mathbf{g}),\\
\mathcal{K}^{(q)}_\chi\mathbf{g}(\mathbf{z}) &\equiv \frac{1}{2}\mathbf{g}(\mathbf{z}) + \chi\,\mathbf{Z}^{(q,2)}(\mathbf{z},\mathbf{g})
\end{aligned} \tag{7.15}$$

for $\mathbf{z} \in S$, where χ is a complex number. Obviously, on the basis of Theorems 7.4 and 7.5, $\mathcal{K}^{(q,j)}$ and $\mathcal{K}^{(q)}_\chi$ are the singular integral operators ($j = 1, 2, 3, 4$).

Let $\boldsymbol{\sigma}^{(q,j)} = (\sigma^{(q,j)}_{lm})_{7\times 7}$ be the symbol of the singular integral operator $\mathcal{K}^{(q,j)}$ ($j = 1, 2, 3, 4$). Taking into account (7.9) and (7.15) we may see that

$$\begin{aligned}
\det\boldsymbol{\sigma}^{(q,1)} &= -\det\boldsymbol{\sigma}^{(q,2)} = -\det\boldsymbol{\sigma}^{(q,3)} = \det\boldsymbol{\sigma}^{(q,4)}\\
&= -\frac{1}{128}\left[1 - \frac{\mu^2}{(\lambda+2\mu)^2}\right] = -\frac{(\lambda+\mu)(\lambda+3\mu)}{128(\lambda+2\mu)^2} < 0.
\end{aligned} \tag{7.16}$$

Hence, the operator $\mathcal{K}^{(q,j)}$ is of the normal type, where $j = 1, 2, 3, 4$.

Let $\boldsymbol{\sigma}^{(q)}_\chi$ and $\operatorname{ind}\mathcal{K}^{(q)}_\chi$ be the symbol and the index of the operator $\mathcal{K}^{(q)}_\chi$, respectively. It may be easily shown that

$$\det\boldsymbol{\sigma}^{(q)}_\chi = \frac{\mu^2\chi^2 - (\lambda+2\mu)^2}{128(\lambda+2\mu)^2}$$

and $\det\boldsymbol{\sigma}^{(q)}_\chi$ vanishes only at two points χ_1 and χ_2 of the complex plane. By virtue of (7.16) and $\det\boldsymbol{\sigma}^{(q)}_1 = \det\boldsymbol{\sigma}^{(q,1)}$ we get $\chi_j \neq 1$ ($j = 1, 2$) and

$$\operatorname{ind}\mathcal{K}^{(q)}_1 = \operatorname{ind}\mathcal{K}^{(s,1)} = \operatorname{ind}\mathcal{K}^{(q)}_0 = 0.$$

Quite similarly we obtain

$$\operatorname{ind}\mathcal{K}^{(q,2)} = -\operatorname{ind}\mathcal{K}^{(q,3)} = 0, \qquad \operatorname{ind}\mathcal{K}^{(q,4)} = -\operatorname{ind}\mathcal{K}^{(q,1)} = 0.$$

Thus, the singular integral operator $\mathcal{K}^{(q,j)}$ $(j = 1, 2, 3, 4)$ is of the normal type with an index equal to zero. Consequently, Fredholm's theorems are valid for the singular integral operator $\mathcal{K}^{(q,j)}$.

7.5 Existence Theorems

By Theorems 7.6 and 7.7 the volume potential $\mathbf{Z}^{(q,3)}(\mathbf{x}, \mathbf{F}, \Omega^{\pm})$ is a regular solution of the nonhomogeneous equation (7.1), where $\mathbf{F} \in C^{0,\nu'}(\Omega^{\pm})$, $0 < \nu' \le 1$ and $\operatorname{supp} \mathbf{F}$ is a finite domain in Ω^-. Therefore, further we will consider problems $(\mathbb{K}_q)^+_{\mathbf{0},\mathbf{f}}$ and $(\mathbb{K}_q)^-_{\mathbf{0},\mathbf{f}}$, where $\mathbb{K} = I, II$.

Now we prove the existence theorems of a regular (classical) solution of the class $\mathfrak{B}$ for these BVPs.

Problem $(I_q)^+_{\mathbf{0},\mathbf{f}}$ We seek a regular solution of the class $\mathfrak{B}$ to this problem in the form of the double-layer potential

$$\mathbf{V}(\mathbf{x}) = \mathbf{Z}^{(q,2)}(\mathbf{x}, \mathbf{g}) \qquad \text{for} \qquad \mathbf{x} \in \Omega^+, \tag{7.17}$$

where $\mathbf{g}$ is the required seven-component vector function.

Obviously, by Theorem 7.5 the vector function $\mathbf{V}$ is a solution of the homogeneous equation

$$\mathcal{A}^{(q)}(\mathbf{D}_{\mathbf{x}}, \omega)\, \mathbf{V}(\mathbf{x}) = \mathbf{0} \tag{7.18}$$

for $\mathbf{x} \in \Omega^+$. Keeping in mind the boundary condition (6.2) and using (7.14), from (7.17) we obtain, for determining the unknown vector $\mathbf{g}$, a singular integral equation

$$\mathcal{K}^{(q,1)}\, \mathbf{g}(\mathbf{z}) = \mathbf{f}(\mathbf{z}) \qquad \text{for} \quad \mathbf{z} \in S. \tag{7.19}$$

We prove that Eq. (7.19) is always solvable for an arbitrary vector $\mathbf{f}$.

Let us consider the associate homogeneous equation

$$\mathcal{K}^{(q,4)}\, \mathbf{h}(\mathbf{z}) = \mathbf{0} \qquad \text{for} \quad \mathbf{z} \in S, \tag{7.20}$$

where $\mathbf{h}$ is the required seven-component vector function. Now we prove that (7.20) has only the trivial solution.

Indeed, let $\mathbf{h}_0$ be a solution of the homogeneous equation (7.20). On the basis of Theorem 7.4 and Eq. (7.20) the vector function $\mathbf{V}_0(\mathbf{x}) = \mathbf{Z}^{(q,1)}(\mathbf{x}, \mathbf{h}_0)$ is a regular solution of the external homogeneous BVP $(II_q)^-_{\mathbf{0},\mathbf{0}}$. Using Theorem 7.3, the problem $(II_q)^-_{\mathbf{0},\mathbf{0}}$ has only the trivial solution, that is

$$\mathbf{V}_0(\mathbf{x}) \equiv \mathbf{0} \qquad \text{for} \qquad \mathbf{x} \in \Omega^-. \tag{7.21}$$

On the other hand, by Theorem 7.4 and (7.21) we get

$$\{\mathbf{V}_0(\mathbf{z})\}^+ = \{\mathbf{V}_0(\mathbf{z})\}^- = \mathbf{0} \qquad \text{for} \qquad \mathbf{z} \in S,$$

i.e., the vector $\mathbf{V}_0(\mathbf{x})$ is a regular solution of problem $(I_q)^+_{\mathbf{0},\mathbf{0}}$. By virtue of Theorem 7.1, the problem $(I_q)^+_{\mathbf{0},\mathbf{0}}$ has only the trivial solution, that is

$$\mathbf{V}_0(\mathbf{x}) \equiv \mathbf{0} \qquad \text{for} \qquad \mathbf{x} \in \Omega^+. \tag{7.22}$$

In view of (7.21), (7.22), and identity (7.13) we obtain

$$\mathbf{h}_0(\mathbf{z}) = \{\mathcal{P}(\mathbf{D}_\mathbf{z}, \mathbf{n})\mathbf{V}_0(\mathbf{z})\}^- - \{\mathcal{P}(\mathbf{D}_\mathbf{z}, \mathbf{n})\mathbf{V}_0(\mathbf{z})\}^+ = \mathbf{0} \qquad \text{for} \qquad \mathbf{z} \in S.$$

Thus, the homogeneous equation (7.20) has only the trivial solution and therefore on the basis of Fredholm's theorem the integral equation (7.19) is always solvable for an arbitrary vector $\mathbf{f}$. We have thereby proved the following.

Theorem 7.8 *If* $S \in C^{2,\nu}$, $\mathbf{f} \in C^{1,\nu'}(S)$, $0 < \nu' < \nu \le 1$, *then a regular solution of the class* $\mathfrak{B}$ *of the internal BVP* $(I_q)^+_{\mathbf{0},\mathbf{f}}$ *exists, is unique, and is represented by double-layer potential (7.17), where* $\mathbf{g}$ *is a solution of the singular integral equation (7.19) which is always solvable for an arbitrary vector* $\mathbf{f}$.

Problem $(II_q)^-_{\mathbf{0},\mathbf{f}}$ We seek a regular solution of the class $\mathfrak{B}$ to this problem in the form

$$\mathbf{V}(\mathbf{x}) = \mathbf{Z}^{(q,1)}(\mathbf{x}, \mathbf{h}) \qquad \text{for} \qquad \mathbf{x} \in \Omega^-, \tag{7.23}$$

where $\mathbf{h}$ is the required seven-component vector function.

Obviously, by virtue of Theorem 7.4 the vector function $\mathbf{V}$ is a solution of (7.18) for $\mathbf{x} \in \Omega^-$. Keeping in mind the boundary condition (6.5) and using (7.13), from (7.23) we obtain, for determining the unknown vector $\mathbf{h}$, a singular integral equation

$$\mathcal{K}^{(q,4)}\mathbf{h}(\mathbf{z}) = \mathbf{f}(\mathbf{z}) \qquad \text{for} \quad \mathbf{z} \in S. \tag{7.24}$$

It has been proved above that the corresponding homogeneous equation (7.20) has only the trivial solution. Hence, it follows that (7.24) is always solvable.

We have thereby proved the following.

Theorem 7.9 *If* $S \in C^{2,\nu}$, $\mathbf{f} \in C^{0,\nu'}(S)$, $0 < \nu' < \nu \le 1$, *then a regular solution of the class* $\mathfrak{B}$ *of the external BVP* $(II_q)^-_{\mathbf{0},\mathbf{f}}$ *exists, is unique, and is represented by single-layer potential (7.23), where* $\mathbf{h}$ *is a solution of the singular integral equation (7.24) which is always solvable for an arbitrary vector* $\mathbf{f}$.

Problem $(I_q)^-_{\mathbf{0},\mathbf{f}}$ We seek a regular solution of the class $\mathfrak{B}$ to this problem in the sum of the double-layer and single-layer potentials

$$\mathbf{V}(\mathbf{x}) = \mathbf{Z}^{(q,2)}(\mathbf{x}, \mathbf{g}) + (1-i)\mathbf{Z}^{(q,1)}(\mathbf{x}, \mathbf{g}) \qquad \text{for} \qquad \mathbf{x} \in \Omega^-, \tag{7.25}$$

where $\mathbf{g}$ is the required seven-component vector function.

Obviously, by Theorems 7.4 and 7.5 the vector function $\mathbf{V}$ is a solution of (7.18) for $\mathbf{x} \in \Omega^-$. Keeping in mind the boundary condition (6.4) and using (7.14), from (7.25) we obtain, for determining the unknown vector $\mathbf{g}$, a singular integral equation

$$\mathcal{K}^{(q,5)}\,\mathbf{g}(\mathbf{z}) \equiv \mathcal{K}^{(q,3)}\,\mathbf{g}(\mathbf{z}) + (1-i)\mathbf{Z}^{(q,1)}(\mathbf{z}, \mathbf{g}) = \mathbf{f}(\mathbf{z}) \qquad \text{for} \quad \mathbf{z} \in S. \tag{7.26}$$

We prove that Eq. (7.26) is always solvable for an arbitrary vector $\mathbf{f}$. Clearly, the singular integral operator $\mathcal{K}^{(q,5)}$ is of the normal type and $\operatorname{ind}\mathcal{K}^{(q,5)} = \operatorname{ind}\mathcal{K}^{(q,3)} = 0$.

Now we prove that the homogeneous equation

$$\mathcal{K}^{(q,5)}\,\mathbf{g}_0(\mathbf{z}) = \mathbf{0} \qquad \text{for} \quad \mathbf{z} \in S. \tag{7.27}$$

has only a trivial solution. Indeed, let $\mathbf{g}_0$ be a solution of the homogeneous equation (7.27). Then the vector

$$\mathbf{V}_0(\mathbf{x}) \equiv \mathbf{Z}^{(q,2)}(\mathbf{x}, \mathbf{g}_0) + (1-i)\mathbf{Z}^{(q,1)}(\mathbf{x}, \mathbf{g}_0) \qquad \text{for} \qquad \mathbf{x} \in \Omega^- \tag{7.28}$$

is a regular solution of problem $(I_q)^-_{\mathbf{0},\mathbf{0}}$. Using Theorem 7.3 we have (7.21).

On the other hand, by Theorems 7.4 and 7.5 from (7.28) we get (6.49), and on the basis of (7.20) from (6.49) it follows the boundary condition (6.50). Hence, we have the relation (6.51), where $\mathbf{V}_0 = (\mathbf{u}_0, \mathbf{p}_0)$, $\mathbf{u}_0$, and $\mathbf{p}_0$ are three- and four-component vector functions, respectively, the operators $\mathcal{P}^{(1)}(\mathbf{D}_\mathbf{z}, \mathbf{n})$ and $\mathcal{P}^{(2)}(\mathbf{D}_\mathbf{z}, \mathbf{n})$ are given in Sect. 3.4.1.

On the other hand, by virtue of (6.51) from (3.18) it follows that

$$\begin{aligned} &\int_{\Omega^+} W_3(\mathbf{V}_0, \mathbf{u}_0)d\mathbf{x} + (1-i)\int_S |\mathbf{u}_0(\mathbf{z})|^2 d_\mathbf{z}S = 0, \\ &\int_{\Omega^+} W_2(\mathbf{V}_0, \mathbf{p}_0)d\mathbf{x} + (1-i)\int_S |\mathbf{p}_0(\mathbf{z})|^2 d_\mathbf{z}S = 0, \end{aligned} \tag{7.29}$$

where W_2 and W_3 are given by (6.13) and (7.4), respectively. The relations (7.29) imply

$$\int_{\Omega^+} \operatorname{Im} W_3(\mathbf{V}_0, \mathbf{u}_0)d\mathbf{x} - \int_S |\mathbf{u}_0(\mathbf{z})|^2 d_\mathbf{z}S = 0,$$

$$\int_{\Omega^+} \operatorname{Re} W_2(\mathbf{V}_0, \mathbf{p}_0)d\mathbf{x} + \int_S |\mathbf{p}_0(\mathbf{z})|^2 d_\mathbf{z}S = 0.$$

Clearly, we may further conclude that

$$\int\limits_{\Omega^+} [\mathrm{Re}\, W_2(\mathbf{V}_0, \mathbf{p}_0) - \omega \mathrm{Im}\, W_3(\mathbf{V}_0, \mathbf{u}_0)]\, d\mathbf{x} + \int\limits_{S} \left[\omega |\mathbf{u}_0(\mathbf{z})|^2 + |\mathbf{p}_0(\mathbf{z})|^2\right] d_{\mathbf{z}} S = 0. \tag{7.30}$$

On the other hand, (7.5) implies

$$\mathrm{Re} W_2(\mathbf{V}_0, \mathbf{p}_0) - \omega \mathrm{Im}\, W_3(\mathbf{V}_0, \mathbf{u}_0) = \mathbf{K}\nabla\mathbf{p}_0 \cdot \nabla\mathbf{p}_0 + \mathbf{d}\mathbf{p}_0 \cdot \mathbf{p}_0 \geq 0. \tag{7.31}$$

Then, employing (7.31), from (7.30) we have

$$\int\limits_{S} \left[\omega |\mathbf{u}_0(\mathbf{z})|^2 + |\mathbf{p}_0(\mathbf{z})|^2\right] d_{\mathbf{z}} S = 0.$$

Clearly, from the last equation we obtain

$$\{\mathbf{V}_0(\mathbf{z})\}^+ = \mathbf{0} \qquad \text{for} \qquad \mathbf{z} \in S. \tag{7.32}$$

Finally, by virtue of (7.21) and (7.32) from the first equation of (6.49) we get $\mathbf{g}_0(\mathbf{z}) \equiv \mathbf{0}$ for $\mathbf{z} \in S$.

Thus, the homogeneous equation (7.27) has only the trivial solution and therefore on the basis of Fredholm's theorem the singular integral equation (7.26) is always solvable for an arbitrary vector **f**. We have thereby proved the following.

Theorem 7.10 *If $S \in C^{2,\nu}$, $\mathbf{f} \in C^{1,\nu'}(S)$, $0 < \nu' < \nu \leq 1$, then a regular solution of the class $\mathfrak{B}$ of the external BVP $(I_q)^-_{\mathbf{0},\mathbf{f}}$ exists, is unique, and is represented by a sum of double-layer and single-layer potentials (7.25), where* **g** *is a solution of the singular integral equation (7.26) which is always solvable for an arbitrary vector* **f**.

Problem $(II_q)^+_{\mathbf{0},\mathbf{f}}$ We seek a regular solution of the class $\mathfrak{B}$ to this problem in the form

$$\mathbf{V}(\mathbf{x}) = \mathbf{Z}^{(q,1)}(\mathbf{x}, \mathbf{g}) \qquad \text{for} \qquad \mathbf{x} \in \Omega^+, \tag{7.33}$$

where **g** is the required seven-component vector function.

Obviously, by virtue of Theorem 7.4 the vector function **V** is a solution of (7.18) for $\mathbf{x} \in \Omega^+$. Keeping in mind the boundary condition (6.3) and using (7.13), from (7.33) we obtain, for determining the unknown vector **g**, the following singular integral equation

$$\mathcal{K}^{(q,2)}\mathbf{g}(\mathbf{z}) = \mathbf{f}(\mathbf{z}) \qquad \text{for} \qquad \mathbf{z} \in S. \tag{7.34}$$

To investigate the solvability of (7.34) we consider the homogeneous equation

$$\mathcal{K}^{(q,2)}\mathbf{g}(\mathbf{z}) = \mathbf{0} \qquad \text{for} \qquad \mathbf{z} \in S. \tag{7.35}$$

The homogeneous adjoint integral equation of (7.35) has the following form

$$\mathcal{K}^{(q,3)}\mathbf{h}(\mathbf{z}) = \mathbf{0} \qquad \text{for} \qquad \mathbf{z} \in S, \tag{7.36}$$

where $\mathbf{h}$ is a seven-component vector function.

We introduce the following seven-component vector functions:

$$\begin{aligned}
&\boldsymbol{\chi}^{(q,1)}(\mathbf{x}) = (1,0,0,0,0,0,0), \qquad \boldsymbol{\chi}^{(q,2)}(\mathbf{x}) = (0,1,0,0,0,0,0),\\
&\boldsymbol{\chi}^{(q,3)}(\mathbf{x}) = (0,0,1,0,0,0,0), \qquad \boldsymbol{\chi}^{(q,4)}(\mathbf{x}) = (0,-x_3,x_2,0,0,0,0),\\
&\boldsymbol{\chi}^{(q,5)}(\mathbf{x}) = (x_3,0,-x_1,0,0,0,0),\\
&\boldsymbol{\chi}^{(q,6)}(\mathbf{x}) = (-x_2,x_1,0,0,0,0,0).
\end{aligned} \tag{7.37}$$

Obviously, $\{\boldsymbol{\chi}^{(q,j)}\}_{j=1}^{6}$ is the system of linearly independent vectors. We have the following.

Lemma 7.1 *The homogeneous equations (7.35) and (7.36) have six linearly independent solutions each and they constitute complete system of solutions.*

Proof By Theorem 7.2 each vector $\boldsymbol{\chi}^{(q,j)}$ $(j = 1, 2, \cdots, 6)$ is a regular solution of problem $(II_q)^+_{\mathbf{0},\mathbf{0}}$, i.e.

$$\begin{aligned}
&\mathcal{A}^{(q)}(\mathbf{D_x},\omega)\,\boldsymbol{\chi}^{(q,j)}(\mathbf{x}) = \mathbf{0} \qquad \text{for} \quad \mathbf{x} \in \Omega^+,\\
&\left\{\mathcal{P}(\mathbf{D_z},\mathbf{n})\boldsymbol{\chi}^{(q,j)}(\mathbf{z})\right\}^+ = \mathbf{0} \qquad \text{for} \quad \mathbf{z} \in S.
\end{aligned} \tag{7.38}$$

On the other hand, by virtue of (7.38) from formula of integral representation of regular vector (3.45) we obtain

$$\boldsymbol{\chi}^{(q,j)}(\mathbf{x}) = \mathbf{Z}^{(q,2)}(\mathbf{x},\boldsymbol{\chi}^{(q,j)}) \qquad \text{for} \quad \mathbf{x} \in \Omega^+. \tag{7.39}$$

On the basis of boundary property of double-layer potential (7.13) from (7.39) we have

$$\boldsymbol{\chi}^{(q,j)}(\mathbf{z}) = \frac{1}{2}\boldsymbol{\chi}^{(q,j)}(\mathbf{z}) + \mathbf{Z}^{(q,2)}(\mathbf{z},\boldsymbol{\chi}^{(q,j)}) \qquad \text{for} \quad \mathbf{z} \in S,\ j = 1, 2, \cdots, 6$$

and therefore, $\boldsymbol{\chi}^{(q,j)}(\mathbf{x})$ is a solution of Eq. (7.36). Using Fredholm's theorem the singular integral equations (7.35) and (7.36) have six linearly independent solutions each.

It will now be shown that $\{\boldsymbol{\chi}^{(q,j)}\}_{j=1}^{6}$ is a complete system of linearly independent solutions of (7.36).

Let $\{\mathbf{h}^{(j)}(\mathbf{z})\}_{j=1}^{m}$ is the complete system of linearly independent solutions of homogeneous equation (7.36), where $m \geq 6$. We construct single-layer potentials $\mathbf{Z}^{(q,1)}(\mathbf{x}, \mathbf{h}^{(j)})$ $(j = 1, 2, \cdots, m)$. By Theorem 7.4 from identity

$$\mathbf{0} = \sum_{j=1}^{m} r_j \mathbf{Z}^{(q,1)}(\mathbf{x}, \mathbf{h}^{(j)}) = \mathbf{Z}^{(q,1)}(\mathbf{x}, \sum_{j=1}^{m} r_j \mathbf{h}^{(j)}) \qquad \text{for} \quad \mathbf{x} \in \Omega^+$$

it follows that

$$\sum_{j=1}^{m} r_j \mathbf{h}^{(j)}(\mathbf{z}) = \mathbf{0}, \tag{7.40}$$

where $r_1, r_2, \cdots, r_m$ are arbitrary constants, and from (7.40) we have $r_1 = r_2 = \cdots = r_m = 0$. Hence, $\left\{\mathbf{Z}^{(q,1)}(\mathbf{x}, \mathbf{h}^{(j)})\right\}_{j=1}^{m}$ is the system of linearly independent vectors.

On the other hand, it is easy to see that each vector $\mathbf{Z}^{(q,1)}(\mathbf{x}, \mathbf{h}^{(j)})$ $(j = 1, 2, \cdots, m)$ is a regular solution of problem $(II_q)^+_{\mathbf{0},\mathbf{0}}$. On the basis of Theorem 7.2, vector $\mathbf{Z}^{(q,1)}(\mathbf{x}, \mathbf{h}^{(j)})$ can be written as follows

$$\mathbf{Z}^{(q,1)}(\mathbf{x}, \mathbf{h}^{(j)}) = \sum_{j=1}^{6} r_{jl} \boldsymbol{\chi}^{(1j)}(\mathbf{x}) \qquad \text{for} \quad \mathbf{x} \in \Omega^+,$$

where r_{jl} $(j = 1, 2, \cdots, m, \ l = 1, 2, \cdots, 6)$ are constants. Hence, each vector of system $\left\{\mathbf{Z}^{(q,1)}(\mathbf{x}, \mathbf{h}^{(j)})\right\}_{j=1}^{m}$ is represented by six linearly independent vectors $\boldsymbol{\chi}^{(q,1)}, \boldsymbol{\chi}^{(q,2)}, \cdots, \boldsymbol{\chi}^{(q,6)}$. Thus, $m = 6$. □

By virtue of Fredholm's third theorem, for Eq. (7.34) to be solvable it is necessary and sufficient that (see the condition (4.30))

$$\int_S \mathbf{f}(\mathbf{z}) \boldsymbol{\chi}^{(q,j)}(\mathbf{z}) d_{\mathbf{z}} S = \mathbf{0} \qquad \text{for} \quad j = 1, 2, \cdots, 6. \tag{7.41}$$

Then, employing (7.37), we may derive from (7.41)

$$\int_S \mathbf{f}^{(q)}(\mathbf{z}) d_{\mathbf{z}} S = \mathbf{0}, \qquad \int_S \left[\mathbf{f}^{(q)}(\mathbf{z}) \times \mathbf{z}\right] d_{\mathbf{z}} S = \mathbf{0}, \tag{7.42}$$

where $\mathbf{f}^{(q)} = (f_1, f_2, f_3)$. We have thereby proved the following.

Theorem 7.11 *If $S \in C^{2,\nu}$, $\mathbf{f} \in C^{0,\nu'}(S)$, $0 < \nu' < \nu \leq 1$,, then the problem $(II_q)^+_{\mathbf{0},\mathbf{f}}$ is solvable only when condition (7.42) is fulfilled. The solution of this problem is represented by a single-layer potential (7.33), where* **g** *is a solution of the solvable singular integral equation (7.34).*

7.6 Solution of the First BVPs by Fredholm's Integral Equations

In Sect. 7.5, the first BVPs $(I_q)^+_{\mathbf{0},\mathbf{f}}$ and $(I_q)^-_{\mathbf{0},\mathbf{f}}$ are investigated by means of singular integral equations. In this section these problems will be also investigated by Fredholm's integral equations. Indeed, the BVPs $(I_q)^+_{\mathbf{0},\mathbf{f}}$ and $(I_q)^-_{\mathbf{0},\mathbf{f}}$ will be reduced to the always solvable Fredholm's integral equations.

7.6.1 Auxiliary Boundary Value Problems

In the sequel, we consider the following auxiliary BVPs:

(a) Find a regular (classical) solution of the class $\mathfrak{B}$ in Ω^+ to system of the homogeneous equations (7.2) for $\mathbf{x} \in \Omega^+$ satisfying the boundary condition

$$\lim_{\Omega^+ \ni \mathbf{x} \to \mathbf{z} \in S} \mathcal{N}(\mathbf{D_x}, \mathbf{n}(\mathbf{z}))\mathbf{V}(\mathbf{x}) \equiv \{\mathcal{N}(\mathbf{D_z}, \mathbf{n})\mathbf{V}(\mathbf{z})\}^+ = \mathbf{f}(\mathbf{z}) \tag{7.43}$$

in the *Problem* $(I_a)^+_{\mathbf{f}}$;

(b) Find a regular (classical) solution of the class $\mathfrak{B}$ to system (7.2) for $\mathbf{x} \in \Omega^-$ satisfying the boundary condition

$$\lim_{\Omega^- \ni \mathbf{x} \to \mathbf{z} \in S} \mathcal{N}(\mathbf{D_x}, \mathbf{n}(\mathbf{z}))\mathbf{V}(\mathbf{x}) \equiv \{\mathcal{N}(\mathbf{D_z}, \mathbf{n})\mathbf{V}(\mathbf{z})\}^- = \mathbf{f}(\mathbf{z}) \tag{7.44}$$

in the *Problem* $(I_a)^-_{\mathbf{f}}$. Here the matrix differential operator $\mathcal{N}(\mathbf{D_x}, \mathbf{n}(\mathbf{x}))$ is defined by (2.61) and **f** is prescribed seven-component vector function. Note that the class $\mathfrak{B}$ of vector functions is given by Definition 3.2 (see Sect. 3.5).

7.6.2 Uniqueness Theorems for the Auxiliary BVPs

We introduce the notation

$$W_1^{(\kappa_0)}(\mathbf{V},\mathbf{u}) = W_0^{(\kappa_0)}(\mathbf{u},\mathbf{u}) - a_\alpha\, p_\alpha\, \mathrm{div}\,\overline{\mathbf{u}},$$

$$W_0^{(\kappa_0)}(\mathbf{u},\mathbf{u}) = \frac{1}{3}(3\lambda + 4\mu - 2\kappa_0)\,|\mathrm{div}\,\mathbf{u}|^2 + \frac{1}{2}(\mu - \kappa_0)\,|\mathrm{curl}\,\mathbf{u}|^2$$

$$+(\mu+\kappa_0)\left[\frac{1}{4}\sum_{l,j=1;\,l\neq j}^{3}\left|\frac{\partial u_j}{\partial x_l} + \frac{\partial u_l}{\partial x_j}\right|^2 + \frac{1}{6}\sum_{l,j=1}^{3}\left|\frac{\partial u_l}{\partial x_l} - \frac{\partial u_j}{\partial x_j}\right|^2\right], \tag{7.45}$$

where κ_0 is given by (2.62).

Obviously, by simple calculation from (3.18) for $\mathbf{u}' = \mathbf{u}$, $\mathbf{p}' = \mathbf{p}$, and $\rho = 0$ we obtain

$$\begin{aligned}
&\int_{\Omega^+}\left[\mathcal{A}^{(q,1)}(\mathbf{D_x})\,\mathbf{V}(\mathbf{x})\cdot\mathbf{u}(\mathbf{x}) + W_1^{(\kappa_0)}(\mathbf{V},\mathbf{u})\right]d\mathbf{x} \\
&\qquad = \int_S \mathcal{N}^{(1)}(\mathbf{D_z},\mathbf{n})\mathbf{V}(\mathbf{z})\cdot\mathbf{u}(\mathbf{z})\,d_{\mathbf{z}}S, \\
&\int_{\Omega^+}\left[\mathcal{A}^{(q,2)}(\mathbf{D_x},\omega)\,\mathbf{V}(\mathbf{x})\cdot\mathbf{p}(\mathbf{x}) + W_2(\mathbf{V},\mathbf{p})\right]d\mathbf{x} \\
&\qquad = \int_S \mathcal{N}^{(2)}(\mathbf{D_z},\mathbf{n})\mathbf{p}(\mathbf{z})\cdot\mathbf{p}(\mathbf{z})\,d_{\mathbf{z}}S.
\end{aligned} \tag{7.46}$$

where $W_2(\mathbf{V},\mathbf{p})$ is given by (6.13),

$$\begin{aligned}
&\mathcal{A}^{(q,1)}(\mathbf{D_x}) = \left(\mathcal{A}_{lj}^{(q,1)}(\mathbf{D_x})\right)_{3\times 7}, \\
&\mathcal{A}^{(q,2)}(\mathbf{D_x},\omega) = \left(\mathcal{A}_{lj}^{(q,2)}(\mathbf{D_x},\omega)\right)_{4\times 7}, \\
&\mathcal{A}_{lj}^{(q,1)}(\mathbf{D_x}) = (\mu\Delta + \rho\omega^2)\delta_{lj} + (\lambda+\mu)\frac{\partial^2}{\partial x_l \partial x_j}, \\
&\mathcal{A}_{l;\alpha+3}^{(q,1)}(\mathbf{D_x}) = -a_\alpha\frac{\partial}{\partial x_l}, \qquad \mathcal{A}_{\alpha+3;j}^{(q,2)}(\mathbf{D_x},\omega) = i\omega a_\alpha\frac{\partial}{\partial x_j}, \\
&\mathcal{A}_{\alpha+3;\beta+3}^{(q,2)}(\mathbf{D_x},\omega) = k_{\alpha\beta}\Delta + c_{\alpha\beta}, \\
&l,j = 1,2,3, \qquad \alpha,\beta = 1,2,3,4.
\end{aligned} \tag{7.47}$$

and

$$\mathcal{N}^{(1)}(\mathbf{D_x},\mathbf{n}) = \left(\mathcal{N}^{(1)}_{lj}(\mathbf{D_x},\mathbf{n})\right)_{3\times 7},$$

$$\mathcal{N}^{(2)}(\mathbf{D_x},\mathbf{n}) = \left(\mathcal{N}^{(2)}_{\alpha j}(\mathbf{D_x},\mathbf{n})\right)_{4\times 7},$$

$$\mathcal{N}^{(1)}_{lj}(\mathbf{D_x},\mathbf{n}) = \mathcal{N}_{lj}(\mathbf{D_x},\mathbf{n}), \tag{7.48}$$

$$\mathcal{N}^{(2)}_{\alpha j}(\mathbf{D_x},\mathbf{n}) = \mathcal{N}_{\alpha+3;\,j}(\mathbf{D_x},\mathbf{n}),$$

$$l = 1,2,3, \qquad \alpha = 1,2,3,4, \qquad j = 1,2,\cdots,7.$$

If we employ the relations (7.47) and (7.48), then from (7.2), (7.43), and (7.44) it follows that

$$\mathcal{A}^{(q,1)}(\mathbf{D_x})\,\mathbf{V}(\mathbf{x}) = \mathbf{0}, \qquad \mathcal{A}^{(q,2)}(\mathbf{D_x})\,\mathbf{V}(\mathbf{x}) = \mathbf{0}, \tag{7.49}$$

$$\{\mathcal{N}^{(1)}(\mathbf{D_z},\mathbf{n})\mathbf{V}(\mathbf{z})\}^+ = \mathbf{f}^{(1)}(\mathbf{z}), \qquad \{\mathcal{N}^{(2)}(\mathbf{D_z},\mathbf{n})\mathbf{V}(\mathbf{z})\}^+ = \mathbf{f}^{(2)}(\mathbf{z}), \tag{7.50}$$

and

$$\{\mathcal{N}^{(1)}(\mathbf{D_z},\mathbf{n})\mathbf{V}(\mathbf{z})\}^- = \mathbf{f}^{(1)}(\mathbf{z}), \qquad \{\mathcal{N}^{(2)}(\mathbf{D_z},\mathbf{n})\mathbf{V}(\mathbf{z})\}^- = \mathbf{f}^{(2)}(\mathbf{z}), \tag{7.51}$$

respectively. Here $\mathbf{f}^{(1)}$ and $\mathbf{f}^{(2)}$ are prescribed three- and four-component vector functions, respectively.

We are now in a position to prove the uniqueness theorems of regular (classical) solutions of the auxiliary BVPs $(I_a)^+_{\mathbf{f}}$ and $(I_a)^-_{\mathbf{f}}$.

We have the following results.

Theorem 7.12 *If the condition*

$$\mu > 0, \qquad 3\lambda + 5\mu > 0 \tag{7.52}$$

is satisfied, then any two solutions of the internal BVP $(I_a)^+_{\mathbf{f}}$ of the class $\mathfrak{B}$ may differ only for an additive vector $\mathbf{V} = (\mathbf{u},\mathbf{p})$, where $\mathbf{p}$ satisfies the condition (7.7),

$$\mathbf{u}(\mathbf{x}) = \mathbf{a}' \qquad \text{for} \quad \mathbf{x} \in \Omega^+, \tag{7.53}$$

and $\mathbf{a}' = (a'_1, a'_2, a'_3)$ is arbitrary complex constant three-component vector.

Proof Suppose that there are two regular solutions of problem $(I_a)^+_{\mathbf{f}}$. Then their difference $\mathbf{V}$ is a regular solution of the internal homogeneous BVP $(I_a)^+_{\mathbf{0}}$. Hence, $\mathbf{V}$ is a regular solution of the system of homogeneous equations (7.49) in Ω^+ satisfying the boundary conditions (7.50) and (7.51) for $\mathbf{f}^{(1)}(\mathbf{z}) = \mathbf{0}$ and $\mathbf{f}^{(2)}(\mathbf{z}) = \mathbf{0}$, i.e.

$$\{\mathcal{N}^{(1)}(\mathbf{D_z},\mathbf{n})\mathbf{V}(\mathbf{z})\}^+=\mathbf{0},\qquad \{\mathcal{N}^{(2)}(\mathbf{D_z},\mathbf{n})\mathbf{V}(\mathbf{z})\}^+=\mathbf{0}. \tag{7.54}$$

On the basis of (7.49) and (7.54), from (7.46) we obtain

$$\int\limits_{\Omega^+} W_1^{(\kappa_0)}(\mathbf{V},\mathbf{u})d\mathbf{x}=0,\qquad \int\limits_{\Omega^+} W_2(\mathbf{V},\mathbf{p})d\mathbf{x}=0. \tag{7.55}$$

Obviously, from (7.45) it follows that

$$\begin{aligned}&\operatorname{Im} W_1^{(\kappa_0)}(\mathbf{V},\mathbf{u})=-\operatorname{Im}\left[\mathbf{ap}\,\mathrm{div}\bar{\mathbf{u}}\right],\\ &\mathrm{Re}W_2(\mathbf{V},\mathbf{p})=k_{\alpha\beta}\nabla p_\beta\cdot\nabla p_\alpha+d_{\alpha\beta}p_\beta\overline{p_\alpha}-\omega\mathrm{Im}\left[\mathbf{ap}\,\mathrm{div}\bar{\mathbf{u}}\right].\end{aligned} \tag{7.56}$$

From (7.56) we can easily verify that

$$\mathrm{Re}W_2(\mathbf{V},\mathbf{p})-\omega\operatorname{Im}W_1(\mathbf{V},\mathbf{u})=k_{\alpha\beta}\nabla p_\beta\cdot\nabla p_\alpha+d_{\alpha\beta}p_\beta\overline{p_\alpha}. \tag{7.57}$$

Then, employing (7.57), we may derive from (7.55)

$$\int\limits_{\Omega^+}\left[k_{\alpha\beta}\nabla p_\beta\cdot\nabla p_\alpha+d_{\alpha\beta}p_\beta\overline{p_\alpha}\right]d\mathbf{x}=0.$$

Clearly, from the last equation we may deduce Eq. (7.6). By virtue of (7.6) and Lemma 6.1 from the second equation of (7.49) we have

$$a_\alpha\,\mathrm{div}\mathbf{u}=b_{\alpha\beta}\,c_\beta$$

for $\mathbf{x}\in\Omega^+$ and $\alpha=1,2,3,4$. Employing the last equation in (7.45) we obtain

$$W_1^{(\kappa_0)}(\mathbf{V},\mathbf{u})=W_0^{(\kappa_0)}(\mathbf{u},\mathbf{u})+b_{\alpha\beta}\,c_\alpha\,\overline{c_\beta}. \tag{7.58}$$

In view of the relation (7.58) from (7.55) it follows that

$$\int\limits_{\Omega^+}\left[W_0^{(\kappa_0)}(\mathbf{u},\mathbf{u})+b_{\alpha\beta}\,c_\alpha\,\overline{c_\beta}\right]d\mathbf{x}=0. \tag{7.59}$$

Furthermore, if we employ (2.62) and (7.52), then we can write

$$\begin{aligned}&3\lambda+4\mu-2\kappa_0=\frac{(\lambda+2\mu)(3\lambda+5\mu)}{\lambda+3\mu}>0,\\ &\mu-\kappa_0=\frac{2\mu^2}{\lambda+3\mu}>0,\qquad \mu+\kappa_0=\frac{2\mu(\lambda+2\mu)}{\lambda+3\mu}>0.\end{aligned} \tag{7.60}$$

Consequently, keeping in mind (7.60) from (7.45) we get

$$W_0^{(\kappa_0)}(\mathbf{u}, \mathbf{u}) \geq 0. \tag{7.61}$$

Now, on the basis of (7.61) and Remark 1.2 from (7.59) it follows that

$$W_0^{(\kappa_0)}(\mathbf{u}, \mathbf{u}) = 0, \qquad b_{\alpha\beta}\, c_\alpha\, \overline{c_\beta} = 0. \tag{7.62}$$

Moreover, from the second equation of (7.62) we may obtain (7.7). On the other hand, if we are using (7.45) and (7.60) from the first equation of (7.62) we may also obtain

$$\operatorname{div}\mathbf{u}(\mathbf{x}) = 0, \qquad \operatorname{curl}\mathbf{u}(\mathbf{x}) = \mathbf{0} \quad \text{for } \mathbf{x} \in \Omega^+. \tag{7.63}$$

Hence, $\mathbf{u}$ is a harmonic vector, i.e., $\Delta\mathbf{u}(\mathbf{x}) = \mathbf{0}$ for $\mathbf{x} \in \Omega^+$.

If we now return to the boundary condition (7.54), then by virtue of the identity

$$\begin{aligned} \mathcal{N}^{(1)}(\mathbf{D}_\mathbf{z}, \mathbf{n})\mathbf{V}(\mathbf{z}) = (\mu + \kappa_0)\frac{\partial\mathbf{u}(\mathbf{z})}{\partial\mathbf{n}(\mathbf{z})} + (\lambda + \mu - \kappa_0)\mathbf{n}(\mathbf{z})\operatorname{div}\mathbf{u}(\mathbf{z}) \\ + \kappa_0\,[\mathbf{n}(\mathbf{z}) \times \operatorname{curl}\mathbf{u}(\mathbf{z})] - a_\alpha\, p_\alpha(\mathbf{z})\,\mathbf{n}(\mathbf{z}) \end{aligned} \tag{7.64}$$

and (7.63) we deduced that

$$\left\{\frac{\partial\mathbf{u}(\mathbf{z})}{\partial\mathbf{n}(\mathbf{z})}\right\}^+ = \mathbf{0}.$$

Therefore, $\mathbf{u}$ is a harmonic vector in Ω^+ satisfying Neumann's internal boundary condition. Consequently, the vector $\mathbf{u}$ is a constant vector, i.e., we have the relation (7.53). Thus, we have the desired result. □

Theorem 7.13 *If the condition (7.52) is satisfied, then the external BVP* $(I_a)_\mathbf{f}^-$ *has one regular solution of the class* $\mathfrak{B}$.

Proof Suppose that there are two regular solutions of problem $(I_a)_\mathbf{f}^-$. Then their difference $\mathbf{V}$ is a regular solution of the internal homogeneous BVP $(I_a)_\mathbf{0}^-$. Hence, $\mathbf{V}$ is a regular solution of the system of homogeneous equations (7.49) in Ω^- satisfying the homogeneous boundary conditions

$$\{\mathcal{N}^{(1)}(\mathbf{D}_\mathbf{z}, \mathbf{n})\mathbf{V}(\mathbf{z})\}^- = \mathbf{0}, \qquad \{\mathcal{N}^{(2)}(\mathbf{D}_\mathbf{z}, \mathbf{n})\mathbf{V}(\mathbf{z})\}^- = \mathbf{0}. \tag{7.65}$$

On the basis of (7.49) and (7.65), from identities

$$\int_{\Omega^-} \left[\mathcal{A}^{(q,1)}(\mathbf{D_x})\,\mathbf{V}(\mathbf{x}) \cdot \mathbf{u}(\mathbf{x}) + W_1^{(\kappa_0)}(\mathbf{V},\mathbf{u}) \right] d\mathbf{x}$$

$$= -\int_S \mathcal{N}^{(1)}(\mathbf{D_z},\mathbf{n})\mathbf{V}(\mathbf{z}) \cdot \mathbf{u}(\mathbf{z})\, d_{\mathbf{z}}S,$$

$$\int_{\Omega^-} \left[\mathcal{A}^{(q,2)}(\mathbf{D_x},\omega)\,\mathbf{V}(\mathbf{x}) \cdot \mathbf{p}(\mathbf{x}) + W_2(\mathbf{V},\mathbf{p}) \right] d\mathbf{x}$$

$$= -\int_S \mathcal{N}^{(2)}(\mathbf{D_z},\mathbf{n})\mathbf{p}(\mathbf{z}) \cdot \mathbf{p}(\mathbf{z})\, d_{\mathbf{z}}S.$$

we obtain

$$\int_{\Omega^-} W_1^{(\kappa_0)}(\mathbf{V},\mathbf{u})d\mathbf{x} = 0, \qquad \int_{\Omega^-} W_2(\mathbf{V},\mathbf{p})d\mathbf{x} = 0. \tag{7.66}$$

In a similar manner as in Theorem 7.12 from (7.66) it follows that

$$p_j(\mathbf{x}) = c_j = \text{const}, \qquad d_{\alpha\beta}\,p_\beta\,\overline{p_\alpha} = 0 \qquad \text{for } \mathbf{x} \in \Omega^-, \\ j = 1,2,3,4. \tag{7.67}$$

On the basis of the conditions at infinity (3.31) from (7.67) we get

$$\mathbf{p}(\mathbf{x}) = \mathbf{0} \qquad \text{for} \quad \mathbf{x} \in \Omega^-. \tag{7.68}$$

Moreover, if we employ (7.68), then by virtue of (7.52) from (7.66) we can write

$$\operatorname{div}\mathbf{u}(\mathbf{x}) = 0, \qquad \operatorname{curl}\mathbf{u}(\mathbf{x}) = \mathbf{0} \qquad \text{for} \quad \mathbf{x} \in \Omega^-,$$

and therefore, we have

$$\Delta\mathbf{u}(\mathbf{x}) = \mathbf{0} \qquad \text{for} \quad \mathbf{x} \in \Omega^-.$$

If we now return to the boundary condition (7.65), then by virtue of the identity (7.64) we deduced that

$$\left\{ \frac{\partial \mathbf{u}(\mathbf{z})}{\partial \mathbf{n}(\mathbf{z})} \right\}^- = \mathbf{0}.$$

Hence, $\mathbf{u}$ is a harmonic vector in Ω^- satisfying Neumann's external boundary condition. On the basis of the conditions at infinity (3.31) we can write $\mathbf{u}(\mathbf{x}) = \mathbf{0}$ for $\mathbf{x} \in \Omega^-$. Thus, we have the desired result. □

7.6.3 Basic Properties of Potential

We introduce the following potential:

$$\mathbf{Z}^{(q,4)}(\mathbf{x}, \mathbf{g}) = \int_S \left\{ \tilde{\mathcal{N}}(\mathbf{D_y}, \mathbf{n}(\mathbf{y}))[\Gamma^{(q)}(\mathbf{x}-\mathbf{y}, \omega)]^\top \right\}^\top \mathbf{g}(\mathbf{y}) d_{\mathbf{y}} S,$$

where $\mathbf{\Gamma}^{(q)}(\mathbf{x}, \omega)$ is the fundamental matrix of the operator $\mathcal{A}^{(q)}(\mathbf{D_x}, \omega)$ and defined in Theorem 2.3, $\mathbf{g}$ is seven-component vector function and the matrix differential operator $\tilde{\mathcal{N}}(\mathbf{D_y}, \mathbf{n}(\mathbf{y}))$ is given by

$$\tilde{\mathcal{N}}(\mathbf{D_y}, \mathbf{n}) = \left(\tilde{\mathcal{N}}_{ms}(\mathbf{D_y}, \mathbf{n})\right)_{7\times 7},$$

$$\tilde{\mathcal{N}}_{lj}(\mathbf{D_y}, \mathbf{n}) = \mu\delta_{lj}\frac{\partial}{\partial \mathbf{n}} + (\lambda+\mu)n_l\frac{\partial}{\partial y_j} + \kappa_0 \mathcal{M}_{lj}(\mathbf{D_y}, \mathbf{n}),$$

$$\tilde{\mathcal{N}}_{l;\alpha+3}(\mathbf{D_y}, \mathbf{n}) = -i\omega a_\alpha n_l, \qquad \tilde{\mathcal{N}}_{\alpha+3;\beta+3}(\mathbf{D_y}, \mathbf{n}) = k_{\alpha\beta}\frac{\partial}{\partial \mathbf{n}}, \tag{7.69}$$

$$\tilde{\mathcal{N}}_{\alpha+3;j}(\mathbf{D_y}, \mathbf{n}) = 0, \qquad l, j = 1, 2, 3, \qquad \alpha, \beta = 1, 2, 3, 4.$$

Clearly, the operator $\tilde{\mathcal{N}}(\mathbf{D_y}, \mathbf{n}(\mathbf{y}))$ may be obtained from the operator $\mathcal{N}(\mathbf{D_y}, \mathbf{n}(\mathbf{y}))$ by replacing a_α by $i\omega a_\alpha (\alpha = 1, 2, 3, 4)$ and vice versa.

Moreover, Theorem 2.17 leads to the following.

Theorem 7.14 *If S is the Liapunov surface of classes $C^{1,\nu}$ $(0 < \nu \le 1)$ and $\mathbf{x}, \mathbf{y} \in S$, then the relations*

$$\left[\tilde{\mathcal{N}}(\mathbf{D_x}, \mathbf{n}(\mathbf{x}))\mathbf{\Gamma}^{(0)}(\mathbf{x}-\mathbf{y})\right]_{lj} = O\left(|\mathbf{x}-\mathbf{y}|^{-2+\nu}\right),$$

$$\left[\tilde{\mathcal{N}}(\mathbf{D_x}, \mathbf{n}(\mathbf{x}))\mathbf{\Gamma}^{(0)}(\mathbf{x}-\mathbf{y})\right]_{l,\beta+3} = O\left(|\mathbf{x}-\mathbf{y}|^{-1}\right),$$

$$\left[\tilde{\mathcal{N}}(\mathbf{D_x}, \mathbf{n}(\mathbf{x}))\mathbf{\Gamma}^{(0)}(\mathbf{x}-\mathbf{y})\right]_{m+3,\beta+3} = O\left(|\mathbf{x}-\mathbf{y}|^{-2+\nu}\right), \tag{7.70}$$

$$\left[\tilde{\mathcal{N}}(\mathbf{D_x}, \mathbf{n}(\mathbf{x}))\mathbf{\Gamma}^{(0)}(\mathbf{x}-\mathbf{y})\right]_{m+3,j} = 0$$

hold in the neighborhood of $\mathbf{x} = \mathbf{y}$, *where* $l, j = 1, 2, 3$, $m, \beta = 1, 2, 3, 4$.

On the basis of Theorems 2.8, 2.12, and 7.14, and the relations (7.69) and (7.70) we have the following consequence.

Theorem 7.15 *If* $S \in C^{m+1,\nu}$, $\mathbf{g} \in C^{m,\nu'}(S)$, $0 < \nu' < \nu \leq 1$, *then:*

(a)

$$\mathbf{Z}^{(q,4)}(\cdot, \mathbf{g}) \in C^{m,\nu'}(\overline{\Omega^{\pm}}) \cap C^{\infty}(\Omega^{\pm}),$$

(b)

$$\mathcal{A}^{(q)}(\mathbf{D_x}, \omega)\, \mathbf{Z}^{(q,4)}(\mathbf{x}, \mathbf{g}) = \mathbf{0} \qquad for \quad \mathbf{x} \in \Omega^{\pm},$$

(c)

$$\left\{\mathbf{Z}^{(q,4)}(\mathbf{z}, \mathbf{g})\right\}^{\pm} = \pm\frac{1}{2}\,\mathbf{g}(\mathbf{z}) + \mathbf{Z}^{(q,4)}(\mathbf{z}, \mathbf{g}), \tag{7.71}$$

for the nonnegative integer m,

(d) $\mathbf{Z}^{(q,4)}(\mathbf{z}, \mathbf{g})$ *is an integral with weak singularity,*

(e)

$$\left\{\mathcal{N}(\mathbf{D_z}, \mathbf{n}(\mathbf{z}))\, \mathbf{Z}^{(q,4)}(\mathbf{z}, \mathbf{g})\right\}^{+} = \left\{\mathcal{N}(\mathbf{D_z}, \mathbf{n}(\mathbf{z}))\, \mathbf{Z}^{(q,4)}(\mathbf{z}, \mathbf{g})\right\}^{-},$$

for the natural number m,

(f)

$$\left\{\mathcal{N}(\mathbf{D_z}, \mathbf{n}(\mathbf{z}))\mathbf{Z}^{(q,1)}(\mathbf{z}, \mathbf{g})\right\}^{\pm} = \mp\frac{1}{2}\,\mathbf{g}(\mathbf{z}) + \mathcal{N}(\mathbf{D_z}, \mathbf{n}(\mathbf{z}))\mathbf{Z}^{(q,1)}(\mathbf{z}, \mathbf{g}), \tag{7.72}$$

where $\mathbf{z} \in S$.

We introduce the notation

$$\begin{aligned}
\mathcal{R}^{(q,1)}\,\mathbf{g}(\mathbf{z}) &\equiv \frac{1}{2}\,\mathbf{g}(\mathbf{z}) + \mathbf{Z}^{(q,4)}(\mathbf{z}, \mathbf{g}),\\
\mathcal{R}^{(q,2)}\mathbf{g}(\mathbf{z}) &\equiv -\frac{1}{2}\,\mathbf{g}(\mathbf{z}) + \mathbf{Z}^{(q,4)}(\mathbf{z}, \mathbf{g}),\\
\mathcal{R}^{(q,3)}\mathbf{g}(\mathbf{z}) &\equiv \frac{1}{2}\,\mathbf{g}(\mathbf{z}) + \mathcal{N}(\mathbf{D_z}, \mathbf{n}(\mathbf{z}))\mathbf{Z}^{(q,1)}(\mathbf{z}, \mathbf{g}),
\end{aligned} \tag{7.73}$$

for $\mathbf{z} \in S$, where $\mathbf{Z}^{(q,1)}(\mathbf{z}, \mathbf{g})$ is a single-layer potential and defined in Sect. 7.3. Obviously, on the basis of Theorem 7.14, $\mathcal{R}^{(q,j)}$ is Fredholm's integral operator ($j = 1, 2, 3$).

7.6.4 Existence Theorems

We are now in a position to prove the existence theorems of regular (classical) solutions of the BVPs $(I_q)^+_{\mathbf{0,f}}$ and $(I_q)^-_{\mathbf{0,f}}$ by always solvable Fredholm's integral equations.

Problem $(I_q)^+_{\mathbf{0,f}}$ We seek a regular solution of the class $\mathfrak{B}$ to this problem in the following form.

$$\mathbf{V}(\mathbf{x}) = \mathbf{Z}^{(q,4)}(\mathbf{x}, \mathbf{g}) \qquad \text{for} \qquad \mathbf{x} \in \Omega^+, \tag{7.74}$$

where $\mathbf{g}$ is the required seven-component vector function.

Obviously, by Theorem 7.15 the vector function $\mathbf{V}$ is a solution of the homogeneous equation (7.18) for $\mathbf{x} \in \Omega^+$. Keeping in mind the boundary condition (6.2) and using (7.71) and (7.73), from (7.74) we obtain, determining the unknown vector $\mathbf{g}$, Fredholm's integral equation

$$\mathcal{R}^{(q,1)}\,\mathbf{g}(\mathbf{z}) = \mathbf{f}(\mathbf{z}) \qquad \text{for} \quad \mathbf{z} \in S. \tag{7.75}$$

We prove that Eq. (7.75) is always solvable for an arbitrary vector $\mathbf{f}$.

Let us consider the associate homogeneous Fredholm's integral equation

$$\mathcal{R}^{(q,3)}\,\mathbf{h}(\mathbf{z}) = \mathbf{0} \qquad \text{for} \quad \mathbf{z} \in S, \tag{7.76}$$

where $\mathbf{h}$ is the required seven-component vector function. Now we prove that (7.76) has only the trivial solution.

Indeed, let $\mathbf{h}_0$ be a solution of the homogeneous equation (7.76). On the basis of Theorem 7.4 and Eqs. (7.72) and (7.76) the vector function $\mathbf{V}_0(\mathbf{x}) = \mathbf{Z}^{(q,1)}(\mathbf{x}, \mathbf{h}_0)$ is a regular solution of the external homogeneous BVP $(I_a)^-_{\mathbf{0}}$. Using Theorem 7.13, the problem $(I_a)^-_{\mathbf{0}}$ has only the trivial solution, that is

$$\mathbf{V}_0(\mathbf{x}) \equiv \mathbf{0} \qquad \text{for} \qquad \mathbf{x} \in \Omega^-. \tag{7.77}$$

On the other hand, by Theorem 7.4 and (7.77) we get

$$\{\mathbf{V}_0(\mathbf{z})\}^+ = \{\mathbf{V}_0(\mathbf{z})\}^- = \mathbf{0} \qquad \text{for} \qquad \mathbf{z} \in S,$$

i.e., the vector $\mathbf{V}_0(\mathbf{x})$ is a regular solution of problem $(I_q)^+_{\mathbf{0,0}}$. By virtue of Theorem 7.1, the problem $(I_q)^+_{\mathbf{0,0}}$ has only the trivial solution, that is

$$\mathbf{V}_0(\mathbf{x}) \equiv \mathbf{0} \qquad \text{for} \qquad \mathbf{x} \in \Omega^+. \tag{7.78}$$

Using (7.77), (7.78), and identity (7.72) we obtain

$$\mathbf{h}_0(\mathbf{z}) = \{\mathcal{N}(\mathbf{D_z}, \mathbf{n})\mathbf{V}_0(\mathbf{z})\}^- - \{\mathcal{N}(\mathbf{D_z}, \mathbf{n})\mathbf{V}_0(\mathbf{z})\}^+ = \mathbf{0} \qquad \text{for} \qquad \mathbf{z} \in S.$$

Thus, the homogeneous equation (7.76) has only the trivial solution and therefore on the basis of Fredholm's theorem the integral equation (7.75) is always solvable for an arbitrary vector **f**. We have thereby proved the following.

Theorem 7.16 *If* $S \in C^{2,\nu}$, $\mathbf{f} \in C^{1,\nu'}(S)$, $0 < \nu' < \nu \leq 1$, *and the condition (7.52) is satisfied, then a regular solution of the class* $\mathfrak{B}$ *of the internal BVP* $(I_q)^+_{\mathbf{0},\mathbf{f}}$ *exists, is unique, and is represented by (7.74), where* **g** *is a solution of Fredholm's integral equation (7.75) which is always solvable for an arbitrary vector* **f**.

Problem $(I_q)^-_{\mathbf{0},\mathbf{f}}$ We seek a regular solution of the class $\mathfrak{B}$ to this problem in the sum of potentials

$$\mathbf{V}(\mathbf{x}) = \mathbf{Z}^{(q,4)}(\mathbf{x}, \mathbf{g}) + (1 - i)\mathbf{Z}^{(q,1)}(\mathbf{x}, \mathbf{g}) \qquad \text{for} \qquad \mathbf{x} \in \Omega^-, \tag{7.79}$$

where **g** is the required seven-component vector function.

Obviously, by Theorems 7.4 and 7.15 the vector function **V** is a solution of (7.18) for $\mathbf{x} \in \Omega^-$. Keeping in mind the boundary condition (6.4) and using (7.71) and (7.73), from (7.79) we obtain, for determining the unknown vector **g**, Fredholm's integral equation

$$\mathcal{R}^{(q,4)}\,\mathbf{g}(\mathbf{z}) \equiv \mathcal{R}^{(q,2)}\,\mathbf{g}(\mathbf{z}) + (1 - i)\mathbf{Z}^{(q,1)}(\mathbf{z}, \mathbf{g}) = \mathbf{f}(\mathbf{z}) \qquad \text{for} \quad \mathbf{z} \in S. \tag{7.80}$$

We prove that Eq. (7.80) is always solvable for an arbitrary vector **f**.

Now we prove that the homogeneous equation

$$\mathcal{R}^{(q,4)}\,\mathbf{g}_0(\mathbf{z}) = \mathbf{0} \qquad \text{for} \quad \mathbf{z} \in S \tag{7.81}$$

has only a trivial solution. Indeed, let $\mathbf{g}_0$ be a solution of the homogeneous equation (7.81). Then the vector

$$\mathbf{V}_0(\mathbf{x}) \equiv \mathbf{Z}^{(q,4)}(\mathbf{x}, \mathbf{g}_0) + (1 - i)\mathbf{Z}^{(q,1)}(\mathbf{x}, \mathbf{g}_0) \qquad \text{for} \qquad \mathbf{x} \in \Omega^- \tag{7.82}$$

is a regular solution of problem $(I_q)^-_{\mathbf{0},\mathbf{0}}$. Using Theorem 7.3 we have (7.77).

On the other hand, by Theorem 7.4 and the relations (7.71) and (7.72) from (7.82) we get

$$\{\mathbf{V}_0(\mathbf{z})\}^- - \{\mathbf{V}_0(\mathbf{z})\}^+ = -\mathbf{g}_0(\mathbf{z}),$$

$$\{\mathcal{N}(\mathbf{D_z}, \mathbf{n})\mathbf{V}_0(\mathbf{z})\}^- - \{\mathcal{N}(\mathbf{D_z}, \mathbf{n})\mathbf{V}_0(\mathbf{z})\}^+ = (1 - i)\mathbf{g}_0(\mathbf{z}) \tag{7.83}$$

for $\mathbf{z} \in S$.

On the basis of (7.77) from (7.83) it follows that

$$\{\mathcal{N}(\mathbf{D_z},\mathbf{n})\mathbf{V}_0(\mathbf{z})+(1-i)\mathbf{V}_0(\mathbf{z})\}^+=\mathbf{0} \qquad \text{for} \quad \mathbf{z}\in S. \tag{7.84}$$

Obviously, the vector $\mathbf{V}_0$ is a solution of Eq. (7.18) in Ω^+ satisfying the boundary condition (7.84). It is easy to see that the relation (7.84) can be written as

$$\begin{aligned} &\left\{\mathcal{N}^{(1)}(\mathbf{D_z},\mathbf{n})\mathbf{V}_0(\mathbf{z})+(1-i)\mathbf{u}_0(\mathbf{z})\right\}^+=\mathbf{0},\\ &\left\{\mathcal{N}^{(2)}(\mathbf{D_z},\mathbf{n})\mathbf{p}_0(\mathbf{z})+(1-i)\mathbf{p}_0(\mathbf{z})\right\}^+=\mathbf{0}, \end{aligned} \tag{7.85}$$

where $\mathbf{u}_0$ and $\mathbf{p}_0$ are three- and four-component vector functions, respectively, $\mathbf{V}_0=(\mathbf{u}_0,\mathbf{p}_0)$, the operators $\mathcal{N}^{(1)}(\mathbf{D_z},\mathbf{n})$ and $\mathcal{N}^{(2)}(\mathbf{D_z},\mathbf{n})$ are given by (7.48).

Moreover, by virtue of (7.85) from (7.46) it follows that

$$\begin{aligned} &\int_{\Omega^+} W_1^{(\kappa_0)}(\mathbf{V}_0,\mathbf{u}_0)d\mathbf{x}+(1-i)\int_S|\mathbf{u}_0(\mathbf{z})|^2 d_{\mathbf{z}}S=0,\\ &\int_{\Omega^+} W_2(\mathbf{V}_0,\mathbf{p}_0)d\mathbf{x}+(1-i)\int_S|\mathbf{p}_0(\mathbf{z})|^2 d_{\mathbf{z}}S=0, \end{aligned} \tag{7.86}$$

where W_2 and $W_1^{(\kappa_0)}$ are given by (6.13) and (7.45), respectively. The relations (7.86) imply

$$\int_{\Omega^+} \operatorname{Im} W_1^{(\kappa_0)}(\mathbf{V}_0,\mathbf{u}_0)d\mathbf{x}-\int_S|\mathbf{u}_0(\mathbf{z})|^2 d_{\mathbf{z}}S=0,$$

$$\int_{\Omega^+} \operatorname{Re} W_2(\mathbf{V}_0,\mathbf{p}_0)d\mathbf{x}+\int_S|\mathbf{p}_0(\mathbf{z})|^2 d_{\mathbf{z}}S=0.$$

Clearly, we may further conclude that

$$\begin{aligned} &\int_{\Omega^+}\left[\operatorname{Re} W_2(\mathbf{V}_0,\mathbf{p}_0)-\omega \operatorname{Im} W_1^{(\kappa_0)}(\mathbf{V}_0,\mathbf{u}_0)\right]d\mathbf{x}\\ &\quad+\int_S\left[\omega|\mathbf{u}_0(\mathbf{z})|^2+|\mathbf{p}_0(\mathbf{z})|^2\right]d_{\mathbf{z}}S=0. \end{aligned} \tag{7.87}$$

Then, the relation (7.57) implies

$$\begin{aligned}&\mathrm{Re}W_2(\mathbf{V}_0,\mathbf{p}_0)-\omega\mathrm{Im}\,W_1^{(\kappa_0)}(\mathbf{V}_0,\mathbf{u}_0)\\&\quad=\mathbf{K}\nabla\mathbf{p}_0\cdot\nabla\mathbf{p}_0+\mathbf{d}\mathbf{p}_0\cdot\mathbf{p}_0\geq 0.\end{aligned}\tag{7.88}$$

Employing (7.88), from (7.87) we have

$$\int_S\left[\omega|\mathbf{u}_0(\mathbf{z})|^2+|\mathbf{p}_0(\mathbf{z})|^2\right]d_{\mathbf{z}}S=0.$$

Clearly, from the last equation we obtain

$$\{\mathbf{V}_0(\mathbf{z})\}^+=\mathbf{0}\qquad\text{for}\qquad\mathbf{z}\in S.\tag{7.89}$$

Finally, by virtue of (7.77) and (7.89) from the first equation of (7.83) we get $\mathbf{g}_0(\mathbf{z})\equiv\mathbf{0}$ for $\mathbf{z}\in S$.

Thus, the homogeneous equation (7.81) has only the trivial solution and therefore on the basis of Fredholm's theorem the integral equation (7.80) is always solvable for an arbitrary vector $\mathbf{f}$. We have thereby proved the following.

Theorem 7.17 *If* $S\in C^{2,\nu}$, $\mathbf{f}\in C^{1,\nu'}(S)$, $0<\nu'<\nu\leq 1$, *and the condition (7.52) is satisfied, then a regular solution of the class* $\mathfrak{B}$ *of the external BVP* $(I_q)^-_{\mathbf{0},\mathbf{f}}$ *exists, is unique, and is represented by a sum of potentials (7.79), where* $\mathbf{g}$ *is a solution of Fredholm's integral equation (7.80) which is always solvable for an arbitrary vector* $\mathbf{f}$.

7.7 On the Problems of Equilibrium in Elasticity

The basic internal and external BVPs of equilibrium in the linear theory of elasticity for quadruple porosity materials are formulated as follows.

Find a regular (classical) solution of the class $\mathfrak{B}$ in Ω^+ to the system (1.41) for $\mathbf{x}\in\Omega^+$ satisfying the boundary condition (6.2) in the *Problem* $(I_e)^+_{\mathbf{F},\mathbf{f}}$, and the boundary condition (6.3) in the *Problem* $(II_e)^+_{\mathbf{F},\mathbf{f}}$, where the matrix differential operators $\boldsymbol{\mathcal{A}}^{(e)}(\mathbf{D}_{\mathbf{x}})$ and $\boldsymbol{\mathcal{P}}(\mathbf{D}_{\mathbf{z}},\mathbf{n})$ are defined by (1.36) and (1.45), respectively, $\mathbf{F}$ and $\mathbf{f}$ are prescribed seven-component vector functions.

Find a regular (classical) solution of the class $\mathfrak{B}$ to the system (1.41) for $\mathbf{x}\in\Omega^-$ satisfying the boundary condition (6.4) in the *Problem* $(I_e)^-_{\mathbf{F},\mathbf{f}}$, and the boundary condition (6.5) in the *Problem* $(II_e)^-_{\mathbf{F},\mathbf{f}}$. Here $\mathbf{F}$ and $\mathbf{f}$ are prescribed seven-component vector functions, and $\mathrm{supp}\,\mathbf{F}$ is a finite domain in Ω^-.

We can investigate these BVPs by the potential method in the following manner: first we can study BVP for the vector $\mathbf{p}$, then we can study BVP for the vector $\mathbf{u}$.

For instance, in the BVP $(I_e)^+_{\mathbf{F},\mathbf{f}}$ we have two BVPs:

(i) The BVP for the vector $\mathbf{p}$

$$\begin{aligned} &(\mathbf{K}\,\Delta - \mathbf{d})\mathbf{p}(\mathbf{x}) = \mathbf{F}^{(2)}(\mathbf{x}) \qquad \text{for} \quad \mathbf{x} \in \Omega^+, \\ &\{\mathbf{p}(\mathbf{z})\}^+ = \mathbf{f}^{(2)}(\mathbf{z}) \qquad \text{for} \quad \mathbf{z} \in S; \end{aligned} \tag{7.90}$$

(ii) The BVP for the vector $\mathbf{u}$

$$\begin{aligned} &\mu\,\Delta\mathbf{u}(\mathbf{x}) + (\lambda + \mu)\,\nabla \mathrm{div}\mathbf{u}(\mathbf{x}) = \tilde{\mathbf{F}}^{(1)}(\mathbf{x}) \qquad \text{for} \quad \mathbf{x} \in \Omega^+, \\ &\{\mathbf{u}(\mathbf{z})\}^+ = \mathbf{f}^{(1)}(\mathbf{z}) \qquad \text{for} \quad \mathbf{z} \in S, \end{aligned} \tag{7.91}$$

where $\tilde{\mathbf{F}}^{(1)}(\mathbf{x}) = \mathbf{F}^{(1)}(\mathbf{x}) + \nabla\,(\mathbf{ap}(\mathbf{x}))$.

Obviously, the BVP (7.90) is the problem of equilibrium of the linear theory for quadruple porosity rigid body and studied in Chap. 5. After solving the BVP (7.90) we can study the BVP (7.91). Clearly, the BVP (7.91) is the displacement problem of equilibrium in the classical theory of elasticity and investigated by potential method in the book by Kupradze et al. [219].

Hence, the BVPs of equilibrium in the linear theory of elasticity for quadruple porosity materials are reduced to the known BVPs of the theories of elasticity and quadruple porosity rigid body.

Chapter 8
Problems of Pseudo-Oscillations in Elasticity

In this chapter, the basic BVPs of pseudo-oscillations in the linear theory of elasticity for quadruple porosity materials are investigated.

Namely, in Sects. 8.1 and 8.2, the basic BVPs of this theory are formulated and the uniqueness theorems of these BVPs are proved, respectively.

Then, in Sects. 8.3 and 8.4, the basic properties of the surface and volume potentials are established and the normal type singular integral operators corresponding BVPs of pseudo-oscillations are studied, respectively.

Finally, in Sect. 8.5, the BVPs of pseudo-oscillations in the linear theory of elasticity for quadruple porosity materials are reduced to the equivalent always solvable singular integral equations and the existence theorems for classical solutions of these BVPs are proved by means of the potential method and the theory of singular integral equations.

8.1 Basic Boundary Value Problems

As mentioned in Sect. 1.5, the system of equations of pseudo-oscillations in the linear theory of elasticity for materials with quadruple porosity can be written in the form

$$
\begin{aligned}
&(\mu\,\Delta - \rho\,\tau^2)\mathbf{u} + (\lambda+\mu)\,\nabla \mathrm{div}\mathbf{u} - \nabla\,(\mathbf{a}\,\mathbf{p}) = \mathbf{F}^{(1)},\\
&(\mathbf{K}\,\Delta + \mathbf{c}')\,\mathbf{p} - \tau\,\mathbf{a}\,\mathrm{div}\mathbf{u} = \mathbf{F}^{(2)},
\end{aligned}
\tag{8.1}
$$

where $\mathbf{u}$ is the displacement vector, $\mathbf{p}$ is the pressure vector; $\mathbf{F}^{(1)}$ and $\mathbf{F}^{(2)}$ are three- and four-component vector functions, respectively; $\mathbf{c}' = -(\tau\mathbf{b} + \mathbf{d})$, $\tau = \tau_1 + i\tau_2$ is a complex number, and τ_1 and τ_2 are real numbers, $\tau_1 > 0$.

M. Svanadze, *Potential Method in Mathematical Theories of Multi-Porosity Media*, Interdisciplinary Applied Mathematics 51,
https://doi.org/10.1007/978-3-030-28022-2_8

Obviously, the systems (8.1) we can rewritten in the matrix form

$$\mathcal{A}(\mathbf{D_x}, i\tau)\,\mathbf{V}(\mathbf{x}) = \mathcal{F}(\mathbf{x}), \tag{8.2}$$

where the matrix $\mathcal{A}(\mathbf{D_x}, i\tau)$ is defined by (1.36), $\mathbf{V} = (\mathbf{u}, \mathbf{p})$ and $\mathcal{F} = (\mathbf{F}^{(1)}, \mathbf{F}^{(2)})$ are seven-component vector functions.

The basic internal and external BVPs of pseudo-oscillations in the linear theory of elasticity for quadruple porosity materials are formulated as follows.

Find a regular (classical) solution of the class $\mathfrak{B}$ in Ω^+ to system (8.2) for $\mathbf{x} \in \Omega^+$ satisfying the boundary condition (6.2) in the *Problem* $(I_l)^+_{\mathbf{F},\mathbf{f}}$, and the boundary condition (6.3) in the *Problem* $(II_l)^+_{\mathbf{F},\mathbf{f}}$, where the matrix differential operator $\mathcal{P}(\mathbf{D_z}, \mathbf{n})$ is given by (1.43); $\mathbf{F}$ and $\mathbf{f}$ are prescribed seven-component vector functions.

Find a regular (classical) solution of the class $\mathfrak{B}$ to system (8.2) for $\mathbf{x} \in \Omega^-$ satisfying the boundary condition (6.4) in the *Problem* $(I_l)^-_{\mathbf{F},\mathbf{f}}$, and the boundary condition (6.5) in the *Problem* $(II_l)^-_{\mathbf{F},\mathbf{f}}$. Here $\mathbf{F}$ and $\mathbf{f}$ are prescribed seven-component vector functions, and $\operatorname{supp}\mathbf{F}$ is a finite domain in Ω^-. Note that the class $\mathfrak{B}$ of vector functions is given by Definition 3.2 (see Sect. 3.5).

8.2 Uniqueness Theorems

In this section we prove uniqueness of regular solutions of BVPs $(\mathbb{K}_l)^+_{\mathbf{F},\mathbf{f}}$ and $(\mathbb{K}_l)^-_{\mathbf{F},\mathbf{f}}$, where $\mathbb{K} = I, II$.

Theorem 8.1 *If the condition (7.9) is satisfied, then the internal BVP* $(\mathbb{K}_l)^+_{\mathbf{F},\mathbf{f}}$ *admits at most one regular (classical) solution of the class* $\mathfrak{B}$, *where* $\mathbb{K} = I, II$.

Proof As usual, we suppose that there are two regular solutions of problem $(\mathbb{K}_l)^+_{\mathbf{F},\mathbf{f}}$. Then their difference $\mathbf{V}$ is a regular solution of the internal homogeneous BVP $(\mathbb{K}_l)^+_{\mathbf{0},\mathbf{0}}$. Hence, $\mathbf{V}$ is a regular solution of the homogeneous system of equations

$$\begin{aligned} &(\mu\,\Delta - \rho\tau^2)\mathbf{u} + (\lambda + \mu)\,\nabla \mathrm{div}\mathbf{u} - \nabla\,(\mathbf{a}\,\mathbf{p}) = \mathbf{0}, \\ &\mathbf{K}\,\Delta\,\mathbf{p} - (\tau\mathbf{b} + \mathbf{d})\,\mathbf{p} - \tau\mathbf{a}\,\mathrm{div}\mathbf{u} = \mathbf{0} \end{aligned} \tag{8.3}$$

in Ω^+ satisfying the homogeneous boundary condition (6.11) in the problem $(I_l)^+_{\mathbf{F},\mathbf{f}}$ and the homogeneous boundary condition (6.24) in the problem $(II_l)^+_{\mathbf{F},\mathbf{f}}$.

On the basis of (6.11), (6.24), and (8.3), from (3.18) for $\mathbf{u}' = \mathbf{u}$, $\mathbf{p}' = \mathbf{p}$, and $\omega = i\tau$ we obtain

$$\int_{\Omega^+} W_4(\mathbf{V}, \mathbf{u})d\mathbf{x} = 0, \qquad \int_{\Omega^+} W_5(\mathbf{V}, \mathbf{p})d\mathbf{x} = 0, \tag{8.4}$$

where

$$W_4(\mathbf{V},\mathbf{u}) = W_0(\mathbf{u},\mathbf{u}) + \rho\tau^2|\mathbf{u}|^2 - a_\alpha\, p_\alpha \operatorname{div}\overline{\mathbf{u}},$$
$$W_5(\mathbf{V},\mathbf{p}) = k_{\alpha\beta}\nabla p_\beta\cdot\nabla p_\alpha + (\tau b_{\alpha\beta} + d_{\alpha\beta})p_\beta\overline{p_\alpha} + \tau\, a_\alpha\overline{p_\alpha}\operatorname{div}\mathbf{u}, \tag{8.5}$$

and $W_0(\mathbf{u},\mathbf{u})$ is defined by (6.13).

Then, employing (6.13), we may derive from (8.5)

$$\operatorname{Re} W_4(\mathbf{V},\mathbf{u}) = W_0(\mathbf{u},\mathbf{u}) + \rho(\tau_1^2-\tau_2^2)|\mathbf{u}|^2 - \operatorname{Re}[\mathbf{ap}\operatorname{div}\bar{\mathbf{u}}],$$
$$\operatorname{Im} W_4(\mathbf{V},\mathbf{u}) = 2\rho\tau_1\tau_2|\mathbf{u}|^2 - \operatorname{Im}[\mathbf{ap}\operatorname{div}\bar{\mathbf{u}}],$$
$$\operatorname{Re} W_5(\mathbf{V},\mathbf{p}) = k_{\alpha\beta}\nabla p_\beta\cdot\nabla p_\alpha + (\tau_1 b_{\alpha\beta}+d_{\alpha\beta})p_\beta\overline{p_\alpha} \tag{8.6}$$
$$+\tau_1\operatorname{Re}[\mathbf{ap}\operatorname{div}\bar{\mathbf{u}}] - \tau_2\operatorname{Im}[\mathbf{ap}\operatorname{div}\bar{\mathbf{u}}],$$
$$\operatorname{Im} W_5(\mathbf{V},\mathbf{p}) = \tau_2 b_{\alpha\beta}p_\beta\overline{p_\alpha} + \tau_2\operatorname{Re}[\mathbf{ap}\operatorname{div}\bar{\mathbf{u}}] + \tau_1\operatorname{Im}[\mathbf{ap}\operatorname{div}\bar{\mathbf{u}}].$$

If $\tau_2 = 0$, then using relations (8.6) we deduce

$$\operatorname{Re} W_4(\mathbf{V},\mathbf{u}) = W_0(\mathbf{u},\mathbf{u}) + \rho\tau_1^2|\mathbf{u}|^2 - \operatorname{Re}[\mathbf{ap}\operatorname{div}\bar{\mathbf{u}}],$$
$$\operatorname{Re} W_5(\mathbf{V},\mathbf{p}) = k_{\alpha\beta}\nabla p_\beta\cdot\nabla p_\alpha + (\tau_1 b_{\alpha\beta}+d_{\alpha\beta})p_\beta\overline{p_\alpha}$$
$$+\tau_1\operatorname{Re}[\mathbf{ap}\operatorname{div}\bar{\mathbf{u}}]$$

and we have

$$\operatorname{Re} W_5(\mathbf{V},\mathbf{p}) + \tau_1\operatorname{Re} W_4(\mathbf{V},\mathbf{u}) = k_{\alpha\beta}\nabla p_\beta\cdot\nabla p_\alpha$$
$$+(\tau_1 b_{\alpha\beta}+d_{\alpha\beta})p_\beta\overline{p_\alpha} + \tau_1 W_0(\mathbf{u},\mathbf{u}) + \rho\tau_1^3|\mathbf{u}|^2 \geq 0. \tag{8.7}$$

On the basis of (7.9) and (8.7) from (8.4) it follows that $\mathbf{u}(\mathbf{x}) \equiv \mathbf{0}$ and $\mathbf{p}(\mathbf{x}) \equiv \mathbf{0}$ for $\mathbf{x} \in \Omega^+$.

If $\tau_2 \neq 0$, then using relations (8.6) we may further conclude that

$$(\tau_1^2+\tau_2^2)\operatorname{Im} W_4(\mathbf{V},\mathbf{u}) + \tau_2\operatorname{Re} W_5(\mathbf{V},\mathbf{p}) - \tau_1\operatorname{Im} W_5(\mathbf{V},\mathbf{p})$$
$$= \tau_2\left[k_{\alpha\beta}\nabla p_\beta\cdot\nabla p_\alpha + d_{\alpha\beta}p_\beta\overline{p_\alpha} + 2\rho\tau_1(\tau_1^2+\tau_2^2)|\mathbf{u}|^2\right]. \tag{8.8}$$

by virtue of relation (8.8) from (8.4) we obtain

$$k_{\alpha\beta}\nabla p_\beta \cdot \nabla p_\alpha + d_{\alpha\beta} p_\beta \overline{p_\alpha} + 2\rho\tau_1(\tau_1^2 + \tau_2^2)|\mathbf{u}|^2 = 0.$$

Clearly, we may derive from the last equation

$$\mathbf{u}(\mathbf{x}) = \mathbf{0}, \qquad d_{\alpha\beta} p_\beta(\mathbf{x})\overline{p_\alpha(\mathbf{x})} = 0, \qquad p_\alpha(\mathbf{x}) = c_\alpha = \text{const} \tag{8.9}$$

for $\mathbf{x} \in \Omega^+$ and $\alpha = 1, 2, 3, 4$. Taking into account (8.9) from (8.6) we deduce that

$$W_5(\mathbf{V}, \mathbf{p}) = \tau b_{\alpha\beta} p_\beta \overline{p_\alpha}. \tag{8.10}$$

On the basis of (8.10) from the second equation of (8.4) it follows that $\mathbf{p}(\mathbf{x}) = \mathbf{0}$. Finally, from this equation and the first equation of (8.9) we get $\mathbf{V}(\mathbf{x}) = \mathbf{0}$ for $\mathbf{x} \in \Omega^+$. Hence, we have desired result. □

In a similar manner as in the previous theorem, on the basis of the conditions at infinity (3.31) we obtain the following.

Theorem 8.2 *If the condition (7.9) is satisfied, then the external BVP* $(\mathbb{K}_l)^-_{\mathbf{F},\mathbf{f}}$ *has one regular solution of the class* $\mathfrak{B}$*, where* $\mathbb{K} = I, II$.

8.3 Basic Properties of Potentials

On the basis of integral representation of a regular vector (3.38) and (3.39) we introduce the following notation:

(i) $\mathbf{Z}^{(l,1)}(\mathbf{x}, \mathbf{g}) = \displaystyle\int_S \mathbf{\Gamma}(\mathbf{x} - \mathbf{y}, i\tau)\mathbf{g}(\mathbf{y}) d_\mathbf{y} S,$

(ii) $\mathbf{Z}^{(l,2)}(\mathbf{x}, \mathbf{g}) = \displaystyle\int_S [\tilde{\mathcal{P}}(\mathbf{D}_\mathbf{y}, \mathbf{n}(\mathbf{y}))\mathbf{\Gamma}^\top(\mathbf{x} - \mathbf{y}, i\tau)]^\top \mathbf{g}(\mathbf{y}) d_\mathbf{y} S,$

(iii) $\mathbf{Z}^{(l,3)}(\mathbf{x}, \boldsymbol{\phi}, \Omega^\pm) = \displaystyle\int_{\Omega^\pm} \mathbf{\Gamma}(\mathbf{x} - \mathbf{y}, i\tau)\boldsymbol{\phi}(\mathbf{y}) d\mathbf{y},$

where $\mathbf{\Gamma}(\mathbf{x}, i\tau)$ is the fundamental matrix of the operator $\mathcal{A}(\mathbf{D}_\mathbf{x}, i\tau)$ and given by Theorem 2.2; the matrix differential operator $\tilde{\mathcal{P}}(\mathbf{D}_\mathbf{y}, \mathbf{n}(\mathbf{y}))$ is defined by (3.22); $\mathbf{g}$ and $\boldsymbol{\phi}$ are seven-component vector functions.

As in the classical theory of elasticity (see, e.g., Kupradze et al. [219]), the vector functions $\mathbf{Z}^{(l,1)}(\mathbf{x}, \mathbf{g})$, $\mathbf{Z}^{(l,2)}(\mathbf{x}, \mathbf{g})$, and $\mathbf{Z}^{(l,3)}(\mathbf{x}, \boldsymbol{\phi}, \Omega^\pm)$ are called *single-layer*, *double-layer*, and *volume potentials* of the linear theory of elasticity for quadruple porosity materials in the Laplace transform space.

On the basis of the properties of fundamental solution $\Gamma(\mathbf{x}, i\tau)$ (see Sect. 2.5) we have the following results.

Theorem 8.3 *If* $S \in C^{m+1,\nu}$, $\mathbf{g} \in C^{m,\nu'}(S)$, $0 < \nu' < \nu \leq 1$, *and* m *is a nonnegative integer, then:*

(a)

$$\mathbf{Z}^{(l,1)}(\cdot, \mathbf{g}) \in C^{0,\nu'}(\mathbb{R}^3) \cap C^{m+1,\nu'}(\overline{\Omega^\pm}) \cap C^\infty(\Omega^\pm),$$

(b)

$$\mathcal{A}(\mathbf{D_x}, i\tau)\, \mathbf{Z}^{(l,1)}(\mathbf{x}, \mathbf{g}) = \mathbf{0},$$

(c)

$$\left\{\mathcal{P}(\mathbf{D_z}, \mathbf{n}(\mathbf{z}))\, \mathbf{Z}^{(l,1)}(\mathbf{z}, \mathbf{g})\right\}^\pm = \mp\frac{1}{2}\,\mathbf{g}(\mathbf{z}) + \mathcal{P}(\mathbf{D_z}, \mathbf{n}(\mathbf{z}))\, \mathbf{Z}^{(l,1)}(\mathbf{z}, \mathbf{g}), \tag{8.11}$$

(d)

$$\mathcal{P}(\mathbf{D_z}, \mathbf{n}(\mathbf{z}))\, \mathbf{Z}^{(l,1)}(\mathbf{z}, \mathbf{g})$$

is a singular integral, where $\mathbf{z} \in S$, $\mathbf{x} \in \Omega^\pm$ *and*

$$\left\{\mathcal{P}(\mathbf{D_z}, \mathbf{n}(\mathbf{z}))\, \mathbf{Z}^{(l,1)}(\mathbf{z}, \mathbf{g})\right\}^\pm \equiv \lim_{\Omega^\pm \ni \mathbf{x} \to \mathbf{z} \in S} \mathcal{P}(\mathbf{D_x}, \mathbf{n}(\mathbf{z}))\, \mathbf{Z}^{(l,1)}(\mathbf{x}, \mathbf{g}),$$

(e) the single-layer potential $\mathbf{Z}^{(l,1)}(\mathbf{x}, \mathbf{g}) \equiv \mathbf{0}$ *for* $\mathbf{x} \in \Omega^+$ *(or* $\mathbf{x} \in \Omega^-$*) if and only if* $\mathbf{g}(\mathbf{z}) \equiv \mathbf{0}$, *where* $\mathbf{z} \in S$.

Theorem 8.4 *If* $S \in C^{m+1,\nu}$, $\mathbf{g} \in C^{m,\nu'}(S)$, $0 < \nu' < \nu \leq 1$, *then:*

(a)

$$\mathbf{Z}^{(l,2)}(\cdot, \mathbf{g}) \in C^{m,\nu'}(\overline{\Omega^\pm}) \cap C^\infty(\Omega^\pm),$$

(b)

$$\mathcal{A}(\mathbf{D_x}, i\tau)\, \mathbf{Z}^{(l,2)}(\mathbf{x}, \mathbf{g}) = \mathbf{0},$$

(c)

$$\left\{\mathbf{Z}^{(l,2)}(\mathbf{z}, \mathbf{g})\right\}^\pm = \pm\frac{1}{2}\,\mathbf{g}(\mathbf{z}) + \mathbf{Z}^{(l,2)}(\mathbf{z}, \mathbf{g}), \tag{8.12}$$

for the nonnegative integer m,

(d) $\mathbf{Z}^{(l,2)}(\mathbf{z}, \mathbf{g})$ *is a singular integral, where* $\mathbf{z} \in S$,

(e)

$$\left\{\mathcal{P}(\mathbf{D_z}, \mathbf{n}(\mathbf{z}))\, \mathbf{Z}^{(l,2)}(\mathbf{z}, \mathbf{g})\right\}^+ = \{\mathcal{P}(\mathbf{D_z}, \mathbf{n}(\mathbf{z}))\, \mathbf{Z}^{(l,2)}(\mathbf{z}, \mathbf{g})\}^-,$$

for the natural number m, *where* $\mathbf{z} \in S$, $\mathbf{x} \in \Omega^\pm$, *and*

$$\left\{\mathbf{Z}^{(l,2)}(\mathbf{z}, \mathbf{g})\right\}^\pm \equiv \lim_{\Omega^\pm \ni \mathbf{x} \to \mathbf{z} \in S} \mathbf{Z}^{(l,2)}(\mathbf{x}, \mathbf{g}).$$

Theorem 8.5 *If* $S \in C^{1,\nu}$, $\boldsymbol{\phi} \in C^{0,\nu'}(\Omega^+)$, $0 < \nu' < \nu \le 1$, *then:*

(a)

$$\mathbf{Z}^{(l,3)}(\cdot, \boldsymbol{\phi}, \Omega^+) \in C^{1,\nu'}(\mathbb{R}^3) \cap C^2(\Omega^+) \cap C^{2,\nu'}\left(\overline{\Omega_0^+}\right),$$

(b)

$$\mathcal{A}(\mathbf{D_x}, i\tau)\, \mathbf{Z}^{(l,3)}(\mathbf{x}, \boldsymbol{\phi}, \Omega^+) = \boldsymbol{\phi}(\mathbf{x}),$$

where $\mathbf{x} \in \Omega^+$, Ω_0^+ *is a domain in* $\mathbb{R}^3$ *and* $\overline{\Omega_0^+} \subset \Omega^+$.

Theorem 8.6 *If* $S \in C^{1,\nu}$, $\text{supp}\boldsymbol{\phi} = \Omega \subset \Omega^-$, $\boldsymbol{\phi} \in C^{0,\nu'}(\Omega^-)$, $0 < \nu' < \nu \le 1$, *then:*

(a)

$$\mathbf{Z}^{(l,3)}(\cdot, \boldsymbol{\phi}, \Omega^-) \in C^{1,\nu'}(\mathbb{R}^3) \cap C^2(\Omega^-) \cap C^{2,\nu'}(\overline{\Omega_0^-}),$$

(b)

$$\mathcal{A}(\mathbf{D_x}, i\tau)\, \mathbf{Z}^{(l,3)}(\mathbf{x}, \boldsymbol{\phi}, \Omega^-) = \boldsymbol{\phi}(\mathbf{x}),$$

where $\mathbf{x} \in \Omega^-$, Ω *is a finite domain in* $\mathbb{R}^3$ *and* $\overline{\Omega_0^-} \subset \Omega^-$.

8.4 Singular Integral Operators

We introduce the notation

$$\mathcal{K}^{(l,1)}\mathbf{g}(\mathbf{z}) \equiv \frac{1}{2}\,\mathbf{g}(\mathbf{z}) + \mathbf{Z}^{(l,2)}(\mathbf{z},\mathbf{g}),$$

$$\mathcal{K}^{(l,2)}\mathbf{g}(\mathbf{z}) \equiv -\frac{1}{2}\,\mathbf{g}(\mathbf{z}) + \mathcal{P}(\mathbf{D}_{\mathbf{z}},\mathbf{n}(\mathbf{z}))\mathbf{Z}^{(l,1)}(\mathbf{z},\mathbf{g}),$$

$$\mathcal{K}^{(l,3)}\mathbf{g}(\mathbf{z}) \equiv -\frac{1}{2}\,\mathbf{g}(\mathbf{z}) + \mathbf{Z}^{(l,2)}(\mathbf{z},\mathbf{g}), \tag{8.13}$$

$$\mathcal{K}^{(l,4)}\mathbf{g}(\mathbf{z}) \equiv \frac{1}{2}\,\mathbf{g}(\mathbf{z}) + \mathcal{P}(\mathbf{D}_{\mathbf{z}},\mathbf{n}(\mathbf{z}))\mathbf{Z}^{(l,1)}(\mathbf{z},\mathbf{g}),$$

$$\mathcal{K}^{(l)}_{\chi}\mathbf{g}(\mathbf{z}) \equiv \frac{1}{2}\,\mathbf{g}(\mathbf{z}) + \chi\,\mathbf{Z}^{(l,2)}(\mathbf{z},\mathbf{g})$$

for $\mathbf{z} \in S$, where χ is a complex number. Obviously, on the basis of Theorems 8.3 and 8.4, $\mathcal{K}^{(l,j)}$ and $\mathcal{K}^{(l)}_{\chi}$ are the singular integral operators $(j = 1, 2, 3, 4)$.

Let $\boldsymbol{\sigma}^{(l,j)} = (\sigma^{(l,j)}_{rm})_{7\times 7}$ be the symbol of the singular integral operator $\mathcal{K}^{(l,j)}$ $(j = 1, 2, 3, 4)$. Taking into account (7.9) and (8.13) we may see that

$$\begin{aligned} \det \boldsymbol{\sigma}^{(l,1)} &= -\det \boldsymbol{\sigma}^{(l,2)} = -\det \boldsymbol{\sigma}^{(l,3)} = \det \boldsymbol{\sigma}^{(l,4)} \\ &= -\frac{1}{128}\left[1 - \frac{\mu^2}{(\lambda+2\mu)^2}\right] = -\frac{(\lambda+\mu)(\lambda+3\mu)}{128(\lambda+2\mu)^2} < 0. \end{aligned} \tag{8.14}$$

Hence, the operator $\mathcal{K}^{(l,j)}$ is of the normal type, where $j = 1, 2, 3, 4$.

Let $\boldsymbol{\sigma}^{(l)}_{\chi}$ and $\operatorname{ind}\mathcal{K}^{(l)}_{\chi}$ be the symbol and the index of the operator $\mathcal{K}^{(l)}_{\chi}$, respectively. It may be easily shown that

$$\det \boldsymbol{\sigma}^{(l)}_{\chi} = \frac{\mu^2\chi^2 - (\lambda+2\mu)^2}{128(\lambda+2\mu)^2}$$

and $\det \boldsymbol{\sigma}^{(l)}_{\chi}$ vanishes only at two points χ_1 and χ_2 of the complex plane. By virtue of (8.14) and $\det \boldsymbol{\sigma}^{(l)}_{1} = \det \boldsymbol{\sigma}^{(l,1)}$ we get $\chi_j \neq 1$ $(j = 1, 2)$ and

$$\operatorname{ind}\mathcal{K}^{(l)}_{1} = \operatorname{ind}\mathcal{K}^{(l,1)} = \operatorname{ind}\mathcal{K}^{(l)}_{0} = 0.$$

Quite similarly we obtain

$$\operatorname{ind}\mathcal{K}^{(l,2)} = -\operatorname{ind}\mathcal{K}^{(l,3)} = 0, \qquad \operatorname{ind}\mathcal{K}^{(l,4)} = -\operatorname{ind}\mathcal{K}^{(l,1)} = 0.$$

Thus, the singular integral operator $\mathcal{K}^{(l,j)}$ $(j = 1, 2, 3, 4)$ is of the normal type with an index equal to zero. Consequently, Fredholm's theorems are valid for the singular integral operator $\mathcal{K}^{(l,j)}$.

8.5 Existence Theorems

By Theorems 8.5 and 8.6 the volume potential $\mathbf{Z}^{(l,3)}(\mathbf{x}, \mathbf{F}, \Omega^{\pm})$ is a regular solution of the nonhomogeneous equation (8.2), where $\mathbf{F} \in C^{0,\nu'}(\Omega^{\pm})$, $0 < \nu' \leq 1$ and $\operatorname{supp} \mathbf{F}$ is a finite domain in Ω^-. Therefore, further we will consider problems $(\mathbb{K}_l)^+_{\mathbf{0},\mathbf{f}}$ and $(\mathbb{K}_l)^-_{\mathbf{0},\mathbf{f}}$, where $\mathbb{K} = I, II$.

Now we prove the existence theorems of a regular (classical) solution of the class $\mathfrak{B}$ for these BVPs.

Problem $(I_l)^+_{\mathbf{0},\mathbf{f}}$ We seek a regular solution of the class $\mathfrak{B}$ to this problem in the form of the double-layer potential

$$\mathbf{V}(\mathbf{x}) = \mathbf{Z}^{(l,2)}(\mathbf{x}, \mathbf{g}) \qquad \text{for} \qquad \mathbf{x} \in \Omega^+, \tag{8.15}$$

where $\mathbf{g}$ is the required seven-component vector function.

Obviously, by Theorem 8.4 the vector function $\mathbf{V}$ is a solution of the homogeneous equation

$$\mathcal{A}(\mathbf{D_x}, i\tau)\, \mathbf{V}(\mathbf{x}) = \mathbf{0} \tag{8.16}$$

for $\mathbf{x} \in \Omega^+$. Keeping in mind the boundary condition (6.2) and using (8.12), from (8.15) we obtain, for determining the unknown vector $\mathbf{g}$, a singular integral equation

$$\mathcal{K}^{(l,1)}\, \mathbf{g}(\mathbf{z}) = \mathbf{f}(\mathbf{z}) \qquad \text{for} \quad \mathbf{z} \in S. \tag{8.17}$$

We prove that Eq. (8.17) is always solvable for an arbitrary vector $\mathbf{f}$.

Let us consider the associate homogeneous equation

$$\mathcal{K}^{(l,4)}\, \mathbf{h}(\mathbf{z}) = \mathbf{0} \qquad \text{for} \quad \mathbf{z} \in S, \tag{8.18}$$

where $\mathbf{h}$ is the required seven-component vector function. Now we prove that (8.18) has only the trivial solution.

Indeed, let $\mathbf{h}_0$ be a solution of the homogeneous equation (8.18). On the basis of Theorem 8.3 and Eq. (8.18) the vector function $\mathbf{V}_0(\mathbf{x}) = \mathbf{Z}^{(l,1)}(\mathbf{x}, \mathbf{h}_0)$ is a regular solution of the external homogeneous BVP $(II_l)^-_{\mathbf{0},\mathbf{0}}$. Using Theorem 8.2, the problem $(II_l)^-_{\mathbf{0},\mathbf{0}}$ has only the trivial solution, that is

$$\mathbf{V}_0(\mathbf{x}) \equiv \mathbf{0} \qquad \text{for} \qquad \mathbf{x} \in \Omega^-. \tag{8.19}$$

By virtue of the proposition (e) of Theorem 8.3 from (8.19) we obtain $\mathbf{h}_0(\mathbf{z}) = \mathbf{0}$ for $\mathbf{z} \in S$. Thus, the homogeneous equation (8.18) has only the trivial solution and therefore on the basis of Fredholm's theorem the integral equation (8.17) is always solvable for an arbitrary vector $\mathbf{f}$. We have thereby proved the following.

Theorem 8.7 *If* $S \in C^{2,\nu}$, $\mathbf{f} \in C^{1,\nu'}(S)$, $0 < \nu' < \nu \leq 1$, *then a regular solution of the class* $\mathfrak{B}$ *of the internal BVP* $(I_l)^+_{\mathbf{0},\mathbf{f}}$ *exists, is unique, and is represented by double-layer potential (8.15), where* $\mathbf{g}$ *is a solution of the singular integral equation (8.17) which is always solvable for an arbitrary vector* $\mathbf{f}$.

Problem $(II_l)^-_{\mathbf{0},\mathbf{f}}$ We seek a regular solution of the class $\mathfrak{B}$ to this problem in the form

$$\mathbf{V}(\mathbf{x}) = \mathbf{Z}^{(l,1)}(\mathbf{x}, \mathbf{h}) \qquad \text{for} \qquad \mathbf{x} \in \Omega^-, \tag{8.20}$$

where $\mathbf{h}$ is the required seven-component vector function.

Obviously, by virtue of Theorem 8.3 the vector function $\mathbf{V}$ is a solution of (8.16) for $\mathbf{x} \in \Omega^-$. Keeping in mind the boundary condition (6.5) and using (8.11), from (8.20) we obtain, for determining the unknown vector $\mathbf{h}$, a singular integral equation

$$\mathcal{K}^{(l,4)}\mathbf{h}(\mathbf{z}) = \mathbf{f}(\mathbf{z}) \qquad \text{for} \quad \mathbf{z} \in S. \tag{8.21}$$

It has been proved above that the corresponding homogeneous equation (8.18) has only the trivial solution. Hence, it follows that (8.21) is always solvable.

We have thereby proved the following.

Theorem 8.8 *If* $S \in C^{2,\nu}$, $\mathbf{f} \in C^{0,\nu'}(S)$, $0 < \nu' < \nu \leq 1$, *then a regular solution of the class* $\mathfrak{B}$ *of the external BVP* $(II_l)^-_{\mathbf{0},\mathbf{f}}$ *exists, is unique, and is represented by single-layer potential (8.20), where* $\mathbf{h}$ *is a solution of the singular integral equation (8.21) which is always solvable for an arbitrary vector* $\mathbf{f}$.

Problem $(II_l)^+_{\mathbf{0},\mathbf{f}}$ We seek a regular solution of the class $\mathfrak{B}$ to this problem in the form

$$\mathbf{V}(\mathbf{x}) = \mathbf{Z}^{(l,1)}(\mathbf{x}, \mathbf{g}) \qquad \text{for} \qquad \mathbf{x} \in \Omega^+, \tag{8.22}$$

where $\mathbf{g}$ is the required seven-component vector function.

Obviously, by virtue of Theorem 8.3 the vector function $\mathbf{V}$ is a solution of (8.16) for $\mathbf{x} \in \Omega^+$. Keeping in mind the boundary condition (6.3) and using (8.11), from (8.22) we obtain, for determining the unknown vector $\mathbf{g}$, the following singular integral equation

$$\mathcal{K}^{(l,2)}\mathbf{g}(\mathbf{z}) = \mathbf{f}(\mathbf{z}) \qquad \text{for} \qquad \mathbf{z} \in S. \tag{8.23}$$

We prove that (8.23) is always solvable for an arbitrary vector $\mathbf{f}$. Let us consider the corresponding homogeneous equation

$$\mathcal{K}^{(l,2)}\mathbf{g}_0(\mathbf{z}) = \mathbf{0} \qquad \text{for} \quad \mathbf{z} \in S, \tag{8.24}$$

where $\mathbf{g}_0$ is the required seven-component vector function. Now we prove that (8.24) has only the trivial solution.

Indeed, let $\mathbf{g}_0$ be a solution of the homogeneous equation (8.24). On the basis of Theorem 8.3 and (8.24) the vector $\mathbf{Z}^{(l,1)}(\mathbf{x}, \mathbf{g}_0)$ is a regular solution of problem $(II_l)^+_{\mathbf{0},\mathbf{0}}$. Using Theorem 8.1, the problem $(II_l)^+_{\mathbf{0},\mathbf{0}}$ has only the trivial solution, that is

$$\mathbf{Z}^{(l,1)}(\mathbf{x}, \mathbf{g}_0) = \mathbf{0} \qquad \text{for} \qquad \mathbf{x} \in \Omega^+. \tag{8.25}$$

By virtue of the proposition (e) of Theorem 8.3 from (8.25) we obtain $\mathbf{g}_0(\mathbf{z}) = \mathbf{0}$.

Thus, the homogeneous equation (8.24) has only a trivial solution and therefore (8.23) is always solvable for an arbitrary vector $\mathbf{f}$.

We have thereby proved the following.

Theorem 8.9 *If* $S \in C^{2,\nu}$, $\mathbf{f} \in C^{0,\nu'}(S)$, $0 < \nu' < \nu \leq 1$,, *then a regular (classical) solution of the class* $\mathfrak{B}$ *of problem* $(II_l)^+_{\mathbf{0},\mathbf{f}}$ *exists, is unique, and is represented by single-layer potential (8.22), where* $\mathbf{g}$ *is a solution of the singular integral equation (8.23) which is always solvable for an arbitrary vector* $\mathbf{f}$.

Problem $(I_l)^-_{\mathbf{0},\mathbf{f}}$ We seek a regular (classical) solution of the class $\mathfrak{B}$ to this problem in the form

$$\mathbf{V}(\mathbf{x}) = \mathbf{Z}^{(l,2)}(\mathbf{x}, \mathbf{h}) \qquad \text{for} \qquad \mathbf{x} \in \Omega^-, \tag{8.26}$$

where $\mathbf{h}$ is the required seven-component vector function.

Obviously, by Theorem 8.4 the vector function $\mathbf{V}$ is a solution of (8.16) for $\mathbf{x} \in \Omega^-$. Keeping in mind the boundary condition (6.4) and using (8.12), from (8.26) we obtain, for determining the unknown vector $\mathbf{h}$, the following integral equation

$$\mathcal{K}^{(l,3)}\mathbf{h}(\mathbf{z}) = \mathbf{f}(\mathbf{z}) \qquad \text{for} \qquad \mathbf{z} \in S. \tag{8.27}$$

It has been proved above that the adjoint homogeneous equation (8.24) of (8.27) has only the trivial solution. Hence, it follows that (8.27) is always solvable.

We have thereby proved the following.

Theorem 8.10 *If* $S \in C^{1,\nu}$, $\mathbf{f} \in C^{1,\nu'}(S)$, $0 < \nu' < \nu \leq 1$, *then a regular (classical) solution of the class* $\mathfrak{B}^{(r)}$ *of problem* $(I_r)^-_{\mathbf{0},\mathbf{f}}$ *exists, is unique, and is represented by double-layer potential (8.26), where* $\mathbf{h}$ *is a solution of the singular integral equation (8.27) which is always solvable for an arbitrary vector* $\mathbf{f}$.

Chapter 9
Problems of Steady Vibrations in Thermoelasticity

We now begin our investigation of the BVPs in the linear theory of thermoelasticity for quadruple porosity materials by means of the potential method and the theory of integral equations. Namely, we investigate the BVPs for the five special cases of equations of general dynamics: (i) equations of steady vibrations, (ii) equations in the Laplace transform space, (iii) equations of steady vibrations in the quasi-static theory, (iv) equations of equilibrium, and (v) equations of steady vibrations for quadruple porosity rigid body.

In this chapter, the basic internal and external BVPs of steady vibrations in the linear theory of thermoelasticity for quadruple porosity materials are investigated by means of the potential method and the theory of singular integral equations.

In Sect. 9.1, the fundamental solution of the system of steady vibrations equations is constructed by means of elementary functions and its basic properties are established.

In Sect. 9.2, the formula of Galerkin-type representation of general solution of the system of steady vibrations equations is given.

In Sect. 9.3, Green's identities for equations of steady vibrations of the considered theory are obtained.

In Sects. 9.4 and 9.5, the basic BVPs of steady vibrations of the linear theory of thermoelasticity for materials with quadruple porosity are formulated and the uniqueness theorems for classical solutions of these BVPs are proved by employing techniques based on Green's identities, respectively.

Then, in Sect. 9.6, the surface and volume potentials of the considered theory are constructed and their basic properties are established.

Furthermore, in Sect. 9.7, some useful singular integral operators are introduced and the basic properties of these operators are presented.

Finally, in Sect. 9.8, the above-mentioned BVPs of steady vibrations of the linear theory of thermoelasticity are reduced to the equivalent always solvable singular integral equations for which Fredholm's theorems are valid. Then the existence

M. Svanadze, *Potential Method in Mathematical Theories of Multi-Porosity Media*, Interdisciplinary Applied Mathematics 51,
https://doi.org/10.1007/978-3-030-28022-2_9

theorems for classical solutions of the BVPs of steady vibrations are proved by means of the potential method.

9.1 Fundamental Solution

In this section the fundamental solution of the system of Eq. (1.15) is constructed explicitly by means of elementary functions. In Sect. 1.6, this system is rewritten in the matrix form (1.31).

We introduce the notation:

$$\begin{gathered}
\mathbf{A}^{(0)}(\mathbf{D_x}) = \left(A^{(0)}_{lj}(\mathbf{D_x})\right)_{8\times 8},\\
A^{(0)}_{lj}(\mathbf{D_x}) = \mu\Delta\delta_{lj} + (\lambda+\mu)\frac{\partial^2}{\partial x_l \partial x_j},\\
A^{(0)}_{\alpha+3;\beta+3}(\mathbf{D_x}) = k_{\alpha\beta}\,\Delta, \qquad A^{(0)}_{88}(\mathbf{D_x}) = k\Delta,\\
A^{(0)}_{lm}(\mathbf{D_x}) = A^{(0)}_{lm}(\mathbf{D_x}) = A^{(0)}_{\alpha+3;8}(\mathbf{D_x}) = A^{(0)}_{8;\alpha+3}(\mathbf{D_x}) = 0,\\
l, j = 1, 2, 3, \qquad \alpha, \beta = 1, 2, 3, 4, \qquad m = 4, 5, \cdots, 8.
\end{gathered} \tag{9.1}$$

Definition 9.1 The matrix differential operator $\mathbf{A}^{(0)}(\mathbf{D_x})$ is called the *principal part* of the operators $\mathbf{A}(\mathbf{D_x}, \omega)$, $\mathbf{A}(\mathbf{D_x}, i\tau)$, $\mathbf{A}^{(q)}(\mathbf{D_x}, \omega)$, and $\mathbf{A}^{(e)}(\mathbf{D_x})$, where $\mathbf{A}(\mathbf{D_x}, \omega)$, $\mathbf{A}(\mathbf{D_x}, i\tau)$, $\mathbf{A}^{(q)}(\mathbf{D_x}, \omega)$, and $\mathbf{A}^{(e)}(\mathbf{D_x})$ are defined by (1.27)–(1.29).

Definition 9.2 The operator $\mathbf{A}(\mathbf{D_x}, \omega)$ ($\mathbf{A}(\mathbf{D_x}, i\tau)$, $\mathbf{A}^{(q)}(\mathbf{D_x}, \omega)$, $\mathbf{A}^{(e)}(\mathbf{D_x})$) is said to be *elliptic* if

$$\det \mathbf{A}^{(0)}(\boldsymbol{\psi}) \neq 0,$$

where $\boldsymbol{\psi} = (\psi_1, \psi_2, \psi_3) \neq \mathbf{0}$.

Obviously, by direct calculation we have

$$\det \mathbf{A}^{(0)}(\boldsymbol{\psi}) = \mu^2\mu_0\, k k_0\, |\boldsymbol{\psi}|^{16},$$

where $\mu_0 = \lambda + 2\mu$, $k_0 = \det\mathbf{K}$.

Hence, $\mathbf{A}(\mathbf{D_x}, \omega)$, $\mathbf{A}(\mathbf{D_x}, i\tau)$, $\mathbf{A}^{(q)}(\mathbf{D_x}, \omega)$, and $\mathbf{A}^{(e)}(\mathbf{D_x})$ are the elliptic differential operators if and only if

$$\mu\,\mu_0\,kk_0 \neq 0. \tag{9.2}$$

In this section we will suppose that the assumption (9.2) holds true.

Definition 9.3 The fundamental solution of system (1.15) (the fundamental matrix of operator $\mathbf{A}(\mathbf{D_x},\omega)$) is the matrix $\mathbf{\Psi}(\mathbf{x},\omega) = \left(\Psi_{lj}(\mathbf{x},\omega)\right)_{8\times 8}$ satisfying the following equation in the class of generalized functions

$$\mathbf{A}(\mathbf{D_x},\omega)\mathbf{\Psi}(\mathbf{x},\omega) = \delta(\mathbf{x})\mathbf{I}_8, \tag{9.3}$$

where $\mathbf{x} \in \mathbb{R}^3$.

Now we will extend the method of explicit construction of a fundamental solution (see Sect. 2.2) to the theory of thermoelasticity for quadruple porosity materials and the matrix $\mathbf{\Psi}(\mathbf{x},\omega)$ will be constructed explicitly by using metaharmonic functions.

We introduce the notation:

1.

$$\begin{aligned}
&\mathbf{B}(\Delta,\omega) = \left(B_{lj}(\Delta,\omega)\right)_{6\times 6}, \qquad B_{11}(\Delta,\omega) = \mu_0\Delta + \rho\omega^2,\\
&B_{1;\alpha+1}(\Delta,\omega) = i\omega a_\alpha\,\Delta, \qquad B_{16}(\Delta,\omega) = i\omega\varepsilon_0 T_0\,\Delta,\\
&B_{\alpha+1;1}(\Delta,\omega) = -a_\alpha, \qquad B_{\alpha+1;\beta+1}(\Delta,\omega) = k_{\alpha\beta}\Delta + c_{\alpha\beta},\\
&B_{\alpha+1;6}(\Delta,\omega) = i\omega T_0\varepsilon_\alpha, \qquad B_{61}(\Delta,\omega) = -\varepsilon_0,\\
&B_{6;\alpha+1}(\Delta,\omega) = i\omega\varepsilon_\alpha, \qquad B_{66}(\Delta,\omega) = k\Delta + i\omega a_0 T_0,\\
&\alpha,\beta = 1,2,3,4.
\end{aligned} \tag{9.4}$$

2.

$$\Theta_1(\Delta,\omega) = \frac{1}{kk_0\mu_0}\det\mathbf{B}(\Delta,\omega) = \prod_{j=1}^{6}(\Delta+\mu_j^2),$$

$$\Theta_2(\Delta,\omega) = \Theta_1(\Delta,\omega)(\Delta+\mu_7^2),$$

where $\mu_1^2, \mu_2^2, \cdots, \mu_6^2$ are the roots of the algebraic equation $\Theta_1(-\xi,\omega) = 0$ (with respect to ξ) and $\mu_7^2 = \dfrac{\rho\omega^2}{\mu}$.

3.

$$\mathbf{\Theta}(\Delta,\omega)=\left(\Theta_{lj}\left(\Delta,\omega\right)\right)_{8\times 8},\qquad \Theta_{11}=\Theta_{22}=\Theta_{33}=\Theta_2,$$
$$\Theta_{44}=\Theta_{55}=\cdots=\Theta_{88}=\Theta_1,\qquad \Theta_{lj}=0,\tag{9.5}$$
$$l,j=1,2,\cdots,8,\qquad l\neq j.$$

4.

$$\mathbf{L}\left(\mathbf{D_x},\omega\right)=\left(L_{lj}\left(\mathbf{D_x},\omega\right)\right)_{8\times 8},$$
$$L_{lj}\left(\mathbf{D_x},\omega\right)=\frac{1}{\mu}\Theta_1(\Delta,\omega)\,\delta_{lj}+m_{11}(\Delta,\omega)\frac{\partial^2}{\partial x_l\partial x_j},$$
$$L_{l;\alpha+2}\left(\mathbf{D_x},\omega\right)=m_{1\alpha}(\Delta,\omega)\frac{\partial}{\partial x_l},\tag{9.6}$$
$$L_{\alpha+2;l}\left(\mathbf{D_x},\omega\right)=m_{\alpha 1}(\Delta,\omega)\frac{\partial}{\partial x_l},$$
$$L_{\alpha+2;\beta+2}\left(\mathbf{D_x},\omega\right)=m_{\alpha\beta}(\Delta,\omega),$$

where

$$m_{r1}(\Delta,\omega)=-\frac{\lambda+\mu}{kk_0\mu\mu_0}B^*_{r1}(\Delta,\omega)$$
$$-\frac{i\omega}{kk_0\mu\mu_0}\left[a_\beta B^*_{r;\beta+1}(\Delta,\omega)+\varepsilon_0 T_0 B^*_{r6}(\Delta,\omega)\right],\tag{9.7}$$
$$m_{r\alpha}(\Delta,\omega)=\frac{1}{kk_0\mu_0}B^*_{r\alpha}(\Delta,\omega),$$
$$l,j=1,2,3,\qquad r=1,2,\cdots,6,\qquad \alpha,\beta=2,3,\cdots,6$$

and B^*_{lj} is the cofactor of element B_{lj} of the matrix $\mathbf{B}$ (see (9.4)).

Quite similarly as in Sect. 2.2 by virtue of (1.27), (9.5)–(9.7) we obtain

$$\mathbf{\Theta}(\Delta,\omega)\mathbf{U}=\mathbf{L}^\top\left(\mathbf{D_x},\omega\right)\mathbf{A}^\top\left(\mathbf{D_x},\omega\right)\mathbf{U},$$

where $\mathbf{U}=(\mathbf{u},\mathbf{p},\theta)$. It is obvious that $\mathbf{L}^\top\mathbf{A}^\top=\mathbf{\Theta}$ and, hence,

$$\mathbf{A}(\mathbf{D_x},\omega)\mathbf{L}(\mathbf{D_x},\omega)=\mathbf{\Theta}(\Delta,\omega).\tag{9.8}$$

We assume that $\mu_l^2 \neq \mu_j^2$, where $l, j = 1, 2, \cdots, 7$ and $l \neq j$. Let

$$\begin{aligned}
&\boldsymbol{\Upsilon}(\mathbf{x}, \omega) = \left(\Upsilon_{lj}(\mathbf{x}, \omega)\right)_{8\times 8}, \\
&\Upsilon_{11}(\mathbf{x}, \omega) = \Upsilon_{22}(\mathbf{x}, \omega) = \Upsilon_{33}(\mathbf{x}, \omega) = \sum_{j=1}^{7} \hat{\eta}_{2j} \vartheta^{(j)}(\mathbf{x}, \omega), \\
&\Upsilon_{44}(\mathbf{x}, \omega) = \Upsilon_{55}(\mathbf{x}, \omega) = \cdots = \Upsilon_{88}(\mathbf{x}, \omega) = \sum_{j=1}^{6} \hat{\eta}_{1j} \vartheta^{(j)}(\mathbf{x}, \omega), \\
&\Upsilon_{lj}(\mathbf{x}, \omega) = 0, \qquad l \neq j, \qquad l, j = 1, 2, \cdots, 8,
\end{aligned} \tag{9.9}$$

where

$$\vartheta^{(j)}(\mathbf{x}, \omega) = -\frac{e^{i\mu_j|\mathbf{x}|}}{4\pi\, |\mathbf{x}|} \tag{9.10}$$

is the fundamental solution of Helmholtz' equation, i.e., $(\Delta + \mu_j^2)\vartheta^{(j)}(\mathbf{x}, \omega) = \delta\,(\mathbf{x})$ and

$$\begin{aligned}
&\hat{\eta}_{1m} = \prod_{l=1,\, l\neq m}^{6} (\mu_l^2 - \mu_m^2)^{-1}, \qquad \hat{\eta}_{2j} = \prod_{l=1,\, l\neq j}^{7} (\mu_l^2 - \mu_j^2)^{-1}, \\
&m = 1, 2, \cdots, 6, \qquad j = 1, 2, \cdots, 7.
\end{aligned} \tag{9.11}$$

Lemma 9.1 *The matrix* $\boldsymbol{\Upsilon}(\mathbf{x}, \omega)$ *is the fundamental solution of the operator* $\boldsymbol{\Theta}(\Delta, \omega)$*, that is,*

$$\boldsymbol{\Theta}(\Delta, \omega)\boldsymbol{\Upsilon}\,(\mathbf{x}, \omega) = \delta\,(\mathbf{x})\,\mathbf{I}_8, \tag{9.12}$$

where $\mathbf{x} \in \mathbb{R}^3$.

Proof It suffices to show that Υ_{11} and Υ_{44} are the fundamental solutions of operators $\Theta_2(\Delta)$ and $\Theta_1(\Delta)$, respectively, i.e.

$$\Theta_2(\Delta, \omega)\Upsilon_{11}\,(\mathbf{x}, \omega) = \delta(\mathbf{x}) \tag{9.13}$$

and

$$\Theta_1(\Delta, \omega)\Upsilon_{44}\,(\mathbf{x}, \omega) = \delta(\mathbf{x}).$$

It is easy to verify that from (9.10) and (9.11) it follows that

$$\sum_{j=1}^{6} \hat{\eta}_{1j} = 0, \qquad \sum_{j=2}^{6} \hat{\eta}_{1j}(\mu_1^2 - \mu_j^2) = 0,$$

$$\sum_{j=3}^{6} \hat{\eta}_{1j}(\mu_1^2 - \mu_j^2)(\mu_2^2 - \mu_j^2) = 0,$$

$$\sum_{j=4}^{6} \hat{\eta}_{1j}(\mu_1^2 - \mu_j^2)(\mu_2^2 - \mu_j^2)(\mu_3^2 - \mu_j^2) = 0,$$

$$\sum_{j=5}^{6} \hat{\eta}_{1j}(\mu_1^2 - \mu_j^2)(\mu_2^2 - \mu_j^2)(\mu_3^2 - \mu_j^2)(\mu_4^2 - \mu_j^2) = 0,$$

$$\hat{\eta}_{16} \prod_{j=1}^{5} (\mu_j^2 - \mu_6^2) = 1,$$

$$(\Delta + \mu_l^2)\vartheta^{(j)}(\mathbf{x}, \omega) = \delta(\mathbf{x}) + (\mu_l^2 - \mu_j^2)\vartheta^{(j)}(\mathbf{x}, \omega),$$

$$l, j = 1, 2, \cdots, 6, \qquad \mathbf{x} \in \mathbb{R}^3.$$

Taking into account these equalities we have

$$\Theta_1(\Delta, \omega)\Upsilon_{44}(\mathbf{x}, \omega) = \prod_{l=2}^{6}(\Delta + \mu_l^2) \sum_{j=1}^{6} \hat{\eta}_{1j} \left[\delta(\mathbf{x}) + (\mu_1^2 - \mu_j^2)\vartheta^{(j)}(\mathbf{x}, \omega)\right]$$

$$= \prod_{l=2}^{6}(\Delta + \mu_l^2) \sum_{j=2}^{6} \hat{\eta}_{1j}(\mu_1^2 - \mu_j^2)\vartheta^{(j)}(\mathbf{x}, \omega)$$

$$= \prod_{l=3}^{6}(\Delta + \mu_l^2) \sum_{j=2}^{6} \hat{\eta}_{1j}(\mu_1^2 - \mu_j^2) \left[\delta(\mathbf{x}) + (\mu_2^2 - \mu_j^2)\vartheta^{(j)}(\mathbf{x}, \omega)\right]$$

$$= \cdots = (\Delta + \mu_6^2)\vartheta^{(6)}(\mathbf{x}, \omega) = \delta(\mathbf{x}).$$

Equation (9.13) is proved quite similarly. □

We introduce the matrix

$$\boldsymbol{\Psi}(\mathbf{x}, \omega) = \mathbf{L}(\mathbf{D}_{\mathbf{x}}, \omega)\,\boldsymbol{\Upsilon}(\mathbf{x}, \omega). \tag{9.14}$$

Using identities (9.8) and (9.12) from (9.14) we get

$$\mathbf{A}(\mathbf{D_x},\omega)\,\boldsymbol{\Psi}(\mathbf{x},\omega) = \mathbf{A}(\mathbf{D_x},\omega)\,\mathbf{L}(\mathbf{D_x},\omega)\,\boldsymbol{\Upsilon}(\mathbf{x},\omega)$$

$$= \boldsymbol{\Theta}(\Delta,\omega)\,\boldsymbol{\Upsilon}(\mathbf{x},\omega) = \delta(\mathbf{x})\,\mathbf{I}_8.$$

Hence, $\boldsymbol{\Psi}(\mathbf{x},\omega)$ is the solution of (9.3). We have thereby proved the following.

Theorem 9.1 *The matrix* $\boldsymbol{\Psi}(\mathbf{x},\omega)$ *defined by (9.14) is the fundamental solution of (1.15), where the matrices* $\mathbf{L}(\mathbf{D_x},\omega)$ *and* $\boldsymbol{\Upsilon}(\mathbf{x},\omega)$ *are given by (9.6) and (9.9), respectively.*

Clearly, each element $\Psi_{lj}(\mathbf{x},\omega)$ of the matrix $\boldsymbol{\Psi}(\mathbf{x},\omega)$ is represented in the following form

$$\Psi_{lj}(\mathbf{x},\omega) = L_{lj}(\mathbf{D_x},\omega)\,\Upsilon_{11}(\mathbf{x},\omega),$$

$$\Psi_{lm}(\mathbf{x},\omega) = L_{lm}(\mathbf{D_x},\omega)\,\Upsilon_{44}(\mathbf{x},\omega), \tag{9.15}$$

$$l = 1, 2, \cdots, 8, \qquad j = 1, 2, 3, \qquad m = 4, 5, \cdots, 8.$$

Obviously, the matrix $\boldsymbol{\Psi}(\mathbf{x},\omega)$ is constructed explicitly by means of seven metaharmonic functions $\vartheta^{(j)}$ $(j = 1, 2, \cdots, 7)$ (see (9.10)).

We now can establish the basic properties of the matrix $\boldsymbol{\Psi}(\mathbf{x},\omega)$. Theorem 9.1 leads to the following consequences.

Theorem 9.2 *If* $\mathbf{F} = \mathbf{0}$, *then the matrix* $\boldsymbol{\Psi}(\mathbf{x},\omega)$ *is a solution of Eq. (1.31), i.e.*

$$\mathbf{A}(\mathbf{D_x},\omega)\boldsymbol{\Psi}(\mathbf{x},\omega) = \mathbf{0} \tag{9.16}$$

at every point $\mathbf{x} \in \mathbb{R}^3$ *except the origin.*

Theorem 9.3 *The relations*

$$\Psi_{lj}(\mathbf{x},\omega) = O\left(|\mathbf{x}|^{-1}\right), \qquad \Psi_{\alpha+3;\beta+3}(\mathbf{x},\omega) = O\left(|\mathbf{x}|^{-1}\right),$$

$$\Psi_{88}(\mathbf{x},\omega) = O\left(|\mathbf{x}|^{-1}\right), \qquad \Psi_{lm}(\mathbf{x},\omega) = O(1),$$

$$\Psi_{mj}(\mathbf{x},\omega) = O(1), \qquad \Psi_{\alpha+3;8}(\mathbf{x},\omega) = O(1),$$

$$\Psi_{8;\alpha+3}(\mathbf{x},\omega) = O(1), \qquad l, j = 1, 2, 3,$$

$$\alpha, \beta = 1, 2, 3, 4, \qquad m = 4, 5, \cdots, 8$$

hold the neighborhood of the origin.

Theorem 9.4 *The relations*

$$\frac{\partial}{\partial x_r}\Psi_{lj}(\mathbf{x},\omega)=O\left(|\mathbf{x}|^{-2}\right),\qquad \frac{\partial}{\partial x_r}\Psi_{\alpha+3;\beta+3}(\mathbf{x},\omega)=O\left(|\mathbf{x}|^{-2}\right),$$

$$\frac{\partial}{\partial x_r}\Psi_{88}(\mathbf{x},\omega)=O\left(|\mathbf{x}|^{-2}\right),\qquad \frac{\partial}{\partial x_r}\Psi_{lm}(\mathbf{x},\omega)=O\left(|\mathbf{x}|^{-1}\right),$$

$$\frac{\partial}{\partial x_r}\Psi_{mj}(\mathbf{x},\omega)=O\left(|\mathbf{x}|^{-1}\right),\qquad \frac{\partial}{\partial x_r}\Psi_{\alpha+3;8}(\mathbf{x},\omega)=O\left(|\mathbf{x}|^{-1}\right),$$

$$\frac{\partial}{\partial x_r}\Psi_{8;\alpha+3}(\mathbf{x},\omega)=O\left(|\mathbf{x}|^{-1}\right),\qquad l,j,r=1,2,3,$$

$$\alpha,\beta=1,2,3,4,\qquad m=4,5,\cdots,8$$

hold the neighborhood of the origin.

Theorem 9.5 *The fundamental matrix of the operator* $\mathbf{A}^{(0)}(\mathbf{D_x})$ *is the matrix* $\mathbf{\Psi}^{(0)}(\mathbf{x})=\left(\Psi^{(0)}_{lj}(\mathbf{x})\right)_{8\times 8}$, *where*

$$\Psi^{(0)}_{lj}(\mathbf{x})=\lambda'\frac{\delta_{lj}}{|\mathbf{x}|}+\mu'\frac{x_l x_j}{|\mathbf{x}|^3},\qquad \Psi^{(0)}_{\alpha+3;\beta+3}(\mathbf{x})=\frac{1}{k_0}k^*_{\alpha\beta}\tilde{\gamma}^{(5)}(\mathbf{x}),$$

$$\Psi^{(0)}_{88}(\mathbf{x})=\frac{1}{k}\tilde{\gamma}^{(5)}(\mathbf{x}),\qquad \Psi^{(0)}_{lm}(\mathbf{x})=\Psi^{(0)}_{ml}(\mathbf{x})=\Psi^{(0)}_{\alpha+3;8}(\mathbf{x})=\Psi^{(0)}_{8;\alpha+3}(\mathbf{x})=0,$$

$$\lambda'=-\frac{\lambda+3\mu}{8\pi\mu\mu_0},\qquad \mu'=-\frac{\lambda+\mu}{8\pi\mu\mu_0},\qquad l,j=1,2,3,$$

$$\alpha,\beta=1,2,3,4,\qquad m=4,5,\cdots,8;$$

the matrix $\mathbf{A}^{(0)}(\mathbf{D_x})$ *is defined by (9.1),* $k^*_{\alpha\beta}$ *is the cofactor of element* $k_{\alpha\beta}$ *of the matrix* $\mathbf{K}$, *and* $\tilde{\gamma}^{(5)}(\mathbf{x})$ *is given by (2.26).*

Theorem 9.5 and the relations (9.15) lead to the following.

Theorem 9.6 *The relations*

$$\Psi_{lj}(\mathbf{x},\omega)-\Psi^{(0)}_{lj}(\mathbf{x})=const+O(|\mathbf{x}|),\qquad l,j=1,2,\cdots,8 \tag{9.17}$$

hold in the neighborhood of the origin.

Thus, on the basis of Theorems 9.3 and 9.6 the matrix $\mathbf{\Psi}^{(0)}(\mathbf{x})$ is the singular part of the fundamental solution $\mathbf{\Psi}(\mathbf{x},\omega)$ in the neighborhood of the origin.

Now we shall construct a solution of the system of steady vibrations equations (1.31) for $\mathbf{F}(\mathbf{x})=\mathbf{0}$ playing a major role in the investigation of the BVPs of the linear theory of thermoelasticity for materials with quadruple porosity by means of the potential method. The following theorem is valid.

Theorem 9.7 *Every column of the matrices*

$$\left[\mathbf{P}^{(\kappa)}(\mathbf{D}_{\mathbf{x}}, \mathbf{n}(\mathbf{x}))\boldsymbol{\Psi}^{\top}(\mathbf{x}-\mathbf{y}, \omega)\right]^{\top},$$

$$\left\{\mathbf{P}^{(\kappa)}(\mathbf{D}_{\mathbf{x}}, \mathbf{n}(\mathbf{x}))[\boldsymbol{\Psi}^{(0)}(\mathbf{x}-\mathbf{y})]^{\top}\right\}^{\top}$$

with respect to the point $\mathbf{x}$ *satisfies the homogeneous equations*

$$\begin{aligned} &\mathbf{A}(\mathbf{D}_{\mathbf{x}}, \omega)\left[\mathbf{P}^{(\kappa)}(\mathbf{D}_{\mathbf{x}}, \mathbf{n}(\mathbf{x}))\boldsymbol{\Psi}^{\top}(\mathbf{x}-\mathbf{y}, \omega)\right]^{\top} = \mathbf{0}, \\ &\mathbf{A}^{(0)}(\mathbf{D}_{\mathbf{x}})\left\{\mathbf{P}^{(\kappa)}(\mathbf{D}_{\mathbf{x}}, \mathbf{n}(\mathbf{x}))[\boldsymbol{\Psi}^{(0)}(\mathbf{x}-\mathbf{y})]^{\top}\right\}^{\top} = \mathbf{0} \end{aligned} \tag{9.18}$$

everywhere in $\mathbb{R}^3$ *except* $\mathbf{x} = \mathbf{y}$, *respectively, the matrix differential operator* $\mathbf{P}^{(\kappa)}(\mathbf{D}_{\mathbf{x}}, \mathbf{n}(\mathbf{x}))$ *is given by (1.46) and* κ *is an arbitrary real number.*

Proof Keeping in mind (9.16) it follows that

$$\begin{aligned} &\left\{\mathbf{A}(\mathbf{D}_{\mathbf{x}}, \omega)\left[\mathbf{P}^{(\kappa)}(\mathbf{D}_{\mathbf{x}}, \mathbf{n}(\mathbf{x}))\boldsymbol{\Psi}^{\top}(\mathbf{x}-\mathbf{y}, \omega)\right]^{\top}\right\}_{lj} \\ &= \sum_{m=1}^{8} A_{lm}(\mathbf{D}_{\mathbf{x}}, \omega)\left[\mathbf{P}^{(\kappa)}(\mathbf{D}_{\mathbf{x}}, \mathbf{n}(\mathbf{x}))\boldsymbol{\Psi}^{\top}(\mathbf{x}-\mathbf{y}, \omega)\right]_{jm} \\ &= \sum_{m,s=1}^{8} A_{lm}(\mathbf{D}_{\mathbf{x}}, \omega)P_{js}^{(\kappa)}(\mathbf{D}_{\mathbf{x}}, \mathbf{n}(\mathbf{x}))\Psi_{ms}(\mathbf{x}-\mathbf{y}, \omega) \\ &= \sum_{m,s=1}^{8} P_{js}^{(\kappa)}(\mathbf{D}_{\mathbf{x}}, \mathbf{n}(\mathbf{x}))A_{lm}(\mathbf{D}_{\mathbf{x}}, \omega)\Psi_{ms}(\mathbf{x}-\mathbf{y}, \omega) = 0 \end{aligned}$$

which is the first formula involved in Eq. (9.18). The other formulas in (9.18) can be proved quite similarly. □

The matrix $\left[\mathbf{P}^{(\kappa)}(\mathbf{D}_{\mathbf{x}}, \mathbf{n}(\mathbf{x}))\boldsymbol{\Psi}^{\top}(\mathbf{x}-\mathbf{y}, \omega)\right]^{\top}$ is called the *singular solution* of the system of steady vibration equation (1.31). Theorem 9.7 leads to the following.

Theorem 9.8 *The relations*

$$\begin{aligned} &\left[\mathbf{P}^{(\kappa)}(\mathbf{D}_{\mathbf{x}}, \mathbf{n}(\mathbf{x}))\boldsymbol{\Psi}^{\top}(\mathbf{x}, \omega)\right]_{lj} = O\left(|\mathbf{x}|^{-2}\right), \\ &\left\{\mathbf{P}^{(\kappa)}(\mathbf{D}_{\mathbf{x}}, \mathbf{n}(\mathbf{x}))[\boldsymbol{\Psi}^{(0)}(\mathbf{x})]^{\top}\right\}_{lj} = O\left(|\mathbf{x}|^{-2}\right), \\ &l, j = 1, 2, \cdots, 8 \end{aligned}$$

hold in the neighborhood of the origin, where κ is an arbitrary real number.

Now we can establish the singular part of the matrix $\mathbf{P}^{(\kappa)}(\mathbf{D_x}, \mathbf{n}(\mathbf{x}))\boldsymbol{\Psi}^\top(\mathbf{x}, \omega)$ in the neighborhood of the origin.

The relations (9.17) lead to the following.

Theorem 9.9 *The relations*

$$\left\{\mathbf{P}^{(\kappa)}(\mathbf{D_x}, \mathbf{n}(\mathbf{x}))\boldsymbol{\Psi}^\top(\mathbf{x}, \omega)\right\}_{lj} - \left\{\mathbf{P}^{(\kappa)}(\mathbf{D_x}, \mathbf{n}(\mathbf{x}))\left[\boldsymbol{\Psi}^{(0)}(\mathbf{x})\right]^\top\right\}_{lj} = O\left(|\mathbf{x}|^{-1}\right), \tag{9.19}$$

$$l, j = 1, 2, \cdots, 8$$

hold in the neighborhood of the origin, where κ is an arbitrary real number.

Thus, the relations (9.19) imply that the matrix $\mathbf{P}^{(\kappa)}(\mathbf{D_x}, \mathbf{n}(\mathbf{x}))\left[\boldsymbol{\Psi}^{(0)}(\mathbf{x})\right]^\top$ is the singular part of the matrix $\mathbf{P}^{(\kappa)}(\mathbf{D_x}, \mathbf{n}(\mathbf{x}))\boldsymbol{\Psi}^\top(\mathbf{x}, \omega)$ in the neighborhood of the origin for arbitrary real number κ.

9.2 Galerkin-Type Solution

The next two theorems provide a Galerkin-type solution to system (1.15).

Theorem 9.10 *Let*

$$\begin{aligned} u_l(\mathbf{x}) &= \frac{1}{\mu}\Theta_1(\Delta, \omega)\, w_l(\mathbf{x}) + m_{11}(\Delta, \omega)\, w_{j,lj}(\mathbf{x}) \\ &\quad + \sum_{\beta=2}^{6} m_{1\beta}(\Delta, \omega)\, w_{\beta+2,l}(\mathbf{x}), \\ p_\alpha(\mathbf{x}) &= m_{\alpha+1;1}(\Delta, \omega) w_{j,j}(\mathbf{x}) + \sum_{\beta=2}^{6} m_{\alpha+1;\beta}(\Delta, \omega)\, w_{\beta+2}(\mathbf{x}), \\ \theta(\mathbf{x}) &= m_{61}(\Delta, \omega) w_{j,j}(\mathbf{x}) + \sum_{\beta=2}^{6} m_{6\beta}(\Delta, \omega)\, w_{\beta+2}(\mathbf{x}), \\ l &= 1, 2, 3, \qquad \alpha = 1, 2, 3, 4, \end{aligned} \tag{9.20}$$

where $\mathbf{w}^{(1)} = (w_1, w_2, w_3) \in C^{14}(\Omega)$, $\mathbf{w}^{(2)} = (w_4, w_5, w_6, w_7) \in C^{12}(\Omega)$, $w_8 \in C^{12}(\Omega)$,

$$\begin{aligned} \Theta_2(\Delta, \omega)\, \mathbf{w}^{(1)}(\mathbf{x}) &= \mathbf{F}^{(1)}(\mathbf{x}), \\ \Theta_1(\Delta, \omega)\, \mathbf{w}^{(2)}(\mathbf{x}) &= \mathbf{F}^{(2)}(\mathbf{x}), \\ \Theta_1(\Delta, \omega)\, w_8(\mathbf{x}) &= F_8(\mathbf{x}), \end{aligned} \tag{9.21}$$

$\mathbf{F}^{(1)} = (F_1, F_2, F_3)$, $\mathbf{F}^{(2)} = (F_4, F_5, F_6, F_7)$, *then* $\mathbf{U} = (\mathbf{u}, \mathbf{p}, \theta)$ *is a solution of the system (1.15).*

Proof By virtue of (9.6) we can rewrite the Eqs. (9.20) and (9.21) in the form

$$\mathbf{U}(\mathbf{x}) = \mathbf{L}(\mathbf{D_x}, \omega)\, \widetilde{\mathbf{w}}(\mathbf{x}) \tag{9.22}$$

and

$$\mathbf{\Theta}(\Delta, \omega)\, \widetilde{\mathbf{w}}(\mathbf{x}) = \mathbf{F}(\mathbf{x}), \tag{9.23}$$

respectively, where $\widetilde{\mathbf{w}} = (w_1, w_2, \cdots, w_8)$, $\mathbf{F} = (F_1, F_2, \cdots, F_8)$. Clearly, on the basis of (9.8), (9.22), and (9.23) it follows that

$$\mathbf{A}(\mathbf{D_x}, \omega)\mathbf{U} = \mathbf{A}(\mathbf{D_x}, \omega)\mathbf{L}(\mathbf{D_x}, \omega)\widetilde{\mathbf{w}} = \mathbf{\Theta}(\Delta, \omega)\widetilde{\mathbf{w}} = \mathbf{F}.$$

Thus, the vector $\mathbf{U}$ is a solution of the system (1.31). □

Theorem 9.11 *If* $\mathbf{U} = (\mathbf{u}, \mathbf{p}, \theta)$ *is a solution of system (1.31) in* Ω*, then* $\mathbf{U}$ *is represented by (9.20), where* $\widetilde{\mathbf{w}} = (w_1, w_2, \cdots, w_8)$ *is a solution of (9.21) and* Ω *is a finite domain in* $\mathbb{R}^3$.

Proof Let $\mathbf{U}$ be a solution of system (1.31). Obviously, if $\mathbf{\Psi}'(\mathbf{x})$ is the fundamental matrix of the operator $\mathbf{L}(\mathbf{D_x}, \omega)$, then the vector function (the volume potential)

$$\widetilde{\mathbf{w}}(\mathbf{x}) = \int_\Omega \mathbf{\Psi}'(\mathbf{x} - \mathbf{y})\mathbf{U}(\mathbf{y})dy$$

is a solution of (9.22).

On the other hand, by virtue of (1.31), (9.8), and (9.22) we have

$$\mathbf{F}(\mathbf{x}) = \mathbf{A}(\mathbf{D_x}, \omega)\mathbf{U}(\mathbf{x}) = \mathbf{A}(\mathbf{D_x}, \omega)\mathbf{L}(\mathbf{D_x}, \omega)\, \widetilde{\mathbf{w}}(\mathbf{x}) = \mathbf{\Theta}(\Delta, \omega)\, \widetilde{\mathbf{w}}(\mathbf{x}).$$

Hence, $\widetilde{\mathbf{w}}$ is a solution of (9.23). □

Thus, on the basis of Theorems 9.10 and 9.11 the completeness of the Galerkin-type solution of system (1.15) is proved.

Remark 9.1 Quite similarly as in Theorem 9.1 we can construct explicitly the fundamental matrix $\mathbf{\Psi}'(\mathbf{x})$ of the operator $\mathbf{L}(\mathbf{D_x}, \omega)$ by elementary functions.

9.3 Green's Formulas

In what follows, we assume that $\mathrm{Im}\mu_j > 0$ for $j = 1, 2, \cdots, 6$ and $\mu_7 > 0$.

Definition 9.4 A vector function $\mathbf{U} = (\mathbf{u}, \mathbf{p}, \theta) = (U_1, U_2, \cdots, U_8)$ is called *regular* of the class $\mathfrak{R}^{(s)}$ in Ω^- (or Ω^+) if

(i)

$$U_l \in C^2(\Omega^-) \cap C^1(\overline{\Omega^-}) \qquad (\text{or } U_l \in C^2(\Omega^+) \cap C^1(\overline{\Omega^+})),$$

(ii)

$$\mathbf{U} = \sum_{j=1}^{7} \mathbf{U}^{(j)}, \qquad \mathbf{U}^{(j)} = (U_1^{(j)}, U_2^{(j)}, \cdots, U_8^{(j)}),$$

$$U_l^{(j)} \in C^2(\Omega^-) \cap C^1(\bar{\Omega}^-),$$

(iii) $(\Delta + \mu_j^2) U_l^{(j)}(\mathbf{x}) = 0$ and

$$\left(\frac{\partial}{\partial |\mathbf{x}|} - i\mu_j\right) U_l^{(j)}(\mathbf{x}) = e^{i\mu_j|\mathbf{x}|} o(|\mathbf{x}|^{-1}) \qquad \text{for} \qquad |\mathbf{x}| \gg 1, \tag{9.24}$$

where $U_r^{(7)} = 0$, $j = 1, 2, \cdots, 7$, $l = 1, 2, \cdots, 8$ and $r = 4, 5, \cdots, 8$.

Obviously, the relation (9.24) implies (for details see Vekua [386])

$$U_l^{(j)}(\mathbf{x}) = e^{i\mu_j|\mathbf{x}|} O(|\mathbf{x}|^{-1}) \qquad \text{for} \qquad |\mathbf{x}| \gg 1, \tag{9.25}$$

where $l = 1, 2, \cdots, 6, \quad j = 1, 2, \cdots, 5$.

Relations (9.24) and (9.25) are Sommerfeld–Kupradze type radiation conditions in the linear theory of thermoelasticity for quadruple porosity materials.

9.3.1 Green's First Identity

In the sequel we use the matrix differential operators:

1.

$$\mathbf{A}^{(1)}(\mathbf{D_x},\omega)=\left(A^{(1)}_{lr}(\mathbf{D_x},\omega)\right)_{3\times 8},\qquad A^{(1)}_{lr}(\mathbf{D_x},\omega)=A_{lr}(\mathbf{D_x},\omega),$$

$$\mathbf{A}^{(2)}(\mathbf{D_x},\omega)=\left(A^{(2)}_{\alpha r}(\mathbf{D_x},\omega)\right)_{4\times 8},\qquad A^{(2)}_{\alpha r}(\mathbf{D_x},\omega)=A_{\alpha+3;r}(\mathbf{D_x},\omega)$$

$$\mathbf{A}^{(3)}(\mathbf{D_x},\omega)=\left(A^{(3)}_{1r}(\mathbf{D_x},\omega)\right)_{1\times 8},\qquad A^{(3)}_{1r}(\mathbf{D_x},\omega)=A_{8r}(\mathbf{D_x},\omega);$$

2.

$$\mathbf{P}^{(1)}(\mathbf{D_x},\mathbf{n})=\left(P^{(1)}_{lr}(\mathbf{D_x},\mathbf{n})\right)_{3\times 8},\qquad P^{(1)}_{lr}(\mathbf{D_x},\mathbf{n})=P_{lr}(\mathbf{D_x},\mathbf{n}),$$

$$\mathbf{P}^{(2)}(\mathbf{D_x},\mathbf{n})=\left(P^{(2)}_{\alpha\beta}(\mathbf{D_x},\mathbf{n})\right)_{4\times 4},\qquad P^{(2)}_{\alpha\beta}(\mathbf{D_x},\mathbf{n})=P_{\alpha+3;\beta+3}(\mathbf{D_x},\mathbf{n}),$$

where $l, j = 1, 2, 3$, $\alpha, \beta = 1, 2, 3, 4$, $r = 1, 2, \cdots, 8$; the matrix differential operators $\mathbf{A} = (A_{lr}(\mathbf{D_x}, \omega))_{8\times 8}$ and $\mathbf{P} = (P_{lr}(\mathbf{D_x}, \mathbf{n}))_{8\times 8}$ are defined by (1.27) and (1.44), respectively.

Let $\mathbf{u}' = (u'_1, u'_2, u'_3)$ and $\mathbf{p}' = (p'_1, p'_2, p'_3, p'_4)$ be three- and four-component vector functions, respectively; $\mathbf{U}' = (\mathbf{u}', \mathbf{p}', \theta')$, where θ' is a scalar function.

We introduce the notation

$$\begin{aligned}
&E_1(\mathbf{U},\mathbf{u}')=W_0(\mathbf{u},\mathbf{u}')-\rho\omega^2\mathbf{u}\cdot\mathbf{u}'-(a_\alpha\,p_\alpha+\varepsilon_0\theta)\,\mathrm{div}\,\overline{\mathbf{u}'},\\
&E_2(\mathbf{U},\mathbf{p}')=k_{\alpha\beta}\nabla p_\alpha\cdot\nabla p'_\beta-c_{\alpha\beta}\,p_\beta\,\overline{p'_\alpha}-i\omega\,a_\alpha\overline{p'_\alpha}\,\mathrm{div}\,\mathbf{u}-i\omega\theta\varepsilon_\alpha\overline{p'_\alpha},\\
&E_3(\mathbf{U},\theta')=k\nabla\theta\cdot\nabla\theta'-i\omega T_0\,(a_0\theta+\varepsilon_0\mathrm{div}\,\mathbf{u}+\varepsilon_\alpha\,p_\alpha)\,\overline{\theta'},\\
&E(\mathbf{U},\mathbf{U}')=E_1(\mathbf{U},\mathbf{u}')+E_2(\mathbf{U},\mathbf{p}')+E_3(\mathbf{U},\theta'),
\end{aligned}\tag{9.26}$$

where $W_0(\mathbf{u}, \mathbf{u}')$ is given by (3.15).

We have the following.

Theorem 9.12 *If* $\mathbf{U} = (\mathbf{u}, \mathbf{p}, \theta)$ *is a regular vector of the class* $\mathfrak{R}^{(s)}$ *in* Ω^+, $u'_j, p'_\alpha, \theta' \in C^1(\Omega^+) \cap C(\overline{\Omega^+})$, $j = 1, 2, 3$, $\alpha = 1, 2, 3, 4$, *then*

$$\int\limits_{\Omega^+}\left[\mathbf{A}(\mathbf{D_x},\omega)\,\mathbf{U}(\mathbf{x})\cdot\mathbf{U}'(\mathbf{x})+E(\mathbf{U},\mathbf{U}')\right]d\mathbf{x}=\int\limits_S \mathbf{P}(\mathbf{D_z},\mathbf{n})\mathbf{U}(\mathbf{z})\cdot\mathbf{U}'(\mathbf{z})\,d_{\mathbf{z}}S,\tag{9.27}$$

where $\mathbf{A}(\mathbf{D_x}, \omega)$, $\mathbf{P}(\mathbf{D_z}, \mathbf{n})$ *and* $E(\mathbf{U}, \mathbf{U}')$ *are defined by (1.27), (1.44), and (9.26), respectively.*

Proof On the basis of identities (3.17), (3.18), and

$$\int_{\Omega^+} \left[\nabla\theta(\mathbf{x}) \cdot \mathbf{u}'(\mathbf{x}) + \theta(\mathbf{x})\, \mathrm{div}\, \overline{\mathbf{u}'(\mathbf{x})}\right] d\mathbf{x} = \int_S \theta(\mathbf{z})\mathbf{n}(\mathbf{z}) \cdot \mathbf{u}'(\mathbf{z})\, d_{\mathbf{z}}S$$

we have

$$\begin{aligned}
&\int_{\Omega^+} \left[\mathbf{A}^{(1)}(\mathbf{D_x}, \omega)\, \mathbf{U}(\mathbf{x}) \cdot \mathbf{u}'(\mathbf{x}) + E_1(\mathbf{U}, \mathbf{u}')\right] d\mathbf{x} \\
&\qquad = \int_S \mathbf{P}^{(1)}(\mathbf{D_z}, \mathbf{n})\mathbf{U}(\mathbf{z}) \cdot \mathbf{u}'(\mathbf{z})\, d_{\mathbf{z}}S, \\
&\int_{\Omega^+} \left[\mathbf{A}^{(2)}(\mathbf{D_x}, \omega)\, \mathbf{U}(\mathbf{x}) \cdot \mathbf{p}'(\mathbf{x}) + E_2(\mathbf{U}, \mathbf{p}')\right] d\mathbf{x} \\
&\qquad = \int_S \mathbf{P}^{(2)}(\mathbf{D_z}, \mathbf{n})\mathbf{p}(\mathbf{z}) \cdot \mathbf{p}'(\mathbf{z})\, d_{\mathbf{z}}S, \\
&\int_{\Omega^+} \left[\mathbf{A}^{(3)}(\mathbf{D_x}, \omega)\, \mathbf{U}(\mathbf{x})\, \overline{\theta'(\mathbf{x})} + E_3(\mathbf{U}, \theta')\right] d\mathbf{x} \\
&\qquad = k \int_S \frac{\partial\theta(\mathbf{z})}{\partial\mathbf{n}(\mathbf{z})}\, \overline{\theta'(\mathbf{z})}\, d_{\mathbf{z}}S.
\end{aligned} \tag{9.28}$$

It is easy to verify that the identity (9.27) may be obtained from (9.28). □

Theorem 9.12 and the radiation conditions (9.24) and (9.25) lead to the following.

Theorem 9.13 *If* $\mathbf{U} = (\mathbf{u}, \mathbf{p}, \theta)$ *and* $\mathbf{U}' = (\mathbf{u}', \mathbf{p}', \theta')$ *are regular vectors of the class* $\mathfrak{R}^{(s)}$ *in* Ω^-, *then*

$$\begin{aligned}
&\int_{\Omega^-} \left[\mathbf{A}(\mathbf{D_x}, \omega)\, \mathbf{U}(\mathbf{x}) \cdot \mathbf{U}'(\mathbf{x}) + E(\mathbf{U}, \mathbf{U}')\right] d\mathbf{x} \\
&\qquad = -\int_S \mathbf{P}(\mathbf{D_z}, \mathbf{n})\mathbf{U}(\mathbf{z}) \cdot \mathbf{U}'(\mathbf{z})\, d_{\mathbf{z}}S,
\end{aligned} \tag{9.29}$$

where $\mathbf{A}(\mathbf{D_x}, \omega)$, $\mathbf{P}(\mathbf{D_z}, \mathbf{n})$, *and* $E(\mathbf{U}, \mathbf{U}')$ *are given by (1.27), (1.44), and (9.26), respectively.*

The formulas (9.27) and (9.29) are Green's first identities of the steady vibrations equations in the linear theory of quadruple porosity thermoelasticity for domains Ω^+ and Ω^-, respectively.

9.3.2 Green's Second Identity

Clearly, the matrix differential operator $\tilde{\mathbf{A}}(\mathbf{D_x}, \omega) = \left(\tilde{A}_{lj}(\mathbf{D_x}, \omega)\right)_{8\times 8}$ is the associate operator of $\mathbf{A}(\mathbf{D_x}, \omega)$, where $\tilde{\mathbf{A}}(\mathbf{D_x}, \omega) = \mathbf{A}^\top(-\mathbf{D_x}, \omega)$. It is easy to see that the associated system of homogeneous equations of (1.15) is

$$
\begin{aligned}
&(\mu\,\Delta + \rho\,\omega^2)\tilde{\mathbf{u}} + (\lambda + \mu)\,\nabla\mathrm{div}\tilde{\mathbf{u}} - i\omega\nabla\,(\mathbf{a}\,\tilde{\mathbf{p}}) - i\omega\varepsilon_0 T_0 \nabla\,\tilde{\theta} = \mathbf{0},\\
&(\mathbf{K}\,\Delta + \mathbf{c})\,\tilde{\mathbf{p}} + \mathbf{a}\,\mathrm{div}\tilde{\mathbf{u}} + i\omega T_0 \boldsymbol{\varepsilon}\,\tilde{\theta} = \mathbf{0}, \qquad (9.30)\\
&(k\Delta + i\omega a_0 T_0)\tilde{\theta} + \varepsilon_0\,\mathrm{div}\,\tilde{\mathbf{u}} + i\omega\boldsymbol{\varepsilon}\,\tilde{\mathbf{p}} = 0,
\end{aligned}
$$

where $\tilde{\theta}$ is a scalar function, and $\tilde{\mathbf{u}}$ and $\tilde{\mathbf{p}}$ are three- and four-component vector functions on $\Omega^\pm$, respectively.

Let $\tilde{\mathbf{U}}_j$ be the j-th column of the matrix $\tilde{\mathbf{U}} = (\tilde{U}_{lj})_{8\times 8}$ $(j = 1, 2, \cdots, 8)$. Quite similarly as in Theorem 3.13 by direct calculation we obtain the following.

Theorem 9.14 *If* $\mathbf{U} = (\mathbf{u}, \mathbf{p}, \theta)$ *and* $\tilde{\mathbf{U}}_j$ $(j = 1, 2, \cdots, 8)$ *are regular vectors of the class* $\mathfrak{R}^{(s)}$ *in* Ω^+, *then*

$$
\begin{aligned}
&\int_{\Omega^+} \left\{[\tilde{\mathbf{A}}(\mathbf{D_y}, \omega)\tilde{\mathbf{U}}(\mathbf{y})]^\top \mathbf{U}(\mathbf{y}) - [\tilde{\mathbf{U}}(\mathbf{y})]^\top \mathbf{A}(\mathbf{D_y}, \omega)\mathbf{U}(\mathbf{y})\right\} d\mathbf{y}\\
&\qquad = \int_S \left\{[\tilde{\mathbf{P}}(\mathbf{D_z}, \mathbf{n})\tilde{\mathbf{U}}(\mathbf{z})]^\top \mathbf{U}(\mathbf{z}) - [\tilde{\mathbf{U}}(\mathbf{z})]^\top \mathbf{P}(\mathbf{D_z}, \mathbf{n})\mathbf{U}(\mathbf{z})\right\} d_{\mathbf{z}}S,
\end{aligned}
\qquad (9.31)
$$

where the matrices $\mathbf{A}(\mathbf{D_x}, \omega)$ *and* $\mathbf{P}(\mathbf{D_z}, \mathbf{n})$ *are given by (1.27) and (1.44), respectively; the operator* $\tilde{\mathbf{P}}(\mathbf{D_z}, \mathbf{n})$ *is defined by*

$$
\begin{aligned}
&\tilde{\mathbf{P}}(\mathbf{D_x}, \mathbf{n}) = \left(\tilde{P}_{lj}(\mathbf{D_x}, \mathbf{n})\right)_{8\times 8}, \qquad \tilde{P}_{lj}(\mathbf{D_x}, \mathbf{n}) = P_{lj}(\mathbf{D_x}, \mathbf{n}),\\
&\tilde{P}_{l;\alpha+3}(\mathbf{D_x}, \mathbf{n}) = -i\omega a_\alpha\, n_l, \qquad \tilde{P}_{l8}(\mathbf{D_x}, \mathbf{n}) = -i\omega\varepsilon_0 T_0\, n_l,\\
&\tilde{P}_{rm}(\mathbf{D_x}, \mathbf{n}) = P_{rm}(\mathbf{D_x}, \mathbf{n}), \qquad l, j = 1, 2, 3, \qquad \alpha = 1, 2, 3, 4,\\
&r = 4, 5, \cdots, 8, \qquad m = 1, 2, \cdots, 8.
\end{aligned}
\qquad (9.32)
$$

Theorem 9.14 and the radiation conditions (9.24) and (9.25) lead to the following.

Theorem 9.15 *If* $\mathbf{U} = (\mathbf{u}, \mathbf{p}, \theta)$ *and* $\tilde{U}_j$ $(j = 1, 2, \cdots, 8)$ *are regular vectors of the class* $\mathfrak{R}^{(s)}$ *in* Ω^-*, then*

$$\int_{\Omega^-} \Big\{[\tilde{\mathbf{A}}(\mathbf{D}_\mathbf{y}, \omega)\tilde{\mathbf{U}}(\mathbf{y})]^\top \mathbf{U}(\mathbf{y}) - [\tilde{\mathbf{U}}(\mathbf{y})]^\top \mathbf{A}(\mathbf{D}_\mathbf{y}, \omega)\mathbf{U}(\mathbf{y})\Big\} d\mathbf{y}$$
$$= -\int_S \Big\{[\tilde{\mathbf{P}}(\mathbf{D}_\mathbf{z}, \mathbf{n})\tilde{\mathbf{U}}(\mathbf{z})]^\top \mathbf{U}(\mathbf{z}) - [\tilde{\mathbf{U}}(\mathbf{z})]^\top \mathbf{P}(\mathbf{D}_\mathbf{z}, \mathbf{n})\mathbf{U}(\mathbf{z})\Big\} d_\mathbf{z}S, \tag{9.33}$$

where the operator $\tilde{\mathbf{P}}(\mathbf{D}_\mathbf{z}, \mathbf{n})$ *is defined by (9.32).*

The formulas (9.31) and (9.33) are Green's second identities of the steady vibrations equations in the linear theory of quadruple porosity thermoelasticity for domains Ω^+ and Ω^-, respectively.

9.3.3 *Green's Third Identity*

Let $\tilde{\mathbf{\Psi}}(\mathbf{x}, \omega)$ be the fundamental solution of the system (9.30) (the fundamental matrix of operator $\tilde{\mathbf{A}}(\mathbf{D}_\mathbf{y}, \omega)$). Obviously, the matrix $\tilde{\mathbf{\Psi}}(\mathbf{x}, \omega)$ satisfies the following condition

$$\tilde{\mathbf{\Psi}}(\mathbf{x}, \omega) = \mathbf{\Psi}^\top(-\mathbf{x}, \omega), \tag{9.34}$$

where $\mathbf{\Psi}(\mathbf{x}, \omega)$ is the fundamental solution of the system (1.15) (see Theorem 9.1).

By virtue of (9.15) and (9.34) from (9.31) and (9.33) we obtain the following results.

Theorem 9.16 *If* $\mathbf{U}$ *is a regular vector of the class* $\mathfrak{R}^{(s)}$ *in* Ω^+*, then*

$$\mathbf{U}(\mathbf{x}) =$$
$$\int_S \Big\{[\tilde{\mathbf{P}}(\mathbf{D}_\mathbf{z}, \mathbf{n})\mathbf{\Psi}^\top(\mathbf{x}-\mathbf{z}, \omega)]^\top \mathbf{U}(\mathbf{z}) - \mathbf{\Psi}(\mathbf{x}-\mathbf{z}, \omega)\,\mathbf{P}(\mathbf{D}_\mathbf{z}, \mathbf{n})\mathbf{U}(\mathbf{z})\Big\} d_\mathbf{z}S$$
$$+ \int_{\Omega^+} \mathbf{\Psi}(\mathbf{x}-\mathbf{y}, \omega)\,\mathbf{A}(\mathbf{D}_\mathbf{y}, \omega)\mathbf{U}(\mathbf{y})d\mathbf{y} \qquad \textit{for} \quad \mathbf{x} \in \Omega^+. \tag{9.35}$$

Theorem 9.17 *If* $\mathbf{U}$ *is a regular vector of the class* $\mathfrak{R}^{(s)}$ *in* Ω^-*, then*

$$\mathbf{U}(\mathbf{x}) =$$

$$-\int_S \left\{[\tilde{\mathbf{P}}(\mathbf{D_z},\mathbf{n})\boldsymbol{\Psi}^\top(\mathbf{x}-\mathbf{z},\omega)]^\top \mathbf{U}(\mathbf{z}) - \boldsymbol{\Psi}(\mathbf{x}-\mathbf{z},\omega)\,\mathbf{P}(\mathbf{D_z},\mathbf{n})\mathbf{U}(\mathbf{z})\right\} d_\mathbf{z}S$$

$$+\int_{\Omega^-} \boldsymbol{\Psi}(\mathbf{x}-\mathbf{y},\omega)\,\mathbf{A}(\mathbf{D_y},\omega)\mathbf{U}(\mathbf{y})d\mathbf{y} \qquad \textit{for} \quad \mathbf{x}\in\Omega^-. \tag{9.36}$$

The formulas (9.35) and (9.36) are Green's third identities of the steady vibrations equations in the linear theory of quadruple porosity thermoelasticity for domains Ω^+ and Ω^-, respectively.

Theorems 9.16 and 9.17 lead to the following.

Corollary 9.1 *If* $\mathbf{U}$ *is a regular solution of the class* $\mathfrak{R}^{(s)}$ *of the homogeneous equation*

$$\mathbf{A}(\mathbf{D_x},\omega)\,\mathbf{U}(\mathbf{x}) = \mathbf{0} \tag{9.37}$$

in Ω^+*, then*

$$\mathbf{U}(\mathbf{x}) =$$

$$\int_S \left\{[\tilde{\mathbf{P}}(\mathbf{D_z},\mathbf{n})\boldsymbol{\Psi}^\top(\mathbf{x}-\mathbf{z},\omega)]^\top \mathbf{U}(\mathbf{z}) - \boldsymbol{\Psi}(\mathbf{x}-\mathbf{z},\omega)\,\mathbf{P}(\mathbf{D_z},\mathbf{n})\mathbf{U}(\mathbf{z})\right\} d_\mathbf{z}S \tag{9.38}$$

for $\mathbf{x}\in\Omega^+$.

Corollary 9.2 *If* $\mathbf{U}$ *is a regular solution of the class* $\mathfrak{R}^{(s)}$ *of (9.37) in* Ω^-*, then*

$$\mathbf{U}(\mathbf{x}) =$$

$$-\int_S \left\{[\tilde{\mathbf{P}}(\mathbf{D_z},\mathbf{n})\boldsymbol{\Psi}^\top(\mathbf{x}-\mathbf{z},\omega)]^\top \mathbf{U}(\mathbf{z}) - \boldsymbol{\Psi}(\mathbf{x}-\mathbf{z},\omega)\,\mathbf{P}(\mathbf{D_z},\mathbf{n})\mathbf{U}(\mathbf{z})\right\} d_\mathbf{z}S \tag{9.39}$$

for $\mathbf{x}\in\Omega^-$.

The formulas (9.38) and (9.39) are the Somigliana-type integral representations of a regular solution of the steady vibrations homogeneous equation (9.37) in the linear theory of quadruple porosity thermoelasticity for domains Ω^+ and Ω^-, respectively.

9.4 Basic Boundary Value Problems

The basic internal and external BVPs of steady vibrations in the linear theory of thermoelasticity for quadruple porosity materials are formulated as follows.

Find a regular (classical) solution of the class $\mathfrak{R}^{(s)}$ in Ω^+ to system

$$\mathbf{A}(\mathbf{D_x}, \omega)\, \mathbf{U}(\mathbf{x}) = \mathbf{F}(\mathbf{x}) \tag{9.40}$$

for $\mathbf{x} \in \Omega^+$ satisfying the boundary condition

$$\lim_{\Omega^+ \ni \mathbf{x} \to \mathbf{z} \in S} \mathbf{U}(\mathbf{x}) \equiv \{\mathbf{U}(\mathbf{z})\}^+ = \mathbf{f}(\mathbf{z}) \tag{9.41}$$

in the *Problem* $(I^s)^+_{\mathbf{F},\mathbf{f}}$, and

$$\lim_{\Omega^+ \ni \mathbf{x} \to \mathbf{z} \in S} \mathbf{P}(\mathbf{D_x}, \mathbf{n}(\mathbf{z}))\mathbf{U}(\mathbf{x}) \equiv \{\mathbf{P}(\mathbf{D_z}, \mathbf{n})\mathbf{U}(\mathbf{z})\}^+ = \mathbf{f}(\mathbf{z}) \tag{9.42}$$

in the *Problem* $(II^s)^+_{\mathbf{F},\mathbf{f}}$, where the matrix differential operators $\mathbf{A}(\mathbf{D_x}, \omega)$ and $\mathbf{P}(\mathbf{D_z}, \mathbf{n})$ are defined by (1.27) and (1.44), respectively; $\mathbf{F}$ and $\mathbf{f}$ are prescribed eight-component vector functions.

Find a regular (classical) solution of the class $\mathfrak{R}^{(s)}$ to system (9.40) for $\mathbf{x} \in \Omega^-$ satisfying the boundary condition

$$\lim_{\Omega^- \ni \mathbf{x} \to \mathbf{z} \in S} \mathbf{U}(\mathbf{x}) \equiv \{\mathbf{U}(\mathbf{z})\}^- = \mathbf{f}(\mathbf{z}) \tag{9.43}$$

in the *Problem* $(I^s)^-_{\mathbf{F},\mathbf{f}}$, and

$$\lim_{\Omega^- \ni \mathbf{x} \to \mathbf{z} \in S} \mathbf{P}(\mathbf{D_x}, \mathbf{n}(\mathbf{z}))\mathbf{U}(\mathbf{x}) \equiv \{\mathbf{P}(\mathbf{D_z}, \mathbf{n})\mathbf{U}(\mathbf{z})\}^- = \mathbf{f}(\mathbf{z}) \tag{9.44}$$

in the *Problem* $(II^s)^-_{\mathbf{F},\mathbf{f}}$. Here $\mathbf{F}$ and $\mathbf{f}$ are prescribed eight-component vector functions, and $\operatorname{supp}\mathbf{F}$ is a finite domain in Ω^-.

9.5 Uniqueness Theorems

In this section we study uniqueness of regular solutions of the BVPs $(\mathbb{K}^s)^+_{\mathbf{F},\mathbf{f}}$ and $(\mathbb{K}^s)^-_{\mathbf{F},\mathbf{f}}$, where $\mathbb{K} = I, II$.

We have the following results.

Theorem 9.18 *If the condition (6.6) is satisfied, then any two solutions of the BVP* $(I^s)^+_{\mathbf{F},\mathbf{f}}$ *of the class* $\mathfrak{R}^{(s)}$ *may differ only for an additive vector* $\mathbf{U} = (\mathbf{u}, \mathbf{p}, \theta)$, *where*

$$\mathbf{p}(\mathbf{x}) = \mathbf{0}, \qquad \theta(\mathbf{x}) = 0 \quad for \quad \mathbf{x} \in \Omega^+, \tag{9.45}$$

and vector $\mathbf{u}$ *is a solution of the following BVP*

$$\left(\Delta + \mu_7^2\right) \mathbf{u}(\mathbf{x}) = \mathbf{0}, \qquad \operatorname{div} \mathbf{u}(\mathbf{x}) = \mathbf{0} \quad for \quad \mathbf{x} \in \Omega^+, \tag{9.46}$$

$$\{\mathbf{u}(\mathbf{z})\}^+ = 0 \quad for \quad \mathbf{z} \in S. \tag{9.47}$$

In addition, the problems $(I^s)^+_{\mathbf{0},\mathbf{0}}$ *and (9.46), (9.47) have the same eigenfrequencies.*

Proof As usual, suppose that there are two regular solutions of problem $(I^s)^+_{\mathbf{F},\mathbf{f}}$ in the class $\mathfrak{R}^{(s)}$. Then their difference $\mathbf{U}$ is a regular solution of the internal homogeneous BVP $(I^s)^+_{\mathbf{0},\mathbf{0}}$. Hence, $\mathbf{U}$ is a regular solution of the homogeneous equation (9.37) in Ω^+ satisfying the homogeneous boundary condition

$$\{\mathbf{U}(\mathbf{z})\}^+ = \mathbf{0} \quad \text{for} \quad \mathbf{z} \in S. \tag{9.48}$$

On the basis of (9.37) and (9.48), from (9.28) we obtain for $\mathbf{u}' = \mathbf{u}$, $\mathbf{p}' = \mathbf{p}$ and $\theta' = \theta$:

$$\begin{aligned} &\int_{\Omega^+} E_1(\mathbf{U}, \mathbf{u}) d\mathbf{x} = 0, \qquad \int_{\Omega^+} E_2(\mathbf{U}, \mathbf{p}) d\mathbf{x} = 0, \\ &\int_{\Omega^+} E_3(\mathbf{U}, \theta) d\mathbf{x} = 0, \end{aligned} \tag{9.49}$$

where

$$\begin{aligned} &E_1(\mathbf{U}, \mathbf{u}) = W_0(\mathbf{u}, \mathbf{u}) - \rho\omega^2 |\mathbf{u}|^2 - (a_\alpha\, p_\alpha + \varepsilon_0 \theta) \operatorname{div} \overline{\mathbf{u}}, \\ &E_2(\mathbf{U}, \mathbf{p}) = k_{\alpha\beta} \nabla p_\beta \cdot \nabla p_\alpha - c_{\alpha\beta}\, p_\beta \overline{p_\alpha} - i\omega (a_\alpha \operatorname{div} \mathbf{u} + \theta \varepsilon_\alpha) \overline{p_\alpha}, \\ &E_3(\mathbf{U}, \theta) = k |\nabla\theta|^2 - i\omega a_0 T_0 |\theta|^2 - i\omega T_0 (\varepsilon_0 \operatorname{div} \mathbf{u} + \varepsilon_\alpha\, p_\alpha) \overline{\theta} \end{aligned} \tag{9.50}$$

and $W_0(\mathbf{u}, \mathbf{u})$ is defined by (3.15).

We may derive from (9.50)

$$\begin{aligned} &\operatorname{Im} E_1(\mathbf{U}, \mathbf{u}) = \operatorname{Im}\left[(a_\alpha \overline{p_\alpha} + \varepsilon_0 \overline{\theta}) \operatorname{div} \mathbf{u}\right], \\ &\operatorname{Re} E_2(\mathbf{U}, \mathbf{p}) = k_{\alpha\beta} \nabla p_\beta \cdot \nabla p_\alpha + d_{\alpha\beta}\, p_\beta \overline{p_\alpha} + \omega \operatorname{Im}\left[(a_\alpha \operatorname{div} \mathbf{u} + \theta \varepsilon_\alpha) \overline{p_\alpha}\right], \\ &\operatorname{Re} E_3(\mathbf{U}, \theta) = k |\nabla\theta|^2 + \omega T_0 \operatorname{Im}\left[(\varepsilon_0 \operatorname{div} \mathbf{u} + \varepsilon_\alpha\, p_\alpha) \overline{\theta}\right]. \end{aligned} \tag{9.51}$$

Obviously, from (9.51) it follows that

$$\begin{aligned}&\operatorname{Re} E_3(\mathbf{U},\theta) + T_0 \operatorname{Re} E_2(\mathbf{U},\mathbf{p}) - \omega T_0 \operatorname{Im} E_1(\mathbf{U},\mathbf{u})\\ &\quad = k|\nabla\theta|^2 + T_0(k_{\alpha\beta}\nabla p_\beta\cdot\nabla p_\alpha + d_{\alpha\beta}p_\beta\overline{p_\alpha}).\end{aligned} \tag{9.52}$$

Taking into account (9.52) from (9.49) we have

$$\int\limits_{\Omega^+}\left[k|\nabla\theta|^2 + T_0(k_{\alpha\beta}\nabla p_\beta\cdot\nabla p_\alpha + d_{\alpha\beta}p_\beta\overline{p_\alpha})\right]d\mathbf{x} = 0.$$

By virtue of (4.10) from the last equation we get

$$k_{\alpha\beta}\nabla p_\beta\cdot\nabla p_\alpha = 0, \qquad d_{\alpha\beta}p_\beta\overline{p_\alpha} = 0, \qquad \nabla\theta = 0, \tag{9.53}$$

and hence, by using Lemma 6.1 from (9.53) we obtain

$$\begin{aligned}&p_\alpha(\mathbf{x}) = c_\alpha = \text{const}, \qquad \theta(\mathbf{x}) = c_5 = \text{const},\\ &d_{\alpha\beta}p_\beta(\mathbf{x}) = 0, \qquad \alpha = 1,2,3,4\end{aligned} \tag{9.54}$$

for $\mathbf{x}\in\Omega^+$. On the basis of homogeneous boundary condition (9.48) from (9.54) it follows Eq. (9.45). Now employing (6.6) and (9.45), we may derive from (9.37) and (9.48) the relations (9.46) and (9.47), respectively.

Finally, it is easy to see that problems $(I^s)^+_{\mathbf{0},\mathbf{0}}$ and (9.46), (9.47) have the same eigenfrequencies. □

Theorem 9.19 *If the conditions (6.6) and*

$$\det\begin{pmatrix} b_{11} & b_{12} & b_{13} & b_{14} & \varepsilon_1\\ b_{12} & b_{22} & b_{23} & b_{24} & \varepsilon_2\\ b_{13} & b_{23} & b_{33} & b_{34} & \varepsilon_3\\ b_{14} & b_{24} & b_{34} & b_{44} & \varepsilon_4\\ \varepsilon_1 & \varepsilon_2 & \varepsilon_3 & \varepsilon_4 & a_0 \end{pmatrix}_{5\times 5} \neq 0 \tag{9.55}$$

are satisfied, then any two solutions of the BVP $(II^s)^+_{\mathbf{F},\mathbf{f}}$ of the class $\mathfrak{R}^{(s)}$ may differ only for an additive vector $\mathbf{U} = (\mathbf{u},\mathbf{p},\theta)$, where $\mathbf{p}$ and θ satisfy the condition (9.45), and the vector $\mathbf{u}$ is a solution of (9.46) satisfying the boundary condition

$$\left\{2\frac{\partial \mathbf{u}(\mathbf{z})}{\partial \mathbf{n}(\mathbf{z})} + [\mathbf{n}(\mathbf{z}) \times \operatorname{curl} \mathbf{u}(\mathbf{z})]\right\}^+ = 0 \qquad for \quad \mathbf{z} \in S. \tag{9.56}$$

In addition, the problems $(II^s)^+_{\mathbf{0},\mathbf{0}}$ *and (9.46), (9.56) have the same eigenfrequencies.*

Proof We suppose that there are two regular solutions of the problem $(II^s)^+_{\mathbf{F},\mathbf{f}}$ in the class $\mathfrak{R}^{(s)}$. Then their difference $\mathbf{U}$ is a regular solution of the homogeneous BVP $(II^s)^+_{\mathbf{0},\mathbf{0}}$. Hence, $\mathbf{U}$ is a regular solution of the homogeneous system (9.37) in Ω^+ satisfying the homogeneous boundary condition

$$\{\mathbf{P}(\mathbf{D}_\mathbf{z}, \mathbf{n})\mathbf{U}(\mathbf{z})\}^+ = \mathbf{0}. \tag{9.57}$$

Quite similarly as in Theorem 9.18, on the basis of (9.37) and (9.57), from (9.28) we have Eq. (9.54). Furthermore, if we employ Eq. (9.54), then from (9.37) we may obtain

$$(\mu\,\Delta + \rho\omega^2)\mathbf{u} + (\lambda + \mu)\,\nabla \operatorname{div} \mathbf{u} = \mathbf{0} \tag{9.58}$$

and

$$\begin{aligned} b_{\alpha\beta}c_\beta + \varepsilon_\alpha c_5 &= -a_\alpha \operatorname{div} \mathbf{u}, \\ \varepsilon_\beta c_\beta + a_0 c_5 &= -\varepsilon_0 \operatorname{div} \mathbf{u} \end{aligned} \tag{9.59}$$

for $\alpha = 1, 2, 3, 4$. Clearly, from (9.59) we deduce that $\operatorname{div} \mathbf{u}(\mathbf{x}) = \text{const}$ for $\mathbf{x} \in \Omega^+$ and consequently, by using (6.6) from (9.58) we may derive the first equation of (9.46). On the other hand, applying the operator div to (9.58) we obtain $\rho\omega^2 \operatorname{div} \mathbf{u}(\mathbf{x}) = 0$; hence, we have the second equation of (9.46).

Then, the system (9.59) may be written as

$$b_{\alpha\beta}c_\beta + \varepsilon_\alpha c_5 = 0, \qquad \varepsilon_\beta c_\beta + a_0 c_5 = 0, \qquad \alpha = 1, 2, 3, 4. \tag{9.60}$$

In view of the condition (9.55), from (9.60) we may deduce Eq. (9.45). In addition, with the help of (9.57) and the second equation of (9.46) we can write the boundary condition (9.56).

Moreover, it is easy to see that problems $(II^s)^+_{\mathbf{0},\mathbf{0}}$ and (9.46), (9.56) have the same eigenfrequencies. □

Theorem 9.20 *If the condition (6.6) is satisfied, then the external BVP* $(\mathbb{K}^s)^-_{\mathbf{F},\mathbf{f}}$ *has one regular solution of the class* $\mathfrak{R}^{(s)}$, *where* $\mathbb{K} = I, II$.

Proof Suppose that there are two regular solutions of problem $(\mathbb{K}^s)^-_{\mathbf{F},\mathbf{f}}$ of the class $\mathfrak{R}^{(s)}$. Then their difference $\mathbf{U}$ is a regular solution of the homogeneous BVP $(\mathbb{K}^s)^-_{\mathbf{0},\mathbf{0}}$. Hence, $\mathbf{U}$ is a regular solution of the homogeneous system (9.37) in Ω^- satisfying the homogeneous boundary condition

$$\{\mathbf{U}(\mathbf{z})\}^- = \mathbf{0} \tag{9.61}$$

in the problem $(I^s)^-_{\mathbf{0},0}$ and

$$\{\mathbf{P}(\mathbf{D_z}, \mathbf{n})\mathbf{U}(\mathbf{z})\}^- = 0 \tag{9.62}$$

in the problem $(II^s)^-_{\mathbf{0},0}$.

Employing (9.37), (9.61), and (9.62) in the following identities (see (9.28))

$$\int_{\Omega^-} \left[\mathbf{A}^{(1)}(\mathbf{D_x}, \omega)\, \mathbf{U}(\mathbf{x}) \cdot \mathbf{u}'(\mathbf{x}) + E_1(\mathbf{U}, \mathbf{u}')\right] d\mathbf{x}$$

$$= -\int_S \mathbf{P}^{(1)}(\mathbf{D_z}, \mathbf{n})\mathbf{U}(\mathbf{z}) \cdot \mathbf{u}'(\mathbf{z})\, d_\mathbf{z}S,$$

$$\int_{\Omega^-} \left[\mathbf{A}^{(2)}(\mathbf{D_x}, \omega)\, \mathbf{U}(\mathbf{x}) \cdot \mathbf{p}'(\mathbf{x}) + E_2(\mathbf{U}, \mathbf{p}')\right] d\mathbf{x}$$

$$= -\int_S \mathbf{P}^{(2)}(\mathbf{D_z}, \mathbf{n})\mathbf{p}(\mathbf{z}) \cdot \mathbf{p}'(\mathbf{z})\, d_\mathbf{z}S,$$

$$\int_{\Omega^-} \left[\mathbf{A}^{(3)}(\mathbf{D_x}, \omega)\, \mathbf{U}(\mathbf{x})\, \overline{\theta'(\mathbf{x})} + E_3(\mathbf{U}, \theta')\right] d\mathbf{x}$$

$$= -k \int_S \frac{\partial \theta(\mathbf{z})}{\partial \mathbf{n}(\mathbf{z})}\, \overline{\theta'(\mathbf{z})}\, d_\mathbf{z}S$$

we may derive

$$\int_{\Omega^-} E_1(\mathbf{U}, \mathbf{u}) d\mathbf{x} = 0, \qquad \int_{\Omega^-} E_2(\mathbf{U}, \mathbf{p}) d\mathbf{x} = 0, \qquad \int_{\Omega^-} E_3(\mathbf{U}, \theta) d\mathbf{x} = 0, \tag{9.63}$$

where $E_1(\mathbf{U}, \mathbf{u})$, $E_2(\mathbf{U}, \mathbf{p})$, and $E_3(\mathbf{U}, \theta)$ are defined by (9.50), and the operators $\mathbf{P}^{(1)}(\mathbf{D_z}, \mathbf{n})$ and $\mathbf{P}^{(2)}(\mathbf{D_z}, \mathbf{n})$ are given in Sect. 9.3.1.

In a similar manner as in Theorem 9.18, from (9.63) we obtain the relation (9.54) for $\mathbf{x} \in \Omega^-$. Using the radiation conditions (9.25) in (9.54) we may deduce that

$$\mathbf{p}(\mathbf{x}) = \mathbf{0}, \qquad \theta(\mathbf{x}) = 0 \quad \text{for} \quad \mathbf{x} \in \Omega^-. \tag{9.64}$$

In addition, combining (6.6) and (9.64) with (9.37) we may further conclude that

$$\left(\Delta + \mu_7^2\right) \mathbf{u}(\mathbf{x}) = \mathbf{0}, \qquad \operatorname{div} \mathbf{u}(\mathbf{x}) = \mathbf{0} \qquad \text{for} \quad \mathbf{x} \in \Omega^-, \tag{9.65}$$

i.e., the vector $\mathbf{u}$ is a solution of Helmholtz equation in Ω^- (see the first equation of (9.65)), satisfying the boundary condition $\{\mathbf{u}(\mathbf{z})\}^- = 0$ or

$$\left\{2\frac{\partial \mathbf{u}(\mathbf{z})}{\partial \mathbf{n}(\mathbf{z})} + [\mathbf{n}(\mathbf{z}) \times \operatorname{curl} \mathbf{u}(\mathbf{z})]\right\}^- = 0$$

for $\mathbf{z} \in S$, and the radiation conditions (9.24) and (9.25). Clearly, we have $\mathbf{u}(\mathbf{x}) \equiv \mathbf{0}$ for $\mathbf{x} \in \Omega^-$. Hence, $\mathbf{U}(\mathbf{x}) \equiv \mathbf{0}$ for $\mathbf{x} \in \Omega^-$ and consequently, a solution to the external BVP $(\mathbb{K}_s)^-_{\mathbf{F},\mathbf{f}}$ is unique in the class $\mathfrak{B}^{(s)}$, where $\mathbb{K} = I, II$. □

9.6 Basic Properties of Potentials

On the basis of integral representation of regular vectors (see (9.35) and (9.36)) we introduce the following notation:

(i) $\mathbf{Q}^{(s,1)}(\mathbf{x}, \mathbf{g}) = \int\limits_S \boldsymbol{\Psi}(\mathbf{x} - \mathbf{y}, \omega)\mathbf{g}(\mathbf{y})d_{\mathbf{y}}S,$

(ii) $\mathbf{Q}^{(s,2)}(\mathbf{x}, \mathbf{g}) = \int\limits_S [\tilde{\mathbf{P}}(\mathbf{D}_{\mathbf{y}}, \mathbf{n}(\mathbf{y}))\boldsymbol{\Psi}^\top(\mathbf{x} - \mathbf{y}, \omega)]^\top \mathbf{g}(\mathbf{y})d_{\mathbf{y}}S,$

(iii) $\mathbf{Q}^{(s,3)}(\mathbf{x}, \boldsymbol{\phi}, \Omega^\pm) = \int\limits_{\Omega^\pm} \boldsymbol{\Psi}(\mathbf{x} - \mathbf{y}, \omega)\boldsymbol{\phi}(\mathbf{y})d\mathbf{y},$

where $\boldsymbol{\Psi}(\mathbf{x}, \omega)$ is the fundamental matrix of the operator $\mathbf{A}(\mathbf{D}_{\mathbf{x}}, \omega)$ and given by (9.14), the matrix differential operator $\tilde{\mathbf{P}}(\mathbf{D}_{\mathbf{y}}, \mathbf{n}(\mathbf{y}))$ is defined by (9.32), and $\mathbf{g}$ and $\boldsymbol{\phi}$ are eight-component vector functions.

As in the classical theory of thermoelasticity (see, e.g., Kupradze et al. [219]), the vector functions $\mathbf{Q}^{(s,1)}(\mathbf{x}, \mathbf{g})$, $\mathbf{Q}^{(s,2)}(\mathbf{x}, \mathbf{g})$, and $\mathbf{Q}^{(s,3)}(\mathbf{x}, \boldsymbol{\phi}, \Omega^\pm)$ are called *single-layer*, *double-layer*, and *volume potentials* in the steady vibration problems of the linear theory of thermoelasticity for quadruple porosity materials.

On the basis of the properties of fundamental solution $\boldsymbol{\Psi}(\mathbf{x}, \omega)$ (see Sect. 9.1) we have the following results.

Theorem 9.21 *If* $S \in C^{m+1,\nu}$, $\mathbf{g} \in C^{m,\nu'}(S)$, $0 < \nu' < \nu \leq 1$, *and* m *is a nonnegative integer, then:*

(a)

$$\mathbf{Q}^{(s,1)}(\cdot, \mathbf{g}) \in C^{0,\nu'}(\mathbb{R}^3) \cap C^{m+1,\nu'}(\overline{\Omega^\pm}) \cap C^\infty(\Omega^\pm),$$

(b)

$$\mathbf{A}(\mathbf{D_x}, \omega)\, \mathbf{Q}^{(s,1)}(\mathbf{x}, \mathbf{g}) = \mathbf{0},$$

(c)

$$\Big\{\mathbf{P}(\mathbf{D_z}, \mathbf{n}(\mathbf{z}))\, \mathbf{Q}^{(s,1)}(\mathbf{z}, \mathbf{g})\Big\}^{\pm} = \mp\frac{1}{2}\, \mathbf{g}(\mathbf{z}) + \mathbf{P}(\mathbf{D_z}, \mathbf{n}(\mathbf{z}))\, \mathbf{Q}^{(s,1)}(\mathbf{z}, \mathbf{g}), \qquad (9.66)$$

(d)

$$\mathbf{P}(\mathbf{D_z}, \mathbf{n}(\mathbf{z}))\, \mathbf{Q}^{(s,1)}(\mathbf{z}, \mathbf{g})$$

is a singular integral, where $\mathbf{z} \in S$, $\mathbf{x} \in \Omega^{\pm}$ *and*

$$\Big\{\mathbf{P}(\mathbf{D_z}, \mathbf{n}(\mathbf{z}))\, \mathbf{Q}^{(s,1)}(\mathbf{z}, \mathbf{g})\Big\}^{\pm} \equiv \lim_{\Omega^{\pm} \ni \mathbf{x} \to \mathbf{z} \in S} \mathbf{P}(\mathbf{D_x}, \mathbf{n}(\mathbf{z}))\, \mathbf{Q}^{(s,1)}(\mathbf{x}, \mathbf{g}),$$

(e) *the single-layer potential* $\mathbf{Q}^{(s,1)}(\mathbf{x}, \mathbf{g}) \equiv \mathbf{0}$ *for* $\mathbf{x} \in \Omega^{+}$ *(or* $\mathbf{x} \in \Omega^{-}$*) if and only if* $\mathbf{g}(\mathbf{z}) \equiv \mathbf{0}$, *where* $\mathbf{z} \in S$.

Theorem 9.22 *If* $S \in C^{m+1,\nu}$, $\mathbf{g} \in C^{m,\nu'}(S)$, $0 < \nu' < \nu \le 1$, *then:*

(a)

$$\mathbf{Q}^{(s,2)}(\cdot, \mathbf{g}) \in C^{m,\nu'}(\overline{\Omega^{\pm}}) \cap C^{\infty}(\Omega^{\pm}),$$

(b)

$$\mathbf{A}(\mathbf{D_x}, \omega)\, \mathbf{Q}^{(s,2)}(\mathbf{x}, \mathbf{g}) = \mathbf{0},$$

(c)

$$\Big\{\mathbf{Q}^{(s,2)}(\mathbf{z}, \mathbf{g})\Big\}^{\pm} = \pm\frac{1}{2}\, \mathbf{g}(\mathbf{z}) + \mathbf{Q}^{(s,2)}(\mathbf{z}, \mathbf{g}), \qquad (9.67)$$

for the nonnegative integer m,

(d) $\mathbf{Q}^{(s,2)}(\mathbf{z}, \mathbf{g})$ *is a singular integral, where* $\mathbf{z} \in S$,

(e)

$$\Big\{\mathbf{P}(\mathbf{D_z}, \mathbf{n}(\mathbf{z}))\, \mathbf{Q}^{(s,2)}(\mathbf{z}, \mathbf{g})\Big\}^{+} = \Big\{\mathbf{P}(\mathbf{D_z}, \mathbf{n}(\mathbf{z}))\, \mathbf{Q}^{(s,2)}(\mathbf{z}, \mathbf{g})\Big\}^{-},$$

for the natural number m, *where* $\mathbf{z} \in S$, $\mathbf{x} \in \Omega^{\pm}$ *and*

$$\Big\{\mathbf{Q}^{(s,2)}(\mathbf{z}, \mathbf{g})\Big\}^{\pm} \equiv \lim_{\Omega^{\pm} \ni \mathbf{x} \to \mathbf{z} \in S} \mathbf{Q}^{(s,2)}(\mathbf{x}, \mathbf{g}).$$

Theorem 9.23 *If* $S \in C^{1,\nu}$, $\boldsymbol{\phi} \in C^{0,\nu'}(\Omega^+)$, $0 < \nu' < \nu \le 1$, *then:*

(a)

$$\mathbf{Q}^{(s,3)}(\cdot, \boldsymbol{\phi}, \Omega^+) \in C^{1,\nu'}(\mathbb{R}^3) \cap C^2(\Omega^+) \cap C^{2,\nu'}\left(\overline{\Omega_0^+}\right),$$

(b)

$$\mathbf{A}(\mathbf{D_x}, \omega)\, \mathbf{Q}^{(s,3)}(\mathbf{x}, \boldsymbol{\phi}, \Omega^+) = \boldsymbol{\phi}(\mathbf{x}),$$

where $\mathbf{x} \in \Omega^+$, Ω_0^+ *is a domain in* $\mathbb{R}^3$ *and* $\overline{\Omega_0^+} \subset \Omega^+$.

Theorem 9.24 *If* $S \in C^{1,\nu}$, $\mathrm{supp}\boldsymbol{\phi} = \Omega \subset \Omega^-$, $\boldsymbol{\phi} \in C^{0,\nu'}(\Omega^-)$, $0 < \nu' < \nu \le 1$, *then:*

(a)

$$\mathbf{Q}^{(s,3)}(\cdot, \boldsymbol{\phi}, \Omega^-) \in C^{1,\nu'}(\mathbb{R}^3) \cap C^2(\Omega^-) \cap C^{2,\nu'}(\overline{\Omega_0^-}),$$

(b)

$$\mathbf{A}(\mathbf{D_x}, \omega)\, \mathbf{Q}^{(s,3)}(\mathbf{x}, \boldsymbol{\phi}, \Omega^-) = \boldsymbol{\phi}(\mathbf{x}),$$

where $\mathbf{x} \in \Omega^-$, Ω *is a finite domain in* $\mathbb{R}^3$ *and* $\overline{\Omega_0^-} \subset \Omega^-$.

9.7 Singular Integral Operators

We introduce the notation

$$\begin{aligned}
\mathcal{H}^{(s,1)}\,\mathbf{g}(\mathbf{z}) &\equiv \frac{1}{2}\,\mathbf{g}(\mathbf{z}) + \mathbf{Q}^{(s,2)}(\mathbf{z}, \mathbf{g}),\\
\mathcal{H}^{(s,2)}\,\mathbf{g}(\mathbf{z}) &\equiv -\frac{1}{2}\,\mathbf{g}(\mathbf{z}) + \mathbf{P}(\mathbf{D_z}, \mathbf{n}(\mathbf{z}))\mathbf{Q}^{(s,1)}(\mathbf{z}, \mathbf{g}),\\
\mathcal{H}^{(s,3)}\mathbf{g}(\mathbf{z}) &\equiv -\frac{1}{2}\,\mathbf{g}(\mathbf{z}) + \mathbf{Q}^{(s,2)}(\mathbf{z}, \mathbf{g}), \qquad (9.68)\\
\mathcal{H}^{(s,4)}\mathbf{g}(\mathbf{z}) &\equiv \frac{1}{2}\,\mathbf{g}(\mathbf{z}) + \mathbf{P}(\mathbf{D_z}, \mathbf{n}(\mathbf{z}))\mathbf{Q}^{(s,1)}(\mathbf{z}, \mathbf{g}),\\
\mathcal{H}_\chi^{(s)}\mathbf{g}(\mathbf{z}) &\equiv \frac{1}{2}\,\mathbf{g}(\mathbf{z}) + \chi\, \mathbf{Q}^{(s,2)}(\mathbf{z}, \mathbf{g})
\end{aligned}$$

for $\mathbf{z} \in S$, where χ is a complex number. Obviously, on the basis of Theorems 9.21 and 9.22, $\mathcal{H}^{(s,j)}$ and $\mathcal{H}_\chi^{(s)}$ are the singular integral operators $(j = 1, 2, 3, 4)$.

Let $\hat{\boldsymbol{\sigma}}^{(s,j)} = (\hat{\sigma}_{lm}^{(s,j)})_{8\times 8}$ be the symbol of the singular integral operator $\mathcal{H}^{(s,j)}$ $(j = 1, 2, 3, 4)$. Taking into account (6.30) and (9.68) we may see that

$$\det \hat{\boldsymbol{\sigma}}^{(s,1)} = -\det \hat{\boldsymbol{\sigma}}^{(s,2)} = -\det \hat{\boldsymbol{\sigma}}^{(s,3)} = \det \hat{\boldsymbol{\sigma}}^{(s,4)}$$
$$= \frac{1}{256}\left[1 - \frac{\mu^2}{(\lambda+2\mu)^2}\right] = \frac{(\lambda+\mu)(\lambda+3\mu)}{128(\lambda+2\mu)^2} > 0. \tag{9.69}$$

Hence, the operator $\mathcal{H}^{(s,j)}$ is of the normal type, where $j = 1, 2, 3, 4$.

Let $\hat{\boldsymbol{\sigma}}_\chi^{(s)}$ and $\operatorname{ind}\mathcal{H}_\chi^{(s)}$ be the symbol and the index of the operator $\mathcal{H}_\chi^{(s)}$, respectively. It may be easily shown that

$$\det \hat{\boldsymbol{\sigma}}_\chi^{(s)} = \frac{(\lambda+2\mu)^2 - \mu^2\chi^2}{256(\lambda+2\mu)^2}$$

and $\det \hat{\boldsymbol{\sigma}}_\chi^{(s)}$ vanishes only at two points χ_1 and χ_2 of the complex plane. By virtue of (9.69) and $\det \hat{\boldsymbol{\sigma}}_1^{(s)} = \det \hat{\boldsymbol{\sigma}}^{(s,1)}$ we get $\chi_j \neq 1$ $(j = 1, 2)$ and

$$\operatorname{ind}\mathcal{H}_1^{(s)} = \operatorname{ind}\mathcal{H}^{(s,1)} = \operatorname{ind}\mathcal{H}_0^{(s)} = 0.$$

Quite similarly we obtain

$$\operatorname{ind}\mathcal{H}^{(s,2)} = -\operatorname{ind}\mathcal{H}^{(s,3)} = 0, \qquad \operatorname{ind}\mathcal{H}^{(s,4)} = -\operatorname{ind}\mathcal{H}^{(s,1)} = 0.$$

Thus, the singular integral operator $\mathcal{H}^{(s,j)}$ $(j = 1, 2, 3, 4)$ is of the normal type with an index equal to zero. Consequently, Fredholm's theorems are valid for the singular integral operator $\mathcal{H}^{(s,j)}$.

9.8 Existence Theorems

By Theorems 9.23 and 9.24 the volume potential $\mathbf{Q}^{(s,3)}(\mathbf{x}, \mathbf{F}, \Omega^\pm)$ is a regular solution of the nonhomogeneous equation (9.40), where $\mathbf{F} \in C^{0,\nu'}(\Omega^\pm)$, $0 < \nu' \leq 1$ and $\operatorname{supp}\mathbf{F}$ is a finite domain in Ω^-. Therefore, further we will consider problems $(\mathbb{K}^s)^+_{\mathbf{0},\mathbf{f}}$ and $(\mathbb{K}^s)^-_{\mathbf{0},\mathbf{f}}$, where $\mathbb{K} = I, II$. Moreover, in this section we assume that the Lamé constants satisfy the condition (6.30).

Now we prove the existence theorems of a regular (classical) solution of the class $\mathfrak{R}^{(s)}$ for these BVPs.

Problem $(I^s)^+_{\mathbf{0},\mathbf{f}}$ We assume that ω is not an eigenfrequency of the BVP $(I^s)^+_{\mathbf{0},\mathbf{0}}$ (see Theorem 9.18). We seek a regular solution of the class $\mathfrak{R}^{(s)}$ to this problem in the form of the double-layer potential

$$\mathbf{U}(\mathbf{x}) = \mathbf{Q}^{(s,2)}(\mathbf{x}, \mathbf{g}) \qquad \text{for} \qquad \mathbf{x} \in \Omega^+, \tag{9.70}$$

where $\mathbf{g}$ is the required eight-component vector function.

Obviously, by Theorem 9.22 the vector function $\mathbf{U}$ is a solution of the homogeneous equation (9.37) for $\mathbf{x} \in \Omega^+$. Keeping in mind the boundary condition (9.41) and using (9.67), from (9.70) we obtain, for determining the vector $\mathbf{g}$, a singular integral equation

$$\mathcal{H}^{(s,1)}\,\mathbf{g}(\mathbf{z}) = \mathbf{f}(\mathbf{z}) \qquad \text{for} \quad \mathbf{z} \in S. \tag{9.71}$$

We prove that Eq. (9.71) is always solvable for an arbitrary vector $\mathbf{f}$.

Let us consider the associate homogeneous equation

$$\mathcal{H}^{(s,4)}\,\mathbf{h}(\mathbf{z}) = \mathbf{0} \qquad \text{for} \quad \mathbf{z} \in S, \tag{9.72}$$

where $\mathbf{h}$ is the required eight-component vector function. Now we prove that (9.72) has only the trivial solution.

Indeed, let $\mathbf{h}_0$ be a solution of the homogeneous equation (9.72). On the basis of Theorem 9.21 and Eq. (9.72) the vector function $\mathbf{U}_0(\mathbf{x}) = \mathbf{Q}^{(s,1)}(\mathbf{x}, \mathbf{h}_0)$ is a regular solution of the external homogeneous BVP $(II^s)^-_{\mathbf{0},\mathbf{0}}$. Using Theorem 9.20, the problem $(II^s)^-_{\mathbf{0},\mathbf{0}}$ has only the trivial solution, that is

$$\mathbf{U}_0(\mathbf{x}) \equiv \mathbf{0} \qquad \text{for} \qquad \mathbf{x} \in \Omega^-. \tag{9.73}$$

On the other hand, by Theorem 9.21 and (9.73) we get

$$\{\mathbf{U}_0(\mathbf{z})\}^+ = \{\mathbf{U}_0(\mathbf{z})\}^- = \mathbf{0} \qquad \text{for} \qquad \mathbf{z} \in S,$$

i.e., the vector $\mathbf{U}_0(\mathbf{x})$ is a regular solution of problem $(I^s)^+_{\mathbf{0},\mathbf{0}}$. By virtue of Theorem 9.18 and the assumption that ω is not an eigenfrequency of the BVP $(I^s)^+_{\mathbf{0},\mathbf{0}}$, the problem $(I^s)^+_{\mathbf{0},\mathbf{0}}$ has only the trivial solution, that is

$$\mathbf{U}_0(\mathbf{x}) \equiv \mathbf{0} \qquad \text{for} \qquad \mathbf{x} \in \Omega^+. \tag{9.74}$$

On the basis of (9.73), (9.74) and identity (9.66) we obtain

$$\mathbf{h}_0(\mathbf{z}) = \{\mathbf{P}(\mathbf{D}_\mathbf{z}, \mathbf{n})\mathbf{U}_0(\mathbf{z})\}^- - \{\mathbf{P}(\mathbf{D}_\mathbf{z}, \mathbf{n})\mathbf{U}_0(\mathbf{z})\}^+ = \mathbf{0} \qquad \text{for} \qquad \mathbf{z} \in S.$$

Thus, the homogeneous equation (9.72) has only the trivial solution and therefore on the basis of Fredholm's theorem the integral equation (9.71) is always solvable for an arbitrary vector **f**. We have thereby proved the following.

Theorem 9.25 *If* $S \in C^{2,\nu}$, $\mathbf{f} \in C^{1,\nu'}(S)$, $0 < \nu' < \nu \leq 1$, *and* ω *is not an eigenfrequency of the BVP* $(I^s)^+_{\mathbf{0},\mathbf{0}}$, *then a regular solution of the class* $\mathfrak{R}^{(s)}$ *of the internal BVP* $(I^s)^+_{\mathbf{0},\mathbf{f}}$ *exists, is unique, and is represented by double-layer potential (9.70), where* **g** *is a solution of the singular integral equation (9.71) which is always solvable for an arbitrary vector* **f**.

Problem $(II^s)^+_{\mathbf{0},\mathbf{f}}$ We assume that ω is not an eigenfrequency of the BVP $(II^s)^+_{\mathbf{0},\mathbf{0}}$ (see Theorem 9.19). We seek a regular solution of the class $\mathfrak{R}^{(s)}$ to this problem in the form of the single-layer potential

$$\mathbf{U}(\mathbf{x}) = \mathbf{Q}^{(s,1)}(\mathbf{x}, \mathbf{g}) \qquad \text{for} \quad \mathbf{x} \in \Omega^+, \tag{9.75}$$

where **g** is the required eight-component vector function.

Obviously, by Theorem 9.21 the vector function **U** is a solution of (9.37) for $\mathbf{x} \in \Omega^+$. Keeping in mind the boundary condition (9.42) and using (9.66), from (9.75) we obtain, for determining the unknown vector **g**, a singular integral equation

$$\mathcal{H}^{(s,2)}\, \mathbf{g}(\mathbf{z}) = \mathbf{f}(\mathbf{z}) \qquad \text{for} \quad \mathbf{z} \in S. \tag{9.76}$$

We prove that Eq. (9.76) is always solvable for an arbitrary vector **f**.

Let us consider the homogeneous equation

$$\mathcal{H}^{(s,2)}\, \mathbf{g}_0(\mathbf{z}) = \mathbf{0} \qquad \text{for} \quad \mathbf{z} \in S, \tag{9.77}$$

where $\mathbf{g}_0$ is the required eight-component vector function. Now we prove that (9.77) has only the trivial solution. On the basis of Theorem 9.21 and Eq. (9.77) the vector function $\mathbf{U}_0(\mathbf{x}) = \mathbf{Q}^{(1)}(\mathbf{x}, \mathbf{g}_0)$ is a regular solution of the internal homogeneous BVP $(II^s)^+_{\mathbf{0},\mathbf{0}}$. Using Theorem 9.19 and the assumption that ω is not an eigenfrequency of the problem $(II^s)^+_{\mathbf{0},\mathbf{0}}$, this problem has only the trivial solution, that is

$$\mathbf{U}_0(\mathbf{x}) \equiv \mathbf{0} \qquad \text{for} \qquad \mathbf{x} \in \Omega^+. \tag{9.78}$$

On the other hand, by Theorem 9.21 and (9.78) we get

$$\{\mathbf{U}_0(\mathbf{z})\}^- = \{\mathbf{U}_0(\mathbf{z})\}^+ = \mathbf{0} \qquad \text{for} \qquad \mathbf{z} \in S,$$

i.e., the vector $\mathbf{U}_0(\mathbf{x})$ is a regular solution of problem $(I^s)^-_{\mathbf{0},\mathbf{0}}$. Using Theorem 9.20 the problem $(I^s)^-_{\mathbf{0},\mathbf{0}}$ has only the trivial solution, that is

$$\mathbf{U}_0(\mathbf{x}) \equiv \mathbf{0} \qquad \text{for} \qquad \mathbf{x} \in \Omega^-. \tag{9.79}$$

By virtue of (9.78), (9.79), and identity (9.66) we obtain

$$\mathbf{g}_0(\mathbf{z}) = \{\mathbf{P}(\mathbf{D_z}, \mathbf{n})\mathbf{U}_0(\mathbf{z})\}^- - \{\mathbf{P}(\mathbf{D_z}, \mathbf{n})\mathbf{U}_0(\mathbf{z})\}^+ = \mathbf{0} \qquad \text{for} \quad \mathbf{z} \in S.$$

Thus, the homogeneous equation (9.77) has only the trivial solution and therefore on the basis of Fredholm's theorem the singular integral equation (9.76) is always solvable for an arbitrary vector **f**. We have thereby proved the following.

Theorem 9.26 *If* $S \in C^{2,\nu}$, $\mathbf{f} \in C^{0,\nu'}(S)$, $0 < \nu' < \nu \le 1$, *and* ω *is not an eigenfrequency of the BVP* $(II^s)^+_{\mathbf{0},\mathbf{0}}$, *then a regular solution of the class* $\mathfrak{R}^{(s)}$ *of the internal BVP* $(II^s)^+_{\mathbf{0},\mathbf{f}}$ *exists, is unique, and is represented by single-layer potential (9.75), where* **g** *is a solution of the singular integral equation (9.76) which is always solvable for an arbitrary vector* **f**.

Problem $(I^s)^-_{\mathbf{0},\mathbf{f}}$ We seek a regular solution of the class $\mathfrak{R}^{(s)}$ to this problem in the sum of the double-layer and single-layer potentials

$$\mathbf{U}(\mathbf{x}) = \mathbf{Q}^{(s,2)}(\mathbf{x}, \mathbf{g}) + (1 - i)\mathbf{Q}^{(s,1)}(\mathbf{x}, \mathbf{g}) \qquad \text{for} \qquad \mathbf{x} \in \Omega^-, \tag{9.80}$$

where **g** is the required eight-component vector function.

Obviously, by Theorems 9.21 and 9.22 the vector function **U** is a solution of (9.37) for $\mathbf{x} \in \Omega^-$. Keeping in mind the boundary condition (9.43) and using (9.66), from (9.80) we obtain, for determining the unknown vector **g**, a singular integral equation

$$\mathcal{H}^{(s,5)}\,\mathbf{g}(\mathbf{z}) \equiv \mathcal{H}^{(s,3)}\,\mathbf{g}(\mathbf{z}) + (1 - i)\mathbf{Q}^{(s,1)}(\mathbf{z}, \mathbf{g}) = \mathbf{f}(\mathbf{z}) \qquad \text{for} \quad \mathbf{z} \in S. \tag{9.81}$$

We prove that Eq. (9.81) is always solvable for an arbitrary vector **f**. Clearly, the singular integral operator $\mathcal{H}^{(s,5)}$ is of the normal type and $\operatorname{ind}\mathcal{H}^{(s,5)} = \operatorname{ind}\mathcal{H}^{(s,3)} = 0$.

Now we prove that the homogeneous equation

$$\mathcal{H}^{(s,5)}\,\mathbf{g}_0(\mathbf{z}) = \mathbf{0} \qquad \text{for} \quad \mathbf{z} \in S \tag{9.82}$$

has only a trivial solution. Indeed, let $\mathbf{g}_0$ be a solution of the homogeneous equation (9.82). Then the vector

$$\mathbf{U}_0(\mathbf{x}) \equiv \mathbf{Q}^{(s,2)}(\mathbf{x}, \mathbf{g}_0) + (1 - i)\mathbf{Q}^{(s,1)}(\mathbf{x}, \mathbf{g}_0) \qquad \text{for} \qquad \mathbf{x} \in \Omega^- \tag{9.83}$$

is a regular solution of problem $(I^s)^-_{\mathbf{0},\mathbf{0}}$. Using Theorem 9.20 we have (9.79).

On the other hand, by Theorems 9.21 and 9.22 from (9.83) we get

$$\{\mathbf{U}_0(\mathbf{z})\}^- - \{\mathbf{U}_0(\mathbf{z})\}^+ = -\mathbf{g}_0(\mathbf{z}),$$

$$\{\mathbf{P}(\mathbf{D_z}, \mathbf{n})\mathbf{U}_0(\mathbf{z})\}^- - \{\mathbf{P}(\mathbf{D_z}, \mathbf{n})\mathbf{U}_0(\mathbf{z})\}^+ = (1 - i)\mathbf{g}_0(\mathbf{z}), \qquad \text{for} \quad \mathbf{z} \in S. \tag{9.84}$$

On the basis of (9.79) from (9.84) it follows that

$$\{\mathbf{P}(\mathbf{D_z}, \mathbf{n})\mathbf{U}_0(\mathbf{z}) + (1 - i)\mathbf{U}_0(\mathbf{z})\}^+ = \mathbf{0} \qquad \text{for} \quad \mathbf{z} \in S. \tag{9.85}$$

Obviously, the vector $\mathbf{U}_0$ is a solution of Eq. (9.37) in Ω^+ satisfying the boundary condition (9.85). It is easy to see that from (9.37) and (9.85) we have

$$\{\mathbf{U}_0(\mathbf{z})\}^+ = \mathbf{0} \qquad \text{for} \qquad \mathbf{z} \in S. \tag{9.86}$$

Finally, by virtue of (9.79) and (9.86) from the first equation of (9.84) we get $\mathbf{g}_0(\mathbf{z}) \equiv \mathbf{0}$ for $\mathbf{z} \in S$.

Thus, the homogeneous equation (9.82) has only the trivial solution and therefore on the basis of Fredholm's theorem the integral equation (9.81) is always solvable for an arbitrary vector **f**. We have thereby proved the following.

Theorem 9.27 *If $S \in C^{2,\nu}$, $\mathbf{f} \in C^{1,\nu'}(S)$, $0 < \nu' < \nu \le 1$, then a regular solution of the class $\mathfrak{R}^{(s)}$ of the external BVP $(I^s)^-_{\mathbf{0},\mathbf{f}}$ exists, is unique, and is represented by sum of double-layer and single-layer potentials (9.80), where* **g** *is a solution of the singular integral equation (9.81) which is always solvable for an arbitrary vector* **f**.

Problem $(II_s)^-_{\mathbf{0},\mathbf{f}}$ We seek a regular solution of the class $\mathfrak{R}^{(s)}$ to this problem in the form

$$\mathbf{U}(\mathbf{x}) = \mathbf{Q}^{(s,1)}(\mathbf{x}, \mathbf{h}) + \hat{\mathbf{U}}(\mathbf{x}) \qquad \text{for} \qquad \mathbf{x} \in \Omega^-, \tag{9.87}$$

where **h** is the required eight-component vector function and $\hat{\mathbf{V}}$ is a regular solution (eight-component vector function) of the equation

$$\mathbf{A}(\mathbf{D_x}, \omega)\, \hat{\mathbf{U}}(\mathbf{x}) = \mathbf{0} \qquad \text{for} \quad \mathbf{x} \in \Omega^-. \tag{9.88}$$

Keeping in mind the boundary condition (9.44) and using the relation (9.66), from (9.87) we obtain the following singular integral equation for determining the unknown vector **h**

$$\mathcal{H}^{(s,4)}\, \mathbf{h}(\mathbf{z}) = \hat{\mathbf{f}}(\mathbf{z}) \qquad \text{for} \quad \mathbf{z} \in S, \tag{9.89}$$

where

$$\hat{\mathbf{f}}(\mathbf{z}) = \mathbf{f}(\mathbf{z}) - \left\{\mathbf{P}(\mathbf{D_z}, \mathbf{n})\hat{\mathbf{U}}(\mathbf{z})\right\}^- . \tag{9.90}$$

We prove that Eq. (9.89) is always solvable for an arbitrary vector $\mathbf{f}$. We assume that the homogeneous equation

$$\mathcal{H}^{(s,4)}\,\mathbf{h}(\mathbf{z}) = \mathbf{0} \tag{9.91}$$

has m linearly independent solutions $\{\mathbf{h}^{(l)}(\mathbf{z})\}_{l=1}^{m}$ that are assumed to the orthonormal.

By Fredholm's theorem the solvability condition of Eq. (9.89) can be written as

$$\int_S \left\{\mathbf{P}(\mathbf{D}_\mathbf{z},\mathbf{n})\hat{\mathbf{U}}(\mathbf{z})\right\}^- \cdot \boldsymbol{\psi}^{(l)}(\mathbf{z})d_\mathbf{z}S = N_l^{(s)}, \tag{9.92}$$

where

$$N_l^{(s)} = \int_S \mathbf{f}(\mathbf{z}) \cdot \boldsymbol{\psi}^{(l)}(\mathbf{z})d_\mathbf{z}S$$

and $\{\boldsymbol{\psi}^{(l)}(\mathbf{z})\}_{l=1}^{m}$ is a complete system of solutions of the homogeneous associated equation of (9.91), i.e.

$$\mathcal{H}^{(s,1)}\,\boldsymbol{\psi}^{(l)}(\mathbf{z}) = \mathbf{0}, \qquad l = 1, 2, \cdots, m.$$

It is easy to see that condition (9.92) takes the form

$$\int_S \mathbf{h}^{(l)}(\mathbf{z}) \cdot \left\{\hat{\mathbf{U}}(\mathbf{z})\right\}^- d_\mathbf{z}S = -N_l^{(s)}, \qquad l = 1, 2, \cdots, m. \tag{9.93}$$

Let the vector $\hat{\mathbf{U}}$ be a solution of (9.88) and satisfies the boundary condition

$$\left\{\hat{\mathbf{U}}(\mathbf{z})\right\}^- = \tilde{\mathbf{f}}(\mathbf{z}), \tag{9.94}$$

where

$$\tilde{\mathbf{f}}(\mathbf{z}) = \sum_{l=1}^{m} N_l^{(s)}\mathbf{h}^{(l)}(\mathbf{z}) \tag{9.95}$$

which is solvable by virtue of Theorem 9.27. Because of the orthonormalization of $\{\mathbf{h}^{(l)}(\mathbf{z})\}_{l=1}^{m}$, the condition (9.93) is fulfilled automatically and the solvability of Eq. (9.89) is proved. Consequently, the existence of regular solution of problem $(II^s)^-_{\mathbf{0},\mathbf{f}}$ is proved too. Thus, the following theorem has been proved.

Theorem 9.28 *If* $S \in C^{2,\nu}$, $\mathbf{f} \in C^{0,\nu'}(S)$, $0 < \nu' < \nu \leq 1$, *then a regular solution of the class* $\mathfrak{R}^{(s)}$ *of the external BVP* $(II^s)^-_{\mathbf{0},\mathbf{f}}$ *exists, is unique, and is represented by sum (9.87), where* $\mathbf{h}$ *is a solution of the singular integral equation (9.89) which is always solvable,* $\hat{\mathbf{V}}$ *is the solution of BVP (9.88), (9.94) which is always solvable; and the vector functions* $\hat{\mathbf{f}}$ *and* $\tilde{\mathbf{f}}$ *are determined by (9.90) and (9.95), respectively.*

Finally, we remark that Theorem 9.28 is proved by the same method used to establish the existence of regular solution of the second external BVP of steady vibrations in the classical theory of thermoelasticity (see Kupradze et al. [219]).

Chapter 10
Problems of Pseudo-Oscillations in Thermoelasticity

In Sect. 10.1, the fundamental solution of the system of equations of pseudo-oscillations in the linear theory of thermoelasticity for quadruple porosity materials is presented and its basic properties are established.

Then, in Sect. 10.2, the formula of Galerkin-type representation of general solution of the above-mentioned system of equations is established.

In Sect. 10.3, Green's identities for the system of equations of pseudo-oscillations in the considered theory are obtained.

Finally, in Sect. 10.4, the basic internal and external BVPs of pseudo-oscillations in the linear theory of thermoelasticity for quadruple porosity materials are formulated.

10.1 Fundamental Solution

In this section the fundamental solution of the system of Eq. (1.16) is constructed explicitly by means of elementary functions. In Sect. 1.6, this system is rewritten in the matrix form (1.32). Obviously, the system (1.16) of equations of pseudo-oscillations in the theory of thermoelasticity for solid with quadruple porosity may be obtained from (1.15) by replacing ω by $i\tau$, where τ is a complex number with $\mathrm{Im}\tau > 0$ (see Sect. 1.4). Clearly, on the basis of theorems 9.1 to 9.9 we have the following results.

Theorem 10.1 *The matrix* $\mathbf{\Psi}(\mathbf{x}, i\tau)$ *defined by*

$$\mathbf{\Psi}(\mathbf{x}, i\tau) = \mathbf{L}(\mathbf{D_x}, i\tau)\,\mathbf{\Upsilon}(\mathbf{x}, i\tau)$$

is the fundamental solution of (1.16), where the matrices $\mathbf{\Psi}(\mathbf{x}, i\tau)$, $\mathbf{L}(\mathbf{D_x}, i\tau)$, *and* $\mathbf{\Upsilon}(\mathbf{x}, i\tau)$ *are given by (9.14), (9.6), and (9.9), respectively.*

M. Svanadze, *Potential Method in Mathematical Theories of Multi-Porosity Media*, Interdisciplinary Applied Mathematics 51,
https://doi.org/10.1007/978-3-030-28022-2_10

Hence, the matrix $\boldsymbol{\Psi}(\mathbf{x}, i\tau)$ is constructed explicitly by means of seven metaharmonic functions $\vartheta^{(j)}$ $(j = 1, 2, \cdots, 7)$ (see (9.10)). Obviously, each element $\Psi_{lj}(\mathbf{x}, i\tau)$ of the matrix $\boldsymbol{\Psi}(\mathbf{x}, i\tau)$ is represented in the following form

$$\Psi_{lj}(\mathbf{x}, i\tau) = L_{lj}(\mathbf{D_x}, i\tau)\,\Upsilon_{11}(\mathbf{x}, i\tau),$$

$$\Psi_{lm}(\mathbf{x}, i\tau) = L_{lm}(\mathbf{D_x}, i\tau)\,\Upsilon_{44}(\mathbf{x}, i\tau),$$

$$l = 1, 2, \cdots, 8, \qquad j = 1, 2, 3, \qquad m = 4, 5, \cdots, 8.$$

Theorem 10.2 *The matrix $\boldsymbol{\Psi}(\mathbf{x}, i\tau)$ is a solution of the equation*

$$\mathbf{A}(\mathbf{D_x}, i\tau)\boldsymbol{\Psi}(\mathbf{x}, i\tau) = \mathbf{0}$$

at every point $\mathbf{x} \in \mathbb{R}^3$ except the origin.

Theorem 10.3 *The relations*

$$\Psi_{lj}(\mathbf{x}, i\tau) = O\left(|\mathbf{x}|^{-1}\right), \qquad \Psi_{\alpha+3;\beta+3}(\mathbf{x}, i\tau) = O\left(|\mathbf{x}|^{-1}\right),$$

$$\Psi_{88}(\mathbf{x}, i\tau) = O\left(|\mathbf{x}|^{-1}\right), \qquad \Psi_{lm}(\mathbf{x}, i\tau) = O(1),$$

$$\Psi_{mj}(\mathbf{x}, i\tau) = O(1), \qquad \Psi_{\alpha+3;8}(\mathbf{x}, i\tau) = O(1),$$

$$\Psi_{8;\alpha+3}(\mathbf{x}, i\tau) = O(1), \qquad l, j = 1, 2, 3,$$

$$\alpha, \beta = 1, 2, 3, 4, \qquad m = 4, 5, \cdots, 8$$

hold the neighborhood of the origin.

Theorem 10.4 *The relations*

$$\frac{\partial}{\partial x_r}\Psi_{lj}(\mathbf{x}, i\tau) = O\left(|\mathbf{x}|^{-2}\right), \qquad \frac{\partial}{\partial x_r}\Psi_{\alpha+3;\beta+3}(\mathbf{x}, i\tau) = O\left(|\mathbf{x}|^{-2}\right),$$

$$\frac{\partial}{\partial x_r}\Psi_{88}(\mathbf{x}, i\tau) = O\left(|\mathbf{x}|^{-2}\right), \qquad \frac{\partial}{\partial x_r}\Psi_{lm}(\mathbf{x}, i\tau) = O\left(|\mathbf{x}|^{-1}\right),$$

$$\frac{\partial}{\partial x_r}\Psi_{mj}(\mathbf{x}, i\tau) = O\left(|\mathbf{x}|^{-1}\right), \qquad \frac{\partial}{\partial x_r}\Psi_{\alpha+3;8}(\mathbf{x}, i\tau) = O\left(|\mathbf{x}|^{-1}\right),$$

$$\frac{\partial}{\partial x_r}\Psi_{8;\alpha+3}(\mathbf{x}, i\tau) = O\left(|\mathbf{x}|^{-1}\right), \qquad l, j, r = 1, 2, 3,$$

$$\alpha, \beta = 1, 2, 3, 4, \qquad m = 4, 5, \cdots, 8$$

hold the neighborhood of the origin.

Theorem 10.5 *The relations*

$$\Psi_{lj}(\mathbf{x}, i\tau) - \Psi_{lj}^{(0)}(\mathbf{x}) = const + O(|\mathbf{x}|), \qquad l, j = 1, 2, \cdots, 8 \tag{10.1}$$

hold in the neighborhood of the origin, where the matrix $\boldsymbol{\Psi}^{(0)}(\mathbf{x})$ *is defined in Theorem 9.5.*

Thus, on the basis of Theorems 10.3 and 10.5 the matrix $\boldsymbol{\Psi}^{(0)}(\mathbf{x})$ is the singular part of the fundamental solution $\boldsymbol{\Psi}(\mathbf{x}, i\tau)$ in the neighborhood of the origin.

Theorem 10.6 *Every column of the matrices*

$$\left[\mathbf{P}^{(\kappa)}(\mathbf{D}_\mathbf{x}, \mathbf{n}(\mathbf{x}))\boldsymbol{\Psi}^\top(\mathbf{x}-\mathbf{y}, i\tau)\right]^\top,$$

$$\left\{\mathbf{P}^{(\kappa)}(\mathbf{D}_\mathbf{x}, \mathbf{n}(\mathbf{x}))[\boldsymbol{\Psi}^{(0)}(\mathbf{x}-\mathbf{y})]^\top\right\}^\top$$

with respect to the point $\mathbf{x}$ *satisfies the homogeneous equations*

$$\begin{aligned} &\mathbf{A}(\mathbf{D}_\mathbf{x}, i\tau)\left[\mathbf{P}^{(\kappa)}(\mathbf{D}_\mathbf{x}, \mathbf{n}(\mathbf{x}))\boldsymbol{\Psi}^\top(\mathbf{x}-\mathbf{y}, i\tau)\right]^\top = \mathbf{0}, \\ &\mathbf{A}^{(0)}(\mathbf{D}_\mathbf{x})\left\{\mathbf{P}^{(\kappa)}(\mathbf{D}_\mathbf{x}, \mathbf{n}(\mathbf{x}))[\boldsymbol{\Psi}^{(0)}(\mathbf{x}-\mathbf{y})]^\top\right\}^\top = \mathbf{0} \end{aligned} \tag{10.2}$$

for $\mathbf{x}, \mathbf{y} \in \mathbb{R}^3$ *and* $\mathbf{x} \neq \mathbf{y}$*, the matrix differential operator* $\mathbf{P}^{(\kappa)}(\mathbf{D}_\mathbf{x}, \mathbf{n}(\mathbf{x}))$ *is given by (1.46) and* κ *is an arbitrary real number.*

Theorem 10.6 leads to the following.

Theorem 10.7 *The relations*

$$\left[\mathbf{P}^{(\kappa)}(\mathbf{D}_\mathbf{x}, \mathbf{n}(\mathbf{x}))\boldsymbol{\Psi}^\top(\mathbf{x}, i\tau)\right]_{lj} = O\left(|\mathbf{x}|^{-2}\right),$$

$$\left\{\mathbf{P}^{(\kappa)}(\mathbf{D}_\mathbf{x}, \mathbf{n}(\mathbf{x}))[\boldsymbol{\Psi}^{(0)}(\mathbf{x})]^\top\right\}_{lj} = O\left(|\mathbf{x}|^{-2}\right),$$

$$l, j = 1, 2, \cdots, 8$$

hold the neighborhood of the origin, where κ *is an arbitrary real number.*

The relations (10.1) lead to the following.

Theorem 10.8 *The relations*

$$\left\{\mathbf{P}^{(\kappa)}(\mathbf{D_x}, \mathbf{n}(\mathbf{x}))\mathbf{\Psi}^\top(\mathbf{x}, i\tau)\right\}_{lj} - \left\{\mathbf{P}^{(\kappa)}(\mathbf{D_x}, \mathbf{n}(\mathbf{x}))\left[\mathbf{\Psi}^{(0)}(\mathbf{x})\right]^\top\right\}_{lj} = O\left(|\mathbf{x}|^{-1}\right), \tag{10.3}$$

$$l, j = 1, 2, \cdots, 8$$

hold in the neighborhood of the origin, where κ is an arbitrary real number.

Thus, the relations (10.3) imply that the matrix $\mathbf{P}^{(\kappa)}(\mathbf{D_x}, \mathbf{n}(\mathbf{x}))\left[\mathbf{\Psi}^{(0)}(\mathbf{x})\right]^\top$ is the singular part of the matrix $\mathbf{P}^{(\kappa)}(\mathbf{D_x}, \mathbf{n}(\mathbf{x}))\mathbf{\Psi}^\top(\mathbf{x}, i\tau)$ in the neighborhood of the origin for arbitrary real number κ.

10.2 Galerkin-Type Solution

The next two theorems provide a Galerkin-type solution to system (1.16). In the same manner as in Theorems 9.10 and 9.11 we obtain the following results.

Theorem 10.9 *Let*

$$u_l(\mathbf{x}) = \frac{1}{\mu}\Theta_1(\Delta, i\tau)\, w_l(\mathbf{x}) + m_{11}(\Delta, i\tau)\, w_{j,lj}(\mathbf{x}) + \sum_{\beta=2}^{6} m_{1\beta}(\Delta, i\tau)\, w_{\beta+2,l}(\mathbf{x}),$$

$$p_\alpha(\mathbf{x}) = m_{\alpha+1;1}(\Delta, i\tau) w_{j,j}(\mathbf{x}) + \sum_{\beta=2}^{6} m_{\alpha+1;\beta}(\Delta, i\tau)\, w_{\beta+2}(\mathbf{x}), \tag{10.4}$$

$$\theta(\mathbf{x}) = m_{61}(\Delta, i\tau) w_{j,j}(\mathbf{x}) + \sum_{\beta=2}^{6} m_{6\beta}(\Delta, i\tau)\, w_{\beta+2}(\mathbf{x}),$$

$$l = 1, 2, 3, \qquad \alpha = 1, 2, 3, 4,$$

then $\mathbf{U} = (\mathbf{u}, \mathbf{p}, \theta)$ *is a solution of the system (1.16), where* $\mathbf{w}^{(1)} = (w_1, w_2, w_3) \in C^{14}(\Omega)$, $\mathbf{w}^{(2)} = (w_4, w_5, w_6, w_7) \in C^{12}(\Omega)$, $w_8 \in C^{12}(\Omega)$,

$$\Theta_2(\Delta, i\tau)\, \mathbf{w}^{(1)}(\mathbf{x}) = \mathbf{F}^{(1)}(\mathbf{x}),$$

$$\Theta_1(\Delta, i\tau)\, \mathbf{w}^{(2)}(\mathbf{x}) = \mathbf{F}^{(2)}(\mathbf{x}), \tag{10.5}$$

$$\Theta_1(\Delta, i\tau)\, w_8(\mathbf{x}) = F_8(\mathbf{x}),$$

$\mathbf{F}^{(1)} = (F_1, F_2, F_3)$, $\mathbf{F}^{(2)} = (F_4, F_5, F_6, F_7)$, and m_{lj} is defined by (9.7).

Theorem 10.10 *If $\mathbf{U} = (\mathbf{u}, \mathbf{p}, \theta)$ is a solution of system (1.32) in Ω, then $\mathbf{U}$ is represented by (10.4), where $\widetilde{\mathbf{w}} = (w_1, w_2, \cdots, w_8)$ is a solution of (10.5) and Ω is a finite domain in $\mathbb{R}^3$.*

Thus, on the basis of Theorems 10.9 and 10.10 the completeness of the Galerkin-type solution of system (1.15) is proved.

10.3 Green's Formulas

10.3.1 Green's First Identity

In what follows, we assume that $\mathrm{Im}\mu_j > 0$ for $j = 1, 2, \cdots, 7$.

Definition 10.1 A vector function $\mathbf{U} = (\mathbf{u}, \mathbf{p}, \theta) = (U_1, U_2, \cdots, U_8)$ is called regular of the class $\mathfrak{R}$ in Ω^- (or Ω^+) if

(i)

$$U_l \in C^2(\Omega^-) \cap C^1(\overline{\Omega^-}) \qquad (\text{or } U_l \in C^2(\Omega^+) \cap C^1(\overline{\Omega^+})),$$

(ii)

$$U_l(\mathbf{x}) = O(|\mathbf{x}|^{-1}) \qquad U_{l,j}(\mathbf{x}) = o(|\mathbf{x}|^{-1}) \qquad \text{for} \qquad |\mathbf{x}| \gg 1, \tag{10.6}$$

where $j = 1, 2, 3$ and $l = 1, 2, \cdots, 8$.

Let $\mathbf{u}' = (u_1', u_2', u_3')$ and $\mathbf{p}' = (p_1', p_2', p_3', p_4')$ be three- and four-component vector functions, respectively; $\mathbf{U}' = (\mathbf{u}', \mathbf{p}', \theta')$, where θ' is a scalar function.

We introduce the notation

$$\begin{aligned}
&\mathcal{E}_1(\mathbf{U},\mathbf{u}') = W_0(\mathbf{u},\mathbf{u}') + \rho\tau^2 \mathbf{u}\cdot\mathbf{u}' - (a_\alpha\, p_\alpha + \varepsilon_0\theta)\,\mathrm{div}\,\overline{\mathbf{u}'},\\
&\mathcal{E}_2(\mathbf{U},\mathbf{p}') = k_{\alpha\beta}\nabla p_\alpha\cdot\nabla p'_\beta - c_{\alpha\beta}\,p_\beta\overline{p'_\alpha} + \tau\, a_\alpha\overline{p'_\alpha}\,\mathrm{div}\,\mathbf{u} + \tau\theta\varepsilon_\alpha\overline{p'_\alpha},\\
&\mathcal{E}_3(\mathbf{U},\theta') = k\nabla\theta\cdot\nabla\theta' + \tau T_0(a_0\theta + \varepsilon_0\,\mathrm{div}\,\mathbf{u} + \varepsilon_\alpha\, p_\alpha)\overline{\theta'},\\
&\mathcal{E}(\mathbf{U},\mathbf{U}') = \mathcal{E}_1(\mathbf{U},\mathbf{u}') + \mathcal{E}_2(\mathbf{U},\mathbf{p}') + \mathcal{E}_3(\mathbf{U},\theta'),
\end{aligned}\tag{10.7}$$

where $W_0(\mathbf{u},\mathbf{u}')$ is given by (3.15).

We have the following.

Theorem 10.11 *If* $\mathbf{U} = (\mathbf{u},\mathbf{p},\theta)$ *is a regular vector of the class* $\mathfrak{R}$ *in* Ω^+, $u'_j, p'_\alpha, \theta' \in C^1(\Omega^+)\cap C(\overline{\Omega^+})$, $j = 1,2,3$, $\alpha = 1,2,3,4$, *then*

$$\int\limits_{\Omega^+}\Big[\mathbf{A}^{(1)}(\mathbf{D_x}, i\tau)\,\mathbf{U}(\mathbf{x})\cdot\mathbf{u}'(\mathbf{x}) + \mathcal{E}_1(\mathbf{U},\mathbf{u}')\Big]\,d\mathbf{x} = \int\limits_S \mathbf{P}^{(1)}(\mathbf{D_z},\mathbf{n})\mathbf{U}(\mathbf{z})\cdot\mathbf{u}'(\mathbf{z})\,d_{\mathbf{z}}S,$$

$$\int\limits_{\Omega^+}\Big[\mathbf{A}^{(2)}(\mathbf{D_x}, i\tau)\,\mathbf{U}(\mathbf{x})\cdot\mathbf{p}'(\mathbf{x}) + \mathcal{E}_2(\mathbf{U},\mathbf{p}')\Big]\,d\mathbf{x} = \int\limits_S \mathbf{P}^{(2)}(\mathbf{D_z},\mathbf{n})\mathbf{p}(\mathbf{z})\cdot\mathbf{p}'(\mathbf{z})\,d_{\mathbf{z}}S,$$

$$\int\limits_{\Omega^+}\Big[\mathbf{A}^{(3)}(\mathbf{D_x}, i\tau)\,\mathbf{U}(\mathbf{x})\,\overline{\theta'(\mathbf{x})} + \mathcal{E}_3(\mathbf{U},\theta')\Big]\,d\mathbf{x} = k\int\limits_S \frac{\partial\theta(\mathbf{z})}{\partial\mathbf{n}(\mathbf{z})}\,\overline{\theta'(\mathbf{z})}\,d_{\mathbf{z}}S,$$

where the matrix differential operators $\mathbf{A}^{(l)}(\mathbf{D_x}, i\tau)$ *and* $\mathbf{P}^{(j)}(\mathbf{D_z},\mathbf{n})$ $(l = 1,2,3,$ $j = 1,2)$ *are defined in Sect. 9.3.1.*

Theorem 10.12 *If* $\mathbf{U} = (\mathbf{u},\mathbf{p},\theta)$ *is a regular vector of the class* $\mathfrak{R}$ *in* Ω^+, $u'_j, p'_\alpha, \theta' \in C^1(\Omega^+)\cap C(\overline{\Omega^+})$, $j = 1,2,3$, $\alpha = 1,2,3,4$, *then*

$$\int_{\Omega^+} \left[\mathbf{A}(\mathbf{D_x}, i\tau)\,\mathbf{U}(\mathbf{x}) \cdot \mathbf{U}'(\mathbf{x}) + \mathcal{E}(\mathbf{U}, \mathbf{U}')\right] d\mathbf{x} = \int_S \mathbf{P}(\mathbf{D_z}, \mathbf{n})\mathbf{U}(\mathbf{z}) \cdot \mathbf{U}'(\mathbf{z})\, d_{\mathbf{z}}S, \tag{10.8}$$

where $\mathbf{A}(\mathbf{D_x}, i\tau)$, $\mathbf{P}(\mathbf{D_z}, \mathbf{n})$ *and* $\mathcal{E}(\mathbf{U}, \mathbf{U}')$ *are defined by (1.27), (1.44), and (10.7), respectively.*

Theorem 10.12 and the condition (10.6) lead to the following.

Theorem 10.13 *If* $\mathbf{U} = (\mathbf{u}, \mathbf{p}, \theta)$ *and* $\mathbf{U}' = (\mathbf{u}', \mathbf{p}', \theta')$ *are regular vectors of the class* $\mathfrak{R}$ *in* Ω^-, *then*

$$\int_{\Omega^-} \left[\mathbf{A}(\mathbf{D_x}, i\tau)\,\mathbf{U}(\mathbf{x}) \cdot \mathbf{U}'(\mathbf{x}) + \mathcal{E}(\mathbf{U}, \mathbf{U}')\right] d\mathbf{x} = -\int_S \mathbf{P}(\mathbf{D_z}, \mathbf{n})\mathbf{U}(\mathbf{z}) \cdot \mathbf{U}'(\mathbf{z})\, d_{\mathbf{z}}S, \tag{10.9}$$

where $\mathbf{A}(\mathbf{D_x}, i\tau)$, $\mathbf{P}(\mathbf{D_z}, \mathbf{n})$ *and* $\mathcal{E}(\mathbf{U}, \mathbf{U}')$ *are given by (1.27), (1.44), and (10.7), respectively.*

The formulas (10.8) and (10.9) are Green's first identities of the pseudo-oscillations equations in the linear theory of quadruple porosity thermoelasticity for domains Ω^+ and Ω^-, respectively.

10.3.2 Green's Second Identity

Clearly, the matrix differential operator $\tilde{\mathbf{A}}(\mathbf{D_x}, i\tau) = \left(\tilde{A}_{lj}(\mathbf{D_x}, i\tau)\right)_{8\times 8}$ is the associate operator of $\mathbf{A}(\mathbf{D_x}, i\tau)$, where $\tilde{\mathbf{A}}(\mathbf{D_x}, i\tau) = \mathbf{A}^\top(-\mathbf{D_x}, i\tau)$. It is easy to see that the associated system of homogeneous equations of (1.16) is

$$\begin{aligned}
&(\mu\,\Delta - \rho\,\tau^2)\tilde{\mathbf{u}} + (\lambda + \mu)\,\nabla\mathrm{div}\tilde{\mathbf{u}} + \tau\nabla\,(\mathbf{a}\,\tilde{\mathbf{p}}) + \tau\varepsilon_0 T_0 \nabla\,\tilde{\theta} = \mathbf{0},\\
&(\mathbf{K}\,\Delta + \mathbf{c}')\,\tilde{\mathbf{p}} + \mathbf{a}\,\mathrm{div}\tilde{\mathbf{u}} - \tau T_0 \boldsymbol{\varepsilon}\,\tilde{\theta} = \mathbf{0},\\
&(k\Delta - \tau a_0 T_0)\tilde{\theta} + \varepsilon_0\,\mathrm{div}\,\tilde{\mathbf{u}} - \tau\boldsymbol{\varepsilon}\,\tilde{\mathbf{p}} = F_8,
\end{aligned} \tag{10.10}$$

where $\tilde{\theta}$ is a scalar function, and $\tilde{\mathbf{u}}$ and $\tilde{\mathbf{p}}$ are three- and four-component vector functions on $\Omega^{\pm}$, respectively.

Let $\tilde{\mathbf{U}}_j$ be the j-th column of the matrix $\tilde{\mathbf{U}} = (\tilde{U}_{lj})_{8\times 8}$ $(j = 1, 2, \cdots, 8)$. By direct calculation we obtain the following.

Theorem 10.14 *If* $\mathbf{U} = (\mathbf{u}, \mathbf{p}, \theta)$ *and* $\tilde{U}_j$ $(j = 1, 2, \cdots, 8)$ *are regular vectors of the class* $\mathfrak{R}$ *in* Ω^+, *then*

$$\int\limits_{\Omega^+} \Big\{[\tilde{\mathbf{A}}(\mathbf{D_y}, i\tau)\tilde{\mathbf{U}}(\mathbf{y})]^\top \mathbf{U}(\mathbf{y}) - [\tilde{\mathbf{U}}(\mathbf{y})]^\top \mathbf{A}(\mathbf{D_y}, i\tau)\mathbf{U}(\mathbf{y})\Big\}\, d\mathbf{y}$$
$$= \int\limits_S \Big\{[\tilde{\mathbf{P}}(\mathbf{D_z}, \mathbf{n})\tilde{\mathbf{U}}(\mathbf{z})]^\top \mathbf{U}(\mathbf{z}) - [\tilde{\mathbf{U}}(\mathbf{z})]^\top \mathbf{P}(\mathbf{D_z}, \mathbf{n})\mathbf{U}(\mathbf{z})\Big\}\, d_{\mathbf{z}}S, \tag{10.11}$$

where the matrices $\mathbf{A}(\mathbf{D_x}, i\tau)$ *and* $\mathbf{P}(\mathbf{D_z}, \mathbf{n})$ *are given by (1.27) and (1.44), respectively; the operator* $\tilde{\mathbf{P}}(\mathbf{D_z}, \mathbf{n})$ *is defined by (9.32).*

Theorem 10.14 and the condition (10.6) lead to the following.

Theorem 10.15 *If* $\mathbf{U} = (\mathbf{u}, \mathbf{p}, \theta)$ *and* $\tilde{U}_j$ $(j = 1, 2, \cdots, 8)$ *are regular vectors of the class* $\mathfrak{R}$ *in* Ω^-, *then*

$$\int\limits_{\Omega^-} \Big\{[\tilde{\mathbf{A}}(\mathbf{D_y}, i\tau)\tilde{\mathbf{U}}(\mathbf{y})]^\top \mathbf{U}(\mathbf{y}) - [\tilde{\mathbf{U}}(\mathbf{y})]^\top \mathbf{A}(\mathbf{D_y}, i\tau)\mathbf{U}(\mathbf{y})\Big\}\, d\mathbf{y}$$
$$= -\int\limits_S \Big\{[\tilde{\mathbf{P}}(\mathbf{D_z}, \mathbf{n})\tilde{\mathbf{U}}(\mathbf{z})]^\top \mathbf{U}(\mathbf{z}) - [\tilde{\mathbf{U}}(\mathbf{z})]^\top \mathbf{P}(\mathbf{D_z}, \mathbf{n})\mathbf{U}(\mathbf{z})\Big\}\, d_{\mathbf{z}}S. \tag{10.12}$$

The formulas (10.11) and (10.12) are Green's second identities of the pseudo-oscillations equations in the linear theory of quadruple porosity thermoelasticity for domains Ω^+ and Ω^-, respectively.

10.3.3 Green's Third Identity

Let $\tilde{\mathbf{\Psi}}(\mathbf{x}, i\tau)$ be the fundamental solution of (10.10) (the fundamental matrix of operator $\tilde{\mathbf{A}}(\mathbf{D_y}, i\tau)$). Obviously, the matrix $\tilde{\mathbf{\Psi}}(\mathbf{x}, i\tau)$ satisfies the following condition

$$\tilde{\mathbf{\Psi}}(\mathbf{x}, i\tau) = \mathbf{\Psi}^\top(-\mathbf{x}, i\tau), \tag{10.13}$$

where $\mathbf{\Psi}(\mathbf{x}, i\tau)$ is the fundamental solution of the system (1.16) (see Theorem 10.1).

By virtue of (10.2), (10.3), and (10.13) from (10.11) and (10.12) we obtain the following results.

Theorem 10.16 *If* **U** *is a regular vector of the class* $\mathfrak{R}$ *in* Ω^+, *then*

$$\mathbf{U}(\mathbf{x}) = \int_S \Big\{[\tilde{\mathbf{P}}(\mathbf{D_z},\mathbf{n})\boldsymbol{\Psi}^\top(\mathbf{x}-\mathbf{z},i\tau)]^\top\mathbf{U}(\mathbf{z}) - \boldsymbol{\Psi}(\mathbf{x}-\mathbf{z},i\tau)\,\mathbf{P}(\mathbf{D_z},\mathbf{n})\mathbf{U}(\mathbf{z})\Big\}\,d_{\mathbf{z}}S + \int_{\Omega^+} \boldsymbol{\Psi}(\mathbf{x}-\mathbf{y},i\tau)\,\mathbf{A}(\mathbf{D_y},i\tau)\mathbf{U}(\mathbf{y})d\mathbf{y} \quad \textit{for} \quad \mathbf{x}\in\Omega^+. \tag{10.14}$$

Theorem 10.17 *If* **U** *is a regular vector of the class* $\mathfrak{R}$ *in* Ω^-, *then*

$$\mathbf{U}(\mathbf{x}) = -\int_S \Big\{[\tilde{\mathbf{P}}(\mathbf{D_z},\mathbf{n})\boldsymbol{\Psi}^\top(\mathbf{x}-\mathbf{z},i\tau)]^\top\mathbf{U}(\mathbf{z}) - \boldsymbol{\Psi}(\mathbf{x}-\mathbf{z},i\tau)\,\mathbf{P}(\mathbf{D_z},\mathbf{n})\mathbf{U}(\mathbf{z})\Big\}\,d_{\mathbf{z}}S + \int_{\Omega^+} \boldsymbol{\Psi}(\mathbf{x}-\mathbf{y},i\tau)\,\mathbf{A}(\mathbf{D_y},i\tau)\mathbf{U}(\mathbf{y})d\mathbf{y} \quad \textit{for} \quad \mathbf{x}\in\Omega^-. \tag{10.15}$$

The formulas (10.14) and (10.15) are Green's third identities of the pseudo-oscillations equations in the linear theory of quadruple porosity thermoelasticity for domains Ω^+ and Ω^-, respectively.

Theorems 10.16 and 10.17 lead to the following.

Corollary 10.1 *If* **U** *is a regular solution of the class* $\mathfrak{R}$ *of the homogeneous equation*

$$\mathbf{A}(\mathbf{D_x},i\tau)\,\mathbf{U}(\mathbf{x}) = \mathbf{0} \tag{10.16}$$

in Ω^+, *then*

$$\mathbf{U}(\mathbf{x}) = \int_S \Big\{[\tilde{\mathbf{P}}(\mathbf{D_z},\mathbf{n})\boldsymbol{\Psi}^\top(\mathbf{x}-\mathbf{z},i\tau)]^\top\mathbf{U}(\mathbf{z}) - \boldsymbol{\Psi}(\mathbf{x}-\mathbf{z},i\tau)\,\mathbf{P}(\mathbf{D_z},\mathbf{n})\mathbf{U}(\mathbf{z})\Big\}\,d_{\mathbf{z}}S \tag{10.17}$$

for $\mathbf{x}\in\Omega^+$.

Corollary 10.2 *If* $\mathbf{U}$ *is a regular solution of the class* $\mathfrak{R}$ *of (10.16) in* Ω^-, *then*

$$\mathbf{U}(\mathbf{x}) =$$

$$-\int_S \Big\{[\tilde{\mathbf{P}}(\mathbf{D_z},\mathbf{n})\boldsymbol{\Psi}^\top(\mathbf{x}-\mathbf{z},i\tau)]^\top \mathbf{U}(\mathbf{z}) - \boldsymbol{\Psi}(\mathbf{x}-\mathbf{z},i\tau)\,\mathbf{P}(\mathbf{D_z},\mathbf{n})\mathbf{U}(\mathbf{z})\Big\}\, d_{\mathbf{z}}S \tag{10.18}$$

for $\mathbf{x}\in\Omega^-$.

The formulas (10.17) and (10.18) are the Somigliana-type integral representations of a regular solution of the pseudo-oscillations homogeneous equation (10.16) in the linear theory of quadruple porosity thermoelasticity for domains Ω^+ and Ω^-, respectively.

10.4 Basic Boundary Value Problems

The basic internal and external BVPs of pseudo-oscillations in the linear theory of thermoelasticity for quadruple porosity materials are formulated as follows.

Find a regular (classical) solution of the class $\mathfrak{R}$ in Ω^+ to system

$$\mathbf{A}(\mathbf{D_x}, i\tau)\,\mathbf{U}(\mathbf{x}) = \mathbf{F}(\mathbf{x}) \tag{10.19}$$

for $\mathbf{x}\in\Omega^+$ satisfying the boundary condition (9.41) in the *Problem* $(I^l)^+_{\mathbf{F},\mathbf{f}}$ and the boundary condition (9.42) in the *Problem* $(II^l)^+_{\mathbf{F},\mathbf{f}}$, where the matrix differential operators $\mathbf{A}(\mathbf{D_x}, i\tau)$ and $\mathbf{P}(\mathbf{D_z},\mathbf{n})$ are defined by (1.27) and (1.44), respectively; $\mathbf{F}$ and $\mathbf{f}$ are prescribed eight-component vector functions.

Find a regular (classical) solution of the class $\mathfrak{R}$ to system (10.19) for $\mathbf{x}\in\Omega^-$ satisfying the boundary condition (9.43) in the *Problem* $(I^l)^-_{\mathbf{F},\mathbf{f}}$ and the boundary condition (9.44) in the *Problem* $(II^l)^-_{\mathbf{F},\mathbf{f}}$. Here $\mathbf{F}$ and $\mathbf{f}$ are prescribed eight-component vector functions, and $\operatorname{supp}\mathbf{F}$ is a finite domain in Ω^-.

In the similar manner as in Chap. 9 we can investigate the BVPs of pseudo-oscillations $(\mathbb{K}^l)^+_{\mathbf{F},\mathbf{f}}$ and $(\mathbb{K}^l)^-_{\mathbf{F},\mathbf{f}}$ by means of the potential method and the theory of singular integral equations, where $\mathbb{K} = I, II$.

Chapter 11
Problems of Quasi-Static in Thermoelasticity

In this chapter, the basic internal and external BVPs of steady vibrations in the quasi-static linear theory of thermoelasticity for quadruple porosity materials are investigated by means of the potential method and the theory of singular integral equations.

In Sect. 11.1, the fundamental solution of the system of steady vibrations equations of the considered theory is constructed by means of elementary functions and its basic properties are established.

In Sects. 11.2 and 11.3, the formula of Galerkin-type representation of general solution for the system of steady vibrations equations is given and Green's identities for the same system of equations are obtained, respectively.

In Sects. 11.4 and 11.5, the basic internal and external BVPs of steady vibrations of the considered theory are formulated and the uniqueness theorems for classical solutions of these BVPs are proved by employing techniques based on Green's identities, respectively.

In Sect. 11.6, the surface and volume potentials are constructed and their basic properties are established.

Then, in Sect. 11.7, some useful singular integral operators are introduced and the basic properties of these operators are established.

Furthermore, in Sect. 11.8, the above-mentioned BVPs of steady vibrations of the linear theory of thermoelasticity are reduced to the equivalent always solvable singular integral equations for which Fredholm's theorems are valid. The existence theorems for classical solutions of the BVPs of steady vibrations are proved by means of the potential method.

Finally, in Sect. 11.9, the first internal and external BVPs are investigated by means of Fredholm's integral equations.

M. Svanadze, *Potential Method in Mathematical Theories of Multi-Porosity Media*,
Interdisciplinary Applied Mathematics 51,
https://doi.org/10.1007/978-3-030-28022-2_11

11.1 Fundamental Solution

In this section the fundamental solution of the system of Eq. (1.17) is constructed explicitly by means of elementary functions. In Sect. 1.6, this system is rewritten in the matrix form (1.33).

In this section we will suppose that the assumption (9.2) holds true.

Definition 11.1 The fundamental solution of system (1.17) (the fundamental matrix of operator $\mathbf{A}^{(q)}(\mathbf{D_x}, \omega)$) is the matrix $\mathbf{\Psi}^{(q)}(\mathbf{x}, \omega) = \left(\Psi_{lj}^{(q)}(\mathbf{x}, \omega)\right)_{8\times 8}$ satisfying the following equation in the class of generalized functions

$$\mathbf{A}^{(q)}(\mathbf{D_x}, \omega)\mathbf{\Psi}^{(q)}(\mathbf{x}, \omega) = \delta(\mathbf{x})\mathbf{I}_8, \tag{11.1}$$

where $\mathbf{x} \in \mathbb{R}^3$.

We introduce the notation:

1.

$$\begin{aligned}
&\mathbf{B}^{(q)}(\Delta, \omega) = \left(B_{lj}^{(q)}(\Delta, \omega)\right)_{6\times 6}, \qquad \tilde{\mathbf{B}}(\Delta, \omega) = \left(\tilde{B}_{lj}(\Delta, \omega)\right)_{6\times 6},\\
&B_{11}^{(q)}(\Delta, \omega) = \mu_0, \qquad \tilde{B}_{11}(\Delta, \omega) = \mu_0\Delta,\\
&B_{1;\alpha+1}^{(q)}(\Delta, \omega) = i\omega a_\alpha, \qquad \tilde{B}_{1;\alpha+1}(\Delta, \omega) = i\omega a_\alpha\,\Delta,\\
&B_{16}^{(q)}(\Delta, \omega) = i\omega\varepsilon_0 T_0, \qquad \tilde{B}_{16}(\Delta, \omega) = i\omega\varepsilon_0 T_0\,\Delta,\\
&B_{\alpha+1;1}^{(q)}(\Delta, \omega) = \tilde{B}_{\alpha+1;1}(\Delta, \omega) = -a_\alpha,\\
&B_{\alpha+1;\beta+1}^{(q)}(\Delta, \omega) = \tilde{B}_{\alpha+1;\beta+1}(\Delta, \omega) = k_{\alpha\beta}\Delta + c_{\alpha\beta},\\
&B_{\alpha+1;6}^{(q)}(\Delta, \omega) = \tilde{B}_{\alpha+1;6}(\Delta, \omega) = i\omega T_0\varepsilon_\alpha,\\
&B_{61}^{(q)}(\Delta, \omega) = \tilde{B}_{61}(\Delta, \omega) = -\varepsilon_0,\\
&B_{6;\alpha+1}^{(q)}(\Delta, \omega) = \tilde{B}_{6;\alpha+1}(\Delta, \omega) = i\omega\varepsilon_\alpha,\\
&B_{66}^{(q)}(\Delta, \omega) = \tilde{B}_{66}(\Delta, \omega) = k\Delta + i\omega a_0 T_0,\\
&\alpha, \beta = 1, 2, 3, 4.
\end{aligned} \tag{11.2}$$

2.

$$\Theta_1^{(q)}(\Delta,\omega) = \frac{1}{kk_0\mu_0}\,\Delta\det\mathbf{B}^{(q)}(\Delta,\omega) = \Delta\prod_{j=1}^{5}(\Delta+\tilde{\mu}_j^2),$$

$$\Theta_2^{(q)}(\Delta,\omega) = \Delta^2\prod_{j=1}^{5}(\Delta+\tilde{\mu}_j^2),$$

where $\tilde{\mu}_1^2,\ \tilde{\mu}_2^2,\cdots,\tilde{\mu}_5^2$ are the roots of the algebraic equation $\det\mathbf{B}^{(q)}(-\xi,\omega) = 0$ (with respect to ξ).

3.

$$\boldsymbol{\Theta}^{(q)}(\Delta,\omega) = \left(\Theta_{lj}^{(q)}(\Delta,\omega)\right)_{8\times 8}, \qquad \Theta_{11}^{(q)} = \Theta_{22}^{(q)} = \Theta_{33}^{(q)} = \Theta_2^{(q)},$$

$$\Theta_{44}^{(q)} = \Theta_{55}^{(q)} = \cdots = \Theta_{88}^{(q)} = \Theta_1^{(q)}, \qquad \Theta_{lj}^{(q)} = 0,$$

$$l, j = 1, 2, \cdots, 8, \qquad l \neq j. \tag{11.3}$$

4.

$$\mathbf{L}^{(q)}(\mathbf{D_x},\omega) = \left(L_{lj}^{(q)}(\mathbf{D_x},\omega)\right)_{8\times 8},$$

$$L_{lj}^{(q)}(\mathbf{D_x},\omega) = \frac{1}{\mu}\Theta_1^{(q)}(\Delta,\omega)\,\delta_{lj} + m_{11}^{(q)}(\Delta,\omega)\frac{\partial^2}{\partial x_l\partial x_j},$$

$$L_{l;r+3}^{(q)}(\mathbf{D_x},\omega) = m_{1;r+1}^{(q)}(\Delta,\omega)\frac{\partial}{\partial x_l}, \tag{11.4}$$

$$L_{m+3;l}^{(q)}(\mathbf{D_x},\omega) = m_{m+1;1}^{(q)}(\Delta,\omega)\frac{\partial}{\partial x_l},$$

$$L_{m+3;r+3}^{(q)}(\mathbf{D_x},\omega) = m_{m+1;r+1}^{(q)}(\Delta,\omega),$$

$$l, j = 1, 2, 3, \qquad m, r = 1, 2, \cdots, 5,$$

where $\tilde{B}_{lj}^*$ is the cofactor of element $\tilde{B}_{lj}$ of the matrix $\tilde{\mathbf{B}}$ (see (11.2)),

$$m_{l1}^{(q)}(\Delta,\omega) = -\frac{\lambda+\mu}{kk_0\mu\mu_0}\tilde{B}_{l1}^*(\Delta,\omega)$$
$$-\frac{i\omega}{kk_0\mu\mu_0}\left[a_\beta \tilde{B}_{l;\beta+1}^*(\Delta,\omega) + \varepsilon_0 T_0 \tilde{B}_{l6}^*(\Delta,\omega)\right],$$
$$m_{l\alpha}^{(q)}(\Delta,\omega) = \frac{1}{kk_0\mu_0}\tilde{B}_{l\alpha}^*(\Delta,\omega), \qquad l=1,2,\cdots,6, \qquad \alpha=2,3,\cdots,6 \tag{11.5}$$

On the basis of (11.3)–(11.5) by direct calculation we obtain

$$\mathbf{A}^{(q)}(\mathbf{D_x},\omega)\mathbf{L}^{(q)}(\mathbf{D_x},\omega) = \mathbf{\Theta}^{(q)}(\Delta,\omega). \tag{11.6}$$

We assume that $\tilde{\mu}_l^2 \neq \tilde{\mu}_j^2 \neq 0$, where $l, j = 1,2,\cdots,5$ and $l \neq j$.

Let

$$\mathbf{\Upsilon}^{(q)}(\mathbf{x},\omega) = \left(\Upsilon_{lj}^{(q)}(\mathbf{x},\omega)\right)_{8\times 8},$$
$$\Upsilon_{11}^{(q)}(\mathbf{x},\omega) = \Upsilon_{22}^{(q)}(\mathbf{x},\omega) = \Upsilon_{33}^{(q)}(\mathbf{x},\omega) = \sum_{j=1}^{7}\check{\eta}_{2j}\check{\vartheta}^{(j)}(\mathbf{x}),$$
$$\Upsilon_{44}^{(q)}(\mathbf{x},\omega) = \Upsilon_{55}^{(q)}(\mathbf{x},\omega) = \cdots = \Upsilon_{88}^{(q)}(\mathbf{x},\omega) = \sum_{j=1}^{6}\check{\eta}_{1j}\check{\vartheta}^{(j)}(\mathbf{x}), \tag{11.7}$$
$$\Upsilon_{lj}^{(q)}(\mathbf{x},\omega) = 0, \qquad l \neq j, \qquad l, j = 1,2,\cdots,8,$$

where the functions $\check{\vartheta}^{(j)}$ $(j = 1,2,\cdots,7)$ are defined by

$$\check{\vartheta}^{(m)}(\mathbf{x}) = -\frac{e^{i\tilde{\mu}_m|\mathbf{x}|}}{4\pi\,|\mathbf{x}|}, \qquad \check{\vartheta}^{(6)}(\mathbf{x}) = -\frac{1}{4\pi\,|\mathbf{x}|},$$
$$\check{\vartheta}^{(7)}(\mathbf{x}) = -\frac{|\mathbf{x}|}{8\pi}, \qquad m = 1,2,\cdots,5 \tag{11.8}$$

and

$$\check{\eta}_{1l} = -\left[\tilde{\mu}_l^2 \prod_{j=1,\, j\neq l}^{5} (\tilde{\mu}_j^2 - \tilde{\mu}_l^2)\right]^{-1}, \qquad \check{\eta}_{2l} = \left[\tilde{\mu}_l^4 \prod_{j=1,\, j\neq l}^{5} (\tilde{\mu}_j^2 - \tilde{\mu}_l^2)\right]^{-1},$$

$$\check{\eta}_{26} = -\left[\tilde{\mu}_1^4\tilde{\mu}_2^4\tilde{\mu}_3^4\tilde{\mu}_4^4\tilde{\mu}_5^4\right]^{-1} \sum_{l=1}^{5} \prod_{j=1,\, j\neq l}^{5} \tilde{\mu}_j^2,$$

$$\check{\eta}_{16} = \check{\eta}_{27} = \prod_{j=1}^{5} \tilde{\mu}_j^{-2}, \qquad l = 1, 2, \cdots, 5.$$

Obviously, the matrix $\boldsymbol{\Upsilon}^{(q)}(\mathbf{x}, \omega)$ is the fundamental solution of the operator $\boldsymbol{\Theta}^{(q)}(\Delta, \omega)$, that is,

$$\boldsymbol{\Theta}^{(q)}(\Delta, \omega)\boldsymbol{\Upsilon}^{(q)}(\mathbf{x}, \omega) = \delta(\mathbf{x})\, \mathbf{I}_8, \tag{11.9}$$

where $\mathbf{x} \in \mathbb{R}^3$.

We introduce the matrix

$$\boldsymbol{\Psi}^{(q)}(\mathbf{x}, \omega) = \mathbf{L}^{(q)}(\mathbf{D}_\mathbf{x}, \omega)\, \boldsymbol{\Upsilon}^{(q)}(\mathbf{x}, \omega). \tag{11.10}$$

Using identities (11.6) and (11.9) from (11.10) we get

$$\begin{aligned}\mathbf{A}^{(q)}(\mathbf{D}_\mathbf{x}, \omega)\, \boldsymbol{\Psi}^{(q)}(\mathbf{x}, \omega) &= \mathbf{A}^{(q)}(\mathbf{D}_\mathbf{x}, \omega)\, \mathbf{L}^{(q)}(\mathbf{D}_\mathbf{x}, \omega)\, \boldsymbol{\Upsilon}^{(q)}(\mathbf{x}, \omega)\\ &= \boldsymbol{\Theta}^{(q)}(\Delta, \omega)\, \boldsymbol{\Upsilon}^{(q)}(\mathbf{x}, \omega) = \delta(\mathbf{x})\, \mathbf{I}_8.\end{aligned}$$

Hence, $\boldsymbol{\Psi}^{(q)}(\mathbf{x}, \omega)$ is the solution of (11.1). We have thereby proved the following.

Theorem 11.1 *The matrix $\boldsymbol{\Psi}^{(q)}(\mathbf{x}, \omega)$ defined by (11.10) is the fundamental solution of (1.17), where the matrices $\mathbf{L}^{(q)}(\mathbf{D}_\mathbf{x}, \omega)$ and $\boldsymbol{\Upsilon}^{(q)}(\mathbf{x}, \omega)$ are given by (11.4) and (11.7), respectively.*

Clearly, each element $\Psi_{lj}^{(q)}(\mathbf{x}, \omega)$ of the matrix $\boldsymbol{\Psi}^{(q)}(\mathbf{x}, \omega)$ is represented in the following form

$$\begin{aligned}&\Psi_{lj}^{(q)}(\mathbf{x}, \omega) = L_{lj}^{(q)}(\mathbf{D}_\mathbf{x}, \omega)\, \Upsilon_{11}^{(q)}(\mathbf{x}, \omega),\\ &\Psi_{lm}^{(q)}(\mathbf{x}, \omega) = L_{lm}^{(q)}(\mathbf{D}_\mathbf{x}, \omega)\, \Upsilon_{44}^{(q)}(\mathbf{x}, \omega),\\ &l = 1, 2, \cdots, 8, \qquad j = 1, 2, 3, \qquad m = 4, 5, \cdots, 8.\end{aligned} \tag{11.11}$$

Obviously, the matrix $\boldsymbol{\Psi}^{(q)}(\mathbf{x}, \omega)$ is constructed explicitly by means of seven elementary functions: metaharmonic functions $\hat{\vartheta}^{(j)}$ $(j = 1, 2, \cdots, 5)$, harmonic function $\check{\vartheta}^{(6)}$, and biharmonic function $\check{\vartheta}^{(7)}$ (see (11.8)).

We now can establish the basic properties of the matrix $\boldsymbol{\Psi}^{(q)}(\mathbf{x}, \omega)$. The relations (11.11) lead to the following results.

Theorem 11.2 *If* $\mathbf{F} = \mathbf{0}$, *then the matrix* $\boldsymbol{\Psi}^{(q)}(\mathbf{x}, \omega)$ *is a solution of (1.33) i.e.*

$$\mathbf{A}^{(q)}(\mathbf{D}_{\mathbf{x}}, \omega)\boldsymbol{\Psi}^{(q)}(\mathbf{x}, \omega) = \mathbf{0} \tag{11.12}$$

at every point $\mathbf{x} \in \mathbb{R}^3$ *except the origin.*

Theorem 11.3 *The relations*

$$\Psi^{(q)}_{lj}(\mathbf{x}, \omega) = O\left(|\mathbf{x}|^{-1}\right), \qquad \Psi^{(q)}_{\alpha+3;\beta+3}(\mathbf{x}, \omega) = O\left(|\mathbf{x}|^{-1}\right),$$

$$\Psi^{(q)}_{88}(\mathbf{x}, \omega) = O\left(|\mathbf{x}|^{-1}\right), \qquad \Psi^{(q)}_{lm}(\mathbf{x}, \omega) = O(1),$$

$$\Psi^{(q)}_{mj}(\mathbf{x}, \omega) = O(1), \qquad \Psi^{(q)}_{\alpha+3;8}(\mathbf{x}, \omega) = O(1),$$

$$\Psi^{(q)}_{8;\alpha+3}(\mathbf{x}, \omega) = O(1), \qquad l, j = 1, 2, 3,$$

$$\alpha, \beta = 1, 2, 3, 4, \qquad m = 4, 5, \cdots, 8$$

hold the neighborhood of the origin.

Theorem 11.4 *The relations*

$$\frac{\partial}{\partial x_r}\Psi^{(q)}_{lj}(\mathbf{x}, \omega) = O\left(|\mathbf{x}|^{-2}\right), \qquad \frac{\partial}{\partial x_r}\Psi^{(q)}_{\alpha+3;\beta+3}(\mathbf{x}, \omega) = O\left(|\mathbf{x}|^{-2}\right),$$

$$\frac{\partial}{\partial x_r}\Psi^{(q)}_{88}(\mathbf{x}, \omega) = O\left(|\mathbf{x}|^{-2}\right), \qquad \frac{\partial}{\partial x_r}\Psi^{(q)}_{lm}(\mathbf{x}, \omega) = O\left(|\mathbf{x}|^{-1}\right),$$

$$\frac{\partial}{\partial x_r}\Psi^{(q)}_{mj}(\mathbf{x}, \omega) = O\left(|\mathbf{x}|^{-1}\right), \qquad \frac{\partial}{\partial x_r}\Psi^{(q)}_{\alpha+3;8}(\mathbf{x}, \omega) = O\left(|\mathbf{x}|^{-1}\right),$$

$$\frac{\partial}{\partial x_r}\Psi^{(q)}_{8;\alpha+3}(\mathbf{x}, \omega) = O\left(|\mathbf{x}|^{-1}\right), \qquad l, j, r = 1, 2, 3,$$

$$\alpha, \beta = 1, 2, 3, 4, \qquad m = 4, 5, \cdots, 8$$

hold the neighborhood of the origin.

Theorem 11.5 *The relations*

$$\Psi^{(q)}_{lj}(\mathbf{x}, \omega) - \Psi^{(0)}_{lj}(\mathbf{x}) = const + O(|\mathbf{x}|), \qquad l, j = 1, 2, \cdots, 8 \tag{11.13}$$

hold in the neighborhood of the origin, where the matrix $\boldsymbol{\Psi}^{(0)}(\mathbf{x})$ *is given in Theorem 9.5.*

Thus, on the basis of Theorems 11.3 and 11.5 the matrix $\boldsymbol{\Psi}^{(0)}(\mathbf{x})$ is the singular part of the fundamental solution $\boldsymbol{\Psi}^{(q)}(\mathbf{x}, \omega)$ in the neighborhood of the origin.

Now we shall construct a solution of the system (1.33) for $\mathbf{F}(\mathbf{x}) = \mathbf{0}$ playing a major role in the investigation of the BVPs of the quasi-static theory of thermoelasticity for materials with quadruple porosity by means of the potential method. The following theorem is valid.

Theorem 11.6 *Every column of the matrix*

$$\left[\mathbf{P}^{(\kappa)}(\mathbf{D}_{\mathbf{x}}, \mathbf{n}(\mathbf{x}))\boldsymbol{\Psi}^{(q)^\top}(\mathbf{x}-\mathbf{y}, \omega)\right]^\top$$

with respect to the point $\mathbf{x}$ *satisfies the homogeneous equations*

$$\mathbf{A}^{(q)}(\mathbf{D}_{\mathbf{x}}, \omega)\left[\mathbf{P}^{(\kappa)}(\mathbf{D}_{\mathbf{x}}, \mathbf{n}(\mathbf{x}))\boldsymbol{\Psi}^{(q)^\top}(\mathbf{x}-\mathbf{y}, \omega)\right]^\top = \mathbf{0} \tag{11.14}$$

everywhere in $\mathbb{R}^3$ *except* $\mathbf{x} = \mathbf{y}$*, respectively, where the matrix differential operator* $\mathbf{P}^{(\kappa)}(\mathbf{D}_{\mathbf{x}}, \mathbf{n}(\mathbf{x}))$ *is given by (1.46) and* κ *is an arbitrary real number.*

The matrix $\left[\mathbf{P}^{(\kappa)}(\mathbf{D}_{\mathbf{x}}, \mathbf{n}(\mathbf{x}))\boldsymbol{\Psi}^{(q)^\top}(\mathbf{x}-\mathbf{y}, \omega)\right]^\top$ is called the *singular solution* of the system of steady vibration equation (1.33). Theorem 11.6 leads to the following.

Theorem 11.7 *The relations*

$$\left[\mathbf{P}^{(\kappa)}(\mathbf{D}_{\mathbf{x}}, \mathbf{n}(\mathbf{x}))\boldsymbol{\Psi}^{(q)^\top}(\mathbf{x}, \omega)\right]_{lj} = O\left(|\mathbf{x}|^{-2}\right),$$

$$l, j = 1, 2, \cdots, 8$$

hold the neighborhood of the origin, where κ *is an arbitrary real number.*

Now we can establish the singular part of the matrix $\mathbf{P}^{(\kappa)}(\mathbf{D}_{\mathbf{x}}, \mathbf{n}(\mathbf{x}))\boldsymbol{\Psi}^{(q)^\top}(\mathbf{x}, \omega)$ in the neighborhood of the origin.

The relations (11.13) lead to the following.

Theorem 11.8 *The relations*

$$\left\{\mathbf{P}^{(\kappa)}(\mathbf{D}_{\mathbf{x}}, \mathbf{n}(\mathbf{x}))\boldsymbol{\Psi}^{(q)^\top}(\mathbf{x}, \omega)\right\}_{lj} - \left\{\mathbf{P}^{(\kappa)}(\mathbf{D}_{\mathbf{x}}, \mathbf{n}(\mathbf{x}))\left[\boldsymbol{\Psi}^{(0)}(\mathbf{x})\right]^\top\right\}_{lj} = O\left(|\mathbf{x}|^{-1}\right), \tag{11.15}$$

$$l, j = 1, 2, \cdots, 8$$

hold in the neighborhood of the origin, where κ is an arbitrary real number.

Thus, the relations (11.15) imply that the matrix $\mathbf{P}^{(\kappa)}(\mathbf{D_x}, \mathbf{n}(\mathbf{x}))\left[\mathbf{\Psi}^{(0)}(\mathbf{x})\right]^\top$ is the singular part of the matrix $\mathbf{P}^{(\kappa)}(\mathbf{D_x}, \mathbf{n}(\mathbf{x}))\mathbf{\Psi}^{(q)^\top}(\mathbf{x}, \omega)$ in the neighborhood of the origin for arbitrary real number κ.

11.2 Galerkin-Type Solution

The next two theorems provide a Galerkin-type solution to system (1.17).

Theorem 11.9 *Let*

$$u_l(\mathbf{x}) = \frac{1}{\mu}\Theta_1^{(q)}(\Delta, \omega)\, w_l(\mathbf{x}) + m_{11}^{(q)}(\Delta, \omega)\, w_{j,lj}(\mathbf{x})$$
$$+ \sum_{\beta=2}^{6} m_{1\beta}^{(q)}(\Delta, \omega)\, w_{\beta+2,l}(\mathbf{x}),$$
$$p_\alpha(\mathbf{x}) = m_{\alpha+1;1}^{(q)}(\Delta, \omega) w_{j,j}(\mathbf{x}) + \sum_{\beta=2}^{6} m_{\alpha+1;\beta}^{(q)}(\Delta, \omega)\, w_{\beta+2}(\mathbf{x}), \tag{11.16}$$
$$\theta(\mathbf{x}) = m_{61}^{(q)}(\Delta, \omega) w_{j,j}(\mathbf{x}) + \sum_{\beta=2}^{6} m_{6\beta}^{(q)}(\Delta, \omega)\, w_{\beta+2}(\mathbf{x}),$$
$$l = 1, 2, 3, \qquad \alpha = 1, 2, 3, 4,$$

where $\mathbf{w}^{(1)} = (w_1, w_2, w_3) \in C^{14}(\Omega)$, $\mathbf{w}^{(2)} = (w_4, w_5, w_6, w_7) \in C^{12}(\Omega)$, $w_8 \in C^{12}(\Omega)$,

$$\Theta_2^{(q)}(\Delta, \omega)\, \mathbf{w}^{(1)}(\mathbf{x}) = \mathbf{F}^{(1)}(\mathbf{x}),$$
$$\Theta_1^{(q)}(\Delta, \omega)\, \mathbf{w}^{(2)}(\mathbf{x}) = \mathbf{F}^{(2)}(\mathbf{x}), \tag{11.17}$$
$$\Theta_1^{(q)}(\Delta, \omega)\, w_8(\mathbf{x}) = F_8(\mathbf{x}),$$

$\mathbf{F}^{(1)} = (F_1, F_2, F_3)$, $\mathbf{F}^{(2)} = (F_4, F_5, F_6, F_7)$, *then* $\mathbf{U} = (\mathbf{u}, \mathbf{p}, \theta)$ *is a solution of the system (1.17).*

Theorem 11.10 *If* $\mathbf{U} = (\mathbf{u}, \mathbf{p}, \theta)$ *is a solution of system (1.33) in* Ω, *then* $\mathbf{U}$ *is represented by (11.16), where* $\widetilde{\mathbf{w}} = (w_1, w_2, \cdots, w_8)$ *is a solution of (11.17) and* Ω *is a finite domain in* $\mathbb{R}^3$.

Theorems 11.9 and 11.10 are proved similarly as Theorems 9.10 and 9.11, respectively. Thus, on the basis of Theorems 11.9 and 11.10 the completeness of the Galerkin-type solution of system (1.17) is proved.

11.3 Green's Formulas

11.3.1 Green's First Identity

In what follows, we assume that $\mathrm{Im}\tilde{\mu}_j > 0$ for $j = 1, 2, \cdots, 5$. In the sequel we use the matrix differential operator:

$$\mathbf{A}^{(q,1)}(\mathbf{D_x}, \omega) = \left(A_{lr}^{(q,1)}(\mathbf{D_x}, \omega)\right)_{3\times 8},$$

$$A_{lj}^{(q,1)}(\mathbf{D_x}, \omega) = \mu\Delta\delta_{lj} + (\lambda + \mu)\frac{\partial^2}{\partial x_l \partial x_j}, \tag{11.18}$$

$$A_{l;\alpha+3}^{(q,1)}(\mathbf{D_x}, \omega) = -a_\alpha \frac{\partial}{\partial x_l}, \qquad A_{l8}^{(q,1)}(\mathbf{D_x}, \omega) = -\varepsilon_0 \frac{\partial}{\partial x_l},$$

where $l, j = 1, 2, 3$, $\alpha = 1, 2, 3, 4$,

Let $\mathbf{u}' = (u_1', u_2', u_3')$ and $\mathbf{p}' = (p_1', p_2', p_3', p_4')$ be three- and four-component vector functions, respectively; $\mathbf{U}' = (\mathbf{u}', \mathbf{p}', \theta')$, where θ' is a scalar function.

We introduce the notation

$$E_1^{(q)}(\mathbf{U}, \mathbf{u}') = W_0(\mathbf{u}, \mathbf{u}') - (a_\alpha\, p_\alpha + \varepsilon_0 \theta)\,\mathrm{div}\,\overline{\mathbf{u}'}, \tag{11.19}$$

where $W_0(\mathbf{u}, \mathbf{u}')$ is given by (3.15).

We have the following.

Theorem 11.11 *If* $\mathbf{U} = (\mathbf{u}, \mathbf{p}, \theta)$ *is a regular vector of the class* $\mathfrak{R}$ *in* Ω^+, $u_j', p_\alpha', \theta' \in C^1(\Omega^+) \cap C(\overline{\Omega^+})$, $j = 1, 2, 3$, $\alpha = 1, 2, 3, 4$, *then*

$$\int_{\Omega^+} \left[\mathbf{A}^{(q)}(\mathbf{D_x}, \omega)\,\mathbf{U}(\mathbf{x}) \cdot \mathbf{U}'(\mathbf{x}) + E^{(q)}(\mathbf{U}, \mathbf{U}')\right] d\mathbf{x}$$
$$= \int_S \mathbf{P}(\mathbf{D_z}, \mathbf{n})\mathbf{U}(\mathbf{z}) \cdot \mathbf{U}'(\mathbf{z})\, d_{\mathbf{z}}S, \tag{11.20}$$

where

$$E^{(q)}(\mathbf{U},\mathbf{U}') = E_1^{(q)}(\mathbf{U},\mathbf{u}') + E_2(\mathbf{U},\mathbf{p}') + E_3(\mathbf{U},\theta'), \tag{11.21}$$

$\mathbf{A}^{(q)}(\mathbf{D_x},\omega)$, $\mathbf{P}(\mathbf{D_z},\mathbf{n})$, E_j $(j = 2, 3)$ *and* $E_1^{(q)}$ *are defined by (1.28), (1.44), (9.26), and (11.19), respectively.*

It is easy to verify that the identity (11.20) may be obtained from the second and third identities of (9.28) and

$$\int_{\Omega^+} \left[\mathbf{A}^{(q,1)}(\mathbf{D_x},\omega)\,\mathbf{U}(\mathbf{x})\cdot\mathbf{u}'(\mathbf{x}) + E_1^{(q)}(\mathbf{U},\mathbf{u}')\right] d\mathbf{x} = \int_S \mathbf{P}^{(1)}(\mathbf{D_z},\mathbf{n})\mathbf{U}(\mathbf{z})\cdot\mathbf{u}'(\mathbf{z})\, d_{\mathbf{z}}S, \tag{11.22}$$

by virtue of (11.19).

Theorem 11.11 and the condition (10.6) lead to the following.

Theorem 11.12 *If* $\mathbf{U} = (\mathbf{u},\mathbf{p},\theta)$ *and* $\mathbf{U}' = (\mathbf{u}',\mathbf{p}',\theta')$ *are regular vectors of the class* $\mathfrak{R}$ *in* Ω^-, *then*

$$\int_{\Omega^-} \left[\mathbf{A}^{(q)}(\mathbf{D_x},\omega)\,\mathbf{U}(\mathbf{x})\cdot\mathbf{U}'(\mathbf{x}) + E^{(q)}(\mathbf{U},\mathbf{U}')\right] d\mathbf{x} = -\int_S \mathbf{P}(\mathbf{D_z},\mathbf{n})\mathbf{U}(\mathbf{z})\cdot\mathbf{U}'(\mathbf{z})\, d_{\mathbf{z}}S, \tag{11.23}$$

where $\mathbf{A}^{(q)}(\mathbf{D_x},\omega)$, $\mathbf{P}(\mathbf{D_z},\mathbf{n})$, *and* $E^{(q)}(\mathbf{U},\mathbf{U}')$ *are given by (1.28), (1.44), and (11.21), respectively.*

The formulas (11.21) and (11.23) are Green's first identities of the steady vibrations equations in the quasi-static linear theory of quadruple porosity thermoelasticity for domains Ω^+ and Ω^-, respectively.

11.3.2 Green's Second Identity

Clearly, the matrix differential operator $\tilde{\mathbf{A}}^{(q)}(\mathbf{D_x},\omega) = \left(\tilde{A}_{lj}^{(q)}(\mathbf{D_x},\omega)\right)_{8\times 8}$ is the associate operator of $\mathbf{A}^{(q)}(\mathbf{D_x},\omega)$, where $\tilde{\mathbf{A}}^{(q)}(\mathbf{D_x},\omega) = \mathbf{A}^{(q)^\top}(-\mathbf{D_x},\omega)$. It is easy to see that the associated system of homogeneous equations of (1.17) is

$$\mu\,\Delta\tilde{\mathbf{u}} + (\lambda+\mu)\,\nabla \mathrm{div}\tilde{\mathbf{u}} - i\omega\nabla\,(\mathbf{a}\,\tilde{\mathbf{p}}) - i\omega\varepsilon_0 T_0 \nabla\,\tilde{\theta} = \mathbf{0},$$

$$(\mathbf{K}\,\Delta + \mathbf{c})\,\tilde{\mathbf{p}} + \mathbf{a}\,\mathrm{div}\tilde{\mathbf{u}} + i\omega T_0 \boldsymbol{\varepsilon}\,\tilde{\theta} = \mathbf{0}, \tag{11.24}$$

$$(k\Delta + i\omega a_0 T_0)\tilde{\theta} + \varepsilon_0\,\mathrm{div}\,\tilde{\mathbf{u}} + i\omega\boldsymbol{\varepsilon}\,\tilde{\mathbf{p}} = F_8,$$

where $\tilde{\theta}$ is a scalar function, and $\tilde{\mathbf{u}}$ and $\tilde{\mathbf{p}}$ are three- and four-component vector functions on $\Omega^{\pm}$, respectively.

Let $\tilde{\mathbf{U}}_j$ be the j-th column of the matrix $\tilde{\mathbf{U}} = (\tilde{U}_{lj})_{8\times 8}$ $(j = 1, 2, \cdots, 8)$. Quite similarly as in Theorem 3.13 by direct calculation we obtain the following.

Theorem 11.13 *If* $\mathbf{U} = (\mathbf{u}, \mathbf{p}, \theta)$ *and* $\tilde{U}_j$ $(j = 1, 2, \cdots, 8)$ *are regular vectors of the class* $\mathfrak{R}$ *in* Ω^+, *then*

$$\int\limits_{\Omega^+} \left\{ [\tilde{\mathbf{A}}^{(q)}(\mathbf{D_y}, \omega)\tilde{\mathbf{U}}(\mathbf{y})]^\top \mathbf{U}(\mathbf{y}) - [\tilde{\mathbf{U}}(\mathbf{y})]^\top \mathbf{A}^{(q)}(\mathbf{D_y}, \omega)\mathbf{U}(\mathbf{y}) \right\} d\mathbf{y}$$
$$= \int\limits_S \left\{ [\tilde{\mathbf{P}}(\mathbf{D_z}, \mathbf{n})\tilde{\mathbf{U}}(\mathbf{z})]^\top \mathbf{U}(\mathbf{z}) - [\tilde{\mathbf{U}}(\mathbf{z})]^\top \mathbf{P}(\mathbf{D_z}, \mathbf{n})\mathbf{U}(\mathbf{z}) \right\} d_{\mathbf{z}}S, \tag{11.25}$$

where the matrices $\mathbf{A}^{(q)}(\mathbf{D_x}, \omega)$ *and* $\mathbf{P}(\mathbf{D_z}, \mathbf{n})$ *are given by (1.28) and (1.44), respectively; the operator* $\tilde{\mathbf{P}}(\mathbf{D_z}, \mathbf{n})$ *is defined by (9.32).*

Theorem 11.13 and the condition (10.6) lead to the following.

Theorem 11.14 *If* $\mathbf{U} = (\mathbf{u}, \mathbf{p}, \theta)$ *and* $\tilde{U}_j$ $(j = 1, 2, \cdots, 8)$ *are regular vectors of the class* $\mathfrak{R}$ *in* Ω^-, *then*

$$\int\limits_{\Omega^-} \left\{ [\tilde{\mathbf{A}}^{(q)}(\mathbf{D_y}, \omega)\tilde{\mathbf{U}}(\mathbf{y})]^\top \mathbf{U}(\mathbf{y}) - [\tilde{\mathbf{U}}(\mathbf{y})]^\top \mathbf{A}^{(q)}(\mathbf{D_y}, \omega)\mathbf{U}(\mathbf{y}) \right\} d\mathbf{y}$$
$$= -\int\limits_S \left\{ [\tilde{\mathbf{P}}(\mathbf{D_z}, \mathbf{n})\tilde{\mathbf{U}}(\mathbf{z})]^\top \mathbf{U}(\mathbf{z}) - [\tilde{\mathbf{U}}(\mathbf{z})]^\top \mathbf{P}(\mathbf{D_z}, \mathbf{n})\mathbf{U}(\mathbf{z}) \right\} d_{\mathbf{z}}S, \tag{11.26}$$

where the operator $\tilde{\mathbf{P}}(\mathbf{D_z}, \mathbf{n})$ *is defined by (9.32).*

The formulas (11.25) and (11.26) are Green's second identities of the steady vibrations equations in the quasi-static linear theory of quadruple porosity thermoelasticity for domains Ω^+ and Ω^-, respectively.

11.3.3 Green's Third Identity

Let $\tilde{\mathbf{\Psi}}^{(q)}(\mathbf{x},\omega)$ be the fundamental solution of the system (11.24) (the fundamental matrix of operator $\tilde{\mathbf{A}}^{(q)}(\mathbf{D}_\mathbf{y},\omega)$). Obviously, the matrix $\tilde{\mathbf{\Psi}}^{(q)}(\mathbf{x},\omega)$ satisfies the following condition

$$\tilde{\mathbf{\Psi}}^{(q)}(\mathbf{x},\omega)=\mathbf{\Psi}^{(q)^\top}(-\mathbf{x},\omega), \tag{11.27}$$

where $\mathbf{\Psi}^{(q)}(\mathbf{x},\omega)$ is defined by (11.10).

By virtue of (11.12), (11.13), (11.15), and (11.27) from (11.25) and (11.26) we have the following results.

Theorem 11.15 *If* $\mathbf{U}$ *is a regular vector of the class* $\mathfrak{R}$ *in* Ω^+*, then*

$$\mathbf{U}(\mathbf{x})=\int\limits_S\left\{[\tilde{\mathbf{P}}(\mathbf{D}_\mathbf{z},\mathbf{n})\mathbf{\Psi}^{(q)^\top}(\mathbf{x}-\mathbf{z},\omega)]^\top\mathbf{U}(\mathbf{z})-\mathbf{\Psi}^{(q)}(\mathbf{x}-\mathbf{z},\omega)\,\mathbf{P}(\mathbf{D}_\mathbf{z},\mathbf{n})\mathbf{U}(\mathbf{z})\right\}d_\mathbf{z}S$$
$$+\int\limits_{\Omega^+}\mathbf{\Psi}^{(q)}(\mathbf{x}-\mathbf{y},\omega)\,\mathbf{A}^{(q)}(\mathbf{D}_\mathbf{y},\omega)\mathbf{U}(\mathbf{y})d\mathbf{y}\qquad \textit{for}\quad \mathbf{x}\in\Omega^+. \tag{11.28}$$

Theorem 11.16 *If* $\mathbf{U}$ *is a regular vector of the class* $\mathfrak{R}$ *in* Ω^-*, then*

$$\mathbf{U}(\mathbf{x})=-\int\limits_S\left\{[\tilde{\mathbf{P}}(\mathbf{D}_\mathbf{z},\mathbf{n})\mathbf{\Psi}^{(q)^\top}(\mathbf{x}-\mathbf{z},\omega)]^\top\mathbf{U}(\mathbf{z})-\mathbf{\Psi}^{(q)}(\mathbf{x}-\mathbf{z},\omega)\,\mathbf{P}(\mathbf{D}_\mathbf{z},\mathbf{n})\mathbf{U}(\mathbf{z})\right\}d_\mathbf{z}S$$
$$+\int\limits_{\Omega^-}\mathbf{\Psi}^{(q)}(\mathbf{x}-\mathbf{y},\omega)\,\mathbf{A}^{(q)}(\mathbf{D}_\mathbf{y},\omega)\mathbf{U}(\mathbf{y})d\mathbf{y}\qquad \textit{for}\quad \mathbf{x}\in\Omega^-. \tag{11.29}$$

The formulas (11.28) and (11.29) are Green's third identities of the steady vibrations equations in the quasi-static linear theory of quadruple porosity thermoelasticity for domains Ω^+ and Ω^-, respectively.

Theorems 11.15 and 11.16 lead to the following.

Corollary 11.1 *If* $\mathbf{U}$ *is a regular solution of the class* $\mathfrak{R}$ *of the homogeneous equation*

$$\mathbf{A}^{(q)}(\mathbf{D}_\mathbf{x},\omega)\,\mathbf{U}(\mathbf{x})=\mathbf{0} \tag{11.30}$$

in Ω^+, *then*

$$\mathbf{U}(\mathbf{x}) = \int_S \left\{ [\tilde{\mathbf{P}}(\mathbf{D_z}, \mathbf{n})\boldsymbol{\Psi}^{(q)^\top}(\mathbf{x}-\mathbf{z}, \omega)]^\top \mathbf{U}(\mathbf{z}) - \boldsymbol{\Psi}^{(q)}(\mathbf{x}-\mathbf{z}, \omega)\, \mathbf{P}(\mathbf{D_z}, \mathbf{n})\mathbf{U}(\mathbf{z}) \right\} d_{\mathbf{z}}S \tag{11.31}$$

for $\mathbf{x} \in \Omega^+$.

Corollary 11.2 *If* $\mathbf{U}$ *is a regular solution of the class* $\mathfrak{R}$ *of (11.30) in* Ω^-, *then*

$$\mathbf{U}(\mathbf{x}) = -\int_S \left\{ [\tilde{\mathbf{P}}(\mathbf{D_z}, \mathbf{n})\boldsymbol{\Psi}^{(q)^\top}(\mathbf{x}-\mathbf{z}, \omega)]^\top \mathbf{U}(\mathbf{z}) - \boldsymbol{\Psi}^{(q)}(\mathbf{x}-\mathbf{z}, \omega)\, \mathbf{P}(\mathbf{D_z}, \mathbf{n})\mathbf{U}(\mathbf{z}) \right\} d_{\mathbf{z}}S \tag{11.32}$$

for $\mathbf{x} \in \Omega^-$.

The formulas (11.31) and (11.32) are the Somigliana-type integral representations of a regular solution of the steady vibrations homogeneous equation (11.30) in the quasi-static linear theory of quadruple porosity thermoelasticity for domains Ω^+ and Ω^-, respectively.

11.4 Basic Boundary Value Problems

The basic internal and external BVPs of steady vibrations in the quasi-static linear theory of thermoelasticity for quadruple porosity materials are formulated as follows.

Find a regular (classical) solution of the class $\mathfrak{R}$ in Ω^+ to system (1.33) for $\mathbf{x} \in \Omega^+$ satisfying the boundary condition (9.41) in the *Problem* $(I^q)^+_{\mathbf{F},\mathbf{f}}$, and the boundary condition (9.42) in the *Problem* $(II^q)^+_{\mathbf{F},\mathbf{f}}$, where the matrix differential operators $\mathbf{A}^{(q)}(\mathbf{D_x}, \omega)$ and $\mathbf{P}(\mathbf{D_z}, \mathbf{n})$ are defined by (1.28) and (1.44), respectively; $\mathbf{F}$ and $\mathbf{f}$ are prescribed eight-component vector functions.

Find a regular (classical) solution of the class $\mathfrak{R}$ to system (1.33) for $\mathbf{x} \in \Omega^-$ satisfying the boundary condition (9.43) in the *Problem* $(I^q)^-_{\mathbf{F},\mathbf{f}}$, and the boundary condition (9.44) in the *Problem* $(II^q)^-_{\mathbf{F},\mathbf{f}}$. Here $\mathbf{F}$ and $\mathbf{f}$ are prescribed eight-component vector functions, and $\operatorname{supp}\mathbf{F}$ is a finite domain in Ω^-.

11.5 Uniqueness Theorems

In this section we study uniqueness of regular solutions of the BVPs $(\mathbb{K}^q)^+_{\mathbf{F},\mathbf{f}}$ and $(\mathbb{K}^q)^-_{\mathbf{F},\mathbf{f}}$, where $\mathbb{K} = I, II$.

We have the following results.

Theorem 11.17 *If the condition (6.6) is satisfied, then the internal BVP $(I^q)^+_{\mathbf{F},\mathbf{f}}$ admits at most one regular (classical) solution of the class $\mathfrak{R}$.*

Proof As usual, suppose that there are two regular solutions of problem $(I^q)^+_{\mathbf{F},\mathbf{f}}$ in the class $\mathfrak{R}$. Then their difference $\mathbf{U}$ is a regular solution of the internal homogeneous BVP $(I^q)^+_{\mathbf{0},\mathbf{0}}$. Hence, $\mathbf{U}$ is a regular solution of the homogeneous equation (11.30) in Ω^+ satisfying the homogeneous boundary condition

$$\{\mathbf{U}(\mathbf{z})\}^+ = \mathbf{0} \qquad \text{for} \quad \mathbf{z} \in S. \tag{11.33}$$

On the basis of (11.30) and (11.33), from (11.22) and the second and third identities of (9.28) we obtain for $\mathbf{u}' = \mathbf{u}$, $\mathbf{p}' = \mathbf{p}$ and $\theta' = \theta$:

$$\int\limits_{\Omega^+} E_1^{(q)}(\mathbf{U}, \mathbf{u}) d\mathbf{x} = 0, \qquad \int\limits_{\Omega^+} E_2(\mathbf{U}, \mathbf{p}) d\mathbf{x} = 0, \qquad \int\limits_{\Omega^+} E_3(\mathbf{U}, \theta) d\mathbf{x} = 0, \tag{11.34}$$

where

$$\begin{aligned} E_1^{(q)}(\mathbf{U}, \mathbf{u}) &= W_0(\mathbf{u}, \mathbf{u}) - (a_\alpha\, p_\alpha + \varepsilon_0 \theta)\,\mathrm{div}\,\overline{\mathbf{u}}, \\ E_2(\mathbf{U}, \mathbf{p}) &= k_{\alpha\beta} \nabla p_\beta \cdot \nabla p_\alpha - c_{\alpha\beta}\, p_\beta \overline{p_\alpha} - i\omega (a_\alpha\,\mathrm{div}\,\mathbf{u} + \theta \varepsilon_\alpha)\,\overline{p_\alpha}, \\ E_3(\mathbf{U}, \theta) &= k|\nabla\theta|^2 - i\omega a_0 T_0 |\theta|^2 - i\omega T_0 (\varepsilon_0\,\mathrm{div}\,\mathbf{u} + \varepsilon_\alpha p_\alpha)\,\overline{\theta}, \end{aligned} \tag{11.35}$$

$W_0(\mathbf{u}, \mathbf{u})$ is defined by (3.15), and E_2 and E_3 are given by (9.26).

Then, employing (6.6), we may derive from (11.35)

$$\begin{aligned} \mathrm{Im}\, E_1^{(q)}(\mathbf{U}, \mathbf{u}) &= \mathrm{Im}\left[(a_\alpha \overline{p_\alpha} + \varepsilon_0 \overline{\theta})\,\mathrm{div}\,\mathbf{u}\right], \\ \mathrm{Re}\, E_2(\mathbf{U}, \mathbf{p}) &= k_{\alpha\beta} \nabla p_\beta \cdot \nabla p_\alpha + d_{\alpha\beta}\, p_\beta \overline{p_\alpha} + \omega \mathrm{Im}\,[(a_\alpha\,\mathrm{div}\,\mathbf{u} + \theta\varepsilon_\alpha)\,\overline{p_\alpha}], \\ \mathrm{Re}\, E_3(\mathbf{U}, \theta) &= k|\nabla\theta|^2 + \omega T_0 \mathrm{Im}\left[(\varepsilon_0\,\mathrm{div}\,\mathbf{u} + \varepsilon_\alpha p_\alpha)\,\overline{\theta}\right]. \end{aligned} \tag{11.36}$$

Obviously, from (11.36) it follows that

$$\begin{aligned} &\mathrm{Re}\, E_3(\mathbf{U}, \theta) + T_0 \mathrm{Re}\, E_2(\mathbf{U}, \mathbf{p}) - \omega T_0 \mathrm{Im}\, E_1^{(q)}(\mathbf{U}, \mathbf{u}) \\ &\quad = k|\nabla\theta|^2 + T_0 (k_{\alpha\beta} \nabla p_\beta \cdot \nabla p_\alpha + d_{\alpha\beta}\, p_\beta \overline{p_\alpha}). \end{aligned} \tag{11.37}$$

Taking into account (11.37) from (11.34) we have

$$\int\limits_{\Omega^+} \left[k|\nabla\theta|^2 + T_0(k_{\alpha\beta}\nabla p_\beta \cdot \nabla p_\alpha + d_{\alpha\beta}\, p_\beta \overline{p_\alpha}) \right] d\mathbf{x} = 0.$$

On the basis of (4.10) from the last equation we get

$$k_{\alpha\beta}\nabla p_\beta \cdot \nabla p_\alpha = 0, \qquad d_{\alpha\beta}\, p_\beta \overline{p_\alpha} = 0, \qquad \nabla\theta = 0, \tag{11.38}$$

and hence, by using Lemma 6.1 from (11.38) we obtain

$$\begin{aligned} &p_\alpha(\mathbf{x}) = c_\alpha = \text{const}, \qquad \theta(\mathbf{x}) = c_5 = \text{const}, \\ &d_{\alpha\beta}\, p_\beta(\mathbf{x}) = 0, \qquad \alpha = 1, 2, 3, 4 \end{aligned} \tag{11.39}$$

for $\mathbf{x} \in \Omega^+$. On the basis of homogeneous boundary condition (11.33) from (11.39) we have

$$\mathbf{p}(\mathbf{x}) = \mathbf{0}, \qquad \theta(\mathbf{x}) = 0 \tag{11.40}$$

for $\mathbf{x} \in \Omega^+$.

Now employing (11.40) and (11.35) from (11.34) we may derive that $W_0(\mathbf{u}, \mathbf{u}) = 0$, and consequently, $\mathbf{u}(\mathbf{x})$ is a rigid displacement vector. On the basis of the homogeneous boundary condition (11.30) it follows that

$$\mathbf{u}(\mathbf{x}) = \mathbf{0} \tag{11.41}$$

for $\mathbf{x} \in \Omega^+$. Hence, from the identities (11.40) and (11.41) we get $\mathbf{U}(\mathbf{x}) = \mathbf{0}$ for $\mathbf{x} \in \Omega^+$. □

Theorem 11.18 *If the condition (7.9) is satisfied and the matrix*

$$\hat{\mathbf{b}} = \begin{pmatrix} b_{11} & b_{12} & b_{13} & b_{14} & \varepsilon_1 \\ b_{12} & b_{22} & b_{23} & b_{24} & \varepsilon_2 \\ b_{13} & b_{23} & b_{33} & b_{34} & \varepsilon_3 \\ b_{14} & b_{24} & b_{34} & b_{44} & \varepsilon_4 \\ \varepsilon_1 & \varepsilon_2 & \varepsilon_3 & \varepsilon_4 & a_0 \end{pmatrix}_{5\times 5}$$

is a positive definite, then any two solutions of the BVP $(II^q)^+_{\mathbf{F},\mathbf{f}}$ *of the class* $\mathfrak{R}$ *may differ only for an additive vector* $\mathbf{U} = (\mathbf{u}, \mathbf{p}, \theta)$*, where* $\mathbf{p}$ *and* θ *satisfy the condition (11.40),* $\mathbf{u}$ *is a rigid displacement vector*

$$\mathbf{u}(\mathbf{x}) = \mathbf{a}' + [\mathbf{b}' \times \mathbf{x}] \qquad \text{for} \quad \mathbf{x} \in \Omega^+, \tag{11.42}$$

and $\mathbf{a}' = (a_1', a_2', a_3')$ *and* $\mathbf{b}' = (b_1', b_2', b_3')$ *are arbitrary complex constant three-component vectors.*

Proof Suppose that there are two regular solutions of problem $(II^q)^+_{\mathbf{F},\mathbf{f}}$ in the class $\mathfrak{R}$. Then their difference $\mathbf{U}$ is a regular solution of the internal homogeneous BVP $(II^q)^+_{\mathbf{0},\mathbf{0}}$. Hence, $\mathbf{U}$ is a regular solution of the homogeneous equation (11.30) in Ω^+ satisfying the homogeneous boundary condition (9.57).

Quite similarly as in Theorem 11.17, on the basis of (11.30) we have Eq. (11.39). Furthermore, if we employ Eq. (11.39), then from (11.30) we may obtain

$$\mu\, \Delta \mathbf{u} + (\lambda + \mu)\, \nabla \mathrm{div}\, \mathbf{u} = \mathbf{0} \tag{11.43}$$

and

$$\begin{aligned} b_{\alpha\beta} c_\beta + \varepsilon_\alpha c_5 &= -a_\alpha\, \mathrm{div}\, \mathbf{u}, \\ \varepsilon_\beta c_\beta + a_0 c_5 &= -\varepsilon_0\, \mathrm{div}\, \mathbf{u} \end{aligned} \tag{11.44}$$

for $\alpha = 1, 2, 3, 4$. Clearly, from (11.44) we deduce that $\mathrm{div}\, \mathbf{u}(\mathbf{x}) = \mathrm{const}$ for $\mathbf{x} \in \Omega^+$ and consequently, from (11.43) we may derive that the displacement $\mathbf{u}(\mathbf{x})$ is a harmonic vector field in Ω^+.

On the other hand, if we employ Eq. (11.44), then from (11.35) we have

$$E_1^{(q)}(\mathbf{U}, \mathbf{u}) = W_0(\mathbf{u}, \mathbf{u}) + (a_0 |c_5|^2 + 2\varepsilon_\alpha \mathrm{Re}(c_\alpha c_5) + b_{\alpha\beta} c_\alpha \overline{c_\beta}).$$

Applying the last relation from the first equation of (11.34) it follows that

$$\int_{\Omega^+} \Big[W_0(\mathbf{u}, \mathbf{u}) + (a_0 |c_5|^2 + 2\varepsilon_\alpha \mathrm{Re}(c_\alpha c_5) + b_{\alpha\beta} c_\alpha \overline{c_\beta}) \Big]\, d\mathbf{x} = 0. \tag{11.45}$$

On the basis of (7.9) and the assumption that $\hat{\mathbf{b}}$ is a positive definite matrix, from (11.45) we get

$$W_0(\mathbf{u}, \mathbf{u}) = 0, \qquad a_0 |c_5|^2 + 2\varepsilon_\alpha \mathrm{Re}(c_\alpha c_5) + b_{\alpha\beta} c_\alpha \overline{c_\beta} = 0. \tag{11.46}$$

Obviously, from the second equation of (11.46) we obtain the relation (11.40).

Moreover, it is well known from the classical theory of elasticity (see Kupradze et al. [219]) that from the first equation of (11.46) we may also obtain (11.42). Thus, we have desired result. □

Quite similarly we can prove the following.

Theorem 11.19 *If the condition (6.6) is satisfied, then the external BVP* $(\mathbb{K}^q)^-_{\mathbf{F},\mathbf{f}}$ *has one regular solution of the class* $\mathfrak{R}$, *where* $\mathbb{K} = I, II$.

11.6 Basic Properties of Potentials

On the basis of integral representation of regular vectors (see (11.28) and (11.29)) we introduce the following notation:

(i) $\mathbf{Q}^{(q,1)}(\mathbf{x},\mathbf{g}) = \int\limits_S \boldsymbol{\Psi}^{(q)}(\mathbf{x}-\mathbf{y},\omega)\mathbf{g}(\mathbf{y})d_{\mathbf{y}}S,$

(ii) $\mathbf{Q}^{(q,2)}(\mathbf{x},\mathbf{g}) = \int\limits_S [\tilde{\mathbf{P}}(\mathbf{D}_{\mathbf{y}},\mathbf{n}(\mathbf{y}))\boldsymbol{\Psi}^{(q)^\top}(\mathbf{x}-\mathbf{y},\omega)]^\top \mathbf{g}(\mathbf{y})d_{\mathbf{y}}S,$

(iii) $\mathbf{Q}^{(q,3)}(\mathbf{x},\boldsymbol{\phi},\Omega^\pm) = \int\limits_{\Omega^\pm} \boldsymbol{\Psi}^{(q)}(\mathbf{x}-\mathbf{y},\omega)\boldsymbol{\phi}(\mathbf{y})d\mathbf{y},$

where $\boldsymbol{\Psi}^{(q)}(\mathbf{x},\omega)$ is the fundamental matrix of the operator $\mathbf{A}^{(q)}(\mathbf{D}_{\mathbf{x}},\omega)$ and given by (11.10); the matrix differential operator $\tilde{\mathbf{P}}(\mathbf{D}_{\mathbf{y}},\mathbf{n}(\mathbf{y}))$ is defined by (9.32); $\mathbf{g}$ and $\boldsymbol{\phi}$ are eight-component vector functions.

As in the classical theory of thermoelasticity (see, e.g., Kupradze et al. [219]), the vector functions $\mathbf{Q}^{(q,1)}(\mathbf{x},\mathbf{g})$, $\mathbf{Q}^{(q,2)}(\mathbf{x},\mathbf{g})$, and $\mathbf{Q}^{(q,3)}(\mathbf{x},\boldsymbol{\phi},\Omega^\pm)$ are called *single-layer, double-layer*, and *volume potentials* in the steady vibration problems of the quasi-static linear theory of thermoelasticity for quadruple porosity materials.

On the basis of the properties of fundamental solution $\boldsymbol{\Psi}^{(q)}(\mathbf{x},\omega)$ (see Sect. 11.1) we have the following consequences.

Theorem 11.20 *If* $S \in C^{m+1,\nu}$, $\mathbf{g} \in C^{m,\nu'}(S)$, $0 < \nu' < \nu \le 1$, *and m is a nonnegative integer, then:*

(a)

$$\mathbf{Q}^{(q,1)}(\cdot,\mathbf{g}) \in C^{0,\nu'}(\mathbb{R}^3) \cap C^{m+1,\nu'}(\overline{\Omega^\pm}) \cap C^\infty(\Omega^\pm),$$

(b)

$$\mathbf{A}^{(q)}(\mathbf{D}_{\mathbf{x}},\omega)\,\mathbf{Q}^{(q,1)}(\mathbf{x},\mathbf{g}) = \mathbf{0},$$

(c)

$$\left\{\mathbf{P}(\mathbf{D_z}, \mathbf{n}(\mathbf{z}))\, \mathbf{Q}^{(q,1)}(\mathbf{z}, \mathbf{g})\right\}^{\pm} = \mp\frac{1}{2}\,\mathbf{g}(\mathbf{z}) + \mathbf{P}(\mathbf{D_z}, \mathbf{n}(\mathbf{z}))\, \mathbf{Q}^{(q,1)}(\mathbf{z}, \mathbf{g}), \tag{11.47}$$

(d)

$$\mathbf{P}(\mathbf{D_z}, \mathbf{n}(\mathbf{z}))\, \mathbf{Q}^{(q,1)}(\mathbf{z}, \mathbf{g})$$

is a singular integral, where $\mathbf{z} \in S$, $\mathbf{x} \in \Omega^{\pm}$ *and*

$$\left\{\mathbf{P}(\mathbf{D_z}, \mathbf{n}(\mathbf{z}))\, \mathbf{Q}^{(q,1)}(\mathbf{z}, \mathbf{g})\right\}^{\pm} \equiv \lim_{\Omega^{\pm}\ni\mathbf{x}\to\ \mathbf{z}\in S} \mathbf{P}(\mathbf{D_x}, \mathbf{n}(\mathbf{z}))\, \mathbf{Q}^{(q,1)}(\mathbf{x}, \mathbf{g}),$$

(e) the single-layer potential $\mathbf{Q}^{(q,1)}(\mathbf{x}, \mathbf{g}) \equiv \mathbf{0}$ *for* $\mathbf{x} \in \Omega^{+}$ *(or* $\mathbf{x} \in \Omega^{-}$*) if and only if* $\mathbf{g}(\mathbf{z}) \equiv \mathbf{0}$*, where* $\mathbf{z} \in S$.

Theorem 11.21 *If* $S \in C^{m+1,\nu}$, $\mathbf{g} \in C^{m,\nu'}(S)$, $0 < \nu' < \nu \le 1$, *then:*

(a)

$$\mathbf{Q}^{(q,2)}(\cdot, \mathbf{g}) \in C^{m,\nu'}(\overline{\Omega^{\pm}}) \cap C^{\infty}(\Omega^{\pm}),$$

(b)

$$\mathbf{A}^{(q)}(\mathbf{D_x}, \omega)\, \mathbf{Q}^{(q,2)}(\mathbf{x}, \mathbf{g}) = \mathbf{0},$$

(c)

$$\left\{\mathbf{Q}^{(q,2)}(\mathbf{z}, \mathbf{g})\right\}^{\pm} = \pm\frac{1}{2}\,\mathbf{g}(\mathbf{z}) + \mathbf{Q}^{(q,2)}(\mathbf{z}, \mathbf{g}), \tag{11.48}$$

for the nonnegative integer m,

(d) $\mathbf{Q}^{(q,2)}(\mathbf{z}, \mathbf{g})$ *is a singular integral, where* $\mathbf{z} \in S$,

(e)

$$\left\{\mathbf{P}(\mathbf{D_z}, \mathbf{n}(\mathbf{z}))\, \mathbf{Q}^{(q,2)}(\mathbf{z}, \mathbf{g})\right\}^{+} = \left\{\mathbf{P}(\mathbf{D_z}, \mathbf{n}(\mathbf{z}))\, \mathbf{Q}^{(q,2)}(\mathbf{z}, \mathbf{g})\right\}^{-},$$

for the natural number m, *where* $\mathbf{z} \in S$, $\mathbf{x} \in \Omega^{\pm}$ *and*

$$\left\{\mathbf{Q}^{(q,2)}(\mathbf{z}, \mathbf{g})\right\}^{\pm} \equiv \lim_{\Omega^{\pm}\ni\mathbf{x}\to\ \mathbf{z}\in S} \mathbf{Q}^{(q,2)}(\mathbf{x}, \mathbf{g}).$$

Theorem 11.22 *If* $S \in C^{1,\nu}$, $\boldsymbol{\phi} \in C^{0,\nu'}(\Omega^+)$, $0 < \nu' < \nu \leq 1$, *then:*

(a)

$$\mathbf{Q}^{(q,3)}(\cdot, \boldsymbol{\phi}, \Omega^+) \in C^{1,\nu'}(\mathbb{R}^3) \cap C^2(\Omega^+) \cap C^{2,\nu'}\left(\overline{\Omega_0^+}\right),$$

(b)

$$\mathbf{A}^{(q)}(\mathbf{D_x}, \omega)\, \mathbf{Q}^{(q,3)}(\mathbf{x}, \boldsymbol{\phi}, \Omega^+) = \boldsymbol{\phi}(\mathbf{x}),$$

where $\mathbf{x} \in \Omega^+$, Ω_0^+ *is a domain in* $\mathbb{R}^3$ *and* $\overline{\Omega_0^+} \subset \Omega^+$.

Theorem 11.23 *If* $S \in C^{1,\nu}$, $\operatorname{supp}\boldsymbol{\phi} = \Omega \subset \Omega^-$, $\boldsymbol{\phi} \in C^{0,\nu'}(\Omega^-)$, $0 < \nu' < \nu \leq 1$, *then:*

(a)

$$\mathbf{Q}^{(q,3)}(\cdot, \boldsymbol{\phi}, \Omega^-) \in C^{1,\nu'}(\mathbb{R}^3) \cap C^2(\Omega^-) \cap C^{2,\nu'}(\overline{\Omega_0^-}),$$

(b)

$$\mathbf{A}^{(q)}(\mathbf{D_x}, \omega)\, \mathbf{Q}^{(q,3)}(\mathbf{x}, \boldsymbol{\phi}, \Omega^-) = \boldsymbol{\phi}(\mathbf{x}),$$

where $\mathbf{x} \in \Omega^-$, Ω *is a finite domain in* $\mathbb{R}^3$ *and* $\overline{\Omega_0^-} \subset \Omega^-$.

11.7 Singular Integral Operators

We introduce the notation

$$\mathcal{H}^{(q,1)}\, \mathbf{g}(\mathbf{z}) \equiv \frac{1}{2}\, \mathbf{g}(\mathbf{z}) + \mathbf{Q}^{(q,2)}(\mathbf{z}, \mathbf{g}),$$

$$\mathcal{H}^{(q,2)}\, \mathbf{g}(\mathbf{z}) \equiv -\frac{1}{2}\, \mathbf{g}(\mathbf{z}) + \mathbf{P}(\mathbf{D_z}, \mathbf{n}(\mathbf{z}))\mathbf{Q}^{(q,1)}(\mathbf{z}, \mathbf{g}),$$

$$\mathcal{H}^{(q,3)}\mathbf{g}(\mathbf{z}) \equiv -\frac{1}{2}\, \mathbf{g}(\mathbf{z}) + \mathbf{Q}^{(q,2)}(\mathbf{z}, \mathbf{g}), \tag{11.49}$$

$$\mathcal{H}^{(q,4)}\mathbf{g}(\mathbf{z}) \equiv \frac{1}{2}\, \mathbf{g}(\mathbf{z}) + \mathbf{P}(\mathbf{D_z}, \mathbf{n}(\mathbf{z}))\mathbf{Q}^{(q,1)}(\mathbf{z}, \mathbf{g}),$$

$$\mathcal{H}_\chi^{(q)}\mathbf{g}(\mathbf{z}) \equiv \frac{1}{2}\, \mathbf{g}(\mathbf{z}) + \chi\, \mathbf{Q}^{(q,2)}(\mathbf{z}, \mathbf{g})$$

for $\mathbf{z} \in S$, where χ is a complex number. Obviously, on the basis of Theorems 11.20 and 11.21, $\mathcal{H}^{(q,j)}$ and $\mathcal{H}_\chi^{(q)}$ are the singular integral operators ($j = 1, 2, 3, 4$).

Let $\hat{\boldsymbol{\sigma}}^{(q,j)} = (\hat{\sigma}_{lm}^{(q,j)})_{8\times 8}$ be the symbol of the singular integral operator $\mathcal{H}^{(q,j)}$ ($j = 1, 2, 3, 4$). Taking into account (7.9), (11.13), and (11.49) we may see that

$$\det \hat{\boldsymbol{\sigma}}^{(q,1)} = -\det \hat{\boldsymbol{\sigma}}^{(q,2)} = -\det \hat{\boldsymbol{\sigma}}^{(q,3)} = \det \hat{\boldsymbol{\sigma}}^{(q,4)}$$
$$= \frac{1}{256}\left[1 - \frac{\mu^2}{(\lambda + 2\mu)^2}\right] = \frac{(\lambda + \mu)(\lambda + 3\mu)}{128(\lambda + 2\mu)^2} > 0. \tag{11.50}$$

Hence, the operator $\mathcal{H}^{(q,j)}$ is of the normal type, where $j = 1, 2, 3, 4$.

Let $\hat{\boldsymbol{\sigma}}_\chi^{(q)}$ and $\operatorname{ind} \mathcal{H}_\chi^{(q)}$ be the symbol and the index of the operator $\mathcal{H}_\chi^{(q)}$, respectively. It may be easily shown that

$$\det \hat{\boldsymbol{\sigma}}_\chi^{(q)} = \frac{(\lambda + 2\mu)^2 - \mu^2\chi^2}{256(\lambda + 2\mu)^2}$$

and $\det \hat{\boldsymbol{\sigma}}_\chi^{(q)}$ vanishes only at two points χ_1 and χ_2 of the complex plane. By virtue of (11.50) and $\det \hat{\boldsymbol{\sigma}}_1^{(q)} = \det \hat{\boldsymbol{\sigma}}^{(q,1)}$ we get $\chi_j \neq 1$ ($j = 1, 2$) and

$$\operatorname{ind} \mathcal{H}_1^{(q)} = \operatorname{ind} \mathcal{H}^{(q,1)} = \operatorname{ind} \mathcal{H}_0^{(q)} = 0.$$

Quite similarly we obtain

$$\operatorname{ind} \mathcal{H}^{(q,2)} = -\operatorname{ind} \mathcal{H}^{(q,3)} = 0, \qquad \operatorname{ind} \mathcal{H}^{(q,4)} = -\operatorname{ind} \mathcal{H}^{(q,1)} = 0.$$

Thus, the singular integral operator $\mathcal{H}^{(q,j)}$ ($j = 1, 2, 3, 4$) is of the normal type with an index equal to zero. Consequently, Fredholm's theorems are valid for the singular integral operator $\mathcal{H}^{(q,j)}$.

11.8 Existence Theorems

By Theorems 11.22 and 11.23 the volume potential $\mathbf{Q}^{(q,3)}(\mathbf{x}, \mathbf{F}, \Omega^\pm)$ is a regular solution of the nonhomogeneous equation (1.33), where $\mathbf{F} \in C^{0,\nu'}(\Omega^\pm)$, $0 < \nu' \leq 1$ and $\operatorname{supp} \mathbf{F}$ is a finite domain in Ω^-. Therefore, further we will consider problems $(\mathbb{K}^q)^+_{\mathbf{0},\mathbf{f}}$ and $(\mathbb{K}^q)^-_{\mathbf{0},\mathbf{f}}$, where $\mathbb{K} = I, II$.

Now we prove the existence theorems of a regular (classical) solution of the class $\mathfrak{R}$ for these BVPs.

Problem $(I^q)^+_{\mathbf{0},\mathbf{f}}$ We seek a regular solution of the class $\mathfrak{R}$ to this problem in the form of the double-layer potential

$$\mathbf{U}(\mathbf{x}) = \mathbf{Q}^{(q,2)}(\mathbf{x}, \mathbf{g}) \qquad \text{for} \qquad \mathbf{x} \in \Omega^+, \tag{11.51}$$

where $\mathbf{g}$ is the required eight-component vector function.

Obviously, by Theorem 11.21 the vector function $\mathbf{U}$ is a solution of the homogeneous equation (11.30) for $\mathbf{x} \in \Omega^+$. Keeping in mind the boundary condition (9.41) and using (11.48), from (11.51) we obtain, for determining the unknown vector $\mathbf{g}$, a singular integral equation

$$\mathcal{H}^{(q,1)}\,\mathbf{g}(\mathbf{z}) = \mathbf{f}(\mathbf{z}) \qquad \text{for} \quad \mathbf{z} \in S. \tag{11.52}$$

We prove that Eq. (11.52) is always solvable for an arbitrary vector $\mathbf{f}$.

Let us consider the associate homogeneous equation

$$\mathcal{H}^{(q,4)}\,\mathbf{h}(\mathbf{z}) = \mathbf{0} \qquad \text{for} \quad \mathbf{z} \in S, \tag{11.53}$$

where $\mathbf{h}$ is the required eight-component vector function. Now we prove that (11.53) has only the trivial solution.

Indeed, let $\mathbf{h}_0$ be a solution of the homogeneous equation (11.53). On the basis of Theorem 11.20 and Eq. (11.53) the vector function $\mathbf{U}_0(\mathbf{x}) = \mathbf{Q}^{(q,1)}(\mathbf{x}, \mathbf{h}_0)$ is a regular solution of the external homogeneous BVP $(II^q)^-_{\mathbf{0},\mathbf{0}}$. Using Theorem 11.19, the problem $(II^q)^-_{\mathbf{0},\mathbf{0}}$ has only the trivial solution, that is

$$\mathbf{U}_0(\mathbf{x}) \equiv \mathbf{0} \qquad \text{for} \qquad \mathbf{x} \in \Omega^-. \tag{11.54}$$

On the other hand, by Theorem 11.20 and (11.54) we get

$$\{\mathbf{U}_0(\mathbf{z})\}^+ = \{\mathbf{U}_0(\mathbf{z})\}^- = \mathbf{0} \qquad \text{for} \qquad \mathbf{z} \in S,$$

i.e., the vector $\mathbf{U}_0(\mathbf{x})$ is a regular solution of problem $(I^q)^+_{\mathbf{0},\mathbf{0}}$. By virtue of Theorem 11.17 the problem $(I^q)^+_{\mathbf{0},\mathbf{0}}$ has only the trivial solution, that is

$$\mathbf{U}_0(\mathbf{x}) \equiv \mathbf{0} \qquad \text{for} \qquad \mathbf{x} \in \Omega^+. \tag{11.55}$$

On the basis of (11.54), (11.55), and identity (11.47) we obtain

$$\mathbf{h}_0(\mathbf{z}) = \{\mathbf{P}(\mathbf{D}_\mathbf{z}, \mathbf{n})\mathbf{U}_0(\mathbf{z})\}^- - \{\mathbf{P}(\mathbf{D}_\mathbf{z}, \mathbf{n})\mathbf{U}_0(\mathbf{z})\}^+ = \mathbf{0} \qquad \text{for} \qquad \mathbf{z} \in S.$$

Thus, the homogeneous equation (11.53) has only the trivial solution and therefore on the basis of Fredholm's theorem the integral equation (11.52) is always solvable for an arbitrary vector $\mathbf{f}$. We have thereby proved the following.

Theorem 11.24 *If $S \in C^{2,\nu}$, $\mathbf{f} \in C^{1,\nu'}(S)$, $0 < \nu' < \nu \leq 1$, then a regular solution of the class $\mathfrak{R}$ of the internal BVP $(I^q)^+_{\mathbf{0},\mathbf{f}}$ exists, is unique, and is* *represented by double-layer potential (11.51), where* **g** *is a solution of the singular integral equation (11.52) which is always solvable for an arbitrary vector* **f**.

Problem $(II^q)^-_{\mathbf{0},\mathbf{f}}$ We seek a regular solution of the class $\mathfrak{R}$ to this problem in the form

$$\mathbf{U}(\mathbf{x}) = \mathbf{Q}^{(q,1)}(\mathbf{x}, \mathbf{h}) \quad \text{for} \quad \mathbf{x} \in \Omega^-, \tag{11.56}$$

where **h** is the required eight-component vector function.

Obviously, by virtue of Theorem 11.20 the vector function **U** is a solution of (11.30) for $\mathbf{x} \in \Omega^-$. Keeping in mind the boundary condition (9.44) and using (11.47), from (11.56) we obtain, for determining the unknown vector **h**, a singular integral equation

$$\mathcal{H}^{(q,4)}\mathbf{h}(\mathbf{z}) = \mathbf{f}(\mathbf{z}) \quad \text{for} \quad \mathbf{z} \in S. \tag{11.57}$$

It has been proved above that the corresponding homogeneous equation (11.53) has only the trivial solution. Hence, it follows that (11.57) is always solvable.

We have thereby proved the following.

Theorem 11.25 *If* $S \in C^{2,\nu}$, $\mathbf{f} \in C^{0,\nu'}(S)$, $0 < \nu' < \nu \leq 1$, *then a regular solution of the class* $\mathfrak{R}$ *of the external BVP* $(II^q)^-_{\mathbf{0},\mathbf{f}}$ *exists, is unique, and is represented by single-layer potential (11.56), where* **h** *is a solution of the singular integral equation (11.57) which is always solvable for an arbitrary vector* **f**.

Problem $(I^q)^-_{\mathbf{0},\mathbf{f}}$ We seek a regular solution of the class $\mathfrak{R}$ to this problem in the sum of the double-layer and single-layer potentials

$$\mathbf{U}(\mathbf{x}) = \mathbf{Q}^{(q,2)}(\mathbf{x}, \mathbf{g}) + (1 - i)\mathbf{Q}^{(q,1)}(\mathbf{x}, \mathbf{g}) \quad \text{for} \quad \mathbf{x} \in \Omega^-, \tag{11.58}$$

where **g** is the required eight-component vector function.

Obviously, by Theorems 11.20 and 11.21 the vector function **U** is a solution of (11.30) for $\mathbf{x} \in \Omega^-$. Keeping in mind the boundary condition (9.43) and using (11.48), from (11.58) we obtain, for determining the unknown vector **g**, a singular integral equation

$$\mathcal{H}^{(q,5)}\,\mathbf{g}(\mathbf{z}) \equiv \mathcal{H}^{(q,3)}\,\mathbf{g}(\mathbf{z}) + (1 - i)\mathbf{Q}^{(q,1)}(\mathbf{z}, \mathbf{g}) = \mathbf{f}(\mathbf{z}) \quad \text{for} \quad \mathbf{z} \in S. \tag{11.59}$$

Clearly, the singular integral operator $\mathcal{H}^{(q,5)}$ is of the normal type and ind $\mathcal{H}^{(q,5)} =$ ind $\mathcal{H}^{(q,3)} = 0$.

Let us consider the homogeneous equation

$$\mathcal{H}^{(q,5)}\,\mathbf{g}_0(\mathbf{z}) = \mathbf{0} \quad \text{for} \quad \mathbf{z} \in S. \tag{11.60}$$

Quite similarly as in Theorem 7.10 we may conclude that (11.60) has only a trivial solution and therefore on the basis of Fredholm's theorem the singular integral equation (11.59) is always solvable for an arbitrary vector **f**. We have thereby proved the following.

Theorem 11.26 *If* $S \in C^{2,\nu}$, $\mathbf{f} \in C^{1,\nu'}(S)$, $0 < \nu' < \nu \leq 1$, *then a regular solution of the class* $\mathfrak{B}$ *of the external BVP* $(I_q)^-_{\mathbf{0},\mathbf{f}}$ *exists, is unique, and is represented by sum of double-layer and single-layer potentials (11.58), where* **g** *is a solution of the singular integral equation (11.59) which is always solvable for an arbitrary vector* **f**.

Problem $(II^q)^+_{\mathbf{0},\mathbf{f}}$ We seek a regular solution of the class $\mathfrak{R}$ to this problem in the form

$$\mathbf{U}(\mathbf{x}) = \mathbf{Q}^{(q,1)}(\mathbf{x}, \mathbf{g}) \qquad \text{for} \qquad \mathbf{x} \in \Omega^+, \tag{11.61}$$

where **g** is the required eight-component vector function.

Obviously, by virtue of Theorem 11.20 the vector function **U** is a solution of (11.30) for $\mathbf{x} \in \Omega^+$. Keeping in mind the boundary condition (9.42) and using (11.47), from (11.61) we obtain, for determining the unknown vector **g**, the following singular integral equation

$$\mathcal{H}^{(q,2)}\mathbf{g}(\mathbf{z}) = \mathbf{f}(\mathbf{z}) \qquad \text{for} \qquad \mathbf{z} \in S. \tag{11.62}$$

To investigate the solvability of (11.62) we consider the homogeneous equation

$$\mathcal{H}^{(q,2)}\mathbf{g}(\mathbf{z}) = \mathbf{0} \qquad \text{for} \qquad \mathbf{z} \in S. \tag{11.63}$$

The homogeneous adjoint integral equation of (11.63) has the following form

$$\mathcal{H}^{(q,3)}\mathbf{h}(\mathbf{z}) = \mathbf{0} \qquad \text{for} \qquad \mathbf{z} \in S, \tag{11.64}$$

where **h** is a eight-component vector function.

We introduce the following eight-component vector functions:

$$\begin{aligned}
&\tilde{\chi}^{(q,1)}(\mathbf{x}) = (1,0,0,0,0,0,0,0), &&\tilde{\chi}^{(q,2)}(\mathbf{x}) = (0,1,0,0,0,0,0,0),\\
&\tilde{\chi}^{(q,3)}(\mathbf{x}) = (0,0,1,0,0,0,0,0), &&\tilde{\chi}^{(q,4)}(\mathbf{x}) = (0,-x_3,x_2,0,0,0,0,0),\\
&\tilde{\chi}^{(q,5)}(\mathbf{x}) = (x_3,0,-x_1,0,0,0,0,0), &&\tilde{\chi}^{(q,6)}(\mathbf{x}) = (-x_2,x_1,0,0,0,0,0,0).
\end{aligned} \tag{11.65}$$

Obviously, the system of vectors $\{\tilde{\chi}^{(q,j)}\}_{j=1}^{6}$ is linearly independent. We have the following.

Lemma 11.1 *The homogeneous equations (11.63) and (11.64) have six linearly independent solutions each and they constitute complete system of solutions.*

Proof By Theorem 11.18 each vector $\tilde{\boldsymbol{\chi}}^{(q,j)}$ $(j = 1, 2, \cdots, 6)$ is a regular solution of problem $(II^q)^+_{\mathbf{0},\mathbf{0}}$, i.e.,

$$\begin{aligned} &\mathbf{A}^{(q)}(\mathbf{D_x}, \omega)\, \tilde{\boldsymbol{\chi}}^{(q,j)}(\mathbf{x}) = \mathbf{0} \qquad \text{for} \quad \mathbf{x} \in \Omega^+, \\ &\left\{\mathbf{P}(\mathbf{D_z}, \mathbf{n})\tilde{\boldsymbol{\chi}}^{(q,j)}(\mathbf{z})\right\}^+ = \mathbf{0} \qquad \text{for} \quad \mathbf{z} \in S. \end{aligned} \tag{11.66}$$

On the other hand, by virtue of (11.66) from formula of integral representation of regular vector (11.28) we obtain

$$\tilde{\boldsymbol{\chi}}^{(q,j)}(\mathbf{x}) = \mathbf{Q}^{(q,2)}(\mathbf{x}, \tilde{\boldsymbol{\chi}}^{(q,j)}) \qquad \text{for} \quad \mathbf{x} \in \Omega^+. \tag{11.67}$$

On the basis of boundary property of double-layer potential (11.48) from (11.67) we have

$$\tilde{\boldsymbol{\chi}}^{(q,j)}(\mathbf{z}) = \frac{1}{2}\tilde{\boldsymbol{\chi}}^{(q,j)}(\mathbf{z}) + \mathbf{Q}^{(q,2)}(\mathbf{z}, \tilde{\boldsymbol{\chi}}^{(q,j)}) \qquad \text{for} \quad \mathbf{z} \in S, \ j = 1, 2, \cdots, 6$$

and therefore, $\tilde{\boldsymbol{\chi}}^{(q,j)}(\mathbf{x})$ is a solution of Eq. (11.64). Using Fredholm's theorem the singular integral equations (11.63) and (11.64) have six linearly independent solutions each.

It will now be shown that $\{\tilde{\boldsymbol{\chi}}^{(q,j)}\}_{j=1}^6$ is a complete system of linearly independent solutions of (11.64).

Let $\{\mathbf{h}^{(j)}(\mathbf{z})\}_{j=1}^m$ is the complete system of linearly independent solutions of homogeneous equation (11.64), where $m \geq 6$. We construct single-layer potentials $\mathbf{Q}^{(q,1)}(\mathbf{x}, \mathbf{h}^{(j)})$ $(j = 1, 2, \cdots, m)$. By Theorem 11.20 from identity

$$\mathbf{0} = \sum_{j=1}^m r_j \mathbf{Q}^{(q,1)}(\mathbf{x}, \mathbf{h}^{(j)}) = \mathbf{Q}^{(q,1)}(\mathbf{x}, \sum_{j=1}^m r_j \mathbf{h}^{(j)}) \qquad \text{for} \quad \mathbf{x} \in \Omega^+$$

it follows that

$$\sum_{j=1}^m r_j \mathbf{h}^{(j)}(\mathbf{z}) = \mathbf{0}, \tag{11.68}$$

where $r_1, r_2, \cdots, r_m$ are arbitrary constants, and from (11.68) we have $r_1 = r_2 = \cdots = r_m = 0$. Hence, $\left\{\mathbf{Q}^{(q,1)}(\mathbf{x}, \mathbf{h}^{(j)})\right\}_{j=1}^m$ is the system of linearly independent vectors.

On the other hand, it is easy to see that each vector $\mathbf{Q}^{(q,1)}(\mathbf{x}, \mathbf{h}^{(j)})$ $(j = 1, 2, \cdots, m)$ is a regular solution of problem $(II^q)^+_{\mathbf{0},\mathbf{0}}$. On the basis of Theorem 11.18, vector $\mathbf{Q}^{(q,1)}(\mathbf{x}, \mathbf{h}^{(j)})$ can be written as follows

$$\mathbf{Q}^{(q,1)}(\mathbf{x}, \mathbf{h}^{(j)}) = \sum_{j=1}^{6} r_{jl} \tilde{\chi}^{(1j)}(\mathbf{x}) \qquad \text{for} \quad \mathbf{x} \in \Omega^+,$$

where r_{jl} $(j = 1, 2, \cdots, m, \ l = 1, 2, \cdots, 6)$ are constants. Hence, each vector of system $\left\{\mathbf{Q}^{(q,1)}(\mathbf{x}, \mathbf{h}^{(j)})\right\}_{j=1}^{m}$ is represented by six linearly independent vectors $\tilde{\chi}^{(q,1)}, \tilde{\chi}^{(q,2)}, \cdots, \tilde{\chi}^{(q,6)}$. Thus, $m = 6$. □

By virtue of the Fredholm third theorem, for Eq. (11.62) to be solvable it is necessary and sufficient that

$$\int_S \mathbf{f}(\mathbf{z}) \tilde{\chi}^{(q,j)}(\mathbf{z}) d_{\mathbf{z}} S = \mathbf{0} \qquad \text{for} \quad j = 1, 2, \cdots, 6. \tag{11.69}$$

Then, employing (11.65), we may derive from (11.69)

$$\int_S \mathbf{f}^{(q)}(\mathbf{z}) d_{\mathbf{z}} S = \mathbf{0}, \qquad \int_S \left[\mathbf{f}^{(q)}(\mathbf{z}) \times \mathbf{z}\right] d_{\mathbf{z}} S = \mathbf{0}, \tag{11.70}$$

where $\mathbf{f}^{(q)} = (f_1, f_2, f_3)$. We have thereby proved the following.

Theorem 11.27 *If $S \in C^{2,\nu}$, $\mathbf{f} \in C^{0,\nu'}(S)$, $0 < \nu' < \nu \leq 1$, then the problem $(II^q)^+_{\mathbf{0},\mathbf{f}}$ is solvable only when condition (11.70) is fulfilled. The solution of this problem is represented by a single-layer potential (11.61), where $\mathbf{g}$ is a solution of the solvable singular integral equation (11.62).*

11.9 Solution of the First BVPs by Fredholm's Integral Equations

In Sect. 11.8, the first BVPs $(I^q)^+_{\mathbf{0},\mathbf{f}}$ and $(I^q)^-_{\mathbf{0},\mathbf{f}}$ are investigated by means of singular integral equations. In this section these problems will be also investigated by Fredholm's integral equations. Indeed, the BVPs $(I^q)^+_{\mathbf{0},\mathbf{f}}$ and $(I^q)^-_{\mathbf{0},\mathbf{f}}$ will be reduced to the always solvable Fredholm's integral equations.

11.9.1 Basic Properties of the Pseudostress Operator in Thermoelasticity

The matrix differential operator $\mathbf{N}(\mathbf{D}_{\mathbf{x}}, \mathbf{n}(\mathbf{x}))$ is introduced at the end of Sect. 1.7 and has the following form

$$
\begin{aligned}
&\mathbf{N}(\mathbf{D_x},\mathbf{n}) = (N_{ms}(\mathbf{D_x},\mathbf{n}))_{8\times 8},\\
&N_{lj}(\mathbf{D_x},\mathbf{n}) = \mu\delta_{lj}\frac{\partial}{\partial \mathbf{n}} + (\lambda+\mu)n_l\frac{\partial}{\partial x_j} + \kappa_0\mathcal{M}_{lj}(\mathbf{D_x},\mathbf{n}),\\
&N_{l;\alpha+3}(\mathbf{D_x},\mathbf{n}) = -a_\alpha\, n_l, \qquad N_{l8}(\mathbf{D_x},\mathbf{n}) = -\varepsilon_0\, n_l, \qquad\qquad (11.71)\\
&N_{\alpha+3;\beta+3}(\mathbf{D_x},\mathbf{n}) = k_{\alpha\beta}\frac{\partial}{\partial \mathbf{n}}, \qquad N_{88}(\mathbf{D_x},\mathbf{n}) = k\frac{\partial}{\partial \mathbf{n}},\\
&N_{\alpha+3;j}(\mathbf{D_x},\mathbf{n}) = N_{\alpha+3;8}(\mathbf{D_x},\mathbf{n}) = N_{8m}(\mathbf{D_x},\mathbf{n}) = 0,
\end{aligned}
$$

where $l, j = 1, 2, 3,\ \alpha, \beta = 1, 2, 3, 4,\ m = 1, 2, \cdots, 7;\ \mathcal{M}_{lj}(\mathbf{D_x},\mathbf{n})$ and κ_0 are denoted by (1.48) and (2.62), respectively.

We now establish an important property of matrix $\mathbf{N}(\mathbf{D_x},\mathbf{n}(\mathbf{x}))\mathbf{\Psi}^{(0)}(\mathbf{x})$, where the matrix $\mathbf{\Psi}^{(0)}(\mathbf{x})$ is introduced in Theorem 9.5.

On the basis of (11.71) by direct calculation for the matrix

$$
\mathbf{N}(\mathbf{D_x},\mathbf{n}(\mathbf{x}))\mathbf{\Psi}^{(0)}(\mathbf{x}) = \left(\left[\mathbf{N}(\mathbf{D_x},\mathbf{n}(\mathbf{x}))\mathbf{\Psi}^{(0)}(\mathbf{x})\right]_{lj}\right)_{8\times 8}
$$

we obtain

$$
\begin{aligned}
&\left[\mathbf{N}(\mathbf{D_x},\mathbf{n}(\mathbf{x}))\mathbf{\Psi}^{(0)}(\mathbf{x})\right]_{lj}\\
&\quad = \frac{x_m n_m(\mathbf{x})\delta_{lj}}{4\pi\,|\mathbf{x}|^3} + \mu'(\kappa_0+\mu)\mathcal{M}_{lm}(\mathbf{D_x},\mathbf{n})\frac{x_m x_j}{|\mathbf{x}|^3},\\
&\left[\mathbf{N}(\mathbf{D_x},\mathbf{n}(\mathbf{x}))\mathbf{\Psi}^{(0)}(\mathbf{x})\right]_{l;\beta+3} = -\frac{a_\alpha k^*_{\alpha\beta}}{k_0}\, n_l(\mathbf{x})\tilde{\gamma}^{(5)}(\mathbf{x}),\\
&\left[\mathbf{N}(\mathbf{D_x},\mathbf{n}(\mathbf{x}))\mathbf{\Psi}^{(0)}(\mathbf{x})\right]_{l8} = -\frac{\varepsilon_0\, n_l}{k}\tilde{\gamma}^{(5)}(\mathbf{x}),\\
&\left[\mathbf{N}(\mathbf{D_x},\mathbf{n}(\mathbf{x}))\mathbf{\Psi}^{(0)}(\mathbf{x})\right]_{m+3;\beta+3} = \delta_{m\beta}\frac{\partial}{\partial \mathbf{n}(\mathbf{x})}\tilde{\gamma}^{(5)}(\mathbf{x}),\\
&\left[\mathbf{N}(\mathbf{D_x},\mathbf{n}(\mathbf{x}))\mathbf{\Psi}^{(0)}(\mathbf{x})\right]_{88} = \frac{\partial}{\partial \mathbf{n}(\mathbf{x})}\tilde{\gamma}^{(5)}(\mathbf{x}),\\
&\left[\mathbf{N}(\mathbf{D_x},\mathbf{n}(\mathbf{x}))\mathbf{\Psi}^{(0)}(\mathbf{x})\right]_{\beta+3;l} = \left[\mathbf{N}(\mathbf{D_x},\mathbf{n}(\mathbf{x}))\mathbf{\Psi}^{(0)}(\mathbf{x})\right]_{\beta+3;8}\\
&\quad = \left[\mathbf{N}(\mathbf{D_x},\mathbf{n}(\mathbf{x}))\mathbf{\Psi}^{(0)}(\mathbf{x})\right]_{8l} = \left[\mathbf{N}(\mathbf{D_x},\mathbf{n}(\mathbf{x}))\mathbf{\Psi}^{(0)}(\mathbf{x})\right]_{8;\beta+3} = 0,
\end{aligned}
$$

where $l, j = 1, 2, 3,\ m, \beta = 1, 2, 3, 4$ and $\tilde{\gamma}^{(5)}$ is given by (2.26).

On the basis of Theorem 2.17 we have the following.

Theorem 11.28 *If S is the Liapunov surface of classes $C^{1,\nu}$ $(0 < \nu \leq 1)$ and $\mathbf{x}, \mathbf{y} \in S$, then the relations*

$$\left[\mathbf{N}(\mathbf{D_x}, \mathbf{n}(\mathbf{x}))\boldsymbol{\Psi}^{(0)}(\mathbf{x}-\mathbf{y})\right]_{lj} = O\left(|\mathbf{x}-\mathbf{y}|^{-2+\nu}\right),$$

$$\left[\mathbf{N}(\mathbf{D_x}, \mathbf{n}(\mathbf{x}))\boldsymbol{\Psi}^{(0)}(\mathbf{x}-\mathbf{y})\right]_{l,m+3} = O\left(|\mathbf{x}-\mathbf{y}|^{-1}\right),$$

$$\left[\mathbf{N}(\mathbf{D_x}, \mathbf{n}(\mathbf{x}))\boldsymbol{\Psi}^{(0)}(\mathbf{x}-\mathbf{y})\right]_{\alpha+3,\beta+3} = O\left(|\mathbf{x}-\mathbf{y}|^{-2+\nu}\right),$$

$$\left[\mathbf{N}(\mathbf{D_x}, \mathbf{n}(\mathbf{x}))\boldsymbol{\Psi}^{(0)}(\mathbf{x}-\mathbf{y})\right]_{88} = O\left(|\mathbf{x}-\mathbf{y}|^{-2+\nu}\right),$$

hold in the neighborhood of $\mathbf{x} = \mathbf{y}$, *where* $l, j = 1, 2, 3,\ \alpha, \beta = 1, 2, 3, 4,\ m = 1, 2, \cdots, 5$.

Theorem 11.28 is proved similarly as Theorem 7.14. Hence, the matrix $\mathbf{N}(\mathbf{D_x}, \mathbf{n}(\mathbf{x}))\boldsymbol{\Psi}^{(0)}(\mathbf{x}-\mathbf{y})$ is weakly singular in the neighborhood of $\mathbf{x} = \mathbf{y}$ for $\mathbf{x}, \mathbf{y} \in S$.

Clearly, Theorem 11.8 leads to the following.

Theorem 11.29 *The relations*

$$\begin{aligned}&\left\{\mathbf{N}(\mathbf{D_x}, \mathbf{n}(\mathbf{x}))[\boldsymbol{\Psi}^{(q)}(\mathbf{x}, \omega)]^\top\right\}_{lj}\\ &\quad - \left\{\mathbf{N}(\mathbf{D_x}, \mathbf{n}(\mathbf{x}))\boldsymbol{\Psi}^{(0)}(\mathbf{x})\right\}_{lj} = O\left(|\mathbf{x}|^{-1}\right),\\ &l, j = 1, 2, \cdots, 8\end{aligned}$$

hold in the neighborhood of the origin.

Thus, the matrix $\mathbf{N}(\mathbf{D_x}, \mathbf{n}(\mathbf{x}))\boldsymbol{\Psi}^{(0)}(\mathbf{x})$ is the singular part of the matrix $\mathbf{N}(\mathbf{D_x}, \mathbf{n}(\mathbf{x}))\left[\boldsymbol{\Psi}^{(q)}(\mathbf{x}, \omega)\right]^\top$ in the neighborhood of the origin.

11.9.2 Auxiliary Boundary Value Problems

In the sequel, we consider the following auxiliary BVPs:

(a) Find a regular (classical) solution of the class $\mathfrak{R}$ in Ω^+ to system of the homogeneous equations (11.30) for $\mathbf{x} \in \Omega^+$ satisfying the boundary condition

$$\lim_{\Omega^+ \ni \mathbf{x} \to \mathbf{z} \in S} \mathbf{N}(\mathbf{D_x}, \mathbf{n}(\mathbf{z}))\mathbf{U}(\mathbf{x}) \equiv \{\mathbf{N}(\mathbf{D_z}, \mathbf{n})\mathbf{U}(\mathbf{z})\}^+ = \mathbf{f}(\mathbf{z}) \qquad (11.72)$$

in the *Problem* $(I^a)^+_{\mathbf{f}}$;

(b) Find a regular (classical) solution of the class $\mathfrak{R}$ to system (7.2) for $\mathbf{x} \in \Omega^-$ satisfying the boundary condition

$$\lim_{\Omega^- \ni \mathbf{x} \to \mathbf{z} \in S} \mathbf{N}(\mathbf{D_x}, \mathbf{n}(\mathbf{z}))\mathbf{U}(\mathbf{x}) \equiv \{\mathbf{N}(\mathbf{D_z}, \mathbf{n})\mathbf{U}(\mathbf{z})\}^- = \mathbf{f}(\mathbf{z}) \tag{11.73}$$

in the *Problem* $(I^a)^-_{\mathbf{f}}$. Here the matrix differential operator $\mathbf{N}(\mathbf{D_x}, \mathbf{n}(\mathbf{x}))$ is defined by (11.67) and $\mathbf{f}$ is prescribed eight-component vector function. Note that the class $\mathfrak{R}$ of vector functions is given by Definition 10.1 (see Sect. 10.3).

11.9.3 Uniqueness Theorems for the Auxiliary BVPs

We introduce the notation

$$E_1^{(\kappa_0)}(\mathbf{U}, \mathbf{u}) = W_0^{(\kappa_0)}(\mathbf{u}, \mathbf{u}) - (a_\alpha\, p_\alpha + \varepsilon_0 \theta)\, \mathrm{div}\, \overline{\mathbf{u}}, \tag{11.74}$$

where $W_0^{(\kappa_0)}(\mathbf{u}, \mathbf{u})$ and κ_0 are given by (7.45) and (2.62), respectively.

Obviously, taking into account (11.18) by simple calculation from (11.22) and (9.28) for $\mathbf{u}' = \mathbf{u}$, $\mathbf{p}' = \mathbf{p}$, $\theta' = \theta$ and $\rho = 0$ we obtain

$$\begin{aligned}
&\int_{\Omega^+} \Big[\mathbf{A}^{(q,1)}(\mathbf{D_x})\, \mathbf{U}(\mathbf{x}) \cdot \mathbf{u}(\mathbf{x}) + E_1^{(\kappa_0)}(\mathbf{U}, \mathbf{u})\Big]\, d\mathbf{x} \\
&\qquad = \int_S \mathbf{N}^{(1)}(\mathbf{D_z}, \mathbf{n})\mathbf{U}(\mathbf{z}) \cdot \mathbf{u}(\mathbf{z})\, d_{\mathbf{z}}S, \\
&\int_{\Omega^+} \Big[\mathbf{A}^{(2)}(\mathbf{D_x}, \omega)\, \mathbf{U}(\mathbf{x}) \cdot \mathbf{p}(\mathbf{x}) + E_2(\mathbf{U}, \mathbf{p})\Big]\, d\mathbf{x} \\
&\qquad = \int_S \mathbf{P}^{(2)}(\mathbf{D_z}, \mathbf{n})\mathbf{p}(\mathbf{z}) \cdot \mathbf{p}(\mathbf{z})\, d_{\mathbf{z}}S, \\
&\int_{\Omega^+} \Big[\mathbf{A}^{(3)}(\mathbf{D_x}, \omega)\, \mathbf{U}(\mathbf{x})\, \overline{\theta(\mathbf{x})} + E_3(\mathbf{U}, \theta)\Big]\, d\mathbf{x} \\
&\qquad = k \int_S \frac{\partial \theta(\mathbf{z})}{\partial \mathbf{n}(\mathbf{z})}\, \overline{\theta(\mathbf{z})}\, d_{\mathbf{z}}S.
\end{aligned} \tag{11.75}$$

where the matrix differential operators $\mathbf{A}^{(2)}(\mathbf{D_x}, \omega)$, $\mathbf{A}^{(3)}(\mathbf{D_x}, \omega)$, and $\mathbf{P}^{(2)}(\mathbf{D_z}, \mathbf{n})$ are defined in Sect. 9.3.1, $E_2(\mathbf{U}, \mathbf{p})$, and $E_3(\mathbf{U}, \theta)$ are given by (11.35),

$$\mathbf{N}^{(1)}(\mathbf{D_x}, \mathbf{n}) = \left(N_{lj}^{(1)}(\mathbf{D_x}, \mathbf{n})\right)_{3\times 8}, \qquad N_{lj}^{(1)}(\mathbf{D_x}, \mathbf{n}) = N_{lj}(\mathbf{D_x}, \mathbf{n}),$$

$$l = 1, 2, 3, \qquad j = 1, 2, \cdots, 8.$$

We are now in a position to prove the uniqueness theorems of regular (classical) solutions of the auxiliary BVPs $(I^a)_{\mathbf{f}}^+$ and $(I^a)_{\mathbf{f}}^-$. We have the following results.

Theorem 11.30 *If the condition*

$$\mu > 0, \qquad 3\lambda + 5\mu > 0 \tag{11.76}$$

is satisfied and the matrix $\hat{\mathbf{b}}$ *is a positive definite, then any two solutions of the internal BVP* $(I^a)_{\mathbf{f}}^+$ *of the class* $\mathfrak{R}$ *may differ only for an additive vector* $\mathbf{U} = (\mathbf{u}, \mathbf{p}, \theta)$, *where* $\mathbf{p}$ *and* θ *satisfy the condition (11.40), and*

$$\mathbf{u}(\mathbf{x}) = \mathbf{a}' \qquad \text{for} \quad \mathbf{x} \in \Omega^+, \tag{11.77}$$

and $\mathbf{a}' = (a_1', a_2', a_3')$ *is arbitrary complex constant three-component vector.*

Proof Suppose that there are two regular solutions of problem $(I^a)_{\mathbf{f}}^+$. Then, employing (11.72), their difference $\mathbf{U}$ is a regular solution of the internal homogeneous BVP $(I^a)_{\mathbf{0}}^+$. Hence, $\mathbf{U}$ is a regular solution of the system of homogeneous equations (11.30) in Ω^+ satisfying the homogeneous boundary condition

$$\{\mathbf{N}(\mathbf{D_z}, \mathbf{n})\mathbf{U}(\mathbf{z})\}^+ = \mathbf{0}. \tag{11.78}$$

On the basis of (11.30) and (11.78), from (11.75) we obtain

$$\begin{aligned} &\int_{\Omega^+} E_1^{(\kappa_0)}(\mathbf{U}, \mathbf{u})d\mathbf{x} = 0, \qquad \int_{\Omega^+} E_2(\mathbf{U}, \mathbf{p})d\mathbf{x} = 0, \\ &\int_{\Omega^+} E_3(\mathbf{U}, \theta)d\mathbf{x} = 0. \end{aligned} \tag{11.79}$$

Obviously, from (11.35) and (11.74) it follows that

$$\begin{aligned} &\operatorname{Im} E_1^{(\kappa_0)}(\mathbf{U}, \mathbf{u}) = \operatorname{Im}\left[(a_\alpha \overline{p_\alpha} + \varepsilon_0 \overline{\theta}) \operatorname{div} \mathbf{u}\right], \\ &\operatorname{Re} E_2(\mathbf{U}, \mathbf{p}) = k_{\alpha\beta} \nabla p_\beta \cdot \nabla p_\alpha + d_{\alpha\beta} p_\beta \overline{p_\alpha} \\ &\qquad\qquad + \omega \operatorname{Im}\left[(a_\alpha \operatorname{div} \mathbf{u} + \theta \varepsilon_\alpha)\, \overline{p_\alpha}\right], \\ &\operatorname{Re} E_3(\mathbf{U}, \theta) = k|\nabla\theta|^2 + \omega T_0 \operatorname{Im}\left[(\varepsilon_0 \operatorname{div} \mathbf{u} + \varepsilon_\alpha p_\alpha)\, \overline{\theta}\right]. \end{aligned} \tag{11.80}$$

From (11.80) we can easily verify that

$$\begin{aligned} &\text{Re}\, E_3(\mathbf{U},\theta) + T_0 \text{Re}\, E_2(\mathbf{U},\mathbf{p}) - \omega T_0 \text{Im}\, E_1^{(\kappa_0)}(\mathbf{U},\mathbf{u}) \\ &\quad = k|\nabla\theta|^2 + T_0(k_{\alpha\beta}\nabla p_\beta \cdot \nabla p_\alpha + d_{\alpha\beta} p_\beta \overline{p_\alpha}). \end{aligned} \tag{11.81}$$

Then, employing (11.81), we may derive from (11.79)

$$\int\limits_{\Omega^+} \left[k|\nabla\theta|^2 + T_0(k_{\alpha\beta}\nabla p_\beta \cdot \nabla p_\alpha + d_{\alpha\beta} p_\beta \overline{p_\alpha}) \right] d\mathbf{x} = 0.$$

On the basis of (4.10) from the last equation we get

$$k_{\alpha\beta}\nabla p_\beta \cdot \nabla p_\alpha = 0, \qquad d_{\alpha\beta} p_\beta \overline{p_\alpha} = 0, \qquad \nabla\theta = 0, \tag{11.82}$$

and hence, by using Lemma 6.1 from (11.82) we obtain

$$\begin{aligned} &p_\alpha(\mathbf{x}) = c_\alpha = \text{const}, \qquad \theta(\mathbf{x}) = c_5 = \text{const}, \\ &d_{\alpha\beta} p_\beta(\mathbf{x}) = 0, \qquad \alpha = 1,2,3,4 \end{aligned} \tag{11.83}$$

for $\mathbf{x} \in \Omega^+$.

Furthermore, if we employ Eq. (11.83), then from (11.30) we may obtain

$$\begin{aligned} b_{\alpha\beta} c_\beta + \varepsilon_\alpha c_5 &= -a_\alpha \,\text{div}\, \mathbf{u}, \\ \varepsilon_\beta c_\beta + a_0 c_5 &= -\varepsilon_0 \,\text{div}\, \mathbf{u} \end{aligned} \tag{11.84}$$

for $\alpha = 1,2,3,4$.

Clearly, if we now employ Eq. (11.84), then from (11.35) we have

$$E_1^{(\kappa_0)}(\mathbf{U},\mathbf{u}) = W_0^{(\kappa_0)}(\mathbf{u},\mathbf{u}) + (a_0|c_5|^2 + 2\varepsilon_\alpha \text{Re}(c_\alpha c_5) + b_{\alpha\beta} c_\alpha \overline{c_\beta}).$$

Applying the last relation from the first equation of (11.79) it follows that

$$\int\limits_{\Omega^+} \left[W_0^{(\kappa_0)}(\mathbf{u},\mathbf{u}) + (a_0|c_5|^2 + 2\varepsilon_\alpha \text{Re}(c_\alpha c_5) + b_{\alpha\beta} c_\alpha \overline{c_\beta}) \right] d\mathbf{x} = 0. \tag{11.85}$$

On the basis of (11.76) and the assumption that $\hat{\mathbf{b}}$ is a positive definite matrix, from (11.85) we get

$$W_0^{(\kappa_0)}(\mathbf{u},\mathbf{u}) = 0, \qquad a_0|c_5|^2 + 2\varepsilon_\alpha \text{Re}(c_\alpha c_5) + b_{\alpha\beta} c_\alpha \overline{c_\beta} = 0. \tag{11.86}$$

Obviously, from the second equation of (11.86) we obtain the relation (11.40).

Moreover, if we are using (7.45) and (11.76) from the first equation of (11.86) we may also obtain

$$\operatorname{div}\mathbf{u}(\mathbf{x}) = 0, \qquad \operatorname{curl}\mathbf{u}(\mathbf{x}) = \mathbf{0} \quad \text{for}\, \mathbf{x} \in \Omega^+. \tag{11.87}$$

Hence, $\mathbf{u}$ is a harmonic vector, i.e., $\Delta\mathbf{u}(\mathbf{x}) = \mathbf{0}$ for $\mathbf{x} \in \Omega^+$.

If we now return to the boundary condition (11.78), then by virtue of the relations (11.40) and (11.87) from identity

$$\begin{aligned}\mathbf{N}^{(1)}(\mathbf{D}_\mathbf{z}, \mathbf{n})\mathbf{U}(\mathbf{z}) &= (\mu + \kappa_0)\frac{\partial \mathbf{u}(\mathbf{z})}{\partial \mathbf{n}(\mathbf{z})} + (\lambda + \mu - \kappa_0)\mathbf{n}(\mathbf{z}) \operatorname{div}\mathbf{u}(\mathbf{z}) \\ &\quad + \kappa\, [\mathbf{n}(\mathbf{z}) \times \operatorname{curl}\mathbf{u}(\mathbf{z})] - [a_\alpha p_\alpha(\mathbf{z}) + \varepsilon_0\theta(\mathbf{z})]\, \mathbf{n}(\mathbf{z})\end{aligned} \tag{11.88}$$

we deduced that

$$\left\{\frac{\partial \mathbf{u}(\mathbf{z})}{\partial \mathbf{n}(\mathbf{z})}\right\}^+ = \mathbf{0}.$$

Therefore, $\mathbf{u}$ is a harmonic vector in Ω^+ satisfying Neumann's internal boundary condition. Consequently, the vector $\mathbf{u}$ is a constant vector, i.e., we have the relation (11.77). Thus, we have desired result. □

Theorem 11.31 *If the condition (11.76) is satisfied, then the external BVP* $(I^a)^-_\mathbf{f}$ *has one regular solution of the class* $\mathfrak{R}$.

Proof Suppose that there are two regular solutions of problem $(I^a)^-_\mathbf{f}$. Then, employing (11.73), their difference $\mathbf{U}$ is a regular solution of the internal homogeneous BVP $(I^a)^-_\mathbf{0}$. Hence, $\mathbf{U}$ is a regular solution of the system of homogeneous equations (11.30) in Ω^- satisfying the homogeneous boundary condition

$$\{\mathbf{N}(\mathbf{D}_\mathbf{z}, \mathbf{n})\mathbf{U}(\mathbf{z})\}^+ = \mathbf{0}. \tag{11.89}$$

On the basis of (11.30) and (11.89), we obtain

$$\begin{aligned}&\int_{\Omega^-} E_1^{(\kappa_0)}(\mathbf{U}, \mathbf{u}) d\mathbf{x} = 0, \qquad \int_{\Omega^-} E_2(\mathbf{U}, \mathbf{p}) d\mathbf{x} = 0, \\ &\int_{\Omega^-} E_3(\mathbf{U}, \theta) d\mathbf{x} = 0.\end{aligned} \tag{11.90}$$

In a similar manner as in Theorem 11.30 from (11.90) it follows that

$$p_j(\mathbf{x}) = c_j = \text{const}, \qquad d_{\alpha\beta} p_\beta \overline{p_\alpha} = 0 \qquad \text{for } \mathbf{x} \in \Omega^-,$$
$$j = 1, 2, 3, 4. \tag{11.91}$$

On the basis of the conditions at infinity (10.6) from (11.91) we get

$$\mathbf{p}(\mathbf{x}) = \mathbf{0} \qquad \theta(\mathbf{x}) = 0 \qquad \text{for} \quad \mathbf{x} \in \Omega^-. \tag{11.92}$$

Moreover, if we employ (11.92) and (11.76) from the first equation of (11.90) we can write

$$\text{div}\, \mathbf{u}(\mathbf{x}) = 0, \qquad \text{curl}\, \mathbf{u}(\mathbf{x}) = \mathbf{0} \qquad \text{for} \quad \mathbf{x} \in \Omega^-,$$

and therefore, we have

$$\Delta \mathbf{u}(\mathbf{x}) = \mathbf{0} \qquad \text{for} \quad \mathbf{x} \in \Omega^-.$$

If we now return to the boundary condition (11.89), then by virtue of the identity (11.88) we deduced that

$$\left\{ \frac{\partial \mathbf{u}(\mathbf{z})}{\partial \mathbf{n}(\mathbf{z})} \right\}^- = \mathbf{0}.$$

Hence, $\mathbf{u}$ is a harmonic vector in Ω^- satisfying Neumann's external boundary condition. On the basis of the conditions at infinity (10.9) we can write $\mathbf{u}(\mathbf{x}) = \mathbf{0}$ for $\mathbf{x} \in \Omega^-$. Thus, we have desired result. □

11.9.4 Basic Properties of Potential

We introduce the following potential:

$$\mathbf{Q}^{(q,4)}(\mathbf{x}, \mathbf{g}) = \int_S \{\tilde{\mathbf{N}}(\mathbf{D}_\mathbf{y}, \mathbf{n}(\mathbf{y}))[\mathbf{\Psi}^{(q)}(\mathbf{x} - \mathbf{y}, \omega)]^\top\}^\top \mathbf{g}(\mathbf{y}) d_\mathbf{y} S,$$

where $\mathbf{\Psi}^{(q)}(\mathbf{x}, \omega)$ is the fundamental matrix of the operator $\mathbf{A}^{(q)}(\mathbf{D}_\mathbf{x}, \omega)$ and defined by (11.10), $\mathbf{g}$ is eight-component vector function, and the matrix differential operator $\tilde{\mathbf{N}}(\mathbf{D}_\mathbf{y}, \mathbf{n}(\mathbf{y}))$ is given by

$$
\begin{aligned}
&\tilde{\mathbf{N}}(\mathbf{D_y}, \mathbf{n}) = \left(\tilde{N}_{ms}(\mathbf{D_y}, \mathbf{n})\right)_{8\times 8},\\
&\tilde{N}_{lj}(\mathbf{D_y}, \mathbf{n}) = \mu\delta_{lj}\frac{\partial}{\partial \mathbf{n}} + (\lambda+\mu)n_l\frac{\partial}{\partial y_j} + \kappa_0\mathcal{M}_{lj}(\mathbf{D_y}, \mathbf{n}),\\
&\tilde{N}_{l;\alpha+3}(\mathbf{D_y}, \mathbf{n}) = -i\omega a_\alpha\, n_l, \qquad \tilde{N}_{l8}(\mathbf{D_y}, \mathbf{n}) = -i\omega T_0\varepsilon_0\, n_l,\\
&\tilde{N}_{\alpha+3;\beta+3}(\mathbf{D_y}, \mathbf{n}) = k_{\alpha\beta}\frac{\partial}{\partial \mathbf{n}}, \qquad \tilde{N}_{88}(\mathbf{D_y}, \mathbf{n}) = k\frac{\partial}{\partial \mathbf{n}},\\
&\tilde{N}_{\alpha+3;j}(\mathbf{D_y}, \mathbf{n}) = \tilde{N}_{\alpha+3;8}(\mathbf{D_y}, \mathbf{n}) = \tilde{N}_{8m}(\mathbf{D_y}, \mathbf{n}) = 0,\\
&l, j = 1, 2, 3, \qquad \alpha, \beta = 1, 2, 3, 4, \qquad m = 1, 2, \cdots, 7.
\end{aligned}
\tag{11.93}
$$

Clearly, the operator $\tilde{\mathbf{N}}(\mathbf{D_y}, \mathbf{n}(\mathbf{y}))$ may be obtained from the operator $\mathbf{N}(\mathbf{D_y}, \mathbf{n}(\mathbf{y}))$ by replacing a_α and ε_0 by $i\omega a_\alpha$ and $i\omega T_0\varepsilon_0$ ($\alpha = 1, 2, 3, 4$), respectively.

Theorem 11.28 and the relation (11.93) lead to the following.

Theorem 11.32 *If S is the Liapunov surface of classes $C^{1,\nu}$ $(0 < \nu \le 1)$ and $\mathbf{x}, \mathbf{y} \in S$, then the relations*

$$
\begin{aligned}
&\left[\tilde{\mathbf{N}}(\mathbf{D_x}, \mathbf{n}(\mathbf{x}))\boldsymbol{\Psi}^{(0)}(\mathbf{x}-\mathbf{y})\right]_{lj} = O\left(|\mathbf{x}-\mathbf{y}|^{-2+\nu}\right),\\
&\left[\tilde{\mathbf{N}}(\mathbf{D_x}, \mathbf{n}(\mathbf{x}))\boldsymbol{\Psi}^{(0)}(\mathbf{x}-\mathbf{y})\right]_{l,m+3} = O\left(|\mathbf{x}-\mathbf{y}|^{-1}\right),\\
&\left[\tilde{\mathbf{N}}(\mathbf{D_x}, \mathbf{n}(\mathbf{x}))\boldsymbol{\Psi}^{(0)}(\mathbf{x}-\mathbf{y})\right]_{\alpha+3,\beta+3} = O\left(|\mathbf{x}-\mathbf{y}|^{-2+\nu}\right),\\
&\left[\tilde{\mathbf{N}}(\mathbf{D_x}, \mathbf{n}(\mathbf{x}))\boldsymbol{\Psi}^{(0)}(\mathbf{x}-\mathbf{y})\right]_{88} = O\left(|\mathbf{x}-\mathbf{y}|^{-2+\nu}\right),
\end{aligned}
$$

hold in the neighborhood of $\mathbf{x} = \mathbf{y}$, where $l, j = 1, 2, 3$, $\alpha, \beta = 1, 2, 3, 4$, $m = 1, 2, \cdots, 5$.

On the basis of Theorems 11.1–11.5 and 11.32 we have the following.

Theorem 11.33 *If $S \in C^{m+1,\nu}$, $\mathbf{g} \in C^{m,\nu'}(S)$, $0 < \nu' < \nu \le 1$, then:*

(a)

$$\mathbf{Q}^{(q,4)}(\cdot, \mathbf{g}) \in C^{m,\nu'}(\overline{\Omega^\pm}) \cap C^\infty(\Omega^\pm),$$

(b)

$$\mathbf{A}^{(q)}(\mathbf{D_x}, \omega)\, \mathbf{Q}^{(q,4)}(\mathbf{x}, \mathbf{g}) = \mathbf{0} \qquad for \quad \mathbf{x} \in \Omega^\pm,$$

(c)

$$\left\{\mathbf{Q}^{(q,4)}(\mathbf{z},\mathbf{g})\right\}^{\pm} = \pm\frac{1}{2}\mathbf{g}(\mathbf{z}) + \mathbf{Q}^{(q,4)}(\mathbf{z},\mathbf{g}), \tag{11.94}$$

for the nonnegative integer m,

(d) $\mathbf{Q}^{(q,4)}(\mathbf{z},\mathbf{g})$ *is an integral with weak singularity,*

(e)

$$\left\{\mathbf{N}(\mathbf{D_z},\mathbf{n}(\mathbf{z}))\,\mathbf{Q}^{(q,4)}(\mathbf{z},\mathbf{g})\right\}^{+} = \left\{\mathbf{N}(\mathbf{D_z},\mathbf{n}(\mathbf{z}))\,\mathbf{Q}^{(q,4)}(\mathbf{z},\mathbf{g})\right\}^{-},$$

for the natural number m,

(f)

$$\left\{\mathbf{N}(\mathbf{D_z},\mathbf{n}(\mathbf{z}))\mathbf{Q}^{(q,1)}(\mathbf{z},\mathbf{g})\right\}^{\pm} = \mp\frac{1}{2}\mathbf{g}(\mathbf{z}) + \mathbf{N}(\mathbf{D_z},\mathbf{n}(\mathbf{z}))\mathbf{Q}^{(q,1)}(\mathbf{z},\mathbf{g}), \tag{11.95}$$

where $\mathbf{z}\in S$.

We introduce the notation

$$\begin{aligned}
\tilde{\mathcal{R}}^{(q,1)}\mathbf{g}(\mathbf{z}) &\equiv \frac{1}{2}\mathbf{g}(\mathbf{z}) + \mathbf{Q}^{(q,4)}(\mathbf{z},\mathbf{g}),\\
\tilde{\mathcal{R}}^{(q,2)}\mathbf{g}(\mathbf{z}) &\equiv -\frac{1}{2}\mathbf{g}(\mathbf{z}) + \mathbf{Q}^{(q,4)}(\mathbf{z},\mathbf{g}),\\
\tilde{\mathcal{R}}^{(q,3)}\mathbf{g}(\mathbf{z}) &\equiv \frac{1}{2}\mathbf{g}(\mathbf{z}) + \mathbf{N}(\mathbf{D_z},\mathbf{n}(\mathbf{z}))\mathbf{Q}^{(q,1)}(\mathbf{z},\mathbf{g}),
\end{aligned} \tag{11.96}$$

for $\mathbf{z}\in S$, where $\mathbf{Q}^{(q,1)}(\mathbf{z},\mathbf{g})$ is a single-layer potential and defined in Sect. 11.6. Obviously, on the basis of Theorem 11.32, $\tilde{\mathcal{R}}^{(q,j)}$ is Fredholm's integral operator $(j = 1, 2, 3)$.

11.9.5 Existence Theorems

We are now in a position to prove the existence theorems of regular (classical) solutions of the BVPs $(I^q)^+_{\mathbf{0},\mathbf{f}}$ and $(I^q)^-_{\mathbf{0},\mathbf{f}}$ by always solvable Fredholm's integral equations.

Problem $(I_q)^+_{\mathbf{0},\mathbf{f}}$ We seek a regular solution of the class $\mathfrak{R}$ to this problem in the following form.

$$\mathbf{U}(\mathbf{x}) = \mathbf{Q}^{(q,4)}(\mathbf{x},\mathbf{g}) \quad \text{for} \quad \mathbf{x}\in\Omega^+, \tag{11.97}$$

where $\mathbf{g}$ is the required eight-component vector function.

Obviously, by Theorem 11.33 the vector function $\mathbf{U}$ is a solution of the homogeneous equation (11.30) for $\mathbf{x} \in \Omega^+$. Keeping in mind the boundary condition (9.41) and using (11.94) and (11.96), from (11.97) we obtain, determining the unknown vector $\mathbf{g}$, Fredholm's integral equation

$$\tilde{\mathcal{R}}^{(q,1)}\, \mathbf{g}(\mathbf{z}) = \mathbf{f}(\mathbf{z}) \qquad \text{for} \quad \mathbf{z} \in S. \tag{11.98}$$

We prove that Eq. (11.98) is always solvable for an arbitrary vector $\mathbf{f}$.

Let us consider the associate homogeneous Fredholm's integral equation

$$\tilde{\mathcal{R}}^{(q,3)}\, \mathbf{h}(\mathbf{z}) = \mathbf{0} \qquad \text{for} \quad \mathbf{z} \in S, \tag{11.99}$$

where $\mathbf{h}$ is the required eight-component vector function. Now we prove that (11.99) has only the trivial solution.

Indeed, let $\mathbf{h}_0$ be a solution of the homogeneous equation (11.99). On the basis of Theorem 11.33 and Eqs. (11.95) and (11.96) the vector function $\mathbf{U}_0(\mathbf{x}) = \mathbf{Q}^{(q,1)}(\mathbf{x}, \mathbf{h}_0)$ is a regular solution of the external homogeneous BVP $(I^a)^-_{\mathbf{0}}$. Using Theorem 11.31, the problem $(I^a)^-_{\mathbf{0}}$ has only the trivial solution, that is

$$\mathbf{U}_0(\mathbf{x}) \equiv \mathbf{0} \qquad \text{for} \qquad \mathbf{x} \in \Omega^-. \tag{11.100}$$

On the other hand, by Theorem 11.33 and (11.100) we get

$$\{\mathbf{U}_0(\mathbf{z})\}^+ = \{\mathbf{U}_0(\mathbf{z})\}^- = \mathbf{0} \qquad \text{for} \qquad \mathbf{z} \in S,$$

i.e., the vector $\mathbf{U}_0(\mathbf{x})$ is a regular solution of problem $(I^q)^+_{\mathbf{0},\mathbf{0}}$. By virtue of Theorem 11.17, the problem $(I^q)^+_{\mathbf{0},\mathbf{0}}$ has only the trivial solution, that is

$$\mathbf{U}_0(\mathbf{x}) \equiv \mathbf{0} \qquad \text{for} \qquad \mathbf{x} \in \Omega^+. \tag{11.101}$$

By virtue of (11.100), (11.101), and identity (11.95) we obtain

$$\mathbf{h}_0(\mathbf{z}) = \{\mathbf{N}(\mathbf{D}_\mathbf{z}, \mathbf{n})\mathbf{U}_0(\mathbf{z})\}^- - \{\mathbf{N}(\mathbf{D}_\mathbf{z}, \mathbf{n})\mathbf{U}_0(\mathbf{z})\}^+ = \mathbf{0} \qquad \text{for} \qquad \mathbf{z} \in S.$$

Thus, the homogeneous equation (11.99) has only the trivial solution and therefore on the basis of Fredholm's theorem the integral equation (11.98) is always solvable for an arbitrary vector $\mathbf{f}$. We have thereby proved the following.

Theorem 11.34 *If $S \in C^{2,\nu}$, $\mathbf{f} \in C^{1,\nu'}(S)$, $0 < \nu' < \nu \le 1$, then a regular solution of the class $\mathfrak{R}$ of the internal BVP $(I^q)^+_{\mathbf{0},\mathbf{f}}$ exists, is unique, and is represented by (11.97), where $\mathbf{g}$ is a solution of Fredholm's integral equation (11.98) which is always solvable for an arbitrary vector $\mathbf{f}$.*

Problem $(I^q)^-_{\mathbf{0},\mathbf{f}}$ We seek a regular solution of the class $\mathfrak{R}$ to this problem in the sum of potentials

$$\mathbf{U}(\mathbf{x}) = \mathbf{Q}^{(q,4)}(\mathbf{x},\mathbf{g}) + (1-i)\mathbf{Q}^{(q,1)}(\mathbf{x},\mathbf{g}) \qquad \text{for} \qquad \mathbf{x}\in\Omega^-, \tag{11.102}$$

where $\mathbf{g}$ is the required eight-component vector function.

Obviously, by Theorems 11.20 and 11.33 the vector function $\mathbf{U}$ is a solution of (11.30) for $\mathbf{x}\in\Omega^-$. Keeping in mind the boundary condition (9.43) and using (11.94) and (11.96), from (11.102) we obtain, for determining the unknown vector $\mathbf{g}$, Fredholm's integral equation

$$\tilde{\mathcal{R}}^{(q,4)}\,\mathbf{g}(\mathbf{z}) \equiv \tilde{\mathcal{R}}^{(q,2)}\,\mathbf{g}(\mathbf{z}) + (1-i)\mathbf{Q}^{(q,1)}(\mathbf{z},\mathbf{g}) = \mathbf{f}(\mathbf{z}) \qquad \text{for}\quad \mathbf{z}\in S. \tag{11.103}$$

We prove that Eq. (11.103) is always solvable for an arbitrary vector $\mathbf{f}$.

Now we prove that the homogeneous equation

$$\tilde{\mathcal{R}}^{(q,4)}\,\mathbf{g}_0(\mathbf{z}) = \mathbf{0} \qquad \text{for}\quad \mathbf{z}\in S \tag{11.104}$$

has only a trivial solution. Indeed, let $\mathbf{g}_0$ be a solution of the homogeneous equation (11.104). Then the vector

$$\mathbf{U}_0(\mathbf{x}) \equiv \mathbf{Q}^{(q,4)}(\mathbf{x},\mathbf{g}_0) + (1-i)\mathbf{Q}^{(q,1)}(\mathbf{x},\mathbf{g}_0) \qquad \text{for} \qquad \mathbf{x}\in\Omega^- \tag{11.105}$$

is a regular solution of problem $(I^q)^-_{\mathbf{0},\mathbf{0}}$. Using Theorem 11.19 we have the relation (11.100).

On the other hand, by Theorems 11.20 and 11.33 from (11.105) we get

$$\begin{aligned} &\{\mathbf{U}_0(\mathbf{z})\}^- - \{\mathbf{U}_0(\mathbf{z})\}^+ = -\mathbf{g}_0(\mathbf{z}),\\ &\{\mathbf{N}(\mathbf{D}_\mathbf{z},\mathbf{n})\mathbf{U}_0(\mathbf{z})\}^- - \{\mathbf{N}(\mathbf{D}_\mathbf{z},\mathbf{n})\mathbf{U}_0(\mathbf{z})\}^+ = (1-i)\mathbf{g}_0(\mathbf{z}) \end{aligned} \tag{11.106}$$

for $\mathbf{z}\in S$. On the basis of (11.100) from (11.106) it follows that

$$\{\mathbf{N}(\mathbf{D}_\mathbf{z},\mathbf{n})\mathbf{U}_0(\mathbf{z}) + (1-i)\mathbf{U}_0(\mathbf{z})\}^+ = \mathbf{0} \qquad \text{for}\quad \mathbf{z}\in S. \tag{11.107}$$

Obviously, the vector $\mathbf{U}_0$ is a solution of Eq. (11.30) in Ω^+ satisfying the boundary condition (11.107).

In the similar manner as in Theorem 7.17, taking into account (11.30) and (11.107) we may further conclude that

$$\{\mathbf{U}_0(\mathbf{z})\}^+ = \mathbf{0} \qquad \text{for} \qquad \mathbf{z}\in S. \tag{11.108}$$

Finally, by virtue of (11.100) and (11.108) from the first equation of (11.106) we get $\mathbf{g}_0(\mathbf{z}) \equiv \mathbf{0}$ for $\mathbf{z}\in S$.

Thus, the homogeneous equation (11.104) has only the trivial solution and therefore on the basis of Fredholm's theorem the integral equation (11.103) is always solvable for an arbitrary vector **f**. We have thereby proved the following.

Theorem 11.35 *If* $S \in C^{2,\nu}$, $\mathbf{f} \in C^{1,\nu'}(S)$, $0 < \nu' < \nu \leq 1$, *then a regular solution of the class* $\mathfrak{R}$ *of the external BVP* $(I^q)^-_{\mathbf{0},\mathbf{f}}$ *exists, is unique, and is represented by sum of potentials (11.102), where* **g** *is a solution of Fredholm's integral equation (11.103) which is always solvable for an arbitrary vector* **f**.

Thus, the homogeneous equation (14.104) has only the trivial solution and therefore on the basis of Fredholm's theorem the integral equation (14.103) is [illegible] for f. We have thereby proved the following.

Theorem [illegible] $f \in C$ [illegible] $0 < \alpha < 1$. Then a regular solution [illegible] of the exterior BVP $(E)_0$ exists, is unique and is represented by sum of potentials (14.102), where g is a solution of [illegible] integral equation (14.103) which is uniquely solvable for an arbitrary vector f.

Chapter 12
Problems of Heat Conduction for Rigid Body

In this chapter, the basic internal and external BVPs of steady vibration in the linear theory of heat conduction for quadruple porosity rigid body are investigated by means of the potential method and the theory of singular integral equations.

In Sect. 12.1, the fundamental solution of the system of steady vibrations equations of heat conduction for quadruple porosity rigid body is constructed by means of elementary functions and its basic properties are established.

In Sect. 12.2, the formula of Galerkin-type representation of general solution of the same system of steady vibrations equations is given.

In Sect. 12.3, Green's identities for equations of steady vibrations of the considered theory are obtained.

In Sects. 12.4 and 12.5, the basic internal and external BVPs of steady vibrations in the linear theory of heat conduction for quadruple porosity rigid body are formulated and the uniqueness theorems for classical solutions of these BVPs are proved by employing techniques based on Green's identities, respectively.

Then, in Sect. 12.6, the surface and volume potentials of the considered theory are constructed and their basic properties are established.

Furthermore, in Sect. 12.7, the above-mentioned BVPs of steady vibrations of the linear theory of heat conduction for quadruple porosity rigid body are reduced to the equivalent always solvable Fredholm integral equations. The existence theorems for classical solutions of the BVPs of steady vibrations are proved by means of the potential method.

Finally, in Sect. 12.8, the BVPs of equilibrium in the linear theory of thermoelasticity for quadruple porosity materials are considered.

M. Svanadze, *Potential Method in Mathematical Theories of Multi-Porosity Media*,
Interdisciplinary Applied Mathematics 51,
https://doi.org/10.1007/978-3-030-28022-2_12

12.1 Fundamental Solution

In this section the fundamental solution of the system of Eq. (1.19) is constructed explicitly by means of elementary functions. In Sect. 1.6, this system is rewritten in the matrix form (1.35).

We introduce the notation:

1.

$$\mathbf{C}(\Delta,\omega) = \big(C_{lj}(\Delta,\omega)\big)_{5\times 5}, \qquad C_{\alpha\beta}(\Delta,\omega) = k_{\alpha\beta}\Delta + c_{\alpha\beta},$$

$$C_{\alpha 5}(\Delta,\omega) = i\omega T_0\varepsilon_\alpha, \qquad C_{5\alpha}(\Delta,\omega) = i\omega\varepsilon_\alpha,$$

$$C_{55}(\Delta,\omega) = k\Delta + i\omega a_0 T_0, \qquad \alpha,\beta = 1,2,3,4.$$

2.

$$\Theta_1^{(r)}(\Delta,\omega) = \frac{1}{kk_0}\det\mathbf{C}(\Delta,\omega) = \prod_{j=1}^{5}(\Delta + \hat{\mu}_j^2),$$

where $\hat{\mu}_1^2, \hat{\mu}_2^2, \cdots, \hat{\mu}_5^2$ are the roots of the algebraic equation $\Theta_1^{(r)}(-\xi,\omega) = 0$ (with respect to ξ).

3.

$$\mathbf{\Theta}^{(r)}(\Delta,\omega) = \left(\Theta_{lj}^{(r)}(\Delta,\omega)\right)_{5\times 5},$$

$$\Theta_{11}^{(r)} = \Theta_{22}^{(r)} = \cdots = \Theta_{55}^{(r)} = \Theta_1^{(r)}, \qquad \Theta_{lj} = 0,$$

$$l,j = 1,2,\cdots,5, \qquad l \neq j.$$

4.

$$\mathbf{L}^{(r)}(\Delta,\omega) = \left(L_{lj}^{(r)}(\Delta,\omega)\right)_{5\times 5},$$

$$L_{lj}^{(r)}(\Delta,\omega) = \frac{1}{kk_0}C_{lj}^*(\Delta,\omega) \tag{12.1}$$

$$l,j = 1,2,\cdots,5,$$

where C_{lj}^* is the cofactor of element C_{lj} of the matrix $\mathbf{C}$. We can easily verify that

$$\mathbf{A}^{(r)}(\mathbf{D_x}, \omega)\mathbf{L}^{(r)}(\Delta, \omega) = \mathbf{\Theta}^{(r)}(\Delta, \omega). \tag{12.2}$$

In what follows we assume that $\text{Im}\hat{\mu}_j > 0$ and $\hat{\mu}_l^2 \neq \hat{\mu}_j^2$, where $l, j = 1, 2, \cdots, 5$ and $l \neq j$. Let

$$\mathbf{\Upsilon}^{(r)}(\mathbf{x}, \omega) = \left(\Upsilon_{lj}^{(r)}(\mathbf{x}, \omega)\right)_{5\times 5},$$

$$\Upsilon_{11}^{(r)}(\mathbf{x}, \omega) = \Upsilon_{22}^{(r)}(\mathbf{x}, \omega) = \cdots = \Upsilon_{55}^{(r)}(\mathbf{x}, \omega) = \sum_{j=1}^{5} \hat{\eta}_j \hat{\vartheta}^{(j)}(\mathbf{x}), \tag{12.3}$$

$$\Upsilon_{lj}^{(r)}(\mathbf{x}, \omega) = 0, \qquad l \neq j, \qquad l, j = 1, 2, \cdots, 5,$$

where

$$\hat{\vartheta}^{(j)}(\mathbf{x}) = -\frac{e^{i\hat{\mu}_j|\mathbf{x}|}}{4\pi\,|\mathbf{x}|} \tag{12.4}$$

is the fundamental solution of Helmholtz' equation, i.e., $(\Delta + \hat{\mu}_j^2)\hat{\vartheta}^{(j)}(\mathbf{x}) = \delta(\mathbf{x})$ and

$$\hat{\eta}_j = \prod_{l=1,\, l\neq j}^{5} (\hat{\mu}_l^2 - \hat{\mu}_j^2)^{-1}, \qquad j = 1, 2, \cdots, 5.$$

It is now easy to prove that the matrix $\mathbf{\Upsilon}^{(r)}(\mathbf{x}, \omega)$ is the fundamental solution of the operator $\mathbf{\Theta}^{(r)}(\Delta, \omega)$, that is,

$$\mathbf{\Theta}^{(r)}(\Delta, \omega)\mathbf{\Upsilon}^{(r)}(\mathbf{x}, \omega) = \delta(\mathbf{x})\,\mathbf{I}_5, \tag{12.5}$$

where $\mathbf{x} \in \mathbb{R}^3$.

We introduce the matrix

$$\mathbf{\Psi}^{(r)}(\mathbf{x}, \omega) = \mathbf{L}^{(r)}(\Delta, \omega)\,\mathbf{\Upsilon}^{(r)}(\mathbf{x}, \omega). \tag{12.6}$$

Using identities (12.2) and (12.5) from (12.6) we get

$$\mathbf{A}^{(r)}(\mathbf{D_x}, \omega)\,\mathbf{\Psi}^{(r)}(\mathbf{x}, \omega) = \mathbf{A}^{(r)}(\mathbf{D_x}, \omega)\,\mathbf{L}^{(r)}(\Delta, \omega)\,\mathbf{\Upsilon}^{(r)}(\mathbf{x}, \omega)$$

$$= \mathbf{\Theta}^{(r)}(\Delta, \omega)\,\mathbf{\Upsilon}^{(r)}(\mathbf{x}, \omega) = \delta(\mathbf{x})\,\mathbf{I}_5.$$

Hence, $\mathbf{\Psi}^{(r)}(\mathbf{x}, \omega)$ is the fundamental matrix of the differential operator $\mathbf{A}^{(r)}(\mathbf{D_x}, \omega)$. We have thereby proved the following.

Theorem 12.1 *The matrix $\mathbf{\Psi}^{(r)}(\mathbf{x},\omega)$ defined by (12.6) is the fundamental solution of the system (1.19), where the matrices $\mathbf{L}^{(r)}(\mathbf{D_x},\omega)$ and $\mathbf{\Upsilon}^{(r)}(\mathbf{x},\omega)$ are given by (12.1) and (12.3), respectively.*

Clearly, each element $\Psi^{(r)}_{lj}(\mathbf{x},\omega)$ of the matrix $\mathbf{\Psi}^{(r)}(\mathbf{x},\omega)$ is represented in the following form

$$\Psi^{(r)}_{lj}(\mathbf{x},\omega) = L^{(r)}_{lj}(\mathbf{D_x},\omega)\,\Upsilon^{(r)}_{11}(\mathbf{x},\omega),$$

$$l,j = 1,2,\cdots,5.$$

Obviously, the matrix $\mathbf{\Psi}^{(r)}(\mathbf{x},\omega)$ is constructed explicitly by means of five metaharmonic functions $\hat{\vartheta}^{(j)}$ $(j = 1,2,\cdots,5)$ (see (12.4)).

We now can establish the basic properties of the matrix $\mathbf{\Psi}^{(r)}(\mathbf{x},\omega)$. Theorem 12.1 leads to the following consequences.

Theorem 12.2 *If $\mathbf{F} = \mathbf{0}$, then the matrix $\mathbf{\Psi}^{(r)}(\mathbf{x},\omega)$ is a solution of Eq. (1.35), i.e.*

$$\mathbf{A}^{(r)}(\mathbf{D_x},\omega)\mathbf{\Psi}^{(r)}(\mathbf{x},\omega) = \mathbf{0} \tag{12.7}$$

at every point $\mathbf{x} \in \mathbb{R}^3$ except the origin.

Theorem 12.3 *The relations*

$$\Psi^{(r)}_{\alpha\beta}(\mathbf{x},\omega) = O\left(|\mathbf{x}|^{-1}\right), \qquad \Psi^{(r)}_{55}(\mathbf{x},\omega) = O\left(|\mathbf{x}|^{-1}\right),$$

$$\Psi^{(r)}_{\alpha 5}(\mathbf{x},\omega) = O(1), \qquad \Psi^{(r)}_{5\alpha}(\mathbf{x},\omega) = O(1),$$

$$\alpha,\beta = 1,2,3,4$$

hold the neighborhood of the origin.

Theorem 12.4 *The relations*

$$\frac{\partial}{\partial x_j}\Psi^{(r)}_{\alpha\beta}(\mathbf{x},\omega) = O\left(|\mathbf{x}|^{-2}\right), \qquad \frac{\partial}{\partial x_j}\Psi^{(r)}_{55}(\mathbf{x},\omega) = O\left(|\mathbf{x}|^{-2}\right),$$

$$\frac{\partial}{\partial x_j}\Psi^{(r)}_{\alpha 5}(\mathbf{x},\omega) = O\left(|\mathbf{x}|^{-1}\right), \qquad \frac{\partial}{\partial x_j}\Psi^{(r)}_{5\alpha}(\mathbf{x},\omega) = O\left(|\mathbf{x}|^{-1}\right),$$

$$\alpha,\beta = 1,2,3,4, \qquad j = 1,2,3$$

hold the neighborhood of the origin.

Theorem 12.5 *The fundamental matrix of the system*

$$\mathbf{K}\,\Delta\mathbf{p} = \mathbf{0}, \qquad k\Delta\,\theta = 0$$

is the matrix $\boldsymbol{\Psi}^{(r0)}(\mathbf{x}) = \left(\Psi^{(r0)}_{lj}(\mathbf{x})\right)_{5\times 5}$, *where*

$$\Psi^{(r0)}_{\alpha\beta}(\mathbf{x}) = \frac{1}{k_0}k^*_{\alpha\beta}\tilde{\gamma}^{(5)}(\mathbf{x}), \qquad \Psi^{(r0)}_{55}(\mathbf{x}) = \frac{1}{k}\tilde{\gamma}^{(5)}(\mathbf{x}),$$

$$\Psi^{(r0)}_{\alpha 5}(\mathbf{x}) = \Psi^{(r0)}_{5\alpha}(\mathbf{x}) = 0, \qquad \alpha, \beta = 1, 2, 3, 4,$$

$k^*_{\alpha\beta}$ *is the cofactor of element* $k_{\alpha\beta}$ *of the matrix* $\mathbf{K}$ *and* $\tilde{\gamma}^{(5)}(\mathbf{x})$ *is given by (2.26).*

Theorem 12.6 *The relations*

$$\Psi^{(r)}_{lj}(\mathbf{x},\omega) - \Psi^{(r0)}_{lj}(\mathbf{x}) = const + O\left(|\mathbf{x}|\right), \qquad l, j = 1, 2, \cdots, 5 \tag{12.8}$$

hold in the neighborhood of the origin.

Thus, on the basis of Theorems 12.3 and the relations (12.8) the matrix $\boldsymbol{\Psi}^{(r0)}(\mathbf{x})$ is the singular part of the fundamental solution $\boldsymbol{\Psi}^{(r)}(\mathbf{x},\omega)$ in the neighborhood of the origin.

Now we shall construct a solution of the system of steady vibrations equations (1.35) for $\mathbf{F}'(\mathbf{x}) = \mathbf{0}$ playing a major role in the investigation of the BVPs of heat conduction in the linear theory of quadruple porosity rigid body by means of the potential method. The following theorem is valid.

Theorem 12.7 *Every column of the matrix*

$$\left[\mathbf{P}^{(r)}(\mathbf{D}_{\mathbf{x}},\mathbf{n}(\mathbf{x}))\boldsymbol{\Psi}^{(r)^\top}(\mathbf{x}-\mathbf{y},\omega)\right]^\top$$

with respect to the point $\mathbf{x}$ *satisfies the homogeneous equation*

$$\mathbf{A}^{(r)}(\mathbf{D}_{\mathbf{x}},\omega)\left[\mathbf{P}^{(r)}(\mathbf{D}_{\mathbf{x}},\mathbf{n}(\mathbf{x}))\boldsymbol{\Psi}^{(r)^\top}(\mathbf{x}-\mathbf{y},\omega)\right]^\top = \mathbf{0}, \tag{12.9}$$

everywhere in $\mathbb{R}^3$ *except* $\mathbf{x} = \mathbf{y}$, *where*

$$\mathbf{P}^{(r)}(\mathbf{D}_{\mathbf{x}},\mathbf{n}) = \left(P^{(r)}_{lj}(\mathbf{D}_{\mathbf{x}},\mathbf{n})\right)_{5\times 5},$$

$$P^{(r)}_{\alpha\beta}(\mathbf{D}_{\mathbf{x}},\mathbf{n}) = k_{\alpha\beta}\frac{\partial}{\partial\mathbf{n}}, \qquad P^{(r)}_{55}(\mathbf{D}_{\mathbf{x}},\mathbf{n}) = k\frac{\partial}{\partial\mathbf{n}}, \tag{12.10}$$

$$P^{(r)}_{\alpha 5}(\mathbf{D}_{\mathbf{x}},\mathbf{n}) = P^{(r)}_{5\alpha}(\mathbf{D}_{\mathbf{x}},\mathbf{n}) = 0, \qquad \alpha, \beta = 1, 2, 3, 4.$$

Theorem 12.7 leads to the following.

Theorem 12.8 *The relations*

$$\left[\mathbf{P}^{(r)}(\mathbf{D_x}, \mathbf{n}(\mathbf{x}))\boldsymbol{\Psi}^{(r)^{\top}}(\mathbf{x}, \omega)\right]_{lj} = O\left(|\mathbf{x}|^{-2}\right),$$

$$\left[\mathbf{P}^{(r)}(\mathbf{D_x}, \mathbf{n}(\mathbf{x}))\boldsymbol{\Psi}^{(r0)}(\mathbf{x})\right]_{lj} = O\left(|\mathbf{x}|^{-2}\right),$$

$$l, j = 1, 2, \cdots, 5$$

hold in the neighborhood of the origin.

Now we can establish the singular part of the matrix $\mathbf{P}^{(r)}(\mathbf{D_x}, \mathbf{n}(\mathbf{x}))\boldsymbol{\Psi}^{(r)^{\top}}(\mathbf{x}, \omega)$ in the neighborhood of the origin. The relations (9.17) lead to the following.

Theorem 12.9 *The relations*

$$\left[\mathbf{P}^{(r)}(\mathbf{D_x}, \mathbf{n}(\mathbf{x}))\boldsymbol{\Psi}^{(r)^{\top}}(\mathbf{x}, \omega)\right]_{lj} - \left[\mathbf{P}^{(r)}(\mathbf{D_x}, \mathbf{n}(\mathbf{x}))\boldsymbol{\Psi}^{(r0)}(\mathbf{x})\right]_{lj} = O\left(|\mathbf{x}|^{-1}\right),$$
$$l, j = 1, 2, \cdots, 5 \tag{12.11}$$

hold in the neighborhood of the origin.

Thus, the relations (12.11) imply that the matrix $\mathbf{P}^{(r)}(\mathbf{D_x}, \mathbf{n}(\mathbf{x}))\boldsymbol{\Psi}^{(r0)}(\mathbf{x})$ is the singular part of the matrix $\mathbf{P}^{(r)}(\mathbf{D_x}, \mathbf{n}(\mathbf{x}))\boldsymbol{\Psi}^{(r)^{\top}}(\mathbf{x}, \omega)$ in the neighborhood of the origin.

12.2 Galerkin-Type Solution

The next two theorems provide a Galerkin-type solution to system (1.19).

Theorem 12.10 *Let*

$$p_\alpha(\mathbf{x}) = \sum_{j=1}^{5} \frac{1}{kk_0} C^*_{\alpha j}(\Delta, \omega) w_j(\mathbf{x}),$$
$$\theta(\mathbf{x}) = \sum_{j=1}^{5} \frac{1}{kk_0} C^*_{5j}(\Delta, \omega) w_j(\mathbf{x}), \tag{12.12}$$

then $\boldsymbol{\vartheta} = (\mathbf{p}, \theta)$ is a solution of the system

$$(k_{\alpha\beta}\,\Delta + c_{\alpha\beta})p_\beta + i\omega\varepsilon_\alpha\,\theta = \varphi_\alpha,$$
$$(k\Delta + i\omega a_0 T_0)\theta + i\omega T_0\varepsilon_j p_j = \varphi_5. \tag{12.13}$$

where $w_j \in C^{10}(\Omega)$ *and*

$$\Theta_1^{(r)}(\Delta,\omega)\,w_j(\mathbf{x}) = \varphi_j(\mathbf{x}), \qquad j = 1, 2, \cdots, 5, \qquad \alpha = 1, 2, 3, 4. \tag{12.14}$$

Proof By virtue of (12.1) Eqs. (12.12) and (12.14) we can rewrite in the form

$$\boldsymbol{\vartheta} = \mathbf{L}^{(r)}(\Delta,\omega)\,\mathbf{w}(\mathbf{x}) \tag{12.15}$$

and

$$\boldsymbol{\Theta}^{(r)}(\Delta,\omega)\,\mathbf{w}(\mathbf{x}) = \boldsymbol{\varphi}(\mathbf{x}), \tag{12.16}$$

respectively, where $\mathbf{w} = (w_1, w_2, \cdots, w_5)$, $\boldsymbol{\varphi} = (\varphi_1, \varphi_2, \cdots, \varphi_5)$. Clearly, on the basis of (12.2), (12.15), and (12.16) it follows that

$$\mathbf{A}^{(r)}(\mathbf{D_x},\omega)\boldsymbol{\vartheta} = \mathbf{A}^{(r)}(\mathbf{D_x},\omega)\mathbf{L}^{(r)}(\Delta,\omega)\mathbf{w} = \boldsymbol{\Theta}^{(r)}(\Delta,\omega)\mathbf{w} = \boldsymbol{\varphi}(\mathbf{x}).$$

Thus, the vector $\boldsymbol{\vartheta}$ is a solution of the system (12.13). □

Theorem 12.11 *If* $\boldsymbol{\vartheta} = (\mathbf{p},\theta)$ *is a solution of system (12.13) in* Ω*, then* $\boldsymbol{\vartheta}$ *is represented by (12.12), where* $\mathbf{w} = (w_1, w_2, \cdots, w_5)$ *is a solution of (12.16) and* Ω *is a finite domain in* $\mathbb{R}^3$.

Proof Let $\boldsymbol{\vartheta}$ be a solution of system (12.13). Obviously, if $\hat{\boldsymbol{\Psi}}(\mathbf{x})$ is the fundamental matrix of the operator $\mathbf{L}^{(r)}(\Delta,\omega)$, then the vector function (the volume potential)

$$\mathbf{w}(\mathbf{x}) = \int_\Omega \hat{\boldsymbol{\Psi}}(\mathbf{x}-\mathbf{y})\boldsymbol{\vartheta}(\mathbf{y})d\mathbf{y}$$

is a solution of (12.15).

On the other hand, by virtue of (12.2), (12.13), and (12.15) we have

$$\boldsymbol{\varphi}(\mathbf{x}) = \mathbf{A}^{(r)}(\mathbf{D_x},\omega)\boldsymbol{\vartheta}(\mathbf{x}) = \mathbf{A}^{(r)}(\mathbf{D_x},\omega)\mathbf{L}^{(r)}(\Delta,\omega)\,\mathbf{w}(\mathbf{x}) = \boldsymbol{\Theta}^{(r)}(\Delta,\omega)\,\mathbf{w}(\mathbf{x}).$$

Hence, $\mathbf{w}$ is a solution of (12.16). □

Thus, on the basis of Theorems 12.10 and 12.11 the completeness of the Galerkin-type solution of system (1.19) is proved.

Remark 12.1 Quite similarly as in Theorem 12.1 we can construct explicitly the fundamental matrix $\hat{\boldsymbol{\Psi}}(\mathbf{x})$ of the operator $\mathbf{L}^{(r)}(\Delta,\omega)$ by elementary functions.

12.3 Green's Formulas

12.3.1 Green's First Identity

Definition 12.1 A vector function $\boldsymbol{\vartheta} = (\mathbf{p}, \theta) = (\vartheta_1, \vartheta_2, \cdots, \vartheta_5)$ is called regular of the class $\mathfrak{R}^{(r)}$ in Ω^- (or Ω^+) if

(i)

$$\vartheta_l \in C^2(\Omega^-) \cap C^1(\overline{\Omega^-}) \qquad (\text{or } U_l \in C^2(\Omega^+) \cap C^1(\overline{\Omega^+})),$$

(ii)

$$\vartheta_l(\mathbf{x}) = O(|\mathbf{x}|^{-1}), \qquad \vartheta_{l,j}(\mathbf{x}) = o(|\mathbf{x}|^{-1}) \qquad \text{for} \qquad |\mathbf{x}| \gg 1, \tag{12.17}$$

where $l = 1, 2, \cdots, 5,\ j = 1, 2, 3.$

In the sequel we use the matrix differential operators:

1.

$$\mathbf{A}^{(r1)}(\mathbf{D_x}, \omega) = \left(A_{lj}^{(r1)}(\mathbf{D_x}, \omega)\right)_{4\times 5}, \qquad \mathbf{A}^{(r2)}(\mathbf{D_x}, \omega) = \left(A_{1j}^{(r2)}(\mathbf{D_x}, \omega)\right)_{1\times 5},$$

$$A_{\alpha\beta}^{(r1)}(\mathbf{D_x}, \omega) = k_{\alpha\beta}\Delta + c_{\alpha\beta}, \qquad A_{\alpha 5}^{(r1)}(\mathbf{D_x}, \omega) = i\omega\varepsilon_\alpha,$$

$$A_{1\alpha}^{(r2)}(\mathbf{D_x}, \omega) = i\omega T_0\varepsilon_\alpha, \qquad A^{(r2)}(\mathbf{D_x}, \omega) = k\Delta + i\omega a_0 T_0;$$

2.

$$\mathbf{P}^{(r1)}(\mathbf{D_x}, \mathbf{n}) = \left(P_{lj}^{(r1)}(\mathbf{D_x}, \mathbf{n})\right)_{4\times 4}, \qquad P_{\alpha\beta}^{(r1)}(\mathbf{D_x}, \mathbf{n}) = k_{\alpha\beta}\frac{\partial}{\partial \mathbf{n}},$$

$$P^{(r2)}(\mathbf{D_x}, \mathbf{n}) = k\frac{\partial}{\partial \mathbf{n}},$$

where $\alpha, \beta = 1, 2, 3, 4.$

Let $\mathbf{p}' = (p_1', p_2', p_3', p_4')$ be four-component vector function, $\boldsymbol{\vartheta}' = (\mathbf{p}', \theta')$, where θ' is a scalar function.

We introduce the notation

$$E_1^{(r)}(\boldsymbol{\vartheta},\mathbf{p}') = k_{\alpha\beta}\nabla p_\alpha \cdot \nabla \overline{p'_\beta} - c_{\alpha\beta} p_\beta \overline{p'_\alpha} - i\omega\theta\varepsilon_\alpha \overline{p'_\alpha},$$

$$E_2^{(r)}(\boldsymbol{\vartheta},\theta') = k\nabla\theta \cdot \nabla\theta' - i\omega T_0\,(a_0\theta + \varepsilon_0 \operatorname{div}\mathbf{u} + \varepsilon_\alpha p_\alpha)\,\overline{\theta'}, \tag{12.18}$$

$$E^{(r)}(\boldsymbol{\vartheta},\boldsymbol{\vartheta}') = E_1^{(r)}(\boldsymbol{\vartheta},\mathbf{p}') + E_2^{(r)}(\boldsymbol{\vartheta},\theta').$$

We have the following.

Theorem 12.12 *If $\boldsymbol{\vartheta} = (\mathbf{p},\theta)$ is a regular vector of the class $\mathfrak{R}^{(r)}$ in Ω^+, $p'_\alpha, \theta' \in C^1(\Omega^+) \cap C(\overline{\Omega^+})$, $\alpha = 1,2,3,4$, then*

$$\int\limits_{\Omega^+} \left[\mathbf{A}^{(r)}(\mathbf{D_x},\omega)\,\boldsymbol{\vartheta}(\mathbf{x}) \cdot \boldsymbol{\vartheta}'(\mathbf{x}) + E^{(r)}(\boldsymbol{\vartheta},\boldsymbol{\vartheta}')\right] d\mathbf{x} = \int\limits_S \mathbf{P}^{(r)}(\mathbf{D_z},\mathbf{n})\boldsymbol{\vartheta}(\mathbf{z}) \cdot \boldsymbol{\vartheta}'(\mathbf{z})\, d_{\mathbf{z}}S, \tag{12.19}$$

where $\mathbf{A}^{(r)}(\mathbf{D_x},\omega)$, $\mathbf{P}^{(r)}(\mathbf{D_z},\mathbf{n})$ and $E^{(r)}(\boldsymbol{\vartheta},\boldsymbol{\vartheta}')$ are defined by (1.30), (12.10), and (12.18), respectively.

Proof By direct calculation we obtain

$$\int\limits_{\Omega^+} \left[\mathbf{A}^{(r1)}(\mathbf{D_x},\omega)\,\boldsymbol{\vartheta} \cdot \mathbf{p}'(\mathbf{x}) + E_1^{(r)}(\boldsymbol{\vartheta},\mathbf{p}')\right] d\mathbf{x} = \int\limits_S \mathbf{P}^{(r1)}(\mathbf{D_z},\mathbf{n})\mathbf{p}(\mathbf{z}) \cdot \mathbf{p}'(\mathbf{z})\, d_{\mathbf{z}}S,$$

$$\int\limits_{\Omega^+} \left[\mathbf{A}^{(r2)}(\mathbf{D_x},\omega)\,\boldsymbol{\vartheta}\, \overline{\theta'(\mathbf{x})} + E_2^{(r)}(\boldsymbol{\vartheta},\theta')\right] d\mathbf{x} = \int\limits_S P^{(r2)}(\mathbf{D_z},\mathbf{n})\theta(\mathbf{z})\, \overline{\theta'(\mathbf{z})}\, d_{\mathbf{z}}S. \tag{12.20}$$

It is easy to verify that the identity (12.19) may be obtained from (12.20). □

Theorem 12.12 and the condition (12.17) lead to the following.

Theorem 12.13 *If $\boldsymbol{\vartheta} = (\mathbf{p},\theta)$ and $\boldsymbol{\vartheta}' = (\mathbf{p}',\theta')$ are regular vectors of the class $\mathfrak{R}^{(r)}$ in Ω^-, then*

$$\int\limits_{\Omega^-} \left[\mathbf{A}^{(r)}(\mathbf{D_x},\omega)\,\boldsymbol{\vartheta}(\mathbf{x}) \cdot \boldsymbol{\vartheta}'(\mathbf{x}) + E^{(r)}(\boldsymbol{\vartheta},\boldsymbol{\vartheta}')\right] d\mathbf{x} = -\int\limits_S \mathbf{P}^{(r)}(\mathbf{D_z},\mathbf{n})\boldsymbol{\vartheta}(\mathbf{z}) \cdot \boldsymbol{\vartheta}'(\mathbf{z})\, d_{\mathbf{z}}S, \tag{12.21}$$

where $\mathbf{A}^{(r)}(\mathbf{D_x},\omega)$, $\mathbf{P}^{(r)}(\mathbf{D_z},\mathbf{n})$, and $E^{(r)}(\mathbf{U},\mathbf{U}')$ are defined by (1.30), (12.10), and (12.18), respectively.

The formulas (12.19) and (12.21) are Green's first identities of the steady vibrations equations of heat conduction in the linear theory of quadruple porosity rigid body for domains Ω^+ and Ω^-, respectively.

12.3.2 Green's Second Identity

Clearly, the matrix differential operator $\tilde{\mathbf{A}}^{(r)}(\mathbf{D_x}, \omega) = \left(\tilde{A}^{(r)}_{lj}(\mathbf{D_x}, \omega)\right)_{5\times 5}$ is the associate operator of $\mathbf{A}^{(r)}(\mathbf{D_x}, \omega)$, where $\tilde{\mathbf{A}}^{(r)}(\mathbf{D_x}, \omega) = \mathbf{A}^{(r)^\top}(-\mathbf{D_x}, \omega)$. It is easy to see that the associated system of homogeneous equations of (1.19) is

$$(\mathbf{K}\,\Delta + \mathbf{c})\,\tilde{\mathbf{p}} + i\omega T_0 \boldsymbol{\varepsilon}\,\tilde{\theta} = \mathbf{0},$$
$$(k\Delta + i\omega a_0 T_0)\tilde{\theta} + i\omega \boldsymbol{\varepsilon}\,\tilde{\mathbf{p}} = 0, \tag{12.22}$$

where $\tilde{\theta}$ is a scalar function and $\tilde{\mathbf{p}}$ is four-component vector function on $\Omega^{\pm}$.

Let $\tilde{\boldsymbol{\vartheta}}_j$ be the j-th column of the matrix $\tilde{\boldsymbol{\vartheta}} = (\tilde{\vartheta}_{lj})_{5\times 5}$ $(j = 1, 2, \cdots, 5)$. We have the following.

Theorem 12.14 *If $\boldsymbol{\vartheta} = (\mathbf{p}, \theta)$ and $\tilde{\boldsymbol{\vartheta}}_j$ $(j = 1, 2, \cdots, 5)$ are regular vectors of the class $\mathfrak{R}^{(r)}$ in Ω^+, then*

$$\int\limits_{\Omega^+} \left\{[\tilde{\mathbf{A}}^{(r)}(\mathbf{D_y}, \omega)\tilde{\boldsymbol{\vartheta}}(\mathbf{y})]^\top \boldsymbol{\vartheta}(\mathbf{y}) - [\tilde{\boldsymbol{\vartheta}}(\mathbf{y})]^\top \mathbf{A}^{(r)}(\mathbf{D_y}, \omega)\boldsymbol{\vartheta}(\mathbf{y})\right\} d\mathbf{y}$$
$$= \int\limits_S \left\{[\mathbf{P}^{(r)}(\mathbf{D_z}, \mathbf{n})\tilde{\boldsymbol{\vartheta}}(\mathbf{z})]^\top \boldsymbol{\vartheta} - [\tilde{\boldsymbol{\vartheta}}(\mathbf{z})]^\top \mathbf{P}^{(r)}(\mathbf{D_z}, \mathbf{n})\boldsymbol{\vartheta}(\mathbf{z})\right\} d_{\mathbf{z}}S, \tag{12.23}$$

where the matrices $\mathbf{A}^{(r)}(\mathbf{D_x}, \omega)$ and $\mathbf{P}^{(r)}(\mathbf{D_z}, \mathbf{n})$ are given by (1.30) and (12.10), respectively.

Theorem 12.14 and the condition (12.17) lead to the following.

Theorem 12.15 *If $\boldsymbol{\vartheta} = (\mathbf{p}, \theta)$ and $\tilde{\boldsymbol{\vartheta}}_j$ $(j = 1, 2, \cdots, 5)$ are regular vectors of the class $\mathfrak{R}^{(r)}$ in Ω^-, then*

$$\int\limits_{\Omega^-} \left\{[\tilde{\mathbf{A}}^{(r)}(\mathbf{D_y}, \omega)\tilde{\boldsymbol{\vartheta}}(\mathbf{y})]^\top \boldsymbol{\vartheta}(\mathbf{y}) - [\tilde{\boldsymbol{\vartheta}}(\mathbf{y})]^\top \mathbf{A}^{(r)}(\mathbf{D_y}, \omega)\boldsymbol{\vartheta}(\mathbf{y})\right\} d\mathbf{y}$$
$$= -\int\limits_S \left\{[\mathbf{P}^{(r)}(\mathbf{D_z}, \mathbf{n})\tilde{\boldsymbol{\vartheta}}(\mathbf{z})]^\top \boldsymbol{\vartheta} - [\tilde{\boldsymbol{\vartheta}}(\mathbf{z})]^\top \mathbf{P}^{(r)}(\mathbf{D_z}, \mathbf{n})\boldsymbol{\vartheta}(\mathbf{z})\right\} d_{\mathbf{z}}S. \tag{12.24}$$

The formulas (12.23) and (12.24) are Green's second identities of the steady vibrations equations of heat conduction in the linear theory of quadruple porosity rigid body for domains Ω^+ and Ω^-, respectively.

12.3.3 Green's Third Identity

Let $\tilde{\boldsymbol{\Psi}}^{(r)}(\mathbf{x},\omega)$ be the fundamental solution of the system (12.22) (the fundamental matrix of operator $\tilde{\mathbf{A}}^{(r)}(\mathbf{D}_{\mathbf{y}},\omega)$). Obviously, the matrix $\tilde{\boldsymbol{\Psi}}^{(r)}(\mathbf{x},\omega)$ satisfies the following condition

$$\tilde{\boldsymbol{\Psi}}^{(r)}(\mathbf{x},\omega) = \boldsymbol{\Psi}^{(r)\top}(-\mathbf{x},\omega), \tag{12.25}$$

where $\boldsymbol{\Psi}^{(r)}(\mathbf{x},\omega)$ is the fundamental solution of the system (1.19) (see Theorem 12.1).

By virtue of (12.7) and (12.25) from (12.23) and (12.24) we obtain the following results.

Theorem 12.16 *If* $\boldsymbol{\vartheta}$ *is a regular vector of the class* $\mathfrak{R}^{(r)}$ *in* Ω^+, *then*

$$\begin{aligned}\boldsymbol{\vartheta}(\mathbf{x}) &= \\ &\int_S \left\{[\mathbf{P}^{(r)}(\mathbf{D}_{\mathbf{z}},\mathbf{n})\boldsymbol{\Psi}^{(r)\top}(\mathbf{x}-\mathbf{z},\omega)]^\top\boldsymbol{\vartheta}(\mathbf{z}) - \boldsymbol{\Psi}^{(r)}(\mathbf{x}-\mathbf{z},\omega)\,\mathbf{P}^{(r)}(\mathbf{D}_{\mathbf{z}},\mathbf{n})\boldsymbol{\vartheta}(\mathbf{z})\right\} d_{\mathbf{z}}S \\ &+ \int_{\Omega^+} \boldsymbol{\Psi}^{(r)}(\mathbf{x}-\mathbf{y},\omega)\,\mathbf{A}^{(r)}(\mathbf{D}_{\mathbf{y}},\omega)\boldsymbol{\vartheta}(\mathbf{y})d\mathbf{y} \qquad \textit{for} \quad \mathbf{x}\in\Omega^+.\end{aligned} \tag{12.26}$$

Theorem 12.17 *If* $\boldsymbol{\vartheta}$ *is a regular vector of the class* $\mathfrak{R}^{(r)}$ *in* Ω^-, *then*

$$\begin{aligned}\boldsymbol{\vartheta}(\mathbf{x}) &= \\ &-\int_S \left\{[\mathbf{P}^{(r)}(\mathbf{D}_{\mathbf{z}},\mathbf{n})\boldsymbol{\Psi}^{(r)\top}(\mathbf{x}-\mathbf{z},\omega)]^\top\boldsymbol{\vartheta}(\mathbf{z}) - \boldsymbol{\Psi}^{(r)}(\mathbf{x}-\mathbf{z},\omega)\,\mathbf{P}^{(r)}(\mathbf{D}_{\mathbf{z}},\mathbf{n})\boldsymbol{\vartheta}(\mathbf{z})\right\} d_{\mathbf{z}}S \\ &+ \int_{\Omega^-} \boldsymbol{\Psi}^{(r)}(\mathbf{x}-\mathbf{y},\omega)\,\mathbf{A}^{(r)}(\mathbf{D}_{\mathbf{y}},\omega)\boldsymbol{\vartheta}(\mathbf{y})d\mathbf{y} \qquad \textit{for} \quad \mathbf{x}\in\Omega^+.\end{aligned} \tag{12.27}$$

The formulas (12.26) and (12.27) are Green's third identities of heat conduction in the linear theory of quadruple porosity rigid body for domains Ω^+ and Ω^-, respectively.

Theorems 12.16 and 12.17 lead to the following.

Corollary 12.1 *If* $\boldsymbol{\vartheta}$ *is a regular solution of the class* $\mathfrak{R}^{(r)}$ *of the homogeneous equation*

$$\mathbf{A}^{(r)}(\mathbf{D_x}, \omega)\,\boldsymbol{\vartheta}\mathbf{x} = \mathbf{0} \tag{12.28}$$

in Ω^+, *then*

$$\boldsymbol{\vartheta}(\mathbf{x}) = \int_S \left\{[\mathbf{P}^{(r)}(\mathbf{D_z}, \mathbf{n})\boldsymbol{\Psi}^{(r)^\top}(\mathbf{x}-\mathbf{z}, \omega)]^\top \boldsymbol{\vartheta}(\mathbf{z}) - \boldsymbol{\Psi}^{(r)}(\mathbf{x}-\mathbf{z}, \omega)\,\mathbf{P}^{(r)}(\mathbf{D_z}, \mathbf{n})\boldsymbol{\vartheta}(\mathbf{z})\right\} d_{\mathbf{z}}S \tag{12.29}$$

for $\mathbf{x} \in \Omega^+$.

Corollary 12.2 *If* $\boldsymbol{\vartheta}$ *is a regular solution of the class* $\mathfrak{R}^{(s)}$ *of (12.28) in* Ω^-, *then*

$$\boldsymbol{\vartheta}(\mathbf{x}) = -\int_S \left\{[\mathbf{P}^{(r)}(\mathbf{D_z}, \mathbf{n})\boldsymbol{\Psi}^{(r)^\top}(\mathbf{x}-\mathbf{z}, \omega)]^\top \boldsymbol{\vartheta}(\mathbf{z}) - \boldsymbol{\Psi}^{(r)}(\mathbf{x}-\mathbf{z}, \omega)\,\mathbf{P}^{(r)}(\mathbf{D_z}, \mathbf{n})\boldsymbol{\vartheta}(\mathbf{z})\right\} d_{\mathbf{z}}S \tag{12.30}$$

for $\mathbf{x} \in \Omega^-$.

The formulas (12.29) and (12.30) are the Somigliana-type integral representations of a regular solution of the steady vibrations homogeneous equation (12.28) in the heat conduction linear theory of quadruple porosity rigid body for domains Ω^+ and Ω^-, respectively.

12.4 Basic Boundary Value Problems

The basic internal and external BVPs of steady vibration of heat conduction in the linear theory of quadruple porosity rigid body are formulated as follows.

Find a regular (classical) solution of the class $\mathfrak{R}^{(r)}$ in Ω^+ to system

$$\mathbf{A}^{(r)}(\mathbf{D_x}, \omega)\boldsymbol{\vartheta}(\mathbf{x}) = \mathbf{F}(\mathbf{x}) \tag{12.31}$$

for $\mathbf{x} \in \Omega^+$ satisfying the boundary condition

$$\lim_{\Omega^+ \ni \mathbf{x} \to \mathbf{z} \in S} \boldsymbol{\vartheta}(\mathbf{x}) \equiv \{\boldsymbol{\vartheta}(\mathbf{z})\}^+ = \mathbf{f}(\mathbf{z}) \tag{12.32}$$

in the *Problem* $(I^r)^+_{\mathbf{F},\mathbf{f}}$, and

$$\{\mathbf{P}^{(r)}(\mathbf{D_z}, \mathbf{n})\boldsymbol{\vartheta}(\mathbf{z})\}^+ = \mathbf{f}(\mathbf{z}) \tag{12.33}$$

in the *Problem* $(II^r)^+_{\mathbf{F},\mathbf{f}}$, where $\mathbf{F}$ and $\mathbf{f}$ are prescribed five-component vector functions.

Find a regular (classical) solution of the class $\mathfrak{R}^{(r)}$ to system (12.31) for $\mathbf{x} \in \Omega^-$ satisfying the boundary condition

$$\lim_{\Omega^- \ni \mathbf{x} \to \mathbf{z} \in S} \boldsymbol{\vartheta}(\mathbf{x}) \equiv \{\boldsymbol{\vartheta}(\mathbf{z})\}^- = \mathbf{f}(\mathbf{z}) \tag{12.34}$$

in the *Problem* $(I^r)^-_{\mathbf{F},\mathbf{f}}$, and

$$\{\mathbf{P}^{(r)}(\mathbf{D}_\mathbf{z}, \mathbf{n})\boldsymbol{\vartheta}(\mathbf{z})\}^- = \mathbf{f}(\mathbf{z}) \tag{12.35}$$

in the *Problem* $(II^r)^-_{\mathbf{F},\mathbf{f}}$. Here $\mathbf{F}$ and $\mathbf{f}$ are prescribed five-component vector functions, and $\operatorname{supp}\mathbf{F}$ is a finite domain in Ω^-.

Furthermore, the BVPs $(I^r)^\pm_{\mathbf{F},\mathbf{f}}$ and $(II^r)^\pm_{\mathbf{F},\mathbf{f}}$ are Dirichlet and Neumann type BVPs in the heat conduction linear theory of quadruple porosity rigid body.

12.5 Uniqueness Theorems

In this section the uniqueness theorems for classical solutions of the BVPs $(I^r)^\pm_{\mathbf{F},\mathbf{f}}$ and $(II^r)^\pm_{\mathbf{F},\mathbf{f}}$ are proved by using Green's identities.

Theorem 12.18 *The internal BVP* $(I^r)^+_{\mathbf{F},\mathbf{f}}$ *admits at most one regular (classical) solution of the class* $\mathfrak{R}^{(r)}$ *in* Ω^+.

Proof Suppose that there are two regular solutions of problem $(I^r)^+_{\mathbf{F},\mathbf{f}}$. Then their difference $\boldsymbol{\vartheta}$ corresponds to zero data ($\mathbf{F} = \mathbf{f} = \mathbf{0}$), i.e., on the basis of (12.31) and (12.32) the vector $\boldsymbol{\vartheta}$ is a regular solution of the system of homogeneous equations (12.28) for $\mathbf{x} \in \Omega^+$ satisfying the boundary condition

$$\{\boldsymbol{\vartheta}(\mathbf{z})\}^+ = 0 \tag{12.36}$$

for $\mathbf{z} \in S$.

On account of (12.18), (12.28), and (12.36) from Green's formulas (12.20) for $\boldsymbol{\vartheta}' = \boldsymbol{\vartheta}$ we may deduce that

$$\int_{\Omega^+} E_1^{(r)}(\boldsymbol{\vartheta}, \mathbf{p})d\mathbf{x} = 0, \qquad \int_{\Omega^+} E_2^{(r)}(\boldsymbol{\vartheta}, \theta)d\mathbf{x} = 0, \tag{12.37}$$

where

$$
\begin{aligned}
E_1^{(r)}(\boldsymbol{\vartheta}, \mathbf{p}) &= k_{\alpha\beta} \nabla p_\alpha \cdot \nabla p_\beta - c_{\alpha\beta} p_\beta \overline{p_\alpha} - i\omega\theta\varepsilon_\alpha \overline{p_\alpha}, \\
E_2^{(r)}(\boldsymbol{\vartheta}, \theta) &= k|\nabla\theta|^2 - i\omega T_0 \left(a_0\theta + \varepsilon_\alpha p_\alpha\right)\overline{\theta}.
\end{aligned}
\tag{12.38}
$$

Then, employing the inequalities (4.10) we may derive from (12.38)

$$
\begin{aligned}
\mathrm{Re}\, E_1^{(r)}(\boldsymbol{\vartheta}, \mathbf{p}) &= k_{\alpha\beta} \nabla p_\beta \cdot \nabla p_\alpha + d_{\alpha\beta} p_\beta \overline{p_\alpha} + \omega \mathrm{Im}\left[\theta\varepsilon_\alpha \overline{p_\alpha}\right], \\
\mathrm{Re}\, E_2^{(r)}(\boldsymbol{\vartheta}, \theta) &= k|\nabla\theta|^2 + \omega T_0 \mathrm{Im}\left[\varepsilon_\alpha p_\alpha \overline{\theta}\right].
\end{aligned}
\tag{12.39}
$$

Obviously, from (12.39) it follows that

$$
\begin{aligned}
&T_0 \mathrm{Re}\, E_1^{(r)}(\boldsymbol{\vartheta}, \mathbf{p}) + \mathrm{Re}\, E_2^{(r)}(\boldsymbol{\vartheta}, \theta) \\
&\qquad = k|\nabla\theta|^2 + T_0(k_{\alpha\beta} \nabla p_\beta \cdot \nabla p_\alpha + d_{\alpha\beta} p_\beta \overline{p_\alpha}).
\end{aligned}
\tag{12.40}
$$

Taking into account (12.40) from (12.37) we have

$$
\int\limits_{\Omega^+} \left[k|\nabla\theta|^2 + T_0(k_{\alpha\beta} \nabla p_\beta \cdot \nabla p_\alpha + d_{\alpha\beta} p_\beta \overline{p_\alpha})\right] d\mathbf{x} = 0.
$$

On the basis of (4.10) from the last equation we get

$$
k_{\alpha\beta} \nabla p_\beta \cdot \nabla p_\alpha = 0, \qquad d_{\alpha\beta} p_\beta \overline{p_\alpha} = 0, \qquad \nabla\theta = 0, \tag{12.41}
$$

and hence, by using Lemma 6.1 from (12.41) we obtain

$$
\begin{aligned}
&p_\alpha(\mathbf{x}) = c_\alpha = \text{const}, \qquad \theta(\mathbf{x}) = c_5 = \text{const}, \\
&d_{\alpha\beta} p_\beta(\mathbf{x}) = 0, \qquad \alpha = 1, 2, 3, 4
\end{aligned}
\tag{12.42}
$$

for $\mathbf{x} \in \Omega^+$. On the basis of homogeneous boundary condition (12.36) from (12.42) we have $\boldsymbol{\vartheta}(\mathbf{x}) = \mathbf{0}$ for $\mathbf{x} \in \Omega^+$. □

Theorem 12.19 *If the condition (9.55) is satisfied, then the internal BVP $(II^r)^+_{\mathbf{F},\mathbf{f}}$ admits at most one regular (classical) solution of the class $\mathfrak{R}^{(r)}$ in Ω^+.*

Proof We suppose that there are two regular solutions of the problem $(II^r)^+_{\mathbf{F},\mathbf{f}}$ in the class $\mathfrak{R}^{(r)}$. Then their difference $\boldsymbol{\vartheta}$ is a regular solution of the homogeneous BVP $(II^r)^+_{\mathbf{0},\mathbf{0}}$. Hence, $\boldsymbol{\vartheta}$ is a regular solution of the homogeneous system of Eq. (12.28) in Ω^+ satisfying the homogeneous boundary condition

$$
\{\mathbf{P}^{(r)}(\mathbf{D}_\mathbf{z}, \mathbf{n})\boldsymbol{\vartheta}(\mathbf{z})\}^+ = \mathbf{0}. \tag{12.43}
$$

Quite similarly as in Theorem 12.18, on the basis of (12.18), (12.28), and (12.36), from (12.20) we have Eqs. (12.42). Furthermore, if we employ Eqs. (12.42), then from (12.28) we may obtain

$$b_{\alpha\beta} c_\beta + \varepsilon_\alpha c_5 = 0, \qquad \varepsilon_\beta c_\beta + a_0 c_5 = 0, \qquad \alpha = 1, 2, 3, 4. \tag{12.44}$$

In view of the condition (9.55), from (12.44) we may deduce $c_j = 0 (j = 1, 2, \cdots, 5)$. Then from (12.42) we have $\boldsymbol{\vartheta}(\mathbf{x}) = \mathbf{0}$ for $\mathbf{x} \in \Omega^+$. □

Quite similar manner as in Theorems 12.18 and 12.19, on the basis of (12.17) we can prove the following.

Theorem 12.20 *The external BVP* $(\mathbb{K}^r)^-_{\mathbf{F},\mathbf{f}}$ *has one regular solution of the class* $\mathfrak{R}^{(r)}$, *where* $\mathbb{K} = I, II$.

12.6 Basic Properties of Potentials

On the basis of integral representation of regular vector (see (12.26) and (12.27)) we introduce the following notation

(i) $\mathbf{Q}^{(r,1)}(\mathbf{x}, \mathbf{g}) = \int_S \boldsymbol{\Psi}^{(r)}(\mathbf{x} - \mathbf{y}, \omega)\mathbf{g}(\mathbf{y}) d_\mathbf{y} S,$

(ii) $\mathbf{Q}^{(r,2)}(\mathbf{x}, \mathbf{g}) = \int_S [\mathbf{P}^{(r)}(\mathbf{D}_\mathbf{y}, \mathbf{n}(\mathbf{y}))\boldsymbol{\Psi}^{(r)^\top}(\mathbf{x} - \mathbf{y}, \omega)]^\top \mathbf{g}(\mathbf{y}) d_\mathbf{y} S,$

(iii) $\mathbf{Q}^{(r,3)}(\mathbf{x}, \boldsymbol{\phi}, \Omega^\pm) = \int_{\Omega^\pm} \boldsymbol{\Psi}^{(r)}(\mathbf{x} - \mathbf{y}, \omega)\boldsymbol{\phi}(\mathbf{y}) d\mathbf{y},$

where $\boldsymbol{\Psi}^{(r)}(\mathbf{x}, \omega)$ is the fundamental matrix of the operator $\mathbf{A}^{(r)}(\mathbf{D}_\mathbf{x}, \omega)$ and given by (12.6); the matrix differential operator $\mathbf{P}^{(r)}(\mathbf{D}_\mathbf{y}, \mathbf{n}(\mathbf{y}))$ is defined by (12.10); $\mathbf{g}$ and $\boldsymbol{\phi}$ are five-component vector functions.

As in mathematical physics (see, e.g., Günther [161], Hsiao and Wendland [169], and Kellogg [195]), the vector functions $\mathbf{Q}^{(r,1)}(\mathbf{x}, \mathbf{g})$, $\mathbf{Q}^{(r,2)}(\mathbf{x}, \mathbf{g})$, and $\mathbf{Q}^{(r,3)}(\mathbf{x}, \boldsymbol{\phi}, \Omega^\pm)$ are called *the single-layer, double-layer*, and *volume potentials* in the steady vibrations problems of the linear theory of heat conduction for quadruple porosity rigid body, respectively.

Obviously, on the basis of Theorems 12.16 and 12.17, the regular vector in Ω^+ and Ω^- is represented by sum of the surface (single-layer and double-layer) and volume potentials as follows

$$\boldsymbol{\vartheta}(\mathbf{x}) = \mathbf{Q}^{(r,2)}\left(\mathbf{x}, \{\boldsymbol{\vartheta}\}^+\right) - \mathbf{Q}^{(r,1)}\left(\mathbf{x}, \{\mathbf{P}^{(r)}\boldsymbol{\vartheta}\}^+\right) + \mathbf{Q}^{(r,3)}\left(\mathbf{x}, \mathbf{A}^{(r)}\boldsymbol{\vartheta}, \Omega^+\right)$$

for $\mathbf{x} \in \Omega^+$ and

$$\boldsymbol{\vartheta}(\mathbf{x}) = -\mathbf{Q}^{(r,2)}\left(\mathbf{x}, \{\boldsymbol{\vartheta}\}^-\right) + \mathbf{Q}^{(r,1)}\left(\mathbf{x}, \{\mathbf{P}^{(r)}\boldsymbol{\vartheta}\}^-\right) + \mathbf{Q}^{(r,3)}\left(\mathbf{x}, \mathbf{A}^{(r)}\boldsymbol{\vartheta}, \Omega^-\right)$$

for $\mathbf{x} \in \Omega^-$, respectively.

We now establish the basic properties of these potentials.

12.6.1 Single-Layer Potential

We have the following results.

Theorem 12.21 *If S is the Liapunov surface of the class $C^{1,\nu}$, $0 < \nu \le 1$ and $\mathbf{g}$ is an integrable vector function on S, then*

(i)

$$\mathbf{Q}^{(r,1)}(\cdot, \mathbf{g}) \in C^\infty(\Omega^\pm),$$

(ii) $\mathbf{Q}^{(r,1)}(\cdot, \mathbf{g})$ *is a solution of the homogeneous equation (12.28) in Ω^+ and Ω^-, i.e.*

$$\mathbf{A}^{(r)}(\mathbf{D}_\mathbf{x}, \omega)\, \mathbf{Q}^{(r,1)}(\mathbf{x}, \mathbf{g}) = \mathbf{0} \tag{12.45}$$

for $\mathbf{x} \in \Omega^\pm$.

Theorem 12.22 *If $S \in C^{1,\nu}$, $0 < \nu \le 1$, $\mathbf{g}$ is an integrable and bounded vector function on S, then*

$$Q_l^{(r,1)}(\mathbf{x}, \mathbf{g}) = O\left(\frac{e^{-\hat{\mu}_0|\mathbf{x}|}}{|\mathbf{x}|}\right), \qquad Q_{l,j}^{(r,1)}(\mathbf{x}, \mathbf{g}) = O\left(\frac{e^{-\hat{\mu}_0|\mathbf{x}|}}{|\mathbf{x}|^2}\right) \tag{12.46}$$

for $|\mathbf{x}| \to +\infty$, where $\mathbf{Q}^{(r,1)} = \left(Q_1^{(r,1)}, Q_2^{(r,1)}, \cdots, Q_5^{(r,1)}\right)$, $j = 1, 2, 3$, $l = 1, 2, \cdots, 5$ and $\hat{\mu}_0 = \min\{\mathrm{Im}\hat{\mu}_1, \mathrm{Im}\hat{\mu}_2, \cdots, \mathrm{Im}\hat{\mu}_5\} > 0$.

Theorem 12.23 *If $S \in C^{1,\nu}$, $0 < \nu \le 1$, $\mathbf{g}$ is an integrable and bounded vector function on S, then*

(i) the integrals $\mathbf{Q}^{(r,1)}(\mathbf{z}, \mathbf{g})$, $\left\{\mathbf{Q}^{(r,1)}(\mathbf{z}, \mathbf{g})\right\}^+$ and $\left\{\mathbf{Q}^{(r,1)}(\mathbf{z}, \mathbf{g})\right\}^-$ exist for $\mathbf{z} \in S$ and

$$\mathbf{Q}^{(r,1)}(\mathbf{z}, \mathbf{g}) = \left\{\mathbf{Q}^{(r,1)}(\mathbf{z}, \mathbf{g})\right\}^+ = \left\{\mathbf{Q}^{(r,1)}(\mathbf{z}, \mathbf{g})\right\}^-, \tag{12.47}$$

(ii)

$$\mathbf{Q}^{(r,1)}(\cdot, \mathbf{g}) \in C(\mathbb{R}^3),$$

(iii)

$$\mathbf{P}^{(r)}(\mathbf{D_z}, \mathbf{n})\, \mathbf{Q}^{(r,1)}(\cdot, \mathbf{g}) \in C^{0,\nu'}(S),$$

where $0 < \nu' < \nu \le 1$ *and*

$$\left\{\mathbf{Q}^{(r,1)}(\mathbf{z}, \mathbf{g})\right\}^{\pm} \equiv \lim_{\Omega^{\pm} \ni \mathbf{x} \to \mathbf{z} \in S} \mathbf{Q}^{(r,1)}(\mathbf{x}, \mathbf{g}).$$

Theorem 12.24 *If* $S \in C^{1,\nu}$, $0 < \nu \le 1$, $\mathbf{g}$ *is a continuous vector function on* S, *then the vector functions* $\mathbf{P}^{(r)}(\mathbf{D_z}, \mathbf{n})\, \mathbf{Q}^{(r,1)}(\mathbf{z}, \mathbf{g})$, $\left\{\mathbf{P}^{(r)}(\mathbf{D_z}, \mathbf{n})\, \mathbf{Q}^{(r,1)}(\mathbf{z}, \mathbf{g})\right\}^{+}$, *and* $\left\{\mathbf{P}^{(r)}(\mathbf{D_z}, \mathbf{n})\, \mathbf{Q}^{(r,1)}(\mathbf{z}, \mathbf{g})\right\}^{-}$ *exist for* $\mathbf{z} \in S$ *and*

$$\left\{\mathbf{P}^{(r)}(\mathbf{D_z}, \mathbf{n})\, \mathbf{Q}^{(r,1)}(\mathbf{z}, \mathbf{g})\right\}^{\pm} = \mp\frac{1}{2}\mathbf{g}(\mathbf{z}) + \mathbf{P}^{(r)}(\mathbf{D_z}, \mathbf{n})\, \mathbf{Q}^{(r,1)}(\mathbf{z}, \mathbf{g}), \qquad (12.48)$$

where

$$\left\{\mathbf{P}^{(r)}(\mathbf{D_z}, \mathbf{n})\, \mathbf{Q}^{(r,1)}(\mathbf{z}, \mathbf{g})\right\}^{\pm} \equiv \lim_{\Omega^{\pm} \ni \mathbf{x} \to \mathbf{z} \in S} \mathbf{P}^{(r)}(\mathbf{D_x}, \mathbf{n})\, \mathbf{Q}^{(r,1)}(\mathbf{x}, \mathbf{g}).$$

We now prove the following useful property of the single-layer potential.

Theorem 12.25 *The single-layer potential* $\mathbf{Q}^{(r,1)}(\mathbf{x}, \mathbf{g}) \equiv \mathbf{0}$ *for* $\mathbf{x} \in \Omega^{+}$ *(or* $\mathbf{x} \in \Omega^{-}$*) if and only if* $\mathbf{g}(\mathbf{z}) \equiv \mathbf{0}$, *where* $\mathbf{z} \in S$.

Proof Obviously, if $\mathbf{g}(\mathbf{z}) \equiv \mathbf{0}$ for $\mathbf{z} \in S$, then $\mathbf{Q}^{(r,1)}(\mathbf{x}, \mathbf{g}) \equiv \mathbf{0}$ for $\mathbf{x} \in \Omega^{\pm}$. Let $\mathbf{Q}^{(r,1)}(\mathbf{x}, \mathbf{g}) \equiv \mathbf{0}$ for $\mathbf{x} \in \Omega^{+}$. Then we have

$$\left\{\mathbf{Q}^{(r,1)}(\mathbf{z}, \mathbf{g})\right\}^{+} = \left\{\mathbf{P}^{(r)}(\mathbf{D_z}, \mathbf{n})\, \mathbf{Q}^{(r,1)}(\mathbf{z}, \mathbf{g})\right\}^{+} = \mathbf{0}. \qquad (12.49)$$

On the basis of (12.49) from (12.47) it follows that

$$\left\{\mathbf{Q}^{(r,1)}(\mathbf{z}, \mathbf{g})\right\}^{-} = \mathbf{0} \qquad (12.50)$$

for $\mathbf{z} \in S$. Furthermore, on the basis of (12.45) the vector function $\mathbf{Q}^{(r,1)}(\mathbf{x}, \mathbf{g})$ is a solution of the homogeneous equation (12.28) for $\mathbf{x} \in \Omega^{-}$ and satisfies the homogeneous boundary condition (12.50), i.e., $\mathbf{Q}^{(r,1)}(\mathbf{x}, \mathbf{g})$ is a regular solution of the external BVP $(I^r)^{-}_{\mathbf{0},\mathbf{0}}$. Keeping in mind Theorem 12.20 we may see that $\mathbf{Q}^{(r,1)}(\mathbf{x}, \mathbf{g}) \equiv \mathbf{0}$ for $\mathbf{x} \in \Omega^{-}$. Then we have

$$\left\{\mathbf{P}^{(r)}(\mathbf{D_z}, \mathbf{n})\, \mathbf{Q}^{(r,1)}(\mathbf{z}, \mathbf{g})\right\}^{-} = \mathbf{0}. \qquad (12.51)$$

Combining (12.48) with (12.49) and (12.51) we may further conclude that $\mathbf{g}(\mathbf{z}) \equiv \mathbf{0}$ for $\mathbf{z} \in S$.

Quite similarly, if $\mathbf{Q}^{(r,1)}(\mathbf{x},\mathbf{g}) \equiv \mathbf{0}$ for $\mathbf{x} \in \Omega^-$, then the vector function $\mathbf{Q}^{(r,1)}(\mathbf{x},\mathbf{g})$ is a regular solution of the internal BVP $(I^r)^+_{\mathbf{0},\mathbf{0}}$. By using Theorem 12.18 we may see that $\mathbf{Q}^{(r,1)}(\mathbf{x},\mathbf{g}) \equiv \mathbf{0}$ for $\mathbf{x} \in \Omega^+$. These equations allow us to show $\mathbf{g}(\mathbf{z}) \equiv \mathbf{0}$ for $\mathbf{z} \in S$. □

12.6.2 Double-Layer Potential

On the basis of Theorem 12.1 and Eq. (12.9) we have the following consequences.

Theorem 12.26 *If $S \in C^{1,\nu}$, $0 < \nu \le 1$, $\mathbf{g}$ is an integrable vector function on S, then*

(i)

$$\mathbf{Q}^{(r,2)}(\cdot,\mathbf{g}) \in C^\infty(\Omega^\pm),$$

(ii) $\mathbf{Q}^{(r,2)}(\cdot,\mathbf{g})$ is a solution of the homogeneous equation (12.28) in Ω^+ and Ω^-, i.e.

$$\mathbf{A}^{(r)}(\mathbf{D}_\mathbf{x},\omega)\,\mathbf{Q}^{(r,2)}(\mathbf{x},\mathbf{g}) = \mathbf{0} \tag{12.52}$$

for $\mathbf{x} \in \Omega^\pm$.

Theorem 12.27 *If $S \in C^{1,\nu}$, $0 < \nu \le 1$, $\mathbf{g}$ is an integrable and bounded vector function on S, $\mathbf{Q}^{(r,2)} = \left(Q_1^{(r,2)}, Q_2^{(r,2)}, \cdots, Q_5^{(r,2)}\right)$, then*

$$Q_l^{(r,2)}(\mathbf{x},\mathbf{g}) = O\left(\frac{e^{-\hat{\mu}_0|\mathbf{x}|}}{|\mathbf{x}|^2}\right), \qquad Q_{l,j}^{(r,2)}(\mathbf{x},\mathbf{g}) = O\left(\frac{e^{-\hat{\mu}_0|\mathbf{x}|}}{|\mathbf{x}|^3}\right) \tag{12.53}$$

for $|\mathbf{x}| \to +\infty$, where $j = 1,2,3$, $l = 1,2,3,4$, and $\hat{\mu}_0$ is defined in Theorem 12.22.

Theorem 12.28 *If $S \in C^{1,\nu}$, $0 < \nu \le 1$, then the matrices $\mathbf{P}^{(r)}(\mathbf{D}_\mathbf{z},\mathbf{n})\boldsymbol{\Psi}^{(r)}(\mathbf{z}-\mathbf{y},\omega)$ and $[\mathbf{P}^{(r)}(\mathbf{D}_\mathbf{y},\mathbf{n})\boldsymbol{\Psi}^{(r)}(\mathbf{z}-\mathbf{y},\omega)]^\top$ are the weakly singular kernels of $\mathbf{P}^{(r)}(\mathbf{D}_\mathbf{z},\mathbf{n})\mathbf{Q}^{(r,1)}(\mathbf{z},\mathbf{g})$ and $\mathbf{Z}^{(r,2)}(\mathbf{z},\mathbf{g})$, respectively, i.e.*

$$\begin{aligned}
\left|[\mathbf{P}^{(r)}(\mathbf{D}_\mathbf{z},\mathbf{n})\boldsymbol{\Psi}^{(r)}(\mathbf{z}-\mathbf{y},\omega)]_{lj}\right| &< \frac{\text{const}}{|\mathbf{z}-\mathbf{y}|^{2-\nu}},\\
\left|[\mathbf{P}^{(r)}(\mathbf{D}_\mathbf{y},\mathbf{n})\boldsymbol{\Psi}^{(r)}(\mathbf{z}-\mathbf{y},\omega)]_{lj}\right| &< \frac{\text{const}}{|\mathbf{z}-\mathbf{y}|^{2-\nu}}
\end{aligned} \tag{12.54}$$

for $\mathbf{z}, \mathbf{y} \in S$, $l, j = 1, 2, \cdots, 5$, *and* const *is a positive number independent on* $\mathbf{z}$ *and* $\mathbf{y}$.

Theorem 12.29 *If* $S \in C^{1,\nu}$, $0 < \nu \leq 1$, $\mathbf{g}$ *is a continuous vector function on* S, *then*

(i) the integrals $\mathbf{Q}^{(r,2)}(\mathbf{z}, \mathbf{g})$, $\left\{\mathbf{Q}^{(r,2)}(\mathbf{z}, \mathbf{g})\right\}^{+}$, *and* $\left\{\mathbf{Q}^{(r,2)}(\mathbf{z}, \mathbf{g})\right\}^{-}$ *exist,*
(ii)

$$\left\{\mathbf{Q}^{(r,2)}(\mathbf{z}, \mathbf{g})\right\}^{\pm} = \pm\frac{1}{2}\mathbf{g}(\mathbf{z}) + \mathbf{Q}^{(r,2)}(\mathbf{z}, \mathbf{g}) \tag{12.55}$$

for $\mathbf{z} \in S$.

Theorem 12.30 *If* $S \in C^{1,\nu}$, $\mathbf{g} \in C^{0,\nu'}(S)$, $0 < \nu' < \nu \leq 1$, *then*

$$\mathbf{Q}^{(r,2)}(\cdot, \mathbf{g}) \in C^{0,\nu'}(S).$$

We now formulate the Liapunov–Tauber type theorem in the linear theory of heat conduction for quadruple porosity rigid body.

Theorem 12.31 *If* $S \in C^{1,\nu}$, $\mathbf{g} \in C^{1,\nu}(S)$, $0 < \nu \leq 1$, *then*

(i) the vector functions $\left\{\mathbf{P}^{(r)}(\mathbf{D_z}, \mathbf{n})\mathbf{Q}^{(r,2)}(\mathbf{z}, \mathbf{g})\right\}^{+}$ *and* $\left\{\mathbf{P}^{(r)}(\mathbf{D_z}, \mathbf{n})\mathbf{Q}^{(r,2)}(\mathbf{z}, \mathbf{g})\right\}^{-}$ *exist,*
(ii) they are elements of the class $C^{1,\nu}(S)$, *and*
(iii)

$$\left\{\mathbf{P}^{(r)}(\mathbf{D_z}, \mathbf{n})\mathbf{Q}^{(r,2)}(\mathbf{z}, \mathbf{g})\right\}^{+} = \left\{\mathbf{P}^{(r)}(\mathbf{D_z}, \mathbf{n})\mathbf{Q}^{(r,2)}(\mathbf{z}, \mathbf{g})\right\}^{-}$$

for $\mathbf{z} \in S$.

12.6.3 Volume Potential

We have the following results.

Theorem 12.32 *If* $S \in C^{1,\nu}$, $\boldsymbol{\phi}$ *is an integrable and bounded vector function in* Ω^{+}, *then*

(i)

$$\mathbf{Q}^{(r,3)}(\cdot, \boldsymbol{\phi}, \Omega^{+}) \in C^{\infty}(\Omega^{-}),$$

(ii)

$$\mathbf{Q}^{(r,3)}(\cdot, \boldsymbol{\phi}, \Omega^{+}) \in C^{1,\nu}(\mathbb{R}^{3}),$$

(iii) $\mathbf{Q}^{(r,3)}(\cdot, \boldsymbol{\phi}, \Omega^+)$ *is a solution of the homogeneous equation (12.28), i.e.*

$$\mathbf{A}^{(r)}(\mathbf{D_x}, \omega)\, \mathbf{Q}^{(r,3)}(\mathbf{x}, \boldsymbol{\phi}, \Omega^+) = \mathbf{0}$$

for $\mathbf{x} \in \Omega^-$.

Theorem 12.33 *If* $S \in C^{1,\nu}$, $0 < \nu \le 1$, $\boldsymbol{\phi} \in C^1(\Omega^+)$, *then*

$$\mathbf{Q}^{(r,3)}(\cdot, \boldsymbol{\phi}, \Omega^+) \in C^2(\Omega^+).$$

Theorem 12.34 *If* $S \in C^{1,\nu}$, $\boldsymbol{\phi}$ *is an integrable and bounded vector function in* Ω^- *and* supp $\boldsymbol{\phi}$ *is a finite domain in* Ω^-, *then*

(i)

$$\mathbf{Q}^{(r,3)}(\cdot, \boldsymbol{\phi}, \Omega^-) \in C^\infty(\Omega^+),$$

(ii)

$$\mathbf{Q}^{(r,3)}(\cdot, \boldsymbol{\phi}, \Omega^-) \in C^{1,\nu}(\mathbb{R}^3),$$

(iii) $\mathbf{Q}^{(r,3)}(\cdot, \boldsymbol{\phi}, \Omega^-)$ *is a solution of the homogeneous equation (12.28), i.e.*

$$\mathbf{A}^{(r)}(\mathbf{D_x}, \omega)\, \mathbf{Q}^{(r,3)}(\mathbf{x}, \boldsymbol{\phi}, \Omega^-) = \mathbf{0}$$

for $\mathbf{x} \in \Omega^+$.

Theorem 12.35 *If* $S \in C^{1,\nu}$, $0 < \nu \le 1$, $\boldsymbol{\phi} \in C^1(\Omega^-)$, *and* supp $\boldsymbol{\phi}$ *is a finite domain in* Ω^-, *then*

$$\mathbf{Q}^{(r,3)}(\cdot, \boldsymbol{\phi}, \Omega^-) \in C^2(\Omega^-).$$

We now formulate Poisson-type theorems in the linear theory of quadruple porosity rigid body for domains Ω^+ and Ω^-.

Theorem 12.36 *If* $S \in C^{1,\nu}$, $0 < \nu \le 1$, $\boldsymbol{\phi} \in C^{0,\nu}(\Omega^+)$, *then*

$$\mathbf{A}^{(r)}(\mathbf{D_x}, \omega)\, \mathbf{Q}^{(r,3)}(\mathbf{x}, \boldsymbol{\phi}, \Omega^+) = \boldsymbol{\phi}(\mathbf{x})$$

for $\mathbf{x} \in \Omega^+$.

Theorem 12.37 *If* $S \in C^{1,\nu}$, $0 < \nu \le 1$, $\boldsymbol{\phi} \in C^{0,\nu}(\Omega^+)$ *and* supp $\boldsymbol{\phi}$ *is a finite domain in* Ω^-, *then*

$$\mathbf{A}^{(r)}(\mathbf{D_x}, \omega)\, \mathbf{Q}^{(r,3)}(\mathbf{x}, \boldsymbol{\phi}, \Omega^-) = \boldsymbol{\phi}(\mathbf{x})$$

for $\mathbf{x} \in \Omega^-$.

12.7 Existence Theorems

We are now in a position to study the existence of regular (classical) solutions of the BVPs $(\mathbb{K}^r)^+_{\mathbf{F},\mathbf{f}}$ and $(\mathbb{K}^r)^-_{\mathbf{F},\mathbf{f}}$ by means of the potential method and the theory of Fredholm integral equations, where $\mathbb{K} = I, II$.

Obviously, by Theorems 12.36 and 12.37 the volume potential $\mathbf{Q}^{(r,3)}(\mathbf{x}, \mathbf{F}, \Omega^\pm)$ is a particular regular solution of the nonhomogeneous equation (12.31), where $\mathbf{F} \in C^{0,\nu'}(\Omega^\pm)$, $0 < \nu' \le 1$; $\operatorname{supp} \mathbf{F}$ is a finite domain in Ω^-. Therefore, in this section we will consider problems $(\mathbb{K}^r)^+_{\mathbf{0},\mathbf{f}}$ and $(\mathbb{K}^r)^-_{\mathbf{0},\mathbf{f}}$, where $\mathbb{K} = I, II$.

We introduce the integral operators

$$
\begin{aligned}
\mathcal{H}^{(r,1)}\mathbf{g}(\mathbf{z}) &\equiv \frac{1}{2}\,\mathbf{g}(\mathbf{z}) + \mathbf{Q}^{(r,2)}(\mathbf{z}, \mathbf{g}),\\
\mathcal{H}^{(r,2)}\mathbf{h}(\mathbf{z}) &\equiv \frac{1}{2}\,\mathbf{h}(\mathbf{z}) + \mathbf{P}^{(r)}(\mathbf{D}_\mathbf{z}, \mathbf{n})\mathbf{Q}^{(r,1)}(\mathbf{z}, \mathbf{h}),\\
\mathcal{H}^{(r,3)}\mathbf{g}(\mathbf{z}) &\equiv -\frac{1}{2}\,\mathbf{g}(\mathbf{z}) + \mathbf{Q}^{(r,2)}(\mathbf{z}, \mathbf{g}),\\
\mathcal{H}^{(r,4)}\mathbf{h}(\mathbf{z}) &\equiv -\frac{1}{2}\,\mathbf{h}(\mathbf{z}) + \mathbf{P}^{(r)}(\mathbf{D}_\mathbf{z}, \mathbf{n})\mathbf{Q}^{(r,1)}(\mathbf{z}, \mathbf{h}),
\end{aligned}
\tag{12.56}
$$

where $\mathbf{g}$ and $\mathbf{h}$ are five-component vector functions on S and $\mathbf{z} \in S$. On the basis of (12.54), the operator $\mathcal{H}^{(r,\alpha)}$ $(\alpha = 1, 2, 3, 4)$ has a weakly singular kernel and consequently, it is a Fredholm operator. Moreover, operators $\mathcal{H}^{(r,j)}$ and $\mathcal{H}^{(r,j+1)}$ are adjoint with respect to each other, where $j = 1, 3$.

Problem $(I^r)^+_{\mathbf{0},\mathbf{f}}$ We seek a regular solution of the class $\mathfrak{R}^{(r)}$ in Ω^+ to this problem in the form

$$
\boldsymbol{\vartheta}(\mathbf{x}) = \mathbf{Q}^{(r,2)}(\mathbf{x}, \mathbf{g}) \qquad \text{for} \qquad \mathbf{x} \in \Omega^+, \tag{12.57}
$$

where $\mathbf{g}$ is the required five-component vector function.

By Eq. (12.52) the vector function $\boldsymbol{\vartheta}$ is a solution of (12.28) for $\mathbf{x} \in \Omega^+$. Keeping in mind the boundary condition (12.32) and using (12.55) and (12.56), from (12.57) we obtain, for determining the unknown vector $\mathbf{g}$, a Fredholm integral equation

$$
\mathcal{H}^{(r,1)}\mathbf{g}(\mathbf{z}) = \mathbf{f}(\mathbf{z}) \qquad \text{for} \quad \mathbf{z} \in S. \tag{12.58}
$$

We prove that (12.58) is always solvable for an arbitrary vector $\mathbf{f}$. Let us consider the adjoint homogeneous equation

$$
\mathcal{H}^{(r,2)}\mathbf{h}_0(\mathbf{z}) = \mathbf{0} \qquad \text{for} \quad \mathbf{z} \in S, \tag{12.59}
$$

where $\mathbf{h}_0$ is the required five-component vector function. Now we prove that (12.59) has only the trivial solution.

Indeed, let $\mathbf{h}_0$ be a solution of the homogeneous equation (12.59). On the basis of (12.45), (12.48), and (12.59) the vector function $\mathbf{Q}^{(r,1)}(\mathbf{x}, \mathbf{h}_0)$ is a regular solution of problem $(II^r)^-_{\mathbf{0},\mathbf{0}}$. Using Theorem 12.20, the problem $(II^r)^-_{\mathbf{0},\mathbf{0}}$ has only the trivial solution, that is

$$\mathbf{Q}^{(r,1)}(\mathbf{x}, \mathbf{h}_0) \equiv \mathbf{0} \qquad \text{for} \qquad \mathbf{x} \in \Omega^-. \tag{12.60}$$

On the other hand, by (12.47) and (12.60) we get

$$\{\mathbf{Q}^{(r,1)}(\mathbf{z}, \mathbf{h}_0)\}^+ = \{\mathbf{Q}^{(r,1)}(\mathbf{z}, \mathbf{h}_0)\}^- = \mathbf{0} \qquad \text{for} \qquad \mathbf{z} \in S,$$

i.e., the vector $\mathbf{Q}^{(r,1)}(\mathbf{x}, \mathbf{h}_0)$ is a regular solution of problem $(I_r)^+_{\mathbf{0},\mathbf{0}}$. By using Theorem 12.18, the problem $(I^r)^+_{\mathbf{0},\mathbf{0}}$ has only the trivial solution, that is

$$\mathbf{Q}^{(r,1)}(\mathbf{x}, \mathbf{h}_0) \equiv \mathbf{0} \qquad \text{for} \qquad \mathbf{x} \in \Omega^+. \tag{12.61}$$

By virtue of (12.60), (12.61), and identity (12.48) we obtain

$$\mathbf{h}_0(\mathbf{z}) = \{\mathbf{P}^{(r)}(\mathbf{D}_\mathbf{z}, \mathbf{n})\mathbf{Q}^{(r,1)}(\mathbf{z}, \mathbf{h}_0)\}^- - \{\mathbf{P}^{(r)}(\mathbf{D}_\mathbf{z}, \mathbf{n})\mathbf{Q}^{(r,1)}(\mathbf{z}, \mathbf{h}_0)\}^+ = \mathbf{0}$$

for $\mathbf{z} \in S$. Thus, the homogeneous equation (12.59) has only the trivial solution and therefore (12.58) is always solvable for an arbitrary vector $\mathbf{f}$.

We have thereby proved the following.

Theorem 12.38 *If $S \in C^{1,\nu}$, $\mathbf{f} \in C^{1,\nu'}(S)$, $0 < \nu' < \nu \le 1$, then a regular (classical) solution of the class $\mathfrak{R}^{(r)}$ of problem $(I^r)^+_{\mathbf{0},\mathbf{f}}$ exists, is unique, and is represented by double-layer potential (12.57), where $\mathbf{g}$ is a solution of the Fredholm integral equation (12.58) which is always solvable for an arbitrary vector* $\mathbf{f}$.

Problem $(II^r)^-_{\mathbf{0},\mathbf{f}}$ We seek a regular solution of the class $\mathfrak{R}^{(r)}$ in Ω^- to this problem in the form

$$\boldsymbol{\vartheta}(\mathbf{x}) = \mathbf{Q}^{(r,1)}(\mathbf{x}, \mathbf{h}) \qquad \text{for} \qquad \mathbf{x} \in \Omega^-, \tag{12.62}$$

where $\mathbf{h}$ is the required five-component vector function.

Obviously, by Theorem 12.21 and the condition (12.46) the vector function $\boldsymbol{\vartheta}$ is a regular solution of (12.28) for $\mathbf{x} \in \Omega^-$. Keeping in mind the boundary condition (12.35) and using (12.48), from (12.62) we obtain, for determining the unknown vector $\mathbf{h}$, a Fredholm integral equation

$$\mathcal{H}^{(r,2)}\mathbf{h}(\mathbf{z}) = \mathbf{f}(\mathbf{z}) \qquad \text{for} \quad \mathbf{z} \in S. \tag{12.63}$$

It has been proved above that the corresponding homogeneous equation (12.59) has only the trivial solution. Hence, it follows that (12.63) is always solvable.

We have thereby proved the following.

Theorem 12.39 *If* $S \in C^{1,\nu}$, $\mathbf{f} \in C^{0,\nu'}(S)$, $0 < \nu' < \nu \le 1$, *then a regular (classical) solution of the class* $\mathfrak{R}^{(r)}$ *of problem* $(II^r)^-_{\mathbf{0},\mathbf{f}}$ *exists, is unique, and is represented by single-layer potential (12.62), where* $\mathbf{h}$ *is a solution of the Fredholm integral equation (12.63) which is always solvable for an arbitrary vector* $\mathbf{f}$.

Problem $(II^r)^+_{\mathbf{0},\mathbf{f}}$ We seek a regular solution of the class $\mathfrak{R}^{(r)}$ in Ω^+ to this problem in the form

$$\boldsymbol{\vartheta}(\mathbf{x}) = \mathbf{Q}^{(r,1)}(\mathbf{x}, \mathbf{g}) \qquad \text{for} \qquad \mathbf{x} \in \Omega^+, \tag{12.64}$$

where $\mathbf{g}$ is the required five-component vector function.

Obviously, by Theorem 12.21 the vector function $\boldsymbol{\vartheta}$ is a solution of (12.28) for $\mathbf{x} \in \Omega^+$. Keeping in mind the boundary condition (12.33) and using (12.48), from (12.64) we obtain, for determining the unknown vector $\mathbf{g}$, the following integral equation

$$\mathcal{H}^{(r,4)}\mathbf{g}(\mathbf{z}) = \mathbf{f}(\mathbf{z}) \qquad \text{for} \qquad \mathbf{z} \in S. \tag{12.65}$$

We prove that (12.65) is always solvable for an arbitrary vector $\mathbf{f}$. Let us consider the corresponding homogeneous equation

$$\mathcal{H}^{(r,4)}\mathbf{g}_0(\mathbf{z}) = \mathbf{0} \qquad \text{for} \quad \mathbf{z} \in S, \tag{12.66}$$

where $\mathbf{g}_0$ is the required five-component vector function. Now we prove that (12.66) has only the trivial solution.

Indeed, let $\mathbf{g}_0$ be a solution of the homogeneous equation (12.66). On the basis of Theorem 12.21, the identity (12.48), Eqs. (12.43) and (12.66) the vector $\mathbf{Q}^{(r,1)}(\mathbf{x}, \mathbf{g}_0)$ is a regular solution of problem $(II^r)^+_{\mathbf{0},\mathbf{0}}$. Using Theorem 12.19, the problem $(II^r)^+_{\mathbf{0},\mathbf{0}}$ has only the trivial solution, that is

$$\mathbf{Q}^{(r,1)}(\mathbf{x}, \mathbf{g}_0) \equiv \mathbf{0} \qquad \text{for} \qquad \mathbf{x} \in \Omega^+. \tag{12.67}$$

By Theorem 12.25 from (12.67) we have $\mathbf{g}_0(\mathbf{z}) = \mathbf{0}$. Thus, the homogeneous equation (12.66) has only a trivial solution and therefore (12.65) is always solvable for an arbitrary vector $\mathbf{f}$.

We have thereby proved the following.

Theorem 12.40 *If* $S \in C^{1,\nu}$, $\mathbf{f} \in C^{0,\nu'}(S)$, $0 < \nu' < \nu \le 1$, *then a regular (classical) solution of the class* $\mathfrak{R}^{(r)}$ *of problem* $(II^r)^+_{\mathbf{0},\mathbf{f}}$ *exists, is unique, and is represented by single-layer potential (12.64), where* $\mathbf{g}$ *is a solution of the Fredholm integral equation (12.65) which is always solvable for an arbitrary vector* $\mathbf{f}$.

Problem $(I^r)^-_{\mathbf{0},\mathbf{f}}$ We seek a regular (classical) solution of the class $\mathfrak{R}^{(r)}$ in Ω^- to this problem in the form

$$\boldsymbol{\vartheta}(\mathbf{x}) = \mathbf{Q}^{(r,2)}(\mathbf{x}, \mathbf{h}) \qquad \text{for} \qquad \mathbf{x} \in \Omega^-, \tag{12.68}$$

where $\mathbf{h}$ is the required five-component vector function.

Obviously, by relations (12.52) and (12.53) the vector function $\boldsymbol{\vartheta}$ is a regular solution of (12.28) for $\mathbf{x} \in \Omega^-$. Keeping in mind the boundary condition (12.34) and using (12.55), from (12.68) we obtain, for determining the unknown vector $\mathbf{h}$, the following integral equation

$$\mathcal{H}^{(r,3)}\mathbf{h}(\mathbf{z}) = \mathbf{f}(\mathbf{z}) \qquad \text{for} \qquad \mathbf{z} \in S. \tag{12.69}$$

It has been proved above that the adjoint homogeneous equation (12.66) has only the trivial solution. Hence, it follows that (12.69) is always solvable.

We have thereby proved the following.

Theorem 12.41 *If $S \in C^{1,\nu}$, $\mathbf{f} \in C^{1,\nu'}(S)$, $0 < \nu' < \nu \leq 1$, then a regular (classical) solution of the class $\mathfrak{R}^{(r)}$ of problem $(I^r)^-_{\mathbf{0},\mathbf{f}}$ exists, is unique, and is represented by double-layer potential (12.68), where $\mathbf{h}$ is a solution of the Fredholm integral equation (12.69) which is always solvable for an arbitrary vector $\mathbf{f}$.*

12.8 On the Problems of Equilibrium in Thermoelasticity

The basic internal and external BVPs of equilibrium in the linear theory of thermoelasticity for quadruple porosity materials are formulated as follows.

Find a regular (classical) solution of the class $\mathfrak{R}$ in Ω^+ to the system (1.34) for $\mathbf{x} \in \Omega^+$ satisfying the boundary condition (9.41) in the *Problem* $(I^e)^+_{\mathbf{F},\mathbf{f}}$, and the boundary condition (9.42) in the *Problem* $(II^e)^+_{\mathbf{F},\mathbf{f}}$, where the matrix differential operators $\mathbf{A}^{(e)}(\mathbf{D_x})$ and $\mathbf{P}(\mathbf{D_z}, \mathbf{n})$ are defined by (1.29) and (1.44), respectively; $\mathbf{F}$ and $\mathbf{f}$ are prescribed eight-component vector functions.

Find a regular (classical) solution of the class $\mathfrak{R}$ to the system (1.34) for $\mathbf{x} \in \Omega^-$ satisfying the boundary condition (9.43) in the *Problem* $(I^e)^-_{\mathbf{F},\mathbf{f}}$, and the boundary condition (9.44) in the *Problem* $(II^e)^-_{\mathbf{F},\mathbf{f}}$. Here $\mathbf{F}$ and $\mathbf{f}$ are prescribed eight-component vector functions, and supp $\mathbf{F}$ is a finite domain in Ω^-.

We can investigate these BVPs by the potential method in the following manner: first we can study the BVPs for the vector $\mathbf{p}$ and function θ, and then we can study BVP for the vector $\mathbf{u}$.

For instance, in the BVP $(I^e)^+_{\mathbf{F},\mathbf{f}}$ we have three BVPs:

(i) The BVP for the vector $\mathbf{p}$

$$(\mathbf{K}\,\Delta - \mathbf{d})\mathbf{p}(\mathbf{x}) = \mathbf{F}^{(2)}(\mathbf{x}) \qquad \text{for} \quad \mathbf{x} \in \Omega^+,$$
$$\{\mathbf{p}(\mathbf{z})\}^+ = \mathbf{f}^{(2)}(\mathbf{z}) \qquad \text{for} \quad \mathbf{z} \in S; \tag{12.70}$$

(ii) The BVP for the function θ

$$k\,\Delta\theta(\mathbf{x}) = \mathbf{F}^{(2)}(\mathbf{x}) \qquad \text{for} \quad \mathbf{x} \in \Omega^+,$$
$$\{\theta(\mathbf{z})\}^+ = f_8(\mathbf{z}) \qquad \text{for} \quad \mathbf{z} \in S; \tag{12.71}$$

(iii) The BVP for the vector $\mathbf{u}$

$$\mu\,\Delta\mathbf{u}(\mathbf{x}) + (\lambda + \mu)\,\nabla \mathrm{div}\mathbf{u}(\mathbf{x}) = \tilde{\mathbf{F}}^{(1)}(\mathbf{x}) \qquad \text{for} \quad \mathbf{x} \in \Omega^+,$$
$$\{\mathbf{u}(\mathbf{z})\}^+ = \mathbf{f}^{(1)}(\mathbf{z}) \qquad \text{for} \quad \mathbf{z} \in S, \tag{12.72}$$

where $\tilde{\mathbf{F}}^{(1)}(\mathbf{x}) = \mathbf{F}^{(1)}(\mathbf{x}) + \nabla\,(\mathbf{ap}(\mathbf{x})) + \varepsilon_0 \nabla\,\theta(\mathbf{x})$.

Obviously, the BVP (12.70) is the problem of equilibrium of the linear theory for quadruple porosity rigid body and studied in Chap. 5. The BVP (12.71) is a well-known Dirichlet problem for Poisson's equation and studied by potential method in the book of Günther [161]. After solving the BVPs (12.70) and (12.71) we can study the BVP (12.72). Clearly, (12.72) is the displacement problem of equilibrium in the classical theory of elasticity and investigated by potential method in the book by Kupradze et al. [219].

Hence, the BVPs of equilibrium in the linear theory of thermoelasticity for quadruple porosity materials are reduced to the known BVPs of mathematical physics, the theories of elasticity, and quadruple porosity rigid body.

Chapter 13
Future Research Perspectives

We have seen that this book focuses only to the 3D linear mathematical theories of elasticity and thermoelasticity for quadruple porosity isotropic materials. Moreover, the classical potential method is extended in these theories and a wide class of the nonclassical problems is extensively investigated.

Furthermore, there are several interesting topics in the theories of elasticity and thermoelasticity for quadruple porosity materials that have not been considered in this book. The scope of the present chapter is to mention a few of these topics briefly. On the basis of results of this book, it is possible:

1. *To develop the potential method in the 2D linear mathematical theories of elasticity and thermoelasticity for quadruple porosity materials.*

It is well known that many plane problems of the classical theory of elasticity are investigated by two elegant methods: the method of complex variable and the potential method. By using these methods the above-mentioned plane problems are reduced to the equivalent one-dimensional singular integral equations. Muskhelishvili [256, 257] developed the theory of one-dimensional singular integral equations and investigated the plane BVPs of the theory of elasticity by virtue of this theory and the method of complex variable.

The classical potential method is extended in the 2D linear theories of elasticity and thermoelasticity by several authors (see Basheleishvili [28], Burchuladze [63], Constanda [103], and the references therein). The potential method is developed for the two-dimensional BVPs of the theory of consolidation with double porosity elastic solids by Basheleishvili and Bitsadze [29]. The thermal stresses in plane strain of elastic solids with single voids are studied by Ieşan and Nappa [180].

It is interesting to extend the potential method in the 2D linear mathematical theories of elasticity and thermoelasticity for quadruple porosity materials.

M. Svanadze, *Potential Method in Mathematical Theories of Multi-Porosity Media*, Interdisciplinary Applied Mathematics 51,
https://doi.org/10.1007/978-3-030-28022-2_13

2. *To develop the potential method in the linear mathematical theories of elasticity and thermoelasticity for quadruple porosity anisotropic materials.*

Theory of consolidation for a porous anisotropic solid is presented by Biot [42]. Straughan [325] generalizes the theories of Gelet et al. [149] and Svanadze [337] of double porosity isotropic solids and gives a system of partial differential equations of an anisotropic double porosity elastic body. Zhao and Chen [404] suggested a fully coupled dual-porosity model for anisotropic materials.

A general model for a triple porosity thermoelastic solid is presented in the linear anisotropic case by Straughan [327]. The same author developed a theory of elasticity for quadruple and quintuple porosity anisotropic materials in [324].

On the other hand, the potential method is extended in the theories of elasticity and thermoelasticity for double and triple porosity isotropic solids by Svanadze [334, 339, 346–352].

It is very important to develop this method in Straughan's theory of quadruple and quintuple porosity anisotropic materials (see [324]) and to prove the existence theorems of classical solutions of a wide class of BVPs.

3. *To establish the basic properties of Rayleigh, Love, harmonic, and acceleration waves in the theories of elasticity and thermoelasticity for quadruple porosity materials.*

Wave propagation in porous media is of interest in various areas of science and engineering. The theory of waves has been studied extensively in soil mechanics, seismology, acoustics, earthquake engineering, ocean engineering, geophysics, and many other disciplines (for details see Albers [9], Allard and Atalla [11], and Carcione [67]).

In this connection, de Boer et al. [130] considered one-dimensional transient wave propagation in fluid-saturated incompressible porous media. The properties of plane and acceleration waves in incompressible saturated poroelastic solids are established by de Boer and Liu [126–128, 229, 230]. Rayleigh waves on an exponentially graded poroelastic half-space are studied by Chirita [80]. Acceleration waves in a nonlinear Biot theory of porous media are investigated by Ciarletta et al. [100]. A model for acoustic waves in a porous medium is studied by Ciarletta and Straughan [92]. Thermo-poroacoustic acceleration waves in elastic materials with voids and second sound are considered by Ciarletta and Straughan [93] and Ciarletta et al. [96]. Chandrasekharaiah [72] established the properties of surface waves in an elastic half-space with voids. Bucur et al. [62] analyzed the behavior of plane harmonic waves and Rayleigh waves in a linear thermoelastic material with voids.

For an extensive review of the works and basic results on the wave propagation in the single porosity solids see the book of Straughan [319] and the review paper by Corapcioglu and Tuncay [104]. The basic results on the linear and nonlinear waves in elastic materials with micro or nanostructure are given in the books of Cattani and Rushchitsky [70] and Rushchitsky [291].

Moreover, wave propagation and attenuation in a double porosity medium are studied by Berryman and Wang [38]. The basic properties of Rayleigh and Love waves in double porosity theory [38] are established by Dai et al. [117] and Dai and Kuang [116], respectively. Acoustic wave propagation in double porosity media is studied by Olny and Boutin [267]. Reflection and transmission of elastic waves from the interface of a fluid-saturated porous solid and a double porosity solid are considered by Dai et al. [118]. Plane harmonic waves for an isotropic linearly elastic body which has a double porosity structure are analyzed by Ba et al. [19], Ciarletta et al. [99], and Svanadze [334, 343]. Plane harmonic waves for Kelvin–Voigt materials with double porosity are considered by Svanadze [367]. Acceleration waves in nonlinear double porosity elasticity are studied by Gentile and Straughan [152]. Puri and Cowin [279] and Singh and Tomar [314] studied the plane waves in linear elastic and thermoelastic materials with voids, respectively. The properties of plane waves in micropolar thermoelastic materials with single voids are established by Singh and Lianngenga [313].

For the basic results and bibliographical materials on wave propagation in the double porosity materials see the book by Straughan [326].

Obviously, further investigation of wave propagation in the theories of triple and quadruple porosity materials is difficult and useful for science and engineering.

4. *To construct explicit classical solutions of the BVPs in the theories of elastic and thermoelastic quadruple porosity domains with special geometry (sphere, half-space, space with spherical cavity, spherical layer, circle, ring, ellipse, etc.).*

The explicit solutions of the BVPs in the theory of elasticity for a circle with double porosity are constructed by Tsagareli and Svanadze [380, 381]. The solutions of the BVPs of the theory of consolidation with double porosity for the half-plane are presented by Basheleishvili and Bitsadze [31]. The explicit solutions of the BVPs for an ellipse with double porosity are constructed by Bitsadze and Zirakashvili [58]. Tsagareli [377] presented an explicit solution of elastostatic BVP for the porous elastic circle with single voids. The solutions of the BVPs in the fully coupled theory of elasticity for a circle with double porosity are obtained by Tsagareli and Bitsadze [379].

The explicit solution of the Dirichlet BVP in the fully coupled theory of elasticity for space, spherical cavity, and spherical layer with double porosity are constructed by Bitsadze and Tsagareli [56, 57]. The solutions of the BVPs of the theory of consolidation with double porosity for a sphere and a half-space are presented by Basheleishvili and Bitsadze [30, 32]. Ieşan and Quintanilla [183] investigated the deformation of an elastic space with a spherical cavity in the theory of thermoelastic solids which double voids. Janjgava [189] solved analytically the BVPs of binary elastic mixture with double porosity structure for the rectangular parallelepiped using the method of separation of variables.

It should be remarked that these explicit solutions are constructed explicitly in the form of absolutely and uniformly convergent series or in quadratures. Clearly, it is interesting to develop the methods of explicit construction of solutions in the theories of elastic and thermoelastic triple and quadruple porosity domains with special geometry.

5. *To develop mathematical models for quadruple porosity rods, plates, and shells.*

The dynamical contact problems describing the mechanical and thermal evolution of an elastic rod with single voids and microtemperatures are investigated by Berti and Naso [39] and Bazarra et al. [34]. A contact problem, between a thermoelastic rod with single voids and microtemperatures and a deformable obstacle, is numerically investigated by Bazarra and Fernández [33]. Existence and uniqueness, as well as an energy decay property, are proved by means of the semigroup arguments in the model involving a strain gradient thermoelastic rod with single voids by Fernández et al. [137] and some numerical simulations are presented.

The theory of micropolar plate made from a material with voids is presented by Tomar [372]. The mathematical models of elastic and thermoelastic plates made from a material with single voids are developed by Bîrsan [50, 55]. Ghiba [154] established the temporal behavior in the bending theory of porous thermoelastic plates with voids. The minimum principles in the dynamic theory of porous elastic plates with voids are established by Scarpetta [303]. The Rayleigh–Lamb waves in an elastic plate with voids are studied by Chandrasekharaiah [73]. The spatial estimates for the harmonic vibrations in rectangular plates with single voids are established by Ghiba [153].

On the basis of Biot [41] theory Kenyon [196] considered the fluid flow through an unconfined cylindrical shell of deformable porous material. The existence and uniqueness of weak solution in the linear theory of elastic shells with single voids are proved by Bîrsan [49]. The problem of radially directed fluid flow through a deformable porous shell is studied by Barry and Aldis [26]. The model of elastic shells made from a material with single voids is considered by Bîrsan [53]. The same author [54] studied several problems of the dynamic theory of thermoelastic Cosserat shells with single voids.

The governing equations of hierarchical models for porous elastic and viscoelastic Kelvin–Voigt prismatic shells are presented by Jaiani [187]. These models are obtained from Ieşan's [177] theory of viscoelastic materials with single voids. Using Vekua's dimension reduction method (for details see Vekua [387], Jaiani [186]), the basic systems of partial differential equations are derived and in the Nth approximation the BVPs are given. Saint-Venant's problem for Cosserat cylindrical shells made from a material with voids is studied by Bîrsan [51]. The same author presented a model for porous thermoelastic shells using the Nunziato–Cowin theory for materials with single voids in [52].

Consequently, nowadays only the theories for rods, plates, and shells made from a material with single porosity or single voids are constructed. Obviously, it is very important to develop mathematical models for multi-porosity rods, plates, and shells.

6. *To obtain numerical solutions in the theories of elasticity and thermoelasticity for quadruple porosity materials by using the boundary and finite element methods and the method of fundamental solutions.*

Since many physical phenomena can be modeled by idealized approaches, the derivation of numerical solutions on various types of problems has attracted considerable attention. We have seen that by virtue of the potential method the elliptic partial differential equations can be reduced to boundary integral equations. The boundary and finite element methods and the method of fundamental solutions are elegant and efficient numerical methods for the solution of these boundary integral equations, and therefore of the basic differential equations. An account of the historical developments as well as references to various contributions may be found in the work of Augustin [18], Sauter and Schwab [295], Steinbach [317], Straughan [319], and Cheng and Cheng [79].

Moreover, there is a large number of papers, where the BVPs of poroelasticity for single porosity materials are studied by using several different numerical techniques. More details on numerical methods in linear poroelasticity can be found in the books of Albers and Kuczma [10], Ehlers and Bluhm [135], in the review papers by Chen and Dargush [77], Schanz [306], and in the references therein.

Clearly, it is important to obtain numerical solutions in the theories of elasticity and thermoelasticity for quadruple porosity materials by using the boundary and finite element methods and the method of fundamental solutions.

7. *To investigate the problems of the nonlinear theories of elasticity and thermoelasticity for quadruple porosity materials.*

The nonlinear deformations in the theory of classical elastic bodies have been intensively studied. The historical development and basic results of the classical nonlinear elasticity are given in the books by Antman [12], Green and Adkins [158], Green and Zerna [160], and Truesdell and Noll [375]. Ball [23] proved the global existence and uniqueness of weak solutions (in Sobolev spaces) for various equilibrium BVPs of the nonlinear elasticity in one, two, and three dimensions by using the method of the calculus of variations. Valent [384] proved the local existence, uniqueness, and analytic dependence on data of weak solutions in the nonlinear elasticity by using the implicit function theorem.

On the other hand, theory of nonlinear elasticity for solids with single porosity is presented by Biot [46, 47] and Van Der Knapp [385]. The nonlinear theory of nonsimple thermoelastic materials with voids is introduced by Ciarletta and Scalia [91] and the continuous dependence upon initial state and supply terms of smooth thermodynamic processes is studied. Nonlinear acoustic wave propagation in the Biot [47] theory for an isotropic porous solids is studied by Tong et al. [374] and Legland et al. [222]. Straughan and Tibullo [328] generalized a theory of Biot for a porous solid based on nonlinear elasticity theory to incorporate temperature effects and studied the acceleration waves in the fully nonlinear theory.

Moreover, Straughan [324] presented the nonlinear models of thermoelasticity for triple and quadruple porosity materials involving quadratic nearest porosity scales connectivity terms, or cubic nearest porosity scales connectivity terms. Nunziato and Cowin [266] introduced a nonlinear theory of elastic materials with single voids. The different nonlinear theory of porous elastic materials with the same structure is developed by Ieşan and Quintanilla [182] and an existence result within the one-dimensional equilibrium theory is obtained.

It is very interesting to prove the existence and uniqueness theorems of the BVPs in the nonlinear theories of poroelasticity and thermoelasticity for single and multiple porosity materials.

8. *To study the structural stability of solutions of the dynamical problems in the theories of elasticity and thermoelasticity for quadruple porosity materials.*

The structural stability of solutions for elastic materials is a classical problem and a subject of great interest of researchers in recent years. In addition, estimates of continuous dependence of solutions upon the constitutive quantities play a central role in obtaining numerical approximations to these solutions.

Knops and Payne [201–203], Quintanilla [280, 281], and Quintanilla and Racke [283, 284] investigated the stability of solutions in the dynamical problems of linear elasticity, thermoelasticity, and heat conduction. Straughan [322] studied the stability and uniqueness in double porosity elasticity. The continuous dependence upon the initial data for the double and triple porosity materials is established by Straughan [326]. The structural stability in local thermal nonequilibrium in a porous medium where the solid and fluid components may have different temperatures is studied by Passarella et al. [273].

However, the structural stability for a rigid body with thermal microstructure is established by Ciarletta et al. [98]. Chirita et al. [87] and Pamplona et al. [270] studied the structural stability of the mathematical model of the linear elastic material with voids. Straughan [321] investigated continuous dependence on the heat source in resonant porous penetrative convection. Global nonlinear stability in porous convection with a thermal nonequilibrium model is studied by Straughan [318]. The continuous dependence upon initial state and supply terms in the nonlinear theory of nonsimple thermoelastic materials with single voids is studied by Ciarletta and Scalia [91].

Moreover, the asymptotic behavior of the solutions in the linear theory of thermo-porous-elasticity for elastic materials with voids when the heat conduction is of type II is established by Leseduarte et al. [223]. The time decay of solutions in one-dimensional theories of porous materials is established by Magaña and Quintanilla [235]. The same authors [236, 237] proved the exponential decay of solutions in one-dimensional generalized porous-thermo-elasticity and in porous-elasticity with quasistatic microvoids. Muñoz-Rivera and Quintanilla [255] established the time polynomial decay of solutions in theory of elastic solids with single voids. The spatial behavior of solution in a cylinder made of an isotropic and homogeneous thermoelastic material with voids is studied by Scalia et al. [298].

More detailed results on the structural stability in the theory of porous media are given in the books of Straughan [319, 323, 326].

Consequently, further investigation of the structural stability of solutions in the theories of elasticity and thermoelasticity for triple and quadruple porosity materials is interesting for mathematical physics and continuum mechanics.

9. *To present mathematical models of thermoelasticity for quadruple porosity materials under LTNE and to develop the potential method in these models.*

Within flow and thermal convection in porous media LTNE (the fluid temperature may be different from the solid skeleton temperature) was introduced in the late 1990s and has been the subject of intense recent attention, see e.g., Nield [258–260] and Rees [286, 287]. The importance of LTNE in porous media with conductive heat transfer is examined by Gandomkar and Gray [146]. Many further references may be found in the book of Straughan [323].

By employing the concept of mixture theory, Aifantis [7] and Aifantis and Beskos [8] proposed a two-temperature model of rigid single porosity body. The global nonlinear stability for convection with a LTNE model of the porous medium equations of Darcy, Forchheimer, or Brinkman is investigated by Straughan [318].

The first theory for double porosity elastic materials with double temperatures is presented by Bai and Roegiers [20]. Modeling of geothermal reservoirs using coupling effects of mechanical, thermal, hydraulic, and chemical concepts is discussed and a brief review of references of this subject is presented by Hayashi et al. [163]. A thermo-hydro-mechanical triple temperature nonlinear model for double porosity elastic media using a effective stress concept is suggested by Khalili and Selvadurai [200], and developed by Gelet [148] and Gelet et al. [150, 151].

Recently, the three-temperature models in local thermal nonequilibrium double porosity thermoelasticity are presented by Franchi et al. [142], Li et al. [224], and Svanadze [355]. The basic results on LTNE effect in the theory of porous media are given in the book of Straughan [323].

The main reason for the increased attention of fluid flows in porous media under LTNE is due to the large number of applications of this area in real life (for details see Straughan [323]). Therefore, it is very important to present mathematical models of thermoelasticity for triple and quadruple porosity materials under LTNE and to develop the potential method in these models.

10. *To present mathematical models for elastic and thermoelastic materials with quadruple voids based on the volume fraction concept and to develop the potential method in these models.*

As mentioned in Sect. 1.1, Nunziato and Cowin [112, 266] and Ieşan [174] presented the theories of elastic and thermoelastic materials with single voids, respectively. Pompei and Scalia [277] studied the BVPs of steady vibrations in the linear theory of thermoelasticity for materials with single voids by using the potential method and the theory of singular integral equations.

However, Ieşan and Quintanilla [183] extended the Nunziato–Cowin [112, 266] mathematical models and developed the theory of thermoelastic deformable materials with double voids structure by using the volume fraction concept. Recently, Svanadze [349, 352] presented the linear model of elasticity for materials with triple voids and investigated the nonclassical BVPs of equilibrium and steady vibrations by using the potential method and the theory of singular integral equations.

Thus it is very interesting to construct mathematical models for elastic and thermoelastic materials with quadruple voids based on the volume fraction concept and to develop the potential method in these models.

11. *To present a mathematical models for viscoelastic and thermoviscoelastic Kelvin–Voigt materials with quadruple porosity and to develop the potential method in these models.*

The basic formulation of viscoelasticity of differential and integral types has been developed by several scientists, including Maxwell (1867), Voigt (1892), Lord Kelvin (1875), and Volterra (1909).

The first theory of deformation of a porous viscoelastic anisotropic solid is presented by Biot [45] and intensively studied by several researches (for details see Carcione [67]). Cowin [109] studied the viscoelastic behavior of linear elastic materials with single voids. Ieşan [177] presented a nonlinear theory of thermoviscoelastic materials with single voids and derived the linearized version of this theory. The potential method in Ieşan's [177] linear theories of viscoelasticity and thermoviscoelasticity for Kelvin–Voigt materials with single voids is developed by Svanadze [361–365]. Scalia [296] established the basic properties of the shock waves in a viscoelastic material with single voids. Tomar et al. [373] studied the time harmonic waves in a thermoviscoelastic material with single voids. Pamplona et al. [271] established the uniqueness and analyticity of solutions in the theory of thermoviscoelasticity for solids with voids. Sharma and Kumar [309] studied the propagation of plane waves and constructed the fundamental solution in the theory of thermoviscoelasticity for medium with voids.

The theory of viscoelasticity for materials with double porosity is presented by Svanadze [338] and studied in the series of papers [366–370]. Recently, a linear theory of viscoelastic materials with double voids is introduced by Ieşan and Quintanilla [184].

Obviously, it is very important to present a mathematical models for viscoelastic and thermoviscoelastic Kelvin–Voigt materials with quadruple porosity and to develop the potential method in these models.

12. *To present hyperbolic and high-order time differential thermoelastic mathematical models for quadruple porosity materials and to develop the potential method in these models.*

It is well known that on the basis of classical Fourier's law of heat conduction (a diffusion phenomenon) the heat transport equation is a parabolic-type partial differential equation that allows an infinite speed for thermal signals. Maxwell (1867) suggested a modification of Fourier's law and postulated wave-type heat flow (a wave phenomenon). This prediction is realistic from a physical point of view (for details, see Cattaneo [69]), and consequently involves hyperbolic-type heat transport equation that allows a finite speed for thermal signals (second sound). Thermoelasticity theories admitting such signals are traditionally called hyperbolic thermoelasticity theories, or generalized thermoelasticity theories, or thermoelasticity theories with second sound. Indeed, Lord and Shulman [234], Green and Lindsay [159], and Roy Choudhuri [290] formulated the generalized thermoelasticity theories involving one, two, and three thermal relaxation times (or phase-lags), respectively. Moreover, Ieşan and Quintanilla [181] presented a linear theory of thermoelastic bodies with microstructure and microtemperatures which permits the transmission of heat as thermal waves at finite speed.

There is a large number of works, where the mathematical models of hyperbolic thermoelasticity are presented and studied by using several different techniques. For the basic results and bibliographical materials on the second sound and hyperbolic thermoelasticity see the books by Straughan [319, 320, 326] and Ignaczak and Ostoja-Starzewski [185], and the review papers by Chandrasekharaiah [71, 76] and Joseph and Preziosi [190, 191].

Recently, Chirita [81] presented a very interesting model on the high-order approximations of the three-phase-lag heat transfer and provided a comprehensive analysis of the model. The well posedness of the initial BVPs of the most general constitutive equation characterizing the time-differential dual-phase-lag model of heat conduction is investigated by Chirita [82, 83] and Chirita and Zampoli [86]. Zampoli [400, 401] proved the uniqueness theorems and obtained some continuous dependence results in the high-order time differential theories of thermoelasticity. Zampoli and Landi [402] investigated the well-posed question for the time differential three-phase-lag model of thermoelasticity. In addition, (see D'Apice and Zampoli [121]) "more generally the delayed thermal effects, which a lagging behavior is able to take into account, play a fundamental role in multi-porous materials, characterized by different levels of porosity, namely macro-, meso- and micro-porosity".

Consequently, now it is very important to present hyperbolic and high-order time differential thermoelastic mathematical models for porous materials and to develop the potential method in these models.

13. *To prove the existence and uniqueness of weak solutions (in Sobolev space) in the linear theories of elasticity and thermoelasticity for quadruple porosity materials.*

It is well known that Sobolev spaces occupy an outstanding place in the theory of partial differential equations (for details see Sobolev [315], Adams and Fournier [2], and Maz'ya [244]).

Fichera [138] and Dafermos [115] established the existence and uniqueness of weak solutions in the classical theories of elasticity and thermoelasticity by functional method, respectively. Ieşan and Quintanilla [183] used the semigroup method to prove the existence of weak solutions in the theory of thermoelastic solids which double voids. For detailed information on the semigroup method see

Goldstein [156] and Liu and Zheng [231]. The existence and uniqueness of a weak solution in the theory of elasticity of dipolar single porosity materials are established by Marin [240].

Clearly, it is great to prove the existence and uniqueness of weak solutions in the linear theories of elasticity and thermoelasticity for multiple porosity materials.

Finally, these topics are very important in mathematical physics and all the results will be used to solve multidisciplinary problems of applied mathematics, continuum mechanics, engineering, technology, medicine, and biology. In addition, finding solutions of equations in the mathematical theories of multi-porosity media is significantly more difficult than for classical elasticity and thermoelasticity of homogeneous materials.

References

1. Abdassah, D., Ershaghi, I.: Triple-porosity systems for representing naturally fractured Reservoirs. SPE Form Eval. (April) 113–127. SPE-13409-PA (1986)
2. Adams, R.A., Fournier, J.J.F.: Sobolev Spaces. Academic Press, New York (2003)
3. Aguilera, R.: Naturally Fractured Reservoirs, 2nd edn. PennWell Books, Tulsa (1995)
4. Aguilera, R.F., Aguilera, R.: A triple - porosity model for petrophysical analysis of naturally fractured reservoirs. Petrophysics **45**, 157–166 (2004)
5. Aguilera, R., Lopez, B.: Evaluation of quintuple porosity in shale petroleum reservoirs. SPE Eastern Regional Meeting, 20–22 August, Pittsburgh. SPE-165681-MS, 28pp. (2013). https://doi.org/10.2118/165681-MS
6. Aifantis, E.C.: Introducing a multi-porous media. Dev. Mech. **9**, 209–211 (1977)
7. Aifantis, E.C.: Further comments on the problem of heat extraction from hot dry rocks. Mech. Res. Commun. **7**, 219–226 (1980)
8. Aifantis, E.C., Beskos, D.E.: Heat extraction from hot dry rocks. Mech. Res. Commun. **7**, 165–170 (1980)
9. Albers, B.: Modeling and Numerical Analysis of Wave Propagation in Saturated and Partially Saturated Porous Media. Shaker Verlag, Maastricht (2009)
10. Albers, B., Kuczma, M. (eds): Continuous Media with Microstructure 2. Springer, Basel (2016)
11. Allard, J.F., Atalla, N.: Propagation of Sound in Porous Media: Modelling Sound Absorbing Materials, 2nd edn. Wiley, West Sussex (2009)
12. Antman, S.S.: Nonlinear Problems of Elasticity. Applied Mathematical Sciences, vol. 107. Springer, Berlin (1995)
13. Aouadi, M.: A theory of thermoelastic diffusion materials with voids. Z. Angew. Math. Phys. **61**, 357–379 (2010)
14. Aouadi, M.: Uniqueness and existence theorems in thermoelasticity with voids without energy dissipation. J. Franklin Inst. **349**, 128–139 (2012)
15. Aouadi, M.: Stability in thermoelastic diffusion theory with voids. Appl. Anal. **91**, 121–139 (2012)
16. Arbogast, T., Douglas, J., Hornung, U.: Derivation of the double porosity model of single phase flow via homogenization theory. SIAM J. Math. Anal. **21**, 823–836 (1990)
17. Arusoaie, A.: Spatial and temporal behavior in the theory of thermoelasticity for solids with double porosity. J. Therm. Stresses **41**, 500–521 (2018)
18. Augustin, M.A.: A Method of Fundamental Solutions in Poroelasticity to Model the Stress Field in Geothermal Reservoirs. Springer, Cham (2015)
19. Ba, J., Carcione, J.M., Nie, J.X.: Biot-Rayleigh theory of wave propagation in double-porosity media. J. Geophys. Res. **116**, B06202 (2011). https://doi.org/10.1029/2010JB008185

M. Svanadze, *Potential Method in Mathematical Theories of Multi-Porosity Media*, Interdisciplinary Applied Mathematics 51, https://doi.org/10.1007/978-3-030-28022-2

20. Bai, M., Roegiers, J.C.: Fluid flow and heat flow in deformable fractured porous media. Int. J. Eng. Sci. **32**, 1615–1633 (1994)
21. Bai, M., Roegiers, J.C.: Triple-porosity analysis of solute transport. J. Cantam. Hydrol. **28**, 189–211 (1997)
22. Bai, M., Elsworth, D., Roegiers, J.C.: Multiporosity/multipermeability approach to the simulation of naturally fractured reservoirs. Water Resour. Res. **29**, 1621–1633 (1993)
23. Ball, J.M.: Convexity conditions and existence theorems in nonlinear elasticity. Arch. Rat. Mech. Anal. **63**, 337–403 (1976)
24. Barenblatt, G.I., Zheltov, Y.P.: Fundamental equations of filtration of homogeneous liquids in fissured rock. Sov. Phys. Dokl. **5**, 522–525 (1960)
25. Barenblatt, G.I., Zheltov, Y.P., Kochina, I.N.: Basic concepts in the theory of seepage of homogeneous liquids in fissured rocks (strata). J. Appl. Math. Mech. **24**, 1286–1303 (1960)
26. Barry, S.I., Aldis, G.K.: Radial flow through deformable porous shells. J. Aust. Math. Soc. Ser. B **34**, 333–354 (1993)
27. Basheleishvili, M.O.: On fundamental solutions of differential equations of an anisotropic elastic body (Russian). Bull. Acad. Sci. Georgian SSR **19**, 393–400 (1957)
28. Basheleishvili, M.O.: Basic plane boundary value problems for nonhomogeneous anisotropic elastic bodies (Russian). Trudy Tbil. Univ. **117**, 279–293 (1966)
29. Basheleishvili, M., Bitsadze, L.: Two-dimensional boundary value problems of the theory of consolidation with double porosity. Mem. Diff. Equ. Math. Phys. **51**, 43–58 (2010)
30. Basheleishvili, M., Bitsadze, L.: Explicit solutions of the BVPs of the theory of consolidation with double porosity for the half-space. Bull. TICMI **14**, 9–15 (2010)
31. Basheleishvili, M., Bitsadze, L.: Explicit solutions of the boundary value problems of the theory of consolidation with double porosity for the half-plane. Georgian Math. J. **19**, 41–48 (2012)
32. Basheleishvili, M., Bitsadze, L.: The basic BVPs of the theory of consolidation with double porosity for the sphere. Bull TICMI **16**, 15–26 (2012)
33. Bazarra, N., Fernández, J.R.: Numerical analysis of a contact problem in poro-thermoelasticity with microtemperatures. ZAMM J. Appl. Math. Mech. **98**, 1190–1209 (2018)
34. Bazarra, N., Berti, A., Fernández, J.R., Naso, M.G.: Analysis of contact problems of porous thermoelastic solids. J. Therm. Stresses **41**, 439–468 (2018)
35. Bazarra, N., Fernández, J.R., Leseduarte, M.C., Magaña, A., Quintanilla, R.: On the thermoelasticity with two porosities: asymptotic behaviour. Math. Mech. Solids (2018, in press). **24**, 2713–2725 (2019)
36. Bear, J.: Modeling Phenomena of Flow and Transport in Porous Media. Springer, Basel (2018)
37. Berryman, J.G., Wang, H.F.: The elastic coefficients of double - porosity models for fluid transport in jointed rock. J. Geophys. Res. B **100**, 24611–24627 (1995)
38. Berryman, J.G., Wang, H.F.: Elastic wave propagation and attenuation in a double-porosity dual-permeability medium. Int. J. Rock Mech. Min. Sci. **37**, 63–78 (2000)
39. Berti, A., Naso, M.G.: A contact problem of a thermoelastic rod with voids and microtemperatures. ZAMM J. Appl. Math. Mech. **97**, 670–685 (2017)
40. Beskos, D.E., Aifantis, E.C.: On the theory of consolidation with double porosity - II. Int. J. Eng. Sci. **24**, 1697–1716 (1986)
41. Biot, M.A.: General theory of three-dimensional consolidation. J. Appl. Phys. **12**, 155–164 (1941)
42. Biot, M.A.: Theory of elasticity and consolidation for a porous anisotropic solid. J. Appl. Phys. **26**, 182–185 (1955)
43. Biot, M.A.: General solutions of the equations of elasticity and consolidation for a porous material. J. Appl. Mech. **28**, 91–96 (1956)
44. Biot, M.A.: Thermoelasticity and irreversible thermodynamics. J. Appl. Phys. **27**, 240–253 (1956)
45. Biot, M.A.: Theory of deformation of a porous viscoelastic anisotropic solid. J. Appl. Phys. **27**, 459–467 (1956)

46. Biot, M.A.: Theory of finite deformations of porous solids. Indiana Univ. Math. J. **21**, 597–620 (1972)
47. Biot, M.A.: Nonlinear and semilinear rheology of porous solids. J. Geophys. Res. **78**, 4924–4937 (1973)
48. Biot, M.A.: Variational Lagrangian-thermodynamics of nonisothermal finite strain mechanics of porous solids and thermomolecular diffusion. Int. J. Solids Struct. **13**, 579–597 (1977)
49. Bîrsan, M.: Existence and uniqueness of weak solution in the linear theory of elastic shells with voids. Libertas Math. **20**, 95–105 (2000)
50. Bîrsan, M.: A bending theory of porous thermoelastic plates. J. Therm. Stresses **26**, 67–90 (2003)
51. Bîrsan, M.: Saint-Venant's problem for Cosserat shells with voids. Int. J. Solids Struct. **42**, 2033–2057 (2005)
52. Bîrsan, M.: On a thermodynamic theory of porous Cosserat elastic shells. J. Therm. Stresses **29**, 879–899 (2006)
53. Bîrsan, M.: On the theory of elastic shells made from a material with voids. Int. J. Solids Struct. **43**, 3106–3123 (2006)
54. Bîrsan, M.: Several results in the dynamic theory of thermoelastic Cosserat shells with voids. Mech. Res. Commun. **33**, 157–176 (2006)
55. Bîrsan, M.: On the bending equations for elastic plates with voids. Math. Mech. Solids **12**, 40–57 (2007)
56. Bitsadze, L., Tsagareli, I.: Solutions of BVPs in the fully coupled theory of elasticity for the space with double porosity and spherical cavity. Math. Methods Appl. Sci. **39**, 2136–2145 (2016)
57. Bitsadze, L., Tsagareli, I.: The solution of the Dirichlet BVP in the fully coupled theory of elasticity for spherical layer with double porosity. Meccanica **51**, 1457–1463 (2016)
58. Bitsadze, L., Zirakashvili, N.: Explicit solutions of the boundary value problems for an ellipse with double porosity. Adv. Math. Phys. **2016**, 1810795, 11 pp. (2016). https://doi.org/10.1155/2016/1810795
59. Bluhm, J., de Boer, R.: The volume fraction concept in the porous media theory. ZAMM J. Appl. Math. Mech. **77**, 563–577 (1997)
60. Bowen, R.M.: Incompressible porous media models by use of the theory of mixtures. Int. J. Eng. Sci. **18**, 1129–1148 (1980)
61. Bowen, R.M.: Compressible porous media models by use of the theory of mixtures. Int. J. Eng. Sci. **20**, 697–735 (1982)
62. Bucur, A.V., Passarella, F., Tibullo, V.: Rayleigh surface waves in the theory of thermoelastic materials with voids. Meccanica **49**, 2069–2078 (2014)
63. Burchuladze, T.V.: Two-dimensional boundary value problems of thermoelasticity (Russian). Trudy Tbil. Mat. Inst. AN GSSR **39**, 5–22 (1971)
64. Burchuladze, T.V., Gegelia, T.G.: The Development of the Potential Methods in the Elasticity Theory (Russian). Metsniereba, Tbilisi (1985)
65. Burchuladze, T., Svanadze, M.: Potential method in the linear theory of binary mixtures for thermoelastic solids. J. Therm. Stresses **23**, 601–626 (2000)
66. Burridge, R., Vargas, C.A.: The fundamental solution in dynamic poroelasticity. Geophys. J. R. Astr. Soc. **58**, 61–90 (1979)
67. Carcione, J.M.: Wave Fields in Real Media: Wave Propagation in Anisotropic, Anelastic, Porous and Electromagnetic Media, 3rd edn. Elsevier, Amsterdam (2015)
68. Casas, P.S., Quintanilla, R.: Exponential decay in one-dimensional porous-thermo-elasticity. Mech. Res. Commun. **32**, 652–658 (2005)
69. Cattaneo, C.: Sulla conduzione del calore. Atti Sem. Mat. Fis. Modena **3**, 83–101 (1948)
70. Cattani, C., Rushchitsky, J.J.: Wavelet and Wave Analysis as Applied to Materials with Micro or Nanostructure. World Scientific, Singapore (2007)
71. Chandrasekharaiah, D.S.: Thermoelasticity with second sound - a review. Appl. Mech. Rev. **39**, 355–376 (1986)

72. Chandrasekharaiah, D.S.: Surface waves in an elastic half-space with voids. Acta Mech. **62**, 77–85 (1986)
73. Chandrasekharaiah, D.S.: Rayleigh-Lamb waves in an elastic plate with voids. J. Appl. Mech. **54**, 509–512 (1987)
74. Chandrasekharaiah, D.S.: Complete solutions in the theory of elastic materials with voids -I. Quart. J. Mech. Appl. Math. **40**, 401–414 (1987)
75. Chandrasekharaiah, D.S.: Complete solutions in the theory of elastic materials with voids -II. Quart. J. Mech. Appl. Math. **42**, 41–54 (1989)
76. Chandrasekharaiah, D.S.: Hyperbolic thermoelasticity: a review of recent literature. Appl. Mech. Rev. **51**, 705–729 (1998)
77. Chen, J., Dargush, G.F.: Boundary element method for dynamic poroelastic and thermoelastic analyses. Int. J. Solids Struct. **32**, 2257–2278 (1995)
78. Cheng, A.H.D.: Poroelasticity. Springer, Basel (2016)
79. Cheng, A.H.D., Cheng, D.T.: Heritage and early history of the boundary element method. Eng. Anal. Bound. Elem. **29**, 268–302 (2005)
80. Chirita, S.: Rayleigh waves on an exponentially graded poroelastic half space. J. Elast. **110**, 185–199 (2013)
81. Chirita, S.: High-order approximations of three-phase-lag heat conduction model: some qualitative results. J. Therm. Stresses **41**, 608–626 (2018)
82. Chirita, S.: On high-order approximations for describing the lagging behavior of heat conduction. Math. Mech. Solids **24**, 1648–1667 (2018)
83. Chirita, S.: High-order effects of thermal lagging in deformable conductors. Int. J. Heat Mass Transf. **127**, 965–974 (2018)
84. Chirita, S., Ghiba, I.D.: Strong ellipticity and progressive waves in elastic materials with voids. Proc. Roy. Soc. Lond. A **466**, 439–458 (2010)
85. Chirita, S., Scalia, A.: On the spatial and temporal behavior in linear thermoelasticity of materials with voids. J. Therm. Stresses **24**, 433–455 (2001)
86. Chirita, S., Zampoli, V.: Spatial behavior of the dual-phase-lag deformable conductors. J. Therm. Stresses **41**, 1276–1296 (2018)
87. Chirita, S., Ciarletta, M., Straughan, B.: Structural stability in porous elasticity. Proc. Roy. Soc. Lond. A **462**, 2593–2605 (2006)
88. Ciarletta, M.: A solution of Galerkin type in the theory of thermoelastic materials with voids. J. Therm. Stresses **14**, 409–417 (1991)
89. Ciarletta, M., Ieşan, D.: Non-Classical Elastic Solids. Longman Scientific and Technical. Wiley, New York (1993)
90. Ciarletta, M., Chirita, S., Passarella, F.: Some results on the spatial behaviour in linear porous elasticity. Arch. Mech. **57**, 43–65 (2005)
91. Ciarletta, M., Scalia, A.: On the nonlinear theory of nonsimple thermoelastic materials with voids. ZAMM J. Appl. Math. Mech. **73**, 67–75 (1993)
92. Ciarletta, M., Straughan, B.: Poroacoustic acceleration waves. Proc. R. Soc. A **462**, 3493–3499 (2006)
93. Ciarletta, M., Straughan, B.: Poroacoustic acceleration waves with second sound. J. Sound Vib. **306**, 725–731 (2007)
94. Ciarletta, M., Straughan, B.: Thermo-poroacoustic acceleration waves in elastic materials with voids. J. Math. Anal. Appl. **333**, 142–150 (2007)
95. Ciarletta, M., Scalia, A., Svanadze, M.: Fundamental solution in the theory of micropolar thermoelasticity for materials with voids. J. Therm. Stresses **30**, 213–229 (2007)
96. Ciarletta, M., Straughan, B., Zampoli, V.: Thermo-poroacoustic acceleration waves in elastic materials with voids without energy dissipation. Int. J. Eng. Sci. **45**, 736–743 (2007)
97. Ciarletta, M., Svanadze, M., Buonano, L.: Plane waves and vibrations in the micropolar thermoelastic materials with voids. Eur. J. Mech. A Solids **28**, 897–903 (2009)
98. Ciarletta, M., Straughan, B., Tibullo, V.: Structural stability for a rigid body with thermal microstructure. Int. J. Eng. Sci. **48**, 592–598 (2010)

99. Ciarletta, M., Passarella, F., Svanadze, M.: Plane waves and uniqueness theorems in the coupled linear theory of elasticity for solids with double porosity. J. Elast. **114**, 55–68 (2014)
100. Ciarletta, M., Straughan, B., Tibullo, V.: Acceleration waves in a nonlinear Biot theory of porous media. Int. J. Non-Linear Mech. **103**, 23–26 (2018)
101. Cleary, M.P.: Fundamental solutions for a fluid-saturated porous solid. Int. J. Solids Struct. **13**, 785–806 (1977)
102. Colton, D., Kress, R.: Inverse Acoustic and Electromagnetic Scattering Theory, 3rd edn. Springer, New York (2013)
103. Constanda, C.: The boundary integral equation method in plane elasticity. Proc. Am. Math. Soc. **123**, 3385–3396 (1995)
104. Corapcioglu, M.Y., Tuncay, K.: Propagation of waves in porous media. In: Corapcioglu, M.Y. (ed.) Advances in Porous Media, vol. 3, pp. 361–440. Elsevier, Amsterdam (1996)
105. Courant, R., Hilbert, D.: Methods of Mathematical Physics, vol. I. Wiley-VCH Verlag, Weinheim (2004)
106. Coussy, O.: Poromechanics. Wiley, Chichester (2004)
107. Coussy, O.: Mechanics and Physics of Porous Solids. Wiley, Chichester (2010)
108. Cowin, S.C.: The stresses around a hole in a linear elastic material with voids. Quart. J. Mech. Appl. Math. **37**, 441–465 (1984)
109. Cowin, S.C.: The viscoelastic behavior of linear elastic materials with voids. J. Elast. **15**, 185–191 (1985)
110. Cowin, S.C.: Bone poroelasticity. J. Biomech. **32**, 217–238 (1999)
111. Cowin, S.C. (ed.): Bone Mechanics Handbook. Informa Healthcare USA, New York (2008)
112. Cowin, S.C., Nunziato, J.W.: Linear elastic materials with voids. J. Elast. **13**, 125–147 (1983)
113. Cowin, S.C., Puri, P.: The classical pressure vessel problems for linear elastic materials with voids. J. Elast. **13**, 157–163 (1983)
114. Cushman, J.H.: The Physics of Fluids in Hierarchical Porous Media: Angstroms to Miles. Springer, Dordrecht (1997)
115. Dafermos, C.M.: On the existence and the asymptotic stability of solutions to the equations of linear thermoelasticity. Arch. Rat. Mech. Anal. **29**, 241–271 (1968)
116. Dai, W.Z., Kuang, Z.B.: Love waves in double porosity media. J. Sound Vib. **296**, 1000–1012 (2006)
117. Dai, Z.J., Kuang, Z.B., Zhao, S.X.: Rayleigh waves in a double porosity half-space. J. Sound Vib. **298**, 319–332 (2006)
118. Dai, Z.J., Kuang, Z.B., Zhao, S.X.: Reflection and transmission of elastic waves from the interface of a fluid-saturated porous solid and a double porosity solid. Transp. Porous Media **65**, 237–264 (2006)
119. d'Alembert, J.R.: Recherches sur la courbe que forme une corde tendue mise en vibration. Mém. Acad. Roy. Sci. Belles-Lett. de Berlin **3**, 214–219 (1747/1749)
120. D'Apice, C., Chirita, S.: Plane harmonic waves in the theory of thermoviscoelastic materials with voids. J. Therm. Stresses **39**, 142–155 (2016)
121. D'Apice, C., Zampoli, V.: Advances on the time differential three-phase-lag heat conduction model and major open issues. AIP Conf. Proc. **1863**, 560056 (2017). https://doi.org/10.1063/1.4992739
122. Das, M.K., Mukherjee, P.P., Muralidhar, K.: Modeling Transport Phenomena in Porous Media with Applications. Springer, Cham (2018)
123. de Boer, R.: Theory of Porous Media: Highlights in the Historical Development and Current State. Springer, Berlin (2000)
124. de Boer, R.: Contemporary progress in porous media theory. Appl. Mech. Rev. **53**, 323–370 (2000)
125. de Boer, R.: Trends in Continuum Mechanics of Porous Media. Springer, Dordrecht (2005)
126. de Boer, R., Liu, Z.: Plane waves in a semi-infinite fluid saturated porous medium. Transp. Porous Media **16**, 147–173 (1994)
127. de Boer, R., Liu, Z.: Propagation of acceleration waves in incompressible saturated porous solids. Transp. Porous Media **21**, 163–173 (1995)

128. de Boer, R., Liu, Z.: Growth and decay of acceleration waves in incompressible saturated poroelastic solids. ZAMM J. Appl. Math. Mech. **76**, 341–347 (1996)
129. de Boer, R., Svanadze, M.: Fundamental solution of the system of equations of steady oscillations in the theory of fluid-saturated porous media. Transp. Porous Media **56**, 39–50 (2004)
130. de Boer, R., Ehlers, W., Liu, Z.: One-dimensional transient wave propagation in fluid-saturated incompressible porous media. Arch. Appl. Mech. **63**, 59–72 (1993)
131. De Cicco, S., Diaco, M.: A theory of thermoelastic materials with voids without energy dissipation. J. Therm. Stresses **25**, 493–503 (2002)
132. Dhaliwal, R.S., Wang, J.: A heat-flux dependent theory of thermoelasticity with voids. Acta Mech. **110**, 33–39 (1995)
133. Dormieux, L., Kondo, D., Ulm, F.-J.: Microporomechanics. Wiley, Chichester (2006)
134. Dragos, L.: Fundamental solutions in micropolar elasticity. Int. J. Eng. Sci. **22**, 265–275 (1984)
135. Ehlers, W., Bluhm, J. (eds): Porous Media: Theory, Experiments and Numerical Applications. Springer, Berlin (2002)
136. Ehrenpreis, L.: Solution of some problems of division. Part I. Division by a polynomial of derivation. Am. J. Math. **76**, 883–903 (1954)
137. Fernández, J.R., Magaña, A., Masid, M., Quintanilla, R.: Analysis for the strain gradient theory of porous thermoelasticity. J. Comput. Appl. Math. **345**, 247–268 (2019)
138. Fichera, G.: Existence theorems in elasticity. In: Truesdel, C. (ed.) Handbuch der Physik, vol. VI a/2. Springer, Berlin (1972)
139. Florea, O.: Spatial behavior in thermoelastodynamics with double porosity structure. Int. J. Appl. Mech. **9**, 1750097, 14pp. (2017). https://doi.org/10.1142/S1758825117500971
140. Florea, O.A.: Harmonic vibrations in thermoelastic dynamics with double porosity structure. Math. Mech. Solids (2018). **24**, 2410–2424 (2019)
141. Fosdick, R., Piccioni, M.D., Puglisi, G.: A note on uniqueness in linear elastostatics. J. Elast. **88**, 79–86 (2007)
142. Franchi, F., Lazzari, B., Nibbi, R., Straughan, B.: Uniqueness and decay in local thermal non-equilibrium double porosity thermoelasticity. Math. Methods Appl. Sci. **41**, 6763–6771 (2018)
143. Fredholm, I.: Sur les équations de l'équilibre d'un corps solide élastique. Acta Math. **23**, 1–42 (1900)
144. Fredholm, I.: Sur une classe d'équations fonctionelles. Acta Math. **27**, 365–390 (1903)
145. Galerkin, B.: Contribution à la solution générale du problème de la théorie de l'élasticité dans le cas de trois dimensions. C. R. Acad. Sci. Paris **190**, 1047–1048 (1930)
146. Gandomkar, A., Gray, K.E.: Local thermal non-equilibrium in porous media with heat conduction. Int. J. Heat Mass Transf. **124**, 1212–1216 (2018)
147. Gegelia, T., Jentsch, L.: Potential methods in continuum mechanics. Georgian Math. J. **1**, 599–640 (1994)
148. Gelet, R.: Thermo-hydro-mechanical study of deformable porous media with double porosity in local thermal nonequilibrium. Ph.D. Thesis, Institut National Polytechnique de Grenoble, France, and The University of New South Wales, Sydney (2011)
149. Gelet, R., Loret, B., Khalili, N.: Borehole stability analysis in a thermoporoelastic dual-porosity medium. Int. J. Rock Mech. Min. Sci. **50**, 65–76 (2012)
150. Gelet, R., Loret, B., Khalili, N.: A thermo-hydromechanical model in local thermal non-equilibrium for fractured HDR reservoirs with double porosity. J. Geop. Res. **117**, B07205 (2012). https://doi.org/10.1029/2012JB009161
151. Gelet, R., Loret, B., Khalili, N.: Thermal recovery from a fractured medium in local thermal non-equilibrium. Int. J. Numer. Anal. Methods Geomech. **37**, 2471–2501 (2013)
152. Gentile, M., Straughan, B.: Acceleration waves in nonlinear double porosity elasticity. Int. J. Eng. Sci. **73**, 10–16 (2013)
153. Ghiba, I.D.: Spatial estimates concerning the harmonic vibrations in rectangular plates with voids. Arch. Mech. **60**, 263–279 (2008)

154. Ghiba, I.D.: On the temporal behaviour in the bending theory of porous thermoelastic plates. ZAMM J. Appl. Math. Mech. **93**, 284–296 (2013)
155. Giraud, G.: Sur une classe generale d'equation a integrales principales. C. R. Acad. Sci. Paris **202**, 2124–2126 (1936)
156. Goldstein, J.A.: Semigroups of Linear Operators and Applications. Oxford University Press, New York (1985)
157. Green, G.: An Essay on the Application of Mathematical Analysis to the Theories of Electricity and Magnetism. Nottingham (1828)
158. Green, A.E., Adkins, J.E.: Large Elastic Deformations, 2nd edn. Clarendon Press, Oxford (1970)
159. Green, A.E., Lindsay, K.A.: Thermoelasicity. J. Elast. **2**, 1–7 (1972)
160. Green, A.E., Zerna, W.: Theoretical Elasticity, 2nd edn. Clarendon Press, Oxford (1968)
161. Günther, N.M.: Potential Theory and its Applications to Basic Problems of Mathematical Physics. Ungar, New York (1967)
162. Gurtin, M.E.: The linear theory of elasticity. In: Truesdell, C. (ed.) Handbuch der Physik, vol. VIa/2, pp. 1–296. Springer, Berlin (1972)
163. Hayashi, K., Willis-Richards, J., Hopkirk, R.J., Niibori, Y.: Numerical models of HDR geothermal reservoirs - a review of current thinking and progress. Geothermics **28**, 507–518 (1999)
164. He, J., Teng, W., Xu, J., Jiang, R., Sun, J.: A quadruple-porosity model for shale gas reservoirs with multiple migration mechanisms. J. Nat. Gas Sci. Eng. **33**, 918–933 (2016)
165. Hetnarski, R.B., Ignaczak, J.: Mathematical Theory of Elasticity, 2nd edn. Taylor and Francis, Abingdon (2011)
166. Holzapfel, G.A., Ogden, R.W. (eds): Biomechanics: Trends in Modeling and Simulation. Springer, Basel (2017)
167. Hörmander, L.: Local and global properties of fundamental solutions. Math. Scand. **5**, 27–39 (1957)
168. Hörmander, L.: The Analysis of Linear Partial Differential Operators, vols. I–IV. Springer, Berlin (1983/1985)
169. Hsiao, G.C., Wendland, W.L.: Boundary Integral Equations. Springer, Berlin (2008)
170. Iacovache, M.: O extindere a metodei lui Galerkin pentru sistemul ecuatiilor elasticităţii. Bul. St. Acad. Rep. Pop. Romîne. Ser. A **1**, 593–596 (1949)
171. Ichikawa, Y., Selvadurai, A.P.S.: Transport Phenomena in Porous Media: Aspects of Micro/Macro Behaviour. Springer, Berlin (2012)
172. Ieşan, D.: Shock waves in micropolar elastic materials with voids. An. St. Univ. Al. I. Cuza Iasi. **81**, 177–186 (1985)
173. Ieşan, D.: Some theorems in the theory of elastic materials with voids. J. Elast. **15**, 215–224 (1985)
174. Ieşan, D.: A theory of thermoelastic materials with voids. Acta Mech. **60**, 67–89 (1986)
175. Ieşan, D.: Thermoelastic Models of Continua. Springer, Dordrecht (2004)
176. Ieşan, D.: Classical and Generalized Models of Elastic Rods. Chapman and Hall/CRC, New York (2008)
177. Ieşan, D.: On a theory of thermoviscoelastic materials with voids. J. Elast. **104**, 369–384 (2011)
178. Ieşan, D.: Method of potentials in elastostatics of solids with double porosity. Int. J. Eng. Sci. **88**, 118–127 (2015)
179. Ieşan, D.: On the prestressed thermoelastic porous materials. J. Therm. Stresses **41**, 1212–1224 (2018)
180. Ieşan, D., Nappa, L.: Thermal stresses in plane strain of porous elastic solids. Meccanica **39**, 125–138 (2004)
181. Ieşan, D., Quintanilla, R.: On thermoelastic bodies with inner structure and microtemperatures. J. Math. Anal. Appl. **354**, 12–23 (2009)
182. Ieşan, D., Quintanilla, R.: Non-linear deformations of porous elastic solids. Int. J. Non-Linear Mech. **49**, 57–65 (2013)

183. Ieşan, D., Quintanilla, R.: On a theory of thermoelastic materials with a double porosity structure. J. Therm. Stresses **37**, 1017–1036 (2014)
184. Ieşan, D., Quintanilla, R.: Viscoelastic materials with a double porosity structure. Comp. Rendus Mécanique **347**, 124–140 (2019)
185. Ignaczak, J., Ostoja-Starzewski, M.: Thermoelasticity with Finite Wave Speeds. Oxford University Press, New York (2010)
186. Jaiani, G.: Cusped Shell-Like Structures. Springer, Heidelberg (2011)
187. Jaiani, G.: Hierarchical models for viscoelastic Kelvin-Voigt prismatic shells with voids. Bull. TICMI **21**, 33–44 (2017)
188. Janjgava, R.: Elastic equilibrium of porous Cosserat media with double porosity. Adv. Math. Phys. **2016**, 4792148, 9pp. (2016). https://doi.org/10.1155/2016/4792148
189. Janjgava, R.: Some three-dimensional boundary value and boundary-contact problems for an elastic mixture with double porosity. Quart. J. Mech. Appl. Math. **71**, 411–425 (2018)
190. Joseph, D.D., Preziosi, L.: Heat waves. Rev. Mod. Phys. **61**, 41–73 (1989)
191. Joseph, D.D., Preziosi, L.: Heat waves: addendum. Rev. Mod. Phys. **62**, 375–391 (1990)
192. Kansal, T.: Generalized theory of thermoelastic diffusion with double porosity. Arch. Mech. **70**, 241–268 (2018)
193. Kansal, T.: Fundamental solution of the system of equations of pseudo oscillations in the theory of thermoelastic diffusion materials with double porosity. Multidisc. Model. Mater. Struct. 20pp. (2018). https://doi.org/10.1108/MMMS-01-2018-0006
194. Kaynia, A.M., Banerjee, P.K.: Fundamental solutions of Biot's equations of dynamic poroelasticity. Int. J. Eng. Sci. **31**, 817–830 (1993)
195. Kellogg, O.D.: Foundations of Potential Theory. Springer, Berlin (1929)
196. Kenyon, D.E.: A mathematical model of water flux through aortic tissue. Bull. Math. Biol. **41**, 79–90 (1979)
197. Khaled, M.Y., Beskos, D.E., Aifantis, E.C.: On the theory of consolidation with double porosity - III: a finite element formulation. Nimer. Anal. Meth. Geomech. **8**, 101–123 (1984)
198. Khalili, N.: Coupling effects in double porosity media with deformable matrix. Geophys. Res. Lett. **30**, 22 (2003). https://doi.org/10.1029/2003GL018544
199. Khalili, N., Habte, M.A., Zargarbashi, S.: A fully coupled flow deformation model for cyclic analysis of unsaturated soils including hydraulic and mechanical hysteresis. Comput. Geotech. **35**, 872–889 (2008)
200. Khalili, N., Selvadurai, A.P.S.: A fully coupled constitutive model for thermo-hydro-mechanical analysis in elastic media with double porosity. Geophys. Res. Lett. **30**(24), 2268 (2003). https://doi.org/10.1029/2003GL018838
201. Knops, R.J., Payne, L.E.: Stability in linear elasticity. Int. J. Solids Struct. **4**, 1233–1242 (1968)
202. Knops, R.J., Payne, L.E.: Continuous data dependence for the equations of classical elasto-dynamics. Math. Proc. Camb. Phil. Soc. **66**, 481–491 (1969)
203. Knops, R.J., Payne, L.E.: On uniqueness and continuous dependence in dynamical problems of linear thermoelasticity. Int. J. Solids Struct. **6**, 1173–1184 (1970)
204. Knops, R.J., Payne, L.E.: Uniqueness Theorems in Linear Elasticity. Springer Tracts in Natural Philosophy, vol. 19. Springer, New York (1971)
205. Kohn, J.J., Nirenberg, L.: An algebra of pseudo-differential operators. Commun. Pure Appl. Math. **18**, 269–305 (1965)
206. Kumar, R., Vohra, R.: State space approach to plane deformation in elastic material with double porosity. Mater. Phys. Mech. **24**, 9–17 (2015)
207. Kumar, R., Vohra, R.: A problem of spherical cavity in an infinite generalized thermoelastic medium with double porosity subjected to moving heat source. Med. J. Model. Simul. **6**, 67–81 (2016)
208. Kumar, R.M., Vohra, R.: Elastodynamic problem for an infinite body having a spherical cavity in the theory of thermoelasticity with double porosity. Mech. Mech. Eng. **21**, 267–289 (2017)
209. Kumar, R.M., Vohra, R.: Vibration analysis of thermoelastic double porous microbeam subjected to laser pulse. Mech. Adv. Mater. Struct. **26**, 471–479 (2017). https://doi.org/10.1080/15376494.2017.1341578

210. Kumar, R., Vohra, R., Gorla, M.G.: Some considerations of fundamental solution in micropolar thermoelastic materials with double porosity. Arch. Mech. **68**, 263–284 (2016)
211. Kumar, R., Vohra, R., Gorla, M.G.: Thermomechanical response in thermoelastic medium with double porosity. J. Solid Mech. **9**, 24–38 (2017)
212. Kumar, R., Vohra, R., Gorla, M.G.: Variational principle and plane wave propagation in thermoelastic medium with double porosity under Lord-Shulman theory. J. Solid Mech. **9**, 423–433 (2017)
213. Kupradze, V.D.: The existence and uniqueness theorems in the diffraction theory (Russian). Doklady AN SSSR **1**, 235–240 (1934)
214. Kupradze, V.D.: Solution of boundary value problems of Helmholtz equations in extraordinary cases (Russian). Doklady AN SSSR **1**, 521–526 (1934)
215. Kupradze, V.D.: Boundary Value Problems of the Oscillation Theory and Integral Equations. M.-L., State Publishing House of technical and theoretical Literature (1950) (Russian). German translation: Kupradze, V.D.: Randwertaufgaben der Schwingungstheorie und Integralgleichongen. Veb Deutscher verlag der Wissenschaften, Berlin (1956)
216. Kupradze, V.D.: Dynamical problems in elasticity. In: Sneddon, I.N., Hill, R. (eds.) Progress in Solid Mechanics, vol. III, pp. 1–259. North Holland, Amsterdam (1963)
217. Kupradze, V.D.: Potential Methods in the Theory of Elasticity. Israel Program for Scientific Translations, Jerusalem (1965)
218. Kupradze, V.D., Burchuladze, T.V.: Boundary value problems of thermoelasticity. Diff. Equ. **5**, 3–43 (1969)
219. Kupradze, V.D., Gegelia, T.G., Basheleishvili, M.O., Burchuladze, T.V.: Three-Dimensional Problems of the Mathematical Theory of Elasticity and Thermoelasticity. North-Holland, Amsterdam (1979)
220. Kythe, P.K.: Fundamental Solutions for Differential Operators and Applications. Birkhäuser, Boston (1996)
221. Laplace, P.S.: Mémoire sur la théorie de l'anneau de Saturne. Mém. Acad. Roy. Sci. Paris 201–234 (1787/1789)
222. Legland, J.-B., Tournat, V., Dazel, O., Novak, A., Gusev, V.: Linear and nonlinear Biot waves in a noncohesive granular medium slab: transfer function, self-action, second harmonic generation. J. Acoust. Soc. Am. **131**, 4292–4303 (2012)
223. Leseduarte, M.C., Magaña, A., Quintanilla, R.: On the time decay of solutions in porous-thermo-elasticity of type II. Discrete Contin. Dynam. Syst. Ser. B **13**, 375–391 (2010)
224. Li, W., Chen, M., Jin, Y., Lu, Y., Gao, J., Meng, H., Zhang, Y., Tan, P.: Effect of local thermal non-equilibrium on thermoporoelastic response of a borehole in dual-porosity media. Appl. Therm. Eng. **142**, 166–183 (2018)
225. Liu, C., Abousleiman, Y.N.: N-porosity and N-permeability generalized wellbore stability analytical solutions and applications: 50th US Rock Mechanics/Geomechanics Symposium, ARMA, 16-417, 9pp. (2016)
226. Liu, C.Q.: Exact solution for the compressible flow equations through a medium with triple-porosity. Appl. Math. Mech. **2**, 457–462 (1981)
227. Liu, J.C., Bodvarsson, G.S., Wu, Y.S.: Analysis of pressure behaviour in fractured lithophysical reservoirs. J. Cantam. Hydrol. **62–63**, 189–211 (2003)
228. Liu, Z.: Multiphysics in Porous Materials. Springer, Basel (2018)
229. Liu, Z., de Boer, R.: Dispersion and attenuation of surface waves in a fluid-saturated porous medium. Transp. Porous Media **29**, 207–233 (1997)
230. Liu, Z., de Boer, R.: Propagation and evolution of wave fronts in two-phase porous media. Transp. Porous Media **34**, 209–225 (1999)
231. Liu, Z., Zheng, S.: Semigroups Associated with Dissipative Systems. Chapman & Hall/CRC Research Notes in Mathematics, vol. 398. Chapman & Hall/CRC, Boca Raton (1999)
232. Lopez, B., Aguilera, R.: Physics-based approach for shale gas numerical simulation: quintuple porosity and gas diffusion from solid kerogen. In: Presented at the SPE Annual Technical Conference and Exhibition, Houston, 28–30 September. SPE-175115-MS, 32pp. (2015). https://doi.org/10.2118/175115-MS

233. Lopez, B., Aguilera, R.: Petrophysical quantification of multiple porosities in shale-petroleum reservoirs with the use of modified pickett plots. SPE Reserv. Eval. Eng. **SPE-171638-PA**, 15pp. (2017). https://doi.org/10.2118/171638-PA
234. Lord, H.W., Shulman, Y.H.: A generalized dynamical theory of thermoelasticity. J. Mech. Phys. Solids **15**, 299–309 (1967)
235. Magaña, A., Quintanilla, R.: On the time decay of solutions in one-dimensional theories of porous materials. Int. J. Solids Struct. **43**, 3414–3427 (2006)
236. Magaña, A., Quintanilla, R.: On the exponential decay of solutions in one-dimensional generalized porous-thermo-elasticity. Asymp. Anal. **49**, 173–187 (2006)
237. Magaña, A., Quintanilla, R.: On the decay of in porous-elasticity with quasistatic microvoids. J. Math. Anal. Appl. **331**, 617–630 (2007)
238. Malgrange, B.: Existence et approximation des solutions des équations aux dérivées partielles et des équations de convolution. Ann. Inst. Fourier **6**, 271–355 (1955/1956)
239. Marin, M.: Some basic theorems in elastostatics of micropolar materials with voids. J. Comput. Appl. Math. **70**, 115–126 (1996)
240. Marin, M.: Weak solutions in elasticity of dipolar porous materials. Math. Prob. Eng. **2008**, 158908, 8pp. (2008). https://doi.org/10.1155/2008/1589082008
241. Marin, M., Nicaise, S.: Existence and stability results for thermoelastic dipolar bodies with double porosity. Cont. Mech. Thermodynam. **28**, 1645–1657 (2016)
242. Marin, M., Vlase, S., Paun, M.: Considerations on double porosity structure for micropolar bodies. AIP Adv. **5**, 037113, 10pp. (2015). https://doi.org/10.1063/1.4914912
243. Masters, I., Pao, W.K.S., Lewis, R.W.: Coupling temperature to a double - porosity model of deformable porous media. Int. J. Numer. Methods Eng. **49**, 421–438 (2000)
244. Maz'ya, V.G: Sobolev Spaces: With Applications to Elliptic Partial Differential Equations, 2nd edn. Grundlehren der mathematischen Wissenschaften, vol. 342. Springer, Berlin (2011)
245. Mehrabian, A.: The poroelastic constants of multiple-porosity solids. Int. J. Eng. Sci. **132**, 97–104 (2018)
246. Mehrabian, A., Abousleiman, Y.N.: Generalized Biot's theory and Mandel's problem of multiple-porosity and multiple-permeability poroelasticity. J. Geophys. Res. Solid Earth. **119**, 2745–2763 (2014)
247. Mehrabian, A., Abousleiman, Y.N.: Multiple-porosity and multiple-permeability poroelasticity: theory and benchmark analytical solution. In: Vandamme, M., Dangla, P., Pereira, J.M., Siavash Ghabezloo, S. (eds.) Poromechanics VI: Proceedings of the Sixth Biot Conference on Poromechanics, pp. 262–271 (2017). https://doi.org/10.1061/9780784480779.032
248. Mehrabian, A., Abousleiman, Y.N.: Theory and analytical solution to Cryer's problem of N-porosity and N-permeability poroelasticity. J. Mech. Phys. Solids **118**, 218–227 (2018)
249. Mikhlin, S.G.: Composition of double singular integrals (Russian). Dokl. Akad. Nauk SSSR **2**(11), 3–6 (1936)
250. Mikhlin, S.G.: Singular integral equations with two independent variables (Russian). Mat. Sbornik **1**(43), 535–552 (1936)
251. Mikhlin, S.G.: An addition to the paper "Singular integral equations with two independent variables" (Russian). Mat. Sbornik **1**(43), 953–954 (1936)
252. Mikhlin, S.G.: Multidimensional Singular Integrals and Integral Equations. Pergamon Press, Oxford (1965)
253. Mikhlin, S.G., Prössdorf, S.: Singular Integral Operators. Springer, Berlin (1986)
254. Moutsopoulos, K.N., Konstantinidis, A.A., Meladiotis, I., Tzimopoulos, C.D., Aifantis, E.C.: Hydraulic behavior and contaminant transport in multiple porosity media. Transp. Porous Media **42**, 265–292 (2001)
255. Muñoz-Rivera, J.E., Quintanilla, R.: On the time polynomial decay in elastic solids with voids. J. Math. Anal. Appl. **338**, 1296–1309 (2008)
256. Muskhelishvili, N.I.: Singular Integral Equations. Noordhoff, Groningen (1953)
257. Muskhelishvili, N.I.: Some Basic Problems of the Mathematical Theory of Elasticity. Noordhoff, Groningen (1953)

258. Nield, D.A.: Effects of local thermal non-equilibrium in steady convection processes in saturated porous media: forced convection in a channel. J. Porous Media **1**, 181–186 (1998)
259. Nield, D.A.: A note on modelling of local thermal non-equilibrium in a structured porous medium. Int. J. Heat Mass Transf. **45**, 4367–4368 (2002)
260. Nield, D.A.: A note on local thermal non-equilibrium in porous media near boundaries and interfaces. Transp. Porous Media **95**, 581–584 (2012)
261. Nield, D.A., Bejan, A.: Convection in Porous Media, 5th edn. Springer, Basel (2017)
262. Nikolaevskij, V.N.: Mechanics of Porous and Fractured Media. World Scientific, Singapore (1990)
263. Nowacki, W.: Thermoelasticity. Pergamon Press, Oxford (1962)
264. Nowacki, W.: Green functions for the thermoelastic medium. Bull. Acad. Polon Sci. Ser. Sci. Tech. **12**, 465–472 (1964)
265. Nowacki, W.: On the completeness of potentials in micropolar elasticity. Arch. Mech. Stos. **21**, 107–122 (1969)
266. Nunziato, J.W., Cowin, S.C.: A nonlinear theory of elastic materials with voids. Arch. Rat. Mech. Anal. **72**, 175–201 (1979)
267. Olny, X., Boutin, C.: Acoustic wave propagation in double porosity media. J. Acoust. Soc. Am. **114**, 73–89 (2003)
268. Ortner, N., Wagner, P.: On the fundamental solution of the operator of dynamic linear thermoelasticity. J. Math. Anal. Appl. **170**, 524–550 (1992)
269. Ortner, N., Wagner, P.: Fundamental Solutions of Linear Partial Differential Operators: Theory and Practice. Springer, Basel (2015)
270. Pamplona, P.X., Muñoz-Rivera, J.E., Quintanilla, R.: Stabilization in elastic solids with voids. J. Math. Anal. Appl. **350**, 37–49 (2009)
271. Pamplona, P.X., Muñoz-Rivera, J.E., Quintanilla, R.: On uniqueness and analyticity in thermoviscoelastic solids with voids. J. Appl. Anal. Comput. **1**, 251–266 (2011)
272. Passarella, F.: Some results in micropolar thermoelasticity. Mech. Res. Commun. **23**, 349–357 (1996)
273. Passarella, F., Straughan, B., Zampoli, V.: Structural stability in local thermal non-equilibrium porous media. Acta Appl. Math. **136**, 43–53 (2015)
274. Passarella, F., Tibullo, V., Zampoli, V.: On the heat-flux dependent thermoelasticity for micropolar porous media. J. Therm. Stresses **34**, 778–794 (2011)
275. Patwardhan, S.D., Famoori, F., Govindarajan, S.K.: Quad-porosity shale systems - a review. World J. Eng. **13**, 529–539 (2016)
276. Poisson, S.D.: M'em. Acad. Sci. Paris **3**, 131–176 (1818)
277. Pompei, A., Scalia, A.: On the steady vibrations of the thermoelastic porous materials. Int. J. Solids Struct. **31**, 2819–2834 (1994)
278. Pride, S.R., Berryman, J.G.: Linear dynamics of double-porosity dual-permeability materials I. Governing equations and acoustic attenuation. Phys. Rev. E **68**, 036603 (2003)
279. Puri, P., Cowin, S.C.: Plane waves in linear elastic material with voids. J. Elast. **15**, 167–183 (1985)
280. Quintanilla, R.: Exponential stability in the dual-phase-lag heat conduction theory. J. Non-Equilibrium Thermodynam. **27**, 217–227 (2002)
281. Quintanilla, R.: Convergence and structural stability in thermoelasticity. Appl. Math. Comput. **135**, 287–300 (2003)
282. Quintanilla, R.: Slow decay for one-dimensional porous dissipation elasticity. Appl. Math. Lett. **16**, 487–491 (2003)
283. Quintanilla, R., Racke, R.: A note on stability in dual-phase-lag heat conduction. Int. J. Heat Mass Transf. **49**, 1209–1213 (2006)
284. Quintanilla, R., Racke, R.: A note on stability in three-phase-lag heat conduction. Int. J. Heat Mass Transf. **51**, 24–29 (2008)
285. Radhika, B.P., Krishnamoorthy, A., Rao, A.U.: A review on consolidation theories and its application. Int. J. Geotech. Eng. 8pp. (2017). https://doi.org/10.1080/19386362.2017.1390899

286. Rees, D.A.S.: Microscopic modelling of the two - temperature model for conduction in heterogeneous media. J. Porous Media **13**, 125–143 (2010)
287. Rees, D.A.S.: The effect of local thermal non-equilibrium on the stability of convection in a vertical porous channel. Transp. Porous Media **87**, 459–464 (2011)
288. Rezaee, R.: Fundamentals of Gas Shale Reservoirs. Wiley, Hoboken (2015)
289. Ricken, T., Bluhm, J.: Remodeling and growth of living tissue: a multiphase theory. Arch. Appl. Mech. **80**, 453–465 (2010)
290. Roy Choudhuri, S.K.: On a thermoelastic three-phase-lag model. J. Therm. Stresses **30**, 231–238 (2007)
291. Rushchitsky, J.J.: Nonlinear Elastic Waves in Materials. Springer, Basel (2014)
292. Russo, R., Starita, G.: Uniqueness in Linear Elastostatics. In: Hetnarski R.B. (ed.) Encyclopedia of Thermal Stresses, pp. 6311–6325. Springer, Dordrecht (2014)
293. Sandru, N.: On some problems of the linear theory of asymmetric elasticity. Int. J. Eng. Sci. **4**, 81–96 (1966)
294. Sang, G., Elsworth, D., Miao, X., Mao, X., Wang, J.: Numerical study of a stress dependent triple porosity model for shale gas reservoirs accommodating gas diffusion in kerogen. J. Nat. Gas Sci. Eng. **32**, 423–438 (2016)
295. Sauter, S.A., Schwab, C.: Boundary Element Method. Springer, Berlin (2011)
296. Scalia, A.: Shock waves in viscoelastic materials with voids. Wave Motion **19**, 125–133 (1994)
297. Scalia, A.: Harmonic oscillations of a rigid punch on a porous elastic layer. J. Appl. Math. Mech. **73**, 344–350 (2009)
298. Scalia, A., Pompei, A., Chirita, S.: On the behavior of steady time-harmonic oscillations in thermoelastic materials with voids. J. Therm. Stresses **27**, 209–226 (2004)
299. Scalia, A., Svanadze, M.: Potential method in the linear theory of thermoelasticity with microtemperatures. J. Therm. Stresses **32**, 1024–1042 (2009)
300. Scalia, A., Svanadze, M.: Basic theorems in thermoelastostatics of bodies with microtemperatures. In: Hetnarski, R.B.(ed), Encyclopedia of Thermal Stresses, 11 vols, pp. 355–365, 1st edn. Springer, Berlin (2014)
301. Scalia, A., Svanadze, M., Tracinà, R.: Basic theorems in the equilibrium theory of thermoelasticity with microtemperatures. J. Therm. Stresses **33**, 721–753 (2010)
302. Scarpetta, E.: On the fundamental solutions in micropolar elasticity with voids. Acta Mech. **82**, 151–158 (1990)
303. Scarpetta, E.: Minimum principles for the bending problem of elastic plates with voids. Int. J. Eng. Sci. **40**, 1317–1327 (2002)
304. Scarpetta, E., Svanadze, M.: Uniqueness theorems in the quasi-static theory of thermoelasticity for solids with double porosity, J. Elast. **120**, 67–86 (2015)
305. Scarpetta, E., Svanadze, M., Zampoli, V.: Fundamental solutions in the theory of thermoelasticity for solids with double porosity. J. Therm. Stresses **37**, 727–748 (2014)
306. Schanz, M.: Poroelastodynamics: linear models, analytical solutions, and numerical methods. Appl. Mech. Rev. **62**, 030803-1–030803-15 (2009)
307. Selvadurai, A.P.S.: The analytical method in geomechanics. Appl. Mech. Rev. ASME **60**, 87–106 (2007)
308. Selvadurai, A.P.S., Suvorov, A.: Thermo-Poroelasticity and Geomechanics. Cambridge University Press, Cambridge (2017)
309. Sharma, K., Kumar, P.: Propagation of plane waves and fundamental solution in thermoviscoelastic medium with voids. J. Therm. Stresses **36**, 94–111 (2013)

310. Sheng, G., Su, Y., Wang, W., Liu, J., Lu, M., Zhang, Q., Ren, L.: A multiple porosity media model for multi-fractured horizontal wells in shale gas reservoirs. J. Nat. Gas Sci. Eng. **27**, 1562–1573 (2015)
311. Showalter, R.E., Visarraga, D.B.: Double-diffusion models from a highly heterogeneous medium. J. Math. Anal. Appl. **295**, 191–210 (2004)
312. Showalter, R.E., Walkington, N.J.: Micro-structure models of diffusion in fissured media. J. Math. Anal. Appl. **155**, 1–20 (1991)
313. Singh, S.S., Lianngenga, R.: Plane waves in micropolar thermoelastic materials with voids. Sci. Technol. J. **4**, 141–151 (2016)
314. Singh, J., Tomar, S.K.: Plane waves in thermo-elastic material with voids. Mech. Mater. **39**, 932–940 (2007)
315. Sobolev, S.L.: Applications of Functional Analysis in Mathematical Physics. American Mathematical Society, Providence (1963)
316. Somigliana, C.: Sopra l'equilibrio di un corpo elastico isotropo. Nuovo Cimento, ser **3**, 17–20 (1885)
317. Steinbach, O.: Numerical Approximation Methods for Elliptic Boundary Value Problems: Finite and Boundary Elements. Springer, New York (2008)
318. Straughan, B.: Global nonlinear stability in porous convection with a thermal non-equilibrium model. Proc. R. Soc. A Math. Phys. **462**(2066), 409–418 (2006)
319. Straughan, B.: Stability and Wave Motion in Porous Media. Springer, New York (2008)
320. Straughan, B.: Heat Waves. Applied Mathematical Sciences, vol. 177. Springer, New York (2011)
321. Straughan, B.: Continuous dependence on the heat source in resonant porous penetrative convection. Stud. Appl. Math. **127**, 302–314 (2011)
322. Straughan, B.: Stability and uniqueness in double porosity elasticity. Int. J. Eng. Sci. **65**, 1–8 (2013)
323. Straughan, B.: Convection with Local Thermal Non-Equilibrium and Microfluidic Effects. Springer, Berlin (2015)
324. Straughan, B.: Modelling questions in multi-porosity elasticity. Meccanica **51**, 2957–2966 (2016)
325. Straughan, B.: Waves and uniqueness in multi-porosity elasticity. J. Therm. Stresses **39**, 704–721 (2016)
326. Straughan, B.: Mathematical Aspects of Multi-Porosity Continua. Springer, Basel (2017)
327. Straughan, B.: Solid mechanics–uniqueness and stability in triple porosity thermoelasticity. Rend. Lincei Mat. Appl. **28**, 191–208 (2017)
328. Straughan, B., Tibullo, V.: Thermal effects on nonlinear acceleration waves in the Biot theory of porous media. Mech. Res. Commun. **94**, 70–73 (2018)
329. Su, B.-L., Sanchez, C., Yang, X.-Y. (eds): Hierarchically Structured Porous Materials: From Nanoscience to Catalysis, Separation, Optics, Energy, and Life Science. Wiley-VCH Verlag, Weinheim (2012)
330. Svanadze, M.: On existence of eigenfrequencies in the theory of two-component elastic mixtures. Quart. J. Mech. Appl. Math. **51**, 427–437 (1998)
331. Svanadze, M.: Fundamental solution in the theory of consolidation with double porosity. J. Mech. Behav. Mater. **16**(1–2), 123–130 (2005)
332. Svanadze, M.: Plane waves and eigenfrequencies in the linear theory of binary mixtures of thermoelastic solids. J. Elast. **92**, 195–207 (2008)
333. Svanadze, M.: Dynamical problems of the theory of elasticity for solids with double porosity. Proc. Appl. Math. Mech. **10**(1), 309–310 (2010)
334. Svanadze, M.: Plane waves and boundary value problems in the theory of elasticity for solids with double porosity. Acta Appl. Math. **122**, 461–471 (2012)
335. Svanadze, M.: The boundary value problems of the fully coupled theory of poroelasticity for materials with double porosity. Proc. Appl. Math. Mech. **12**(1), 279–282 (2012)
336. Svanadze, M.: Fundamental solution in the linear theory of consolidation for elastic solids with double porosity. J. Math. Sci. **195**, 258–268 (2013)

337. Svanadze, M.: Uniqueness theorems in the theory of thermoelasticity for solids with double porosity. Meccanica **49**, 2099–2108 (2014)
338. Svanadze, M.: On the theory of viscoelasticity for materials with double porosity. Disc. Contin. Dynam. Syst. Ser. B **19**, 2335–2352 (2014)
339. Svanadze, M.: Boundary value problems in the theory of thermoporoelasticity for materials with double porosity. Proc. Appl. Math. Mech. **14**(1), 327–328 (2014)
340. Svanadze, M.: Large existence of solutions in thermoelasticity theory of steady vibrations. In: Hetnarski, R.B. (ed), Encyclopedia of Thermal Stresses, 11 vols., 1st edn., pp. 2677–2687. Springer, Berlin (2014)
341. Svanadze, M.: Potentials in thermoelasticity theory. In: Hetnarski, R.B. (ed.), Encyclopedia of Thermal Stresses, 11 vols., 1st edn., pp. 4013–4023. Springer, Berlin (2014)
342. Svanadze, M.: External boundary value problems of steady vibrations in the theory of rigid bodies with a double porosity structure. Proc. Appl. Math. Mech. **15**(1), 365–366 (2015)
343. Svanadze, M.: Plane waves, uniqueness theorems and existence of eigenfrequencies in the theory of rigid bodies with a double porosity structure. In: Albers, B., Kuczma, M. (eds.) Continuous Media with Microstructure, vol. 2, pp. 287–306. Springer, Basel (2016)
344. Svanadze, M.: On the linear theory of thermoelasticity for triple porosity materials. In: Ciarletta, M., Tibullo, V., Passarella, F. (eds), Proceedings of 11th International Congress Thermal Stresses, 5–9 June, 2016, Salerno, Italy, pp. 259–262 (2016)
345. Svanadze, M.: Fundamental solutions in the theory of elasticity for triple porosity materials. Meccanica **51**, 1825–1837 (2016)
346. Svanadze, M.: Boundary value problems in the theory of thermoelasticity for triple porosity materials. In: Proceedings of ASME2016. 50633; Vol. 9: Mechanics of Solids, Structures and Fluids; NDE, Diagnosis, and Prognosis, V009T12A079. November 11, 2016, IMECE2016-65046 (2016). https://doi.org/10.1115/IMECE2016-65046
347. Svanadze, M.: Boundary value problems of steady vibrations in the theory of thermoelasticity for materials with double porosity structure. Arch. Mech. **69**, 347–370 (2017)
348. Svanadze, M.: External boundary value problems in the quasi static theory of thermoelasticity for triple porosity materials. Proc. Appl. Math. Mech. **17**(1), 471–472 (2017)
349. Svanadze, M.: Steady vibrations problems in the theory of elasticity for materials with double voids. Acta Mech. **229**, 1517–1536 (2018)
350. Svanadze, M.: Potential method in the theory of elasticity for triple porosity materials. J. Elast. **130**, 1–24 (2018)
351. Svanadze, M.: Potential method in the linear theory of triple porosity thermoelasticity. J. Math. Anal. Appl. **461**, 1585–1605 (2018)
352. Svanadze, M.: On the linear equilibrium theory of elasticity for materials with triple voids. Quart. J. Mech. Appl. Math. **71**, 329–248 (2018)
353. Svanadze, M.: External boundary value problems in the quasi static theory of thermoelasticity for materials with triple voids. Proc. Appl. Math. Mech. **18**(1), e201800171 (2018)
354. Svanadze, M.: Fundamental solutions in the linear theory of thermoelasticity for solids with triple porosity. Math. Mech. Solids **24**, 919–938 (2019)
355. Svanadze, M.: On the linear theory of double porosity thermoelasticity under local thermal non-equilibrium. J. Therm. Stresses **42**, 890–913 (2019)
356. Svanadze, M.: Potential method in the theory of thermoelasticity for materials with triple voids. Arch. Mech. **71**, 113–136 (2019)
357. Svanadze, M., de Boer, R.: On the representations of solutions in the theory of fluid-saturated porous media. Quart. J. Mech. Appl. Math. **58**, 551–562 (2005)
358. Svanadze, M., De Cicco, S.: Fundamental solutions in the full coupled linear theory of elasticity for solid with double porosity. Arch. Mech. **65**, 367–390 (2013)
359. Svanadze, M., Scalia, A.: Mathematical problems in the coupled linear theory of bone poroelasticity. Comput. Math. Appl. **66**, 1554–1566 (2013)
360. Svanadze, M., Scalia, A.: Potential method in the theory of thermoelasticity with microtemperatures for microstretch solids. Trans. Nanjing Univ. Aeron. Astron. **31**, 159–163 (2014)

361. Svanadze, M.M.: Steady vibrations problem in the theory of viscoelasticity for Kelvin-Voigt materials with voids. Proc. Appl. Math. Mech. **12**, 283–284 (2012)
362. Svanadze, M.M.: Potential method in the linear theory of viscoelastic materials with voids. J. Elast. **114**, 101–126 (2014)
363. Svanadze, M.M.: On the solutions of equations of the linear thermoviscoelasticity theory for Kelvin-Voigt materials with voids. J. Therm. Stresses **37**, 253–269 (2014)
364. Svanadze, M.M.: Potential method in the theory of thermoviscoelasticity for materials with voids. J. Therm. Stresses **37**, 905–927 (2014)
365. Svanadze, M.M.: Potential method in the steady vibrations problems of the theory of thermoviscoelasticity for Kelvin-Voigt materials with voids. Proc. Appl. Math. Mech. **14**, 347–348 (2014)
366. Svanadze, M.M.: External boundary value problems in the quasi static theory of viscoelasticity for Kelvin-Voigt materials with double porosity. Proc. Appl. Math. Mech. **16**(1), 497–498 (2016)
367. Svanadze, M.M.: Plane waves and problems of steady vibrations in the theory of viscoelasticity for Kelvin-Voigt materials with double porosity. Arch. Mech. **68**, 441–458 (2016)
368. Svanadze, M.M.: Fundamental solution and uniqueness theorems in the linear theory of thermoviscoelasticity for solids with double porosity. J. Therm. Stresses **40**, 1339–1352 (2017)
369. Svanadze, M.M.: External boundary value problems in the quasi static theory of thermoviscoelasticity for Kelvin-Voigt materials with double porosity. Proc. Appl. Math. Mech. **17**(1), 469–470 (2017)
370. Svanadze, M.M.: Fundamental solutions and uniqueness theorems in the theory of viscoelasticity for materials with double porosity. Trans. A. Razmadze Math. Inst. **172**, 276–292 (2018)
371. Thomson, W. (Lord Kelvin): On the equations of equilibrium of an elastic solid. Camb. Dubl. Math. J. **3**, 87–89 (1848)
372. Tomar, S.K.: Wave propagation in a micropolar elastic plate with voids. J. Vib. Control **11**, 849–863 (2005)
373. Tomar, S.K., Bhagwan, J., Steeb, H.: Time harmonic waves in a thermo-viscoelastic material with voids. J. Vib. Control **20**, 1119–1136 (2014)
374. Tong, L., Liu, Y., Geng, D., Lai, S.: Nonlinear wave propagation in porous materials based on the Biot theory. J. Acoust. Soc. Am. **142**, 756–770 (2017)
375. Truesdell, C., Noll, W.: The Non-Linear Field Theories of Mechanics. Handbuch der Physik, Band III/3, Flügge, S. (ed.). Springer, Berlin (1965)
376. Truesdell, C., Toupin, R.: The Classical Field Theories. Handbuch der Physik, Band III/1, Flügge, S. (ed.). Springer, Berlin (1960)
377. Tsagareli, I.: Explicit solution of elastostatic boundary value problems for the elastic circle with voids. Adv. Math. Phys. **2018**, 6275432, 6pp. (2018). https://doi.org/10.1155/2018/6275432
378. Tsagareli, I., Bitsadze, L.: Explicit solution of one boundary value problem in the full coupled theory of elasticity for solids with double porosity. Acta Mech. **26**, 1409–1418 (2015)
379. Tsagareli, I., Bitsadze, L.: Explicit solutions on some problems in the fully coupled theory of elasticity for a circle with double porosity. Bull. TICMI **20**, 11–23 (2016)
380. Tsagareli, I., Svanadze, M.M.: Explicit solution of the boundary value problems of the theory of elasticity for solids with double porosity. Proc. Appl. Math. Mech. **10**, 337–338 (2010)
381. Tsagareli, I., Svanadze, M.M.: Explicit solution of the problems of elastostatics for an elastic circle with double porosity. Mech. Res. Commun. **46**, 76–80 (2012)
382. Unger, D.J., Aifantis, E.C.: Completeness of solutions in the double porosity theory. Acta Mech. **75**, 269–274 (1988)
383. Vafai, K.: Porous Media: Applications in Biological Systems and Biotechnology. CRC Press, Boca Raton (2011)
384. Valent, T.: Boundary Value Problems of Finite Elasticity. Local Theorems on Existence, Uniqueness, and Analytic Dependence on Data. Springer Tracts in Natural Philosophy, vol. 31. Springer, Berlin (1988)

385. Van Der Knapp, W.: Nonlinear behavior of elastic porous media. Pet. Trans. AIME **216**, 179–187 (1959)
386. Vekua, I.N.: On metaharmonic functions (Russian). Proc. Tbilisi Math. Inst. Acad. Sci. Georgian SSR **12**, 105–174 (1943). Eng. Trans: Vekua, I.N.: Lecture Notes of TICMI **14**, 1–62 (2013)
387. Vekua, I.N.: Shell Theory: General Methods of Construction. Pitman Advanced Publishing Program, Boston (1985)
388. Verruijt, A.: The completeness of Biot's solution of the coupled thermoelasticity problem. Quart. Appl. Math. **26**, 485–490 (1969)
389. Verruijt, A.: Theory and Problems of Poroelasticity. Delft University of Technology, Delft (2015)
390. Vladimirov, V.S.: Equations of Mathematical Physics. Marcel Dekker, New York (1971)
391. Volterra, V.: Sur les vibrations des corps élastiques isotropes. Acta Math. **18**, 161–232 (1894)
392. Wang, H.F.: Theory of Linear Poro-Elasticity with Applications to Geomechanics and Hydrogeology. Princeton University Press, Princeton (2000)
393. Wang, M.Z., Xu, B.X., Gao, C.F.: Recent general solutions in linear elasticity and their applications. Appl. Mech. Rev. **61**, 030803, 20pp. (2008). https://doi.org/10.1115/1.2909607
394. Warren, J.R., Root, P.J.: The behaviour of naturally fractured reservoirs. Soc. Pet. Eng. J. **228**, 245–255 (1963)
395. Wei, Z., Zhang, D.: Coupled fluid - flow and geomechanics for triple - porosity/dual - permeability modelling of coalbed methane recovery. Int. J. Rock Mech. Min. Sci. **47**, 1242–1253 (2008)
396. Wilson, R.K., Aifantis, E.C.: On the theory of consolidation with double porosity - I. Int. J. Eng. Sci. **20**, 1009–1035 (1982)
397. Wu, Y.-S.: Multiphase Fluid Flow in Porous and Fractured Reservoirs. Elsevier, Amsterdam (2016)
398. Wu, Y.-S., Liu, H.H., Bodavarsson, G.S.: A triple-continuum approach for modelling flow and transport processes in fractured rock. J. Contam. Hydrol. **73**, 145–179 (2004)
399. Zaman, S.I.: A comprehensive review of the boundary integral formulations of acoustic scattering problems. Journal of Scientific Research Science and Technology, Special Edition, pp. 281–310 S.Q.U. Oman (2000)
400. Zampoli, V.: Uniqueness theorems about high-order time differential thermoelastic models. Ricerche Mat. **67**, 929–950 (2018)
401. Zampoli, V.: Some continuous dependence results about high-order time differential thermoelastic models. J. Therm. Stresses **41**, 827–846 (2018)
402. Zampoli, V., Landi, A.: A domain of influence result about the time differential three-phase-lag thermoelastic model. J. Therm. Stresses **40**, 108–120 (2017)
403. Zhang, W., Xu, J., Jiang, R., Cui, Y., Qiao, J., Kang, C., Lu, Q.: Employing a quad-porosity numerical model to analyze the productivity of shale gas reservoir. J. Petrol. Sci. Eng. **157**, 1046–1055 (2017)
404. Zhao, Y., Chen, M.: Fully coupled dual-porosity model for anisotropic formations. Int. J. Rock Mech. Min. Sci. **43**, 1128–1133 (2006)
405. Zou, M., Wei, C., Yu, H., Song, L.: Modelling and application of coalbed methane recovery performance based on a triple porosity/dual permeability model. J. Nat. Gas Sci. Eng. **22**, 679–688 (2015)

Index

M. Svanadze, *Potential Method in Mathematical Theories of Multi-Porosity Media*, Interdisciplinary Applied Mathematics 51, https://doi.org/10.1007/978-3-030-28022-2

Printed in the United States
By Bookmasters

Printed in the United States
By Bookmasters